SI Prefixes

Multiple	Exponential Form	Prefix	SI Symbol
1 000 000 000	10^9	giga	G
1 000 000	10^6	mega	M
1 000	10^3	kilo	k
Submultiple			
0.001	10^{-3}	milli	m
0.000 001	10^{-6}	micro	μ
0.000 000 001	10^{-9}	nano	n

Conversion Factors (FPS) to (SI)

Quantity	Unit of Measurement (FPS)	Equals	Unit of Measurement (SI)
Force	lb		4.4482 N
Mass	slug		14.5938 kg
Length	ft		0.3048 m

Conversion Factors (FPS)

$$1 \text{ ft} = 12 \text{ in. (inches)}$$
$$1 \text{ mi. (mile)} = 5{,}280 \text{ ft}$$
$$1 \text{ kip (kilopound)} = 1{,}000 \text{ lb}$$
$$1 \text{ ton} = 2{,}000 \text{ lb}$$

ENGINEERING MECHANICS

Statics

SIXTH EDITION

R. C. HIBBELER

Macmillan Publishing Company
New York

Maxwell Macmillan Canada
Toronto

Maxwell Macmillan International
New York   Oxford   Singapore   Sydney

Macmillan Publishing Company
866 Third Avenue, New York, New York 10022

Macmillan Publishing Company is part
of the Maxwell Communication Group of Companies.

Maxwell Macmillan Canada, Inc.
1200 Eglinton Avenue East
Suite 200
Don Mills, Ontario M3C 3N1

LIBRARY OF CONGRESS CATALOGING-IN-PUBLICATION DATA

Hibbeler, R. C.
    Engineering mechanics—statics/R. C. Hibbeler.—6th ed.
        p. cm.
    Companion volume to Engineering mechanics—dynamics. These two works are also
published together as Engineering mechanics—statics and dynamics.
    Includes index.
    ISBN 0-02-354685-9
    1. Statics. I. Title.
TA351.H5 1992
620.1′053—dc20   91-42329
                                                                                          CIP

Printing: 2 3 4 5 6 7 8   Year: 2 3 4 5 6 7 8 9 0 1

TO THE STUDENT

With the hope that this work
will stimulate an interest in Engineering Mechanics
and provide an acceptable guide to its understanding.

Preface

The purpose of this book is to provide the student with a clear and thorough presentation of the theory and applications of engineering mechanics. To achieve this objective the author has by no means worked alone on this task, for to a large extent this book has been shaped by the comments and suggestions of more than a hundred reviewers in the teaching profession and many of the author's students who have used its previous editions.

Much has been done during the preparation of this new edition. Former users of the book will notice that the art work has been greatly enhanced in order to gain a more realistic and understandable sense of the material. There are more problems in this edition than ever before, and most of them are new. Although the contents of the book have remained in the same order, some topics have been expanded, examples have been replaced with others, and the explanations of many topics have been improved by a careful rewording of selected sentences. The hallmark of the book remains the same, however, and that is, where necessary, a strong emphasis is placed on drawing a free-body diagram, and also the importance of selecting an appropriate coordinate system and associated sign convention for vector components is stressed when the equations of mechanics are applied.

Organization and Approach. The contents of each chapter are organized into well-defined sections. Selected groups of sections contain an explanation of specific topics, illustrative example problems, and a set of homework problems. The topics within each section are placed into subgroups defined by boldface titles. The purpose of this is to present a structured method for introducing each new definition or concept, and to make the book convenient for later reference and review.

A ''procedure for analysis'' is given at the end of many sections of the book in order to provide the student with a review or summary of the material and a logical and orderly method to follow when applying the theory. As in the previous editions, the example problems are solved using this outlined method in order to clarify its numerical application. It is to be understood, however, that once the relevant principles have been mastered and enough confidence and judgment have been obtained, the student can then develop his or her own procedures for solving problems. In most cases, it is felt that the first step in any procedure should require drawing a diagram. By doing so, the student forms the habit of tabulating the necessary data while focusing on the physical aspects of the problem and its associated geometry. If this step is correctly performed, applying the relevant equations of mechanics becomes somewhat methodical, since the data can be taken directly from the diagram. This step is particularly important when solving problems involving equilibrium, and for this reason, drawing a free-body diagram is strongly emphasized throughout the book.

Since mathematics provides a systematic means of applying the principles of mechanics, the student is expected to have prior knowledge of algebra, geometry, trigonometry, and, for complete coverage, some calculus. Vector analysis is introduced at points where it is most applicable. Its use often provides a convenient means for presenting concise derivations of the theory, and it makes possible a simple and systematic solution of many complicated three-dimensional problems. Occasionally, the example problems are solved using more than one method of analysis so that the student develops the ability to use mathematics as a tool whereby the solution of any problem may be carried out in the most direct and effective manner.

Problems. Numerous problems in the book depict realistic situations encountered in engineering practice. It is hoped that this realism will both stimulate the student's interest in engineering mechanics and provide a means for developing the skill to reduce any such problem from its physical description to a model or symbolic representation to which the principles of mechanics may be applied. As in the previous edition, an effort has been made to include some problems which may be solved using a numerical procedure executed on either a desk-top computer or a programmable pocket calculator. Suitable numerical techniques along with associated computer programs are given in Appendix B. The intent here is to broaden the student's capacity for using other forms of mathematical analysis *without* sacrificing the time needed to focus on the application of the principles of mechanics. Problems of this type which either can or must be solved using numerical procedures are identified by a ''square'' symbol (■) preceding the problem number.

Throughout the text there is an approximate balance of problems using either SI or FPS units. Furthermore, in any set, an attempt has been made to arrange the problems in order of increasing difficulty.* The answers to all but

---

*Review problems, at the end of each chapter, are presented in random order.

every fourth problem are listed in the back of the book. To alert the user to a problem without a reported answer, an asterisk (*) is placed before the problem number.

Contents. The book is divided into 11 chapters, in which the principles introduced are first applied to simple situations. Most often, each principle is applied first to a particle, then to a rigid body subjected to a coplanar system of forces, and finally to the general case of three-dimensional force systems acting on a rigid body.

The text begins in Chapter 1 with an introduction to mechanics and a discussion of units. The notion of a vector and the properties of a concurrent force system are introduced in Chapter 2. This theory is then applied to the equilibrium of particles in Chapter 3. Chapter 4 contains a general discussion of both concentrated and distributed force systems and the methods used to simplify them. The principles of rigid-body equilibrium are developed in Chapter 5 and then applied to specific problems involving the equilibrium of trusses, frames, and machines in Chapter 6, and to the analysis of internal forces in beams and cables in Chapter 7. Applications to problems involving frictional forces are discussed in Chapter 8, and topics related to the center of gravity and centroid are treated in Chapter 9. If time permits, sections concerning more advanced topics, indicated by stars (★), may be covered. Most of these topics are included in Chapter 10 (area and mass moments of inertia) and Chapter 11 (virtual work and potential energy). Note that this material also provides a suitable reference for basic principles when it is discussed in more advanced courses.

At the discretion of the instructor, some of the material may be presented in a different sequence with no loss in continuity. For example, it is possible to introduce the concept of a force and all the necessary methods of vector analysis by first covering Chapter 2 and Sec. 4.1. Then, after covering the rest of Chapter 4 (force and moment systems), the equilibrium methods in Chapters 3 and 5 can be discussed.

Acknowledgments. I have endeavored to write this book so that it will appeal to both the student and instructor. Through the years many people have helped in its development and I should like to acknowledge their valued suggestions and comments. Specifically, I wish to personally thank the following individuals who have contributed to this edition, namely, Professor Jafar Al-Abdulla, University of Wisconsin; Professor Ziad Bayasi, Bradley University; Professor William A. Best, Lafayette College; Professor James Devine, University of South Florida; Professor E. S. Doderer, Trinity University, San Antonio; Professor Brenda Gile, Wichita State University; Professor Zouheir Hashem, University of Texas at Arlington; Professor Henry Haslach, Jr., University of Maryland; Lt. Commander David Shikada, U.S. Naval Academy; Professor Larry Stauffer, University of Idaho, Moscow; Professor Ralph Stephens, University of Iowa; Professor Poojitha Yapa, Clarkson University; and Professor Kingman Yee, Lawrence Technological University. A particular note of thanks is given to Professors Edward Hornsey, University of Missouri, Rolla and Will Lidell, Jr., Auburn University at Montgomery, and a former graduate student of mine, Mr. Kai Beng Yap, for their help in checking the solutions to the problems.

Many thanks are also extended to all my students and to members of the teaching profession who have freely taken the time to send me their suggestions and comments. Since the list is too long to mention, it is hoped that those who have given help in this manner will accept this anonymous recognition. Furthermore, I greatly appreciate the freedom and support given to me by my editors and the staff at Macmillan, notably David Johnstone, Gary Ostedt, Dora Rizzuto, Anna Yip, and Sandy Moore. Lastly, I should like to acknowledge the assistance of my wife, Conny, during the time it has taken to prepare the manuscript for publication.

Russell Charles Hibbeler

Contents

6

Structural Analysis

7

Internal Forces

8

Friction

9

Center of Gravity and Centroid

10

Moments of Inertia   455

11

Virtual Work   501

APPENDIXES

A

Mathematical Expressions

B

Numerical and Computer Analysis

Answers

Index

STATICS

1 General Principles

This chapter provides an introduction to many of the fundamental concepts in mechanics. It includes a discussion of models or idealizations that are used to apply the theory, a statement of Newton's laws of motion, upon which this subject is based, and a general review of the principles for applying the SI system of units. Standard procedures for performing numerical calculations are then discussed. At the end of the chapter we will present a general guide that should be followed for solving problems.

1.1 Mechanics

Mechanics can be defined as that branch of the physical sciences concerned with the state of rest or motion of bodies that are subjected to the action of forces. In general, this subject is subdivided into three branches: *rigid-body mechanics, deformable-body mechanics,* and *fluid mechanics.* This book treats only rigid-body mechanics, since it forms a suitable basis for the design and analysis of many types of structural, mechanical, or electrical devices encountered in engineering. Also, rigid-body mechanics provides part of the necessary background for the study of the mechanics of deformable bodies and the mechanics of fluids.

Rigid-body mechanics is divided into two areas: statics and dynamics. *Statics* deals with the equilibrium of bodies, that is, those that are either at rest or move with a constant velocity; whereas *dynamics* is concerned with the accelerated motion of bodies. Although statics can be considered as a special case of dynamics, in which the acceleration is zero, statics deserves separate treatment in engineering education, since many objects are designed with the intention that they remain in equilibrium.

Historical Development. The subject of statics developed very early in history, because the principles involved could be formulated simply from measurements of geometry and force. For example, the writings of Archimedes (287–212 B.C.) deal with the principle of the lever. Studies of the pulley, inclined plane, and wrench are also recorded in ancient writings—at times when the requirements of engineering were limited primarily to building construction.

Since the principles of dynamics depend on an accurate measurement of time, this subject developed much later. Galileo Galilei (1564–1642) was one of the first major contributors to this field. His work consisted of experiments using pendulums and falling bodies. The most significant contributions in dynamics, however, were made by Isaac Newton (1642–1727), who is noted for his formulation of the three fundamental laws of motion and the law of universal gravitational attraction. Shortly after these laws were postulated, important techniques for their application were developed by Euler, D'Alembert, Lagrange, and others.

1.2 Fundamental Concepts

Before beginning our study of rigid-body mechanics, it is important to understand the meaning of certain fundamental concepts and principles.

Basic Quantities. The following four quantities are used throughout rigid-body mechanics.

Length. *Length* is needed to locate the position of a point in space and thereby describe the size of a physical system. Once a standard unit of length is defined, one can then quantitatively define distances and geometric properties of a body as multiples of the unit length.

Time. *Time* is conceived as a succession of events. Although the principles of statics are time independent, this quantity does play an important role in the study of dynamics.

Mass. *Mass* is a property of matter by which we can compare the action of one body with that of another. This property manifests itself as a gravitational attraction between two bodies and provides a quantitative measure of the resistance of matter to a change in velocity.

Force. In general, *force* is considered as a "push" or "pull" exerted by one body on another. This interaction can occur when there is either direct contact between the bodies, such as a person pushing on a wall, or it can occur through a distance when the bodies are physically separated. Examples of the latter type include gravitational, electrical, and magnetic forces. In any case, a force is completely characterized by its magnitude, direction, and point of application.

Idealizations. Models or idealizations are used in mechanics in order to simplify application of the theory. A few of the more important idealizations will now be defined. Others that are noteworthy will be discussed at points where they are needed.

Particle. A *particle* has a mass but a size that can be neglected. For example, the size of the earth is insignificant compared to the size of its orbit, and therefore the earth can be modeled as a particle when studying its orbital motion. When a body is idealized as a particle, the principles of mechanics reduce to a rather simplified form, since the geometry of the body will not be involved in the analysis of the problem.

Rigid Body. A *rigid body* can be considered as a combination of a large number of particles in which all the particles remain at a fixed distance from one another both before and after applying a load. As a result, the material properties of any body that is assumed to be rigid will not have to be considered when analyzing the forces acting on the body. In most cases the actual deformations occurring in structures, machines, mechanisms, and the like are relatively small, and the rigid-body assumption is suitable for analysis.

Concentrated Force. A *concentrated force* represents the effect of a loading which is assumed to act at a point on a body. We can represent this effect by a concentrated force, provided the area over which the load is applied is very small compared to the overall size of the body.

Newton's Three Laws of Motion. The entire subject of rigid-body mechanics is formulated on the basis of Newton's three laws of motion, the validity of which is based on experimental observation. They apply to the motion of a particle as measured from a nonaccelerating reference frame and may be briefly stated as follows.

First Law. A particle originally at rest, or moving in a straight line with constant velocity, will remain in this state provided the particle is *not* subjected to an unbalanced force.

Second Law. A particle acted upon by an *unbalanced force* **F** experiences an acceleration **a** that has the same direction as the force and a magnitude that is directly proportional to the force.* If **F** is applied to a particle of mass m, this law may be expressed mathematically as

$$\mathbf{F} = m\mathbf{a} \qquad (1-1)$$

Third Law. The mutual forces of action and reaction between two particles are equal, opposite, and collinear.

---

*Stated another way, the unbalanced force acting on the particle is proportional to the time rate of change of the particle's linear momentum.

Newton's Law of Gravitational Attraction. Shortly after formulating his three laws of motion, Newton postulated a law governing the gravitational attraction between any two particles. Stated mathematically,

$$F = G\frac{m_1 m_2}{r^2} \tag{1-2}$$

where

F = force of gravitation between the two particles

G = universal constant of gravitation; according to experimental evidence, $G = 66.73(10^{-12})$ m^3/(kg · s^2)

m_1, m_2 = mass of each of the two particles

r = distance between the two particles

Weight. According to Eq. 1–2, any two particles or bodies have a mutual attractive (gravitational) force acting between them. In the case of a particle located at or near the surface of the earth, however, the only gravitational force having any sizable magnitude is that between the earth and the particle. Consequently, this force, termed the *weight,* will be the only gravitational force considered in our study of mechanics.

From Eq. 1–2, we can develop an approximate expression for finding the weight W of a particle having a mass $m_1 = m$. If we assume the earth to be a nonrotating sphere of constant density and having a mass m_2, then if r is the distance between the earth's center and the particle, we have

$$W = G\frac{mm_2}{r^2}$$

Letting $g = Gm_2/r^2$ yields

$$\boxed{W = mg} \tag{1-3}$$

By comparison with Eq. 1–1, we term g the acceleration due to gravity. Since it depends on r, it can be seen that the weight of a body is *not* an absolute quantity. Instead, its magnitude is determined from where the measurement was made. For most engineering calculations, however, g is determined at sea level and at a latitude of 45°, which is considered the "standard location."

1.3 Units of Measurement

The four basic quantities—length, time, mass, and force—are not all independent from one another; in fact, they are *related* by Newton's second law of motion, $\mathbf{F} = m\mathbf{a}.$ Hence, the *units* used to define force, mass, length, and time cannot *all* be selected arbitrarily. The equality $\mathbf{F} = m\mathbf{a}$ is maintained only if three of the four units, called *base units,* are *arbitrarily defined* and the fourth unit is *derived* from the equation.

SI Units. The International System of units, abbreviated SI after the French "Système International d'Unités," is a modern version of the metric system which has received worldwide recognition. As shown in Table 1–1, the SI system specifies length in meters (m), time in seconds (s), and mass in kilograms (kg). The unit of force, called a newton (N), is *derived* from $F = m\mathbf{a}$. Thus, 1 newton is equal to a force required to give 1 kilogram of mass an acceleration of 1 m/s^2 (N = kg · m/s^2).

If the weight of a body located at the "standard location" is to be determined in newtons, then Eq. 1–3 must be applied. Here $g = 9.806\ 65$ m/s^2; however, for calculations, the value $g = 9.81$ m/s^2 will be used. Thus,

$$W = mg \qquad (g = 9.81 \text{ m/s}^2) \qquad (1\text{--}4)$$

Therefore, a body of mass 1 kg has a weight of 9.81 N, a 2-kg body weighs 19.62 N, and so on.

U.S. Customary. In the U.S. Customary system of units (FPS) length is measured in feet (ft), force in pounds (lb), and time in seconds (s), Table 1–1. The unit of mass, called a *slug,* is *derived* from $F = m\mathbf{a}$. Hence, 1 slug is equal to the amount of matter accelerated at 1 ft/s^2 when acted upon by a force of 1 lb (slug = lb · s^2/ft).

In order to determine the mass of a body having a weight measured in pounds, we must apply Eq. 1–3. If the measurements are made at the "standard location," then $g = 32.2$ ft/s^2 will be used for calculations. Therefore,

$$m = \frac{W}{g} \qquad (g = 32.2 \text{ ft/s}^2) \qquad (1\text{--}5)$$

And so a body weighing 32.2 lb has a mass of 1 slug, a 64.4-lb body has a mass of 2 slugs, and so on.

Table 1–1 Systems of Units

Name	SI	FPS
Length	meter (m)	foot (ft)
Time	second (s)	second (s)
Mass	kilogram (kg)	slug* $\left(\dfrac{\text{lb} \cdot \text{s}}{\text{ft}}\right)$
Force	newton* (N) $\left(\dfrac{\text{kg} \cdot \text{m}}{\text{s}^2}\right)$	pound (lb)

*Derived unit.

Conversion of Units. In some cases it may be necessary to convert from one system of units to another. In this regard, Table 1–2 provides a set of direct conversion factors between FPS and SI units for the basic quantities. Also, in the FPS system, recall that 1 ft = 12 in. (inches), 5280 ft = 1 mi (mile), 1000 lb = 1 kip (kilo-pound), and 2000 lb = 1 ton.

Table 1–2 Conversion Factors

Quantity	Unit of Measurement (FPS)	Equals	Unit of Measurement (SI)
Force	lb		4.448 2 N
Mass	slug		14.593 8 kg
Length	ft		0.304 8 m

1.4 The International System of Units

The SI system of units is used extensively in this book since it is intended to become the worldwide standard for measurement. Consequently, the rules for its use and some of its terminology relevant to mechanics will now be presented.

Prefixes. When a numerical quantity is either very large or very small, the units used to define its size may be modified by using a prefix. Some of the prefixes used in the SI system are shown in Table 1–3. Each represents a multiple or submultiple of a unit which, if applied successively, moves the decimal point of a numerical quantity to every third place.* For example, 4 000 000 N = 4 000 kN (kilo-newton) = 4 MN (mega-newton), or 0.005 m = 5 mm (milli-meter). Notice that the SI system does not include the multiple deca (10) or the submultiple centi (0.01), which form part of the metric system. Except for some volume and area measurements, the use of these prefixes is to be avoided in science and engineering.

Table 1–3 Prefixes

	Exponential Form	Prefix	SI Symbol
Multiple			
1 000 000 000	10^9	giga	G
1 000 000	10^6	mega	M
1 000	10^3	kilo	k
Submultiple			
0.001	10^{-3}	milli	m
0.000 001	10^{-6}	micro	μ
0.000 000 001	10^{-9}	nano	n

*The kilogram is the only base unit that is defined with a prefix.

Rules for Use. The following rules are given for the proper use of the various SI symbols:

1. A symbol is *never* written with a plural "s," since it may be confused with the unit for second (s).

2. Symbols are always written in lowercase letters, with the following exceptions: symbols for the two largest prefixes shown in Table 1–3, giga and mega, are capitalized as G and M, respectively; and symbols named after an individual are also capitalized, e.g., N.

3. Quantities defined by several units which are multiples of one another are separated by a *dot* to avoid confusion with prefix notation, as indicated by $N = kg \cdot m/s^2 = kg \cdot m \cdot s^{-2}$. Also, m · s (meter-second), whereas ms (milli-second).

4. The exponential power represented for a unit having a prefix refers to both the unit *and* its prefix. For example, $\mu N^2 = (\mu N)^2 = \mu N \cdot \mu N$. Likewise, mm^2 represents $(mm)^2 = mm \cdot mm$.

5. Physical constants or numbers having several digits on either side of the decimal point should be reported with a *space* between every three digits rather than with a comma; e.g., 73 569.213 427. In the case of four digits on either side of the decimal, the spacing is optional; e.g., 8537 or 8 537. Furthermore, always try to use decimals and avoid fractions; that is, write 15.25, *not* $15\frac{1}{4}$.

6. When performing calculations, represent the numbers in terms of their *base or derived units* by converting all prefixes to powers of 10. The final result should then be expressed using a *single prefix*. Also, after calculation, it is best to keep numerical values between 0.1 and 1000; otherwise, a suitable prefix should be chosen. For example,

$$(50 \text{ kN})(60 \text{ nm}) = [50(10^3) \text{ N}][60(10^{-9}) \text{ m}]$$
$$= 3000(10^{-6}) \text{ N} \cdot \text{m} = 3(10^{-3}) \text{ N} \cdot \text{m} = 3 \text{ mN} \cdot \text{m}$$

7. Compound prefixes should not be used; e.g., kμs (kilo-micro-second) should be expressed as ms (milli-second) since $1 \text{ k}\mu\text{s} = 1(10^3)(10^{-6}) \text{ s} = 1(10^{-3}) \text{ s} = 1 \text{ ms}$.

8. With the exception of the base unit the kilogram, in general avoid the use of a prefix in the denominator of composite units. For example, do not write N/mm, but rather kN/m; also, m/mg should be written as mm/kg.

9. Although not expressed in multiples of 10, the minute, hour, etc., are retained for practical purposes as multiples of the second. Furthermore, plane angular measurement is made using radians (rad). In this book, however, degrees will often be used, where $180° = \pi$ rad.

1.5 Numerical Calculations

Numerical work in engineering practice is most often performed by using hand-held calculators and computers. It is important, however, that the answers to any problem be reported with both justifiable accuracy and appropriate significant figures. In this section we will discuss these topics together with some other important aspects involved in all engineering calculations.

Dimensional Homogeneity. The terms of any equation used to describe a physical process must be *dimensionally homogeneous;* that is, each term must be expressed in the same units. Provided this is the case, all the terms of an equation can then be combined if numerical values are substituted for the variables. Consider, for example, the equation $s = vt + \frac{1}{2}at^2$, where, in SI units, s is the position in meters, m, t is time in seconds, s, v is velocity in m/s, and a is acceleration in m/s^2. Regardless of how this equation is evaluated, it maintains its dimensional homogeneity. In the form stated each of the three terms is expressed in meters [m, (m/s̸)s̸, (m/s̸^{2})s̸^{2}], or solving for a, $a = 2s/t^2 - 2v/t$, the terms are each expressed in units of m/s^2 [m/s^2, m/s^2, (m/s)/s].

Since problems in mechanics involve the solution of dimensionally homogeneous equations, the fact that all terms of an equation are represented by a consistent set of units can be used as a partial check for algebraic manipulations of an equation.

Significant Figures. The accuracy of a number is specified by the number of significant figures it contains. A *significant figure* is any digit, including a zero, provided it is not used to specify the location of the decimal point for the number. For example, the numbers 5604 and 34.52 each have four significant figures. When numbers begin or end with zeros, however, it is difficult to tell how many significant figures are in the number. Consider the number 40. Does it have one (4), or perhaps two (40) significant figures? In order to clarify this situation, the number should be reported using powers of 10. There are two ways of doing this. The format for *scientific notation* specifies one digit to the left of the decimal point, with the remaining digits to the right; for example, 40 expressed to one significant figure would be $4(10^1)$. Using *engineering notation,* which is preferred here, the exponent is displayed in multiples of three in order to facilitate conversion of SI units to those having an appropriate prefix. Thus, 40 expressed to one significant figure would be $0.04(10^3)$. Likewise, 2500 and 0.00546 expressed to three significant figures would be $2.50(10^3)$ and $5.46(10^{-3})$.

Rounding Off Numbers.   For numerical calculations, the accuracy obtained from the solution of a problem generally can never be better than the accuracy of the problem data. This is what is to be expected, but often hand-held calculators or computers involve more figures in the answer than the number of significant figures used for the data. For this reason, a calculated result should always be "rounded off" to an appropriate number of significant figures.

To ensure accuracy, the following rules for rounding off a number to n significant figures apply:

1. If the $n + 1$ digit is *less than 5,* the $n + 1$ digit and others following it are dropped. For example, 2.326 and 0.451 rounded off to $n = 2$ significant figures would be 2.3 and 0.45.

2. If the $n + 1$ digit is equal to 5 with zeros following it, then round off the nth digit to an *even number*. For example, 1245 and 0.8655 rounded off to $n = 3$ significant figures become 1240 and 0.866.

3. If the $n + 1$ digit is *greater than 5* or equal to 5 with any nonzero digits following it, then increase the nth digit by 1 and drop the $n + 1$ digit and others following it. For example, 0.723 87 and 565.500 3 rounded off to $n = 3$ significant figures become 0.724 and 566.

Calculations.   As a general rule, to ensure accuracy of a final result when performing calculations with numbers of unequal accuracy, always retain one extra significant figure in the more accurate numbers than in the least accurate number *before* beginning the computations. Then round off the final result so that it has the same number of significant figures as the least accurate number. If possible, try to work out the computations so that numbers which are approximately equal are not subtracted, since accuracy is often lost from this calculation.

Most of the example problems in this book are solved with the assumption that any measured *data* are accurate to three significant figures.* Consequently, intermediate calculations are often worked out to four significant figures and the answers are generally reported to *three* significant figures.

The following examples illustrate application of the principles just discussed as related to the proper use and conversion of units.

---

*Of course, some numbers, such as π, e, or numbers used in derived formulas, are exact and are therefore accurate to an infinite number of significant figures.

Example 1–1

Convert 2 km/h to m/s.

SOLUTION

Since 1 km = 1000 m and 1 h = 3600 s, the factors of conversion are arranged in the following order, so that a cancellation of the units can be applied:

$$2 \text{ km/h} = \frac{2 \text{ km}}{\text{h}}\left(\frac{1000 \text{ m}}{\text{km}}\right)\left(\frac{1 \text{ h}}{3600 \text{ s}}\right)$$

$$= \frac{2000 \text{ m}}{3600 \text{ s}} = 0.556 \text{ m/s} \qquad\qquad Ans.$$

Example 1–2

Convert the quantity 300 lb · s to appropriate SI units.

SOLUTION

Using Table 1–2, 1 lb = 4.448 2 N.

$$300 \text{ lb} \cdot \text{s} = 300 \text{ lb} \cdot \text{s}\left(\frac{4.448 \ 2 \text{ N}}{\text{lb}}\right)$$

$$= 1334.5 \text{ N} \cdot \text{s} = 1.33 \text{ kN} \cdot \text{s} \qquad\qquad Ans.$$

Example 1–3

Evaluate each of the following and express with SI units having an appropriate prefix: (a) (50 mN)(6 GN), (b) (400 mm)(0.6 MN)^2, (c) 45 MN^3/900 Gg.

SOLUTION

First convert each number to base units, perform the indicated operations, then choose an appropriate prefix (see Rule 6 on p. 7).

Part (a)

$$(50 \text{ mN})(6 \text{ GN}) = [50(10^{-3}) \text{ N}][6(10^{9}) \text{ N}]$$
$$= 300(10^{6}) \text{ N}^2$$
$$= 300(10^{6}) \text{ N}^2\left(\frac{1 \text{ kN}}{10^3 \text{ N}}\right)\left(\frac{1 \text{ kN}}{10^3 \text{ N}}\right)$$
$$= 300 \text{ kN}^2 \qquad\qquad Ans.$$

Note carefully the convention kN^2 = (kN)^2 = 10^6 N^2 (Rule 4 on p. 7).

Part (b)

$$(400 \text{ mm})(0.6 \text{ MN})^2 = [400(10^{-3}) \text{ m}][0.6(10^{6}) \text{ N}]^2$$
$$= [400(10^{-3}) \text{ m}][0.36(10^{12}) \text{ N}^2]$$
$$= 144(10^{9}) \text{ m} \cdot \text{N}^2$$
$$= 144 \text{ Gm} \cdot \text{N}^2 \qquad\qquad Ans.$$

We can also write

$$144(10^{9}) \text{ m} \cdot \text{N}^2 = 144(10^{9}) \text{ m} \cdot \text{N}^2\left(\frac{1 \text{ MN}}{10^6 \text{ N}}\right)\left(\frac{1 \text{ MN}}{10^6 \text{ N}}\right)$$
$$= 0.144 \text{ m} \cdot \text{MN}^2$$

Part (c)

$$45 \text{ MN}^3/900 \text{ Gg} = \frac{45(10^{6} \text{ N})^3}{900(10^{6}) \text{ kg}}$$
$$= 0.05(10^{12}) \text{ N}^3/\text{kg}$$
$$= 0.05(10^{12}) \text{ N}^3\left(\frac{1 \text{ kN}}{10^3 \text{ N}}\right)^3 \frac{1}{\text{kg}}$$
$$= 0.05(10^{3}) \text{ kN}^3/\text{kg}$$
$$= 50 \text{ kN}^3/\text{kg} \qquad\qquad Ans.$$

Here we have used rules 4 and 8 of p. 7.

1.6 General Procedure for Analysis

The most effective way of learning the principles of engineering mechanics is to *solve problems*. To be successful at this, it is important always to present the work in a *logical* and *orderly manner,* as suggested by the following sequence of steps:

1. Read the problem carefully and try to correlate the actual physical situation with the theory studied.

2. Draw any necessary diagrams and tabulate the problem data.

3. Apply the relevant principles, generally in mathematical form.

4. Solve the necessary equations algebraically as far as practical, then, making sure they are dimensionally homogeneous, use a consistent set of units and complete the solution numerically. Report the answer with no more significant figures than the accuracy of the given data.

5. Study the answer with technical judgment and common sense to determine whether or not it seems reasonable.

6. Once the solution has been completed, review the problem. Try to think of other ways of obtaining the same solution.

In applying this general procedure, do the work as neatly as possible. Being neat generally stimulates clear and orderly thinking, and vice versa.

PROBLEMS

1–1. Round off the following numbers to three significant figures: (*a*) 3.45555 m, (*b*) 45.556 s, (*c*) 5555 N, (*d*) 4525 kg.

1–2. If a car is traveling at 55 mi/h, determine its speed in kilometers per hour and meters per second.

1–3. If a car weighs 3500 lb, determine its mass and express the result using SI units.

*1–4. Convert 63 ft$^2 \cdot$ s to m$^2 \cdot$ s.

1–5. Represent each of the following combinations of units in the correct SI form using an appropriate prefix: (*a*) Mg/mm, (*b*) mN/μs, (*c*) μm $\cdot$ Mg.

1–6. Represent each of the following combinations of units in the correct SI form using an appropriate prefix: (*a*) m/ms, (*b*) μkm, (*c*) ks/mg, (*d*) km $\cdot$ μN.

1–7. Represent each of the following quantities in the correct SI form using an appropriate prefix: (*a*) 0.000431 kg, (*b*) 35.3(10^3) N, (*c*) 0.00532 km.

*1–8. Evaluate each of the following to three significant figures and express each answer in SI units using an appropriate prefix: (*a*) (212 mN)^2, (*b*) (52 800 ms)^2, (*c*) [548(10^6)]$^{1/2}$ ms.

1–9. Evaluate each of the following to three significant figures and express each answer in SI units using an appropriate prefix: (*a*) 0.631 Mm/(8.60 kg)^2, (*b*) (35 mm)^2(48 kg)^3.

1–10. Evaluate (204 mm)(0.004 57 kg)/(34.6 N) and express the answer in SI units using an appropriate prefix.

1–11. Evaluate each of the following and express each answer in SI units using an appropriate prefix: (*a*) (684 μm)/43 ms, (*b*) (28 ms)(0.0458 Mm)/(348 mg), (*c*) (2.68 mm)(426 Mg).

*1–12. Evaluate each of the following and express each answer in SI units using an appropriate prefix: (*a*) 354 mg(45 km)/(0.035 6 kN), (*b*) (.004 53 Mg)(201 ms), (*c*) 435 MN/23.2 mm.

1–13. Convert each of the following and express the answer using an appropriate prefix: (*a*) 175 lb/ft^3 to kN/m^3, (*b*) 6 ft/h to mm/s, (*c*) 835 lb $\cdot$ ft to kN $\cdot$ m.

1–14. The specific weight (wt./vol.) of brass is 520 lb/ft^3. Determine its density (mass/vol.) in SI units. Use an appropriate prefix.

1–15. Determine the mass in kilograms of an object that has a weight of (*a*) 20 mN, (*b*) 150 kN, (*c*) 60 MN. Express each answer using an appropriate prefix.

*1–16. Using Table 1–3, determine your own mass in kilograms, your weight in newtons, and your height in meters.

1–17. A rocket has a mass of 250(10^3) slugs on earth. Specify (*a*) its mass in SI units, and (*b*) its weight in SI units. If the rocket is on the moon, where the acceleration due to gravity is $g_m = 5.30$ ft/s^2, determine (*c*) its weight in SI units, and (*d*) its mass in SI units.

1–18. The density (mass/volume) of aluminum is 5.26 slug/ft^3. Determine its density in SI units. Use an appropriate prefix.

1–19. A concrete column has a diameter of 350 mm and a length of 2 m. If the density (mass/volume) of concrete is 2.45 Mg/m^3, determine the weight of the column in pounds.

*1–20. Two particles have a mass of 350 kg and 250 kg, respectively. If they are 4 m apart, determine the force of gravitational attraction between them.

2 Force Vectors

In this chapter we will introduce the concept of a concentrated force and give the procedures for adding forces, resolving them into components, and projecting them along an axis. Since force is a vector quantity, we must use the rules of vector algebra whenever forces are considered. We will begin our study by defining scalar and vector quantities and then develop some of the basic rules of vector algebra.

2.1 Scalars and Vectors

Most of the physical quantities in mechanics can be expressed mathematically by means of scalars and vectors.

Scalar. A quantity characterized by a positive or negative number is called a *scalar*. Mass, volume, and length are scalar quantities often used in statics. In this book, scalars are indicated by letters in italic type, such as the scalar A. The mathematical operations involving scalars follow the same rules as those of elementary algebra.

Vector. A *vector* is a quantity that has both a magnitude, direction, and sense, and obeys the parallelogram law of addition. This law, which will be described later, utilizes a form of construction that accounts for the combined magnitude and direction of the vector. Vector quantities commonly used in statics are position, force, and moment vectors.

Fig. 2–1

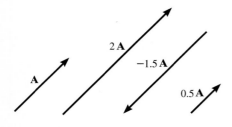

Fig. 2–2

A vector is represented graphically by an arrow, which is used to define its magnitude, direction, and sense. The *magnitude* of the vector is indicated by the length of the arrow, the *direction* is defined by the angle between a reference axis and the arrow's line of action, and the *sense* is indicated by the arrowhead. For example, the vector **A** shown in Fig. 2–1 has a magnitude of 4 units, a direction which is 20° measured clockwise from the horizontal axis, and a sense which is upward and to the right. The point *O* is called the *tail* of the vector, the point *P* is the *tip* or *head*.

For handwritten work, a vector is generally represented by a letter with an arrow written over it, such as $\vec{A}$. The magnitude is designated $|\vec{A}|$ or simply A. In this book vectors will be symbolized in boldface type; for example, **A** is used to designate the vector "A". Its magnitude, which is always a positive quantity, is symbolized in italic type, written as $|A|$, or simply A when it is understood that A is a positive scalar.

2.2 Vector Operations

Scalar Multiplication and Division

Fig. 2–3

Multiplication and Division of a Vector by a Scalar. The product of vector **A** and scalar a, yielding a**A,** is defined as a vector having a magnitude $|aA|$. The *sense* of a**A** is the *same* as **A** provided a is *positive;* it is *opposite* to **A** if a is *negative*. Consequently, the negative of a vector is formed by multiplying the vector by the scalar (-1), Fig. 2–2. Division of a vector by a scalar can be defined using the laws of multiplication, since $\mathbf{A}/a = (1/a)\mathbf{A}$, $a \neq 0$. Graphic examples of these operations are shown in Fig. 2–3.

Vector Addition. Two vectors **A** and **B** of the same type, Fig. 2–4*a*, may be added to form a "resultant" vector $\mathbf{R} = \mathbf{A} + \mathbf{B}$ by using the *parallelogram law*. To do this, **A** and **B** are joined at their tails, Fig. 2–4*b*. Parallel dashed lines drawn from the head of each vector intersect at a common point, thereby forming the adjacent sides of a parallelogram. As shown, the resultant **R** is the diagonal of the parallelogram, which extends from the tails of **A** and **B** to the intersection of the dashed lines.

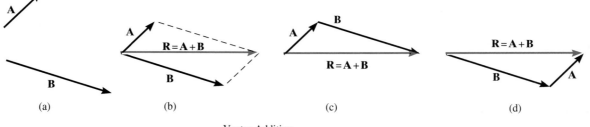

Vector Addition

Fig. 2–4

We can also add **B** to **A** using a *triangle construction,* which is a special case of the parallelogram law, whereby vector **B** is added to vector **A** in a "head-to-tail" fashion, i.e., by connecting the head of **A** to the tail of **B,** Fig. 2–4c. The resultant **R** extends from the tail of **A** to the head of **B.** In a similar manner, **R** can also be obtained by adding **A** to **B,** Fig. 2–4d. By comparison, it is seen that vector addition is commutative; in other words, the vectors can be added in either order, i.e., **R** = **A** + **B** = **B** + **A.**

As a special case, if the two vectors **A** and **B** are *collinear,* i.e., both have the same line of action, the parallelogram law reduces to an *algebraic* or *scalar addition R = A + B,* as shown in Fig. 2–5.

$R = A+B$

Fig. 2–5

Vector Subtraction. The resultant *difference* between two vectors **A** and **B** of the same type may be expressed as

$$\mathbf{R'} = \mathbf{A} - \mathbf{B} = \mathbf{A} + (-\mathbf{B})$$

This vector sum is shown graphically in Fig. 2–6. Subtraction is therefore defined as a special case of addition, so the rules of vector addition also apply to vector subtraction.

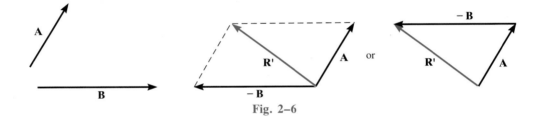

Fig. 2–6

Resolution of a Vector. A vector may be resolved into two "components" having known lines of action by using the parallelogram law. For example, if **R** in Fig. 2–7a is to be resolved into components acting along the lines *a* and *b*, one starts at the *head* of **R** and extends a dashed line *parallel* to *a* until it intersects *b*. Likewise, a dashed line parallel to *b* is drawn from the *head* of **R** to the point of intersection with *a*, Fig. 2–7a. The two components **A** and **B** are then drawn such that they extend from the tail of **R** to the points of intersection, as shown in Fig. 2–7b.

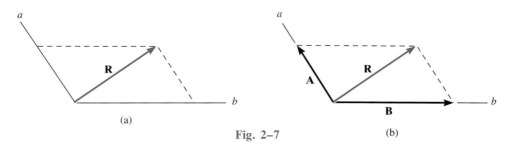

(a)

(b)

Fig. 2–7

2.3 Vector Addition of Forces

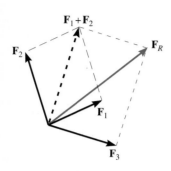

Fig. 2–8

Experimental evidence has shown that a force is a vector quantity since it has a specified magnitude, direction, and sense and it adds according to the parallelogram law. Two common problems in statics involve finding either the resultant force, knowing its components, or resolving a known force into two components. As described in Sec. 2–2, both of these problems require application of the parallelogram law.

If more than two forces are to be added, successive applications of the parallelogram law can be carried out in order to obtain the resultant force. For example, if three forces F_1, F_2, F_3 act at point O, Fig. 2–8, the resultant of any two of the forces is found—say, $F_1 + F_2$—and then this resultant is added to the third force, yielding the resultant of all three forces; i.e., $F_R = (F_1 + F_2) + F_3$. Using the parallelogram law to add more than two forces, as shown here, often requires extensive geometric and trigonometric calculation to determine the numerical values for the magnitude and direction of the resultant. Instead, problems of this type are easily solved by using the "rectangular-component method," which is explained in Sec. 2.4.

PROCEDURE FOR ANALYSIS

Problems that involve the addition of two forces and contain at most *two unknowns* can be solved by using the following procedure:

Parallelogram Law.   Make a sketch showing the vector addition using the parallelogram law. If possible, determine the interior angles of the parallelogram from the geometry of the problem. Recall that the sum total of these angles is 360°. Unknown angles, along with known and unknown force magnitudes, should be clearly labeled on this sketch. Redraw a half portion of the constructed parallelogram to illustrate the triangular head-to-tail addition of the components.

Trigonometry.   By using trigonometry, the two unknowns can be determined from the data listed on the triangle. If the triangle does *not* contain a 90° angle, the law of sines and/or the law of cosines may be used for the solution. These formulas are given in Fig. 2–9 for the triangle shown.

The following examples illustrate this method numerically.

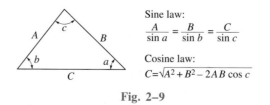

Sine law:
$$\frac{A}{\sin a} = \frac{B}{\sin b} = \frac{C}{\sin c}$$

Cosine law:
$$C = \sqrt{A^2 + B^2 - 2AB \cos c}$$

Fig. 2–9

Example 2–1

The screw eye in Fig. 2–10a is subjected to two forces, $\mathbf{F}_1$ and $\mathbf{F}_2$. Determine the magnitude and direction of the resultant force.

(a)

(b)

SOLUTION

Parallelogram Law. The parallelogram law of addition is shown in Fig. 2–10b. The two unknowns are the magnitude of $\mathbf{F}_R$ and the angle θ (theta). From Fig. 2–10b, the vector triangle, Fig. 2–10c, is constructed.

Trigonometry. F_R is determined by using the law of cosines:

$$F_R = \sqrt{(100)^2 + (150)^2 - 2(100)(150)\cos 115°}$$
$$= \sqrt{10\,000 + 22\,500 - 30\,000(-0.4226)} = 212.6 \text{ N}$$
$$= 213 \text{ N} \qquad\qquad Ans.$$

The angle θ is determined by applying the law of sines, using the computed value of F_R.

$$\frac{150}{\sin \theta} = \frac{212.6}{\sin 115°}$$

$$\sin \theta = \frac{150}{212.6}(0.9063)$$

$$\theta = 39.8°$$

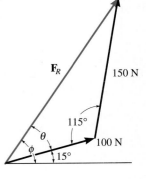

(c)

Fig. 2–10

Thus, the direction ϕ (phi) of $\mathbf{F}_R$, measured from the horizontal, is

$$\phi = 39.8° + 15.0° = 54.8° \quad \measuredangle^\phi \qquad\qquad Ans.$$

Example 2–2

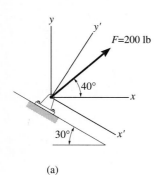

(a)

Resolve the 200-lb force shown acting on the pin, Fig. 2–11a, into components in the (a) x and y directions, and (b) x' and y directions.

SOLUTION

In each case the parallelogram law is used to resolve $\mathbf{F}$ into its two components, and then the vector triangle is constructed to determine the numerical results by trigonometry.

Part (a).   The vector addition $\mathbf{F} = \mathbf{F}_x + \mathbf{F}_y$ is shown in Fig. 2–11b. In particular, note that the length of the components is scaled along the x and y axes by first constructing dashed lines parallel to the axes in accordance with the parallelogram law. From the vector triangle, Fig. 2–11c,

$$F_x = 200 \cos 40° = 153 \text{ lb} \qquad \textit{Ans.}$$
$$F_y = 200 \sin 40° = 129 \text{ lb} \qquad \textit{Ans.}$$

(d)

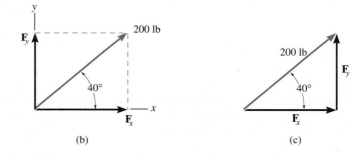

(b)   (c)

Fig. 2–11

Part (b).   The vector addition $\mathbf{F} = \mathbf{F}_{x'} + \mathbf{F}_y$ is shown in Fig. 2–11d. Note carefully how the parallelogram is constructed. Applying the law of sines and using the data listed on the vector triangle, Fig. 2–11e, yields

$$\frac{F_{x'}}{\sin 50°} = \frac{200}{\sin 60°}$$

$$F_{x'} = 200\left(\frac{\sin 50°}{\sin 60°}\right) = 177 \text{ lb} \qquad \textit{Ans.}$$

$$\frac{F_y}{\sin 70°} = \frac{200}{\sin 60°}$$

$$F_y = 200\left(\frac{\sin 70°}{\sin 60°}\right) = 217 \text{ lb} \qquad \textit{Ans.}$$

(e)

Example 2–3

The force **F** acting on the frame shown in Fig. 2–12*a* has a magnitude of 500 N and is to be resolved into two components acting along struts *AB* and *AC*. Determine the angle θ, measured *below* the horizontal, so that the component $\mathbf{F}_{AC}$ is directed from *A* toward *C* and has a magnitude of 400 N.

$F = 500$ N

(a)

Fig. 2–12

$F_{AC} = 400$ N

(b)

400 N

(c)

SOLUTION

By using the parallelogram law, the vector addition of the two components yielding the resultant is shown in Fig. 2–12*b*. Note carefully how the resultant force is resolved into the two components $\mathbf{F}_{AB}$ and $\mathbf{F}_{AC}$, which have specified lines of action. The corresponding vector triangle is shown in Fig. 2–12*c*. The angle ϕ can be determined by using the law of sines:

$$\frac{400}{\sin \phi} = \frac{500}{\sin 60°}$$

$$\sin \phi = \left(\frac{400}{500}\right) \sin 60°$$

$$\phi = 43.9°$$

Hence,

$$\theta = 180° - 60° - 43.9° = 76.1° \qquad \mathbf{Ans.}$$

Using this value for θ, apply the law of cosines and show that $\mathbf{F}_{AB}$ has a magnitude of 560 N.

Notice that **F** can also be directed at an angle θ *above* the horizontal, as shown in Fig. 2–12*d*, and still produce the required component $\mathbf{F}_{AC}$. Show that in this case $\theta = 16.1°$ and $F_{AB} = 161$ N.

(d)

Example 2–4

The ring shown in Fig. 2–13a is subjected to two forces, $\mathbf{F}_1$ and $\mathbf{F}_2$. If it is required that the resultant force have a magnitude of 1 kN and be directed vertically downward, determine (a) the magnitudes of $\mathbf{F}_1$ and $\mathbf{F}_2$ provided $\theta = 30°$, and (b) the magnitudes of $\mathbf{F}_1$ and $\mathbf{F}_2$ if F_2 is to be a minimum.

(a)

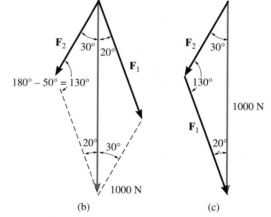

(b)　　　　　(c)

SOLUTION

Part (a). A sketch of the vector addition according to the parallelogram law is shown in Fig. 2–13b. From the vector triangle constructed in Fig. 2–13c, the unknown magnitudes F_1 and F_2 can be determined by using the law of sines.

$$\frac{F_1}{\sin 30°} = \frac{1000}{\sin 130°}$$

$$F_1 = 653 \text{ N} \qquad\qquad \textit{Ans.}$$

$$\frac{F_2}{\sin 20°} = \frac{1000}{\sin 130°}$$

$$F_2 = 446 \text{ N} \qquad\qquad \textit{Ans.}$$

Part (b). If θ is not specified, then by the vector triangle, Fig. 2–13d, $\mathbf{F}_2$ may be added to $\mathbf{F}_1$ in various ways to yield the resultant 1000-N force. In particular, the *minimum* length or magnitude of $\mathbf{F}_2$ will occur when its line of action is *perpendicular to* $\mathbf{F}_1$. Any other direction, such as OA or OB, yields a larger value for F_2. Hence, when $\theta = 90° - 20° = 70°$, F_2 is minimum. From the triangle shown in Fig. 2–13e, it is seen that

$$F_1 = 1000 \sin 70° = 940 \text{ N} \qquad \textit{Ans.}$$

$$F_2 = 1000 \sin 20° = 342 \text{ N} \qquad \textit{Ans.}$$

(d)　　　　　(e)

Fig. 2–13

PROBLEMS

2–1. Determine the magnitude of the resultant force $\mathbf{F}_R =$ $\mathbf{F}_1 + \mathbf{F}_2$ and its orientation θ, measured counterclockwise from the positive x axis.

2–2. Determine the magnitude of the resultant force $\mathbf{F}_R =$ $\mathbf{F}_1 - \mathbf{F}_2$ and its orientation θ, measured counterclockwise from the positive x axis.

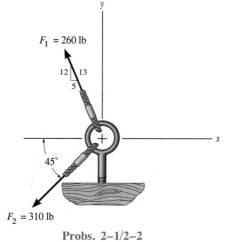

Probs. 2–1/2–2

2–5. Determine the magnitude of the resultant force $\mathbf{F}_R =$ $\mathbf{F}_1 + \mathbf{F}_2$ and its orientation θ, measured counterclockwise from the positive u axis.

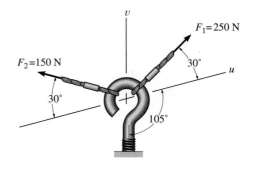

Prob. 2–5

2–3. Determine the magnitude of the resultant force $\mathbf{F}_R =$ $\mathbf{F}_1 + \mathbf{F}_2$ and its orientation θ, measured clockwise from the positive x axis.

*2–4. Determine the magnitude of the resultant force $\mathbf{F}_R =$ $\mathbf{F}_1 + \mathbf{F}_3$ and its orientation θ, measured counterclockwise from the positive x axis.

Probs. 2–3/2–4

2–6. Resolve the force $\mathbf{F}_1$ into components acting along the u and v axes and determine the magnitudes of the components.

2–7. Resolve the force $\mathbf{F}_2$ into components acting along the u and v axes and determine the magnitudes of the components.

Probs. 2–6/2–7

*2–8. The screw eye is subjected to the two forces shown. **F** is directed along the inclined plane and the 450-N force is horizontal. Determine the magnitude of **F** so that the resultant force **F**$_R$ is directed normal to the plane, i.e., along the positive n axis. What is the magnitude of **F**$_R$?

Prob. 2–8

2–9. The plate is subjected to the forces acting on members A and B as shown. If $\theta = 60°$, determine the magnitude of the resultant of these forces and its direction measured clockwise from the positive x axis.

2–10. Determine the angle θ for connecting member A to the plate so that the resultant angle of **F**$_A$ and **F**$_B$ is directed horizontally to the right. What is the magnitude of the resultant force?

Probs. 2–9/2–10

2–11. Wire B is subjected to a force of 2 kN and is directed at 60° from the horizontal. Determine the force **F**$_A$ in wire A when $\theta = 45°$ if the resultant force of the two wires is directed vertically downward along wire C. Also calculate the magnitude of the resultant force.

*2–12. Wire B is subjected to a force of 2 kN and is directed at 60° from the horizontal. If the resultant force of wires A and B is to be 3 kN, directed vertically downward along wire C, determine the force **F**$_A$ in wire A and the corresponding angle θ.

Probs. 2–11/2–12

2–13. The mast of the sailboat is subjected to the force of the two cables. If the force in cable AB is 80 lb, determine the required force in AC so that the resultant force caused by both cables is directed vertically downward along the axis AD of the mast. Also, calculate this resultant force.

Prob. 2–13

2–14. Determine the design angle θ $(0° \leq \theta \leq 90°)$ for member AB so that the 400-lb horizontal force has a component of 500 lb directed from A toward C. What is the component of force acting along member AB? Take $\phi = 40°$.

2–15. Determine the design angle ϕ $(0° \leq \phi \leq 90°)$ between members AB and AC so that the 400-lb horizontal force has a component of 600 lb which acts up to the right, in the same direction as from B toward A. Also calculate the magnitude of the force component along AC. Take $\theta = 30°$.

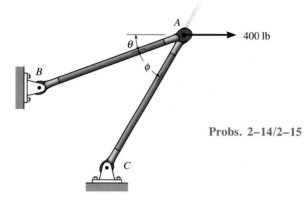

400 lb

Probs. 2–14/2–15

*2–16. The cable exerts a force of 600 N on the frame. Resolve this force into components acting (*a*) along the *x* and *y* axes and (*b*) along the *u* and *v* axes. What is the magnitude of each component?

2–17. The cable exerts a force of 600 N on the frame. Resolve this force into components acting (*a*) along the *x* and *v* axes and (*b*) along the *y* and *u* axes. What is the magnitude of each component?

Probs. 2–16/2–17

600 N

2–18. Determine the orientation θ of the 500-N force such that when the force is resolved into two components acting along members AB and AC, the component of force along AC is 300 N, directed from A toward C. What is the magnitude of the force component acting along AB?

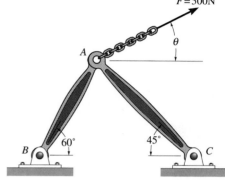

$F = 500\text{N}$

Prob. 2–18

2–19. The horizontal force $F = 500$ N acts to the left at A on the two-member frame. Determine the magnitudes of the two components of $\mathbf{F}$ directed along the axes of members AB and AC.

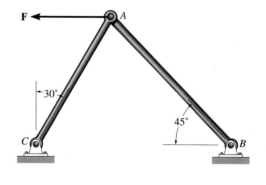

Prob. 2–19

*2–20. Two forces have the same magnitude F and produce a resultant force having a magnitude F_R when the angle between the tails of the two forces is 2ϕ. Show that the resultant force has a magnitude $4F_R$ when the angle between the two forces is 2θ, where $\theta = \cos^{-1} (4 \cos \phi)$.

2–21. Two forces $\mathbf{F}_1$ and $\mathbf{F}_2$ act on the pin. If each force acts at an angle θ with the horizontal and has a magnitude F, determine the magnitude of the resultant force $\mathbf{F}_R$. Plot the value of F_R versus θ.

Prob. 2–21

2–22. Two forces, having a magnitude of 20 lb and 15 lb, act on the rod. If the resultant force has a magnitude of 30 lb, determine the angle θ between the forces.

Prob. 2–22

2–23. The hipbone H is connected to the femur F at A using three different muscles, which exert the forces shown on the femur. Determine the resultant force on the femur and specify its orientation θ measured counterclockwise from the positive x axis.

Prob. 2–23

*** 2–24.** Three chains act on the bracket such that they create a resultant force having a magnitude of 500 lb. If two of the chains are subjected to known forces, as shown, determine the orientation θ of the third chain, measured clockwise from the positive x axis, so that the magnitude of force $\mathbf{F}$ in this chain is a *minimum*. All forces lie in the x–y plane. What is the magnitude of $\mathbf{F}$? *Hint:* First find the resultant of the two known forces. Force $\mathbf{F}$ acts in this direction.

Prob. 2–24

2–25. The jet aircraft is being towed by two trucks B and C. Determine the magnitudes of the two towing forces $\mathbf{F}_B$ and $\mathbf{F}_C$ if the resultant force has a magnitude $F_R = 10$ kN and is directed along the positive x axis. Set $\theta = 15°$.

2–26. If the resultant $\mathbf{F}_R$ of the two forces acting on the jet aircraft is to be directed along the positive x axis and have a magnitude of 10 kN, determine the angle θ of the cable attached to the truck at B such that the force $\mathbf{F}_B$ in this cable is a *minimum*. What is the magnitude of force in each cable when this occurs?

Probs. 2–25/2–26

2.4 Addition of a System of Coplanar Forces

When the resultant of more than two forces has to be obtained, it is easier to find the components of each force along specified axes, add these components algebraically, and then form the resultant, rather than form the resultant of the forces by successive application of the parallelogram law as discussed in Sec. 2.3. In this section we will resolve each force into its rectangular components $\mathbf{F}_x$ and $\mathbf{F}_y$, which lie along the x and y axes, respectively, Fig. 2–14a. Although the axes are shown here to be horizontal and vertical, they may in general be directed at any inclination, as long as they remain perpendicular to one another, Fig. 2–14b. In either case, by the parallelogram law, we require:

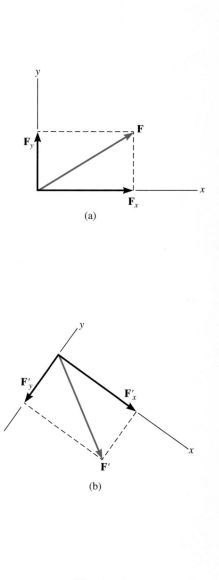

(a)

and

$$\mathbf{F} = \mathbf{F}_x + \mathbf{F}_y$$

$$\mathbf{F}' = \mathbf{F}'_x + \mathbf{F}'_y$$

As shown in Fig. 2–14, the sense of direction of each component is represented *graphically* by the *arrowhead*. For *analytical* work, however, we must establish a notation for representing the directional sense of the rectangular components for each coplanar vector. This can be done in one of two ways.

Scalar Notation. Since the x and y axes have designated positive and negative directions, the magnitude and directional sense of the rectangular components of a force can be expressed in terms of *algebraic scalars*. For example, the components of $\mathbf{F}$ in Fig. 2–14a can be represented by positive scalars F_x and F_y since their sense of direction is along the *positive x* and *y* axes, respectively. In a similar manner, the components of $\mathbf{F}'$ in Fig. 2–14b are F'_x and $-F'_y$. Here the y component is negative, since $\mathbf{F}'_y$ is directed along the negative y axis. It is important to keep in mind that this scalar notation is to be used only for computational purposes, not for graphical representations in figures. Throughout the text, the *head of a vector arrow* in any figure indicates the sense of the vector *graphically;* algebraic signs are not used for this purpose. Thus, the vectors in Figs. 2–14a and 2–14b are designated by using boldface (vector) notation.* Whenever italic symbols are written near vector arrows in figures, they indicate the *magnitude* of the vector, which is *always* a *positive* quantity.

(b)

Fig. 2–14

---

*Negative signs are used only in figures with boldface notation when showing equal but opposite pairs of vectors as in Fig. 2–2.

(a)

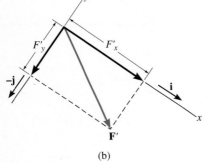

(b)

Fig. 2–15

Cartesian Vector Notation. It is also possible to represent the components of a force in terms of Cartesian unit vectors. By doing this the methods of vector algebra are easier to apply, and we will see that this becomes particularly advantageous for solving problems in three dimensions. In two dimensions the *Cartesian unit vectors* **i** and **j** are used to designate the *directions* of the x and y axes, respectively, Fig. 2–15a.* These vectors have a dimensionless magnitude of unity, and their sense (or arrowhead) will be described analytically by a plus or minus sign, depending on whether they are pointing along the positive or negative x or y axis.

As shown in Fig. 2–15a, the *magnitude* of each component of **F** is *always a positive quantity*, which is represented by the (positive) scalars F_x and F_y. Therefore, having established notation to represent the magnitude and the direction of each component, we can express **F** in Fig. 2–15a as a *Cartesian vector*, i.e.,

$$\mathbf{F} = F_x\mathbf{i} + F_y\mathbf{j}$$

And in the same way, **F'** in Fig. 2–15b can be expressed as

$$\mathbf{F'} = F_x'\mathbf{i} + F_y'(-\mathbf{j})$$

or simply

$$\mathbf{F'} = F_x'\mathbf{i} - F_y'\mathbf{j}$$

Coplanar Force Resultants. Either of the two methods just described for representing the rectangular components of a force can be used to determine the resultant of several *coplanar forces*. To do this, each force is first resolved into its x and y components and then the respective components are added using *scalar algebra* since they are collinear. The resultant force is then formed by adding the resultants of the x and y components using the parallelogram law. For example, consider the three forces in Fig. 2–16a, which have x and y components as shown in Fig. 2–16b. To solve this problem using *Cartesian vector notation,* each force is first represented as a Cartesian vector, i.e.,

$$\mathbf{F}_1 = F_{1x}\mathbf{i} + F_{1y}\mathbf{j}$$
$$\mathbf{F}_2 = -F_{2x}\mathbf{i} + F_{2y}\mathbf{j}$$
$$\mathbf{F}_3 = F_{3x}\mathbf{i} - F_{3y}\mathbf{j}$$

*For handwritten work, unit vectors are usually indicated using a circumflex, e.g., $\hat{\imath}$ and $\hat{\mathbf{j}}$.

The vector resultant is therefore

$$
\begin{aligned}
\mathbf{F}_R &= \mathbf{F}_1 + \mathbf{F}_2 + \mathbf{F}_3 \\
&= F_{1x}\mathbf{i} + F_{1y}\mathbf{j} - F_{2x}\mathbf{i} + F_{2y}\mathbf{j} + F_{3x}\mathbf{i} - F_{3y}\mathbf{j} \\
&= (F_{1x} - F_{2x} + F_{3x})\mathbf{i} + (F_{1y} + F_{2y} - F_{3y})\mathbf{j} \\
&= (F_{Rx})\mathbf{i} + (F_{Ry})\mathbf{j}
\end{aligned}
$$

If *scalar notation* is used, then, from Fig. 2–16b, since x is positive to the right and y is positive upward, we have

$(\xrightarrow{+})$ $$F_{Rx} = F_{1x} - F_{2x} + F_{3x}$$
$(+\uparrow)$ $$F_{Ry} = F_{1y} + F_{2y} - F_{3y}$$

These results are the *same* as the $\mathbf{i}$ and $\mathbf{j}$ components of $\mathbf{F}_R$ determined above.

In the general case, the x and y components of the resultant of any number of coplanar forces can be represented symbolically by the algebraic sum of the x and y components of all the forces, i.e.,

$$
\boxed{\begin{aligned}
F_{Rx} &= \Sigma F_x \\
F_{Ry} &= \Sigma F_y
\end{aligned}}
\qquad (2\text{--}1)
$$

When applying these equations it is important to use the sign convention established for the components; and that is, components having a directional sense along the positive coordinate axes are considered positive scalars, whereas those having a directional sense along the negative coordinate axes are considered negative scalars. If this convention is followed, then the signs of the resultant components will specify the sense of these components. For example, a positive result indicates that the component has a directional sense which is in the positive coordinate direction.

Once the resultant components are determined, they may be sketched along the x and y axes in their proper directions, and the resultant force can be determined from vector addition, as shown in Fig. 2–16c. From this sketch, the magnitude of $\mathbf{F}_R$ is then found from the Pythagorean theorem; that is,

$$F_R = \sqrt{F_{Rx}^2 + F_{Ry}^2}$$

Also, the direction angle θ, which specifies the orientation of the force, is determined from trigonometry.

$$\theta = \tan^{-1}\left|\frac{F_{Ry}}{F_{Rx}}\right|$$

The above concepts are illustrated numerically in the following examples.

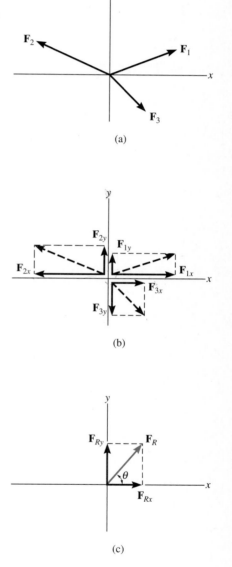

(a)

(b)

(c)

Fig. 2–16

■ Example 2–5 ■

Determine the x and y components of $\mathbf{F}_1$ and $\mathbf{F}_2$ shown in Fig. 2–17a. Express each force as a Cartesian vector.

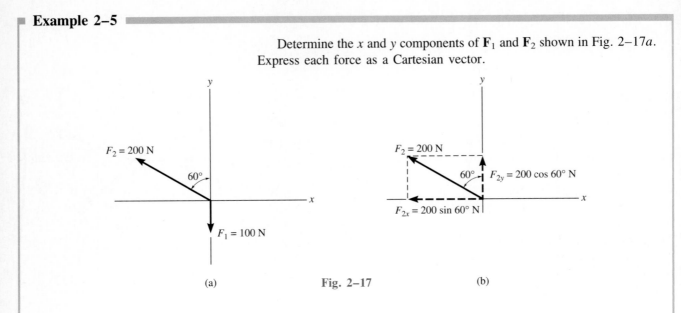

(a)   Fig. 2–17   (b)

SOLUTION

Scalar Notation.   Since $\mathbf{F}_1$ acts along the negative y axis, and the magnitude of $\mathbf{F}_1$ is 100 N, the components written in scalar form are

$$F_{1x} = 0, \qquad F_{1y} = -100 \text{ N} \hspace{3cm} Ans.$$

or, alternatively,

$$F_{1x} = 0, \qquad F_{1y} = 100 \text{ N} \downarrow \hspace{2.5cm} Ans.$$

By the parallelogram law, $\mathbf{F}_2$ is resolved into x and y components, Fig. 2–17b. The magnitude of each component is determined by trigonometry. Since $\mathbf{F}_{2x}$ acts in the $-x$ direction, and $\mathbf{F}_{2y}$ acts in the $+y$ direction, we have

$$F_{2x} = -200 \sin 60° \text{ N} = -173 \text{ N} = 173 \text{ N} \leftarrow \hspace{1cm} Ans.$$
$$F_{2y} = 200 \cos 60° \text{ N} = 100 \text{ N} = 100 \text{ N} \uparrow \hspace{1cm} Ans.$$

Cartesian Vector Notation.   Having computed the magnitudes of the components of $\mathbf{F}_2$, Fig. 2–17b, we can express each force as a Cartesian vector.

$$\mathbf{F}_1 = 0\mathbf{i} + 100 \text{ N}(-\mathbf{j})$$
$$= \{-100\mathbf{j}\} \text{ N} \hspace{3cm} Ans.$$

and

$$\mathbf{F}_2 = 200 \sin 60° \text{ N}(-\mathbf{i}) + 200 \cos 60° \text{ N}(\mathbf{j})$$
$$= \{-173\mathbf{i} + 100\mathbf{j}\} \text{ N} \hspace{2.5cm} Ans.$$

Example 2–6

Determine the x and y components of the force $\mathbf{F}$ shown in Fig. 2–18a.

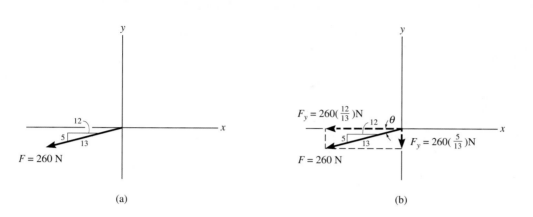

(a)   (b)

Fig. 2–18

SOLUTION

The force is resolved into its x and y components as shown in Fig. 2–18b. Here the *slope* of the line of action for the force is indicated. From this "slope triangle" we could obtain the direction angle θ, e.g., $\theta = \tan^{-1}\left(\frac{5}{12}\right)$, and then proceed to determine the magnitudes of the components in the same manner as for $\mathbf{F}_2$ in Example 2–5. An easier method, however, consists of using proportional parts of similar triangles, i.e.,

$$\frac{F_x}{260} = \frac{12}{13} \qquad F_x = 260\left(\frac{12}{13}\right) = 240 \text{ N}$$

Similarly,

$$F_y = 260\left(\frac{5}{13}\right) = 100 \text{ N}$$

Notice that the magnitude of the *horizontal component*, F_x, was obtained by multiplying the force magnitude by the ratio of the *horizontal leg* of the slope triangle divided by the hypotenuse; whereas the magnitude of the *vertical component*, F_y, was obtained by multiplying the force magnitude by the ratio of the *vertical leg* divided by the hypotenuse. Hence, using scalar notation,

$$F_x = -240 \text{ N} = 240 \text{ N} \leftarrow \qquad\qquad Ans.$$
$$F_y = -100 \text{ N} = 100 \text{ N} \downarrow \qquad\qquad Ans.$$

If $\mathbf{F}$ is expressed as a Cartesian vector, we have

$$\mathbf{F} = \{-240\mathbf{i} - 100\mathbf{j}\} \text{ N} \qquad\qquad Ans.$$

Example 2–7

(a)

(b)

(c)

Fig. 2–19

The link in Fig. 2–19*a* is subjected to two forces **F**₁ and **F**₂. Determine the magnitude and orientation of the resultant force.

SOLUTION I

Scalar Notation.   This problem can be solved by using the parallelogram law; however, here we will resolve each force into its *x* and *y* components, Fig. 2–19*b*, and sum these components algebraically. Indicating the "positive" sense of the *x* and *y* force components alongside Eqs. 2–1, we have

$$\xrightarrow{+} F_{Rx} = \Sigma F_x; \qquad F_{Rx} = 600 \cos 30° - 400 \sin 45°$$
$$= 236.8 \text{ N} \rightarrow$$

$$+ \uparrow F_{Ry} = \Sigma F_y; \qquad F_{Ry} = 600 \sin 30° + 400 \cos 45°$$
$$= 582.8 \text{ N} \uparrow$$

The resultant force, shown in Fig. 2–19*c*, has a *magnitude* of

$$F_R = \sqrt{(236.8)^2 + (582.8)^2}$$
$$= 629 \text{ N} \qquad\qquad\qquad \textit{Ans.}$$

From the vector addition, Fig. 2–19*c*, the direction angle θ is

$$\theta = \tan^{-1}\left(\frac{582.8}{236.8}\right) = 67.9° \qquad\qquad \textit{Ans.}$$

SOLUTION II

Cartesian Vector Notation.   From Fig. 2–19*b*, each force expressed as a Cartesian vector is

$$\mathbf{F}_1 = 600 \cos 30°\mathbf{i} + 600 \sin 30°\mathbf{j}$$
$$\mathbf{F}_2 = -400 \sin 45°\mathbf{i} + 400 \cos 45°\mathbf{j}$$

Thus

$$\mathbf{F}_R = \mathbf{F}_1 + \mathbf{F}_2 = (600 \cos 30° - 400 \sin 45°)\mathbf{i}$$
$$+ (600 \sin 30° + 400 \cos 45°)\mathbf{j}$$
$$= \{236.8\mathbf{i} + 582.8\mathbf{j}\} \text{ N}$$

The magnitude and direction of **F**₍R₎ are determined in the same manner as shown above.

   Comparing the two methods of solution, it is seen that use of scalar notation is more efficient, since the scalar components can be found *directly,* without first having to express each force as a Cartesian vector before adding the components. Cartesian vector analysis, however, will later be shown to be more advantageous for solving three-dimensional problems.

Example 2–8

The end of the boom O in Fig. 2–20a is subjected to three concurrent and coplanar forces. Determine the magnitude and orientation of the resultant force.

(a)

$F_1 = -400 i + 0 j$

$F_2 \quad 250 \sin 45 + 250 \cos 45$

$F_3 \quad -200 \cos 36.8 + 200 \sin 36.8$

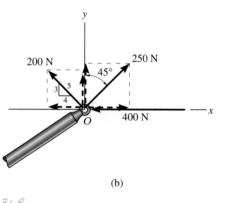

(b)

$\cos \theta = \dfrac{a}{h}$

$\sin \theta = \dfrac{o}{h} \qquad \theta = h \sin \theta$

SOLUTION

Each force is resolved into its x and y components as shown in Fig. 2–20b. Summing the x components, we have

$$\xrightarrow{+} F_{Rx} = \Sigma F_x; \qquad F_{Rx} = -400 + 250 \sin 45° - 200(\tfrac{4}{5})$$
$$= -383.2 \text{ N} = 383.2 \text{ N} \leftarrow$$

The negative sign indicates that F_{Rx} acts to the left, i.e., in the negative x direction as noted by the small arrow. Summing the y components yields

$$+ \uparrow F_{Ry} = \Sigma F_y; \qquad F_{Ry} = 250 \cos 45° + 200(\tfrac{3}{5})$$
$$= 296.8 \text{ N} \uparrow$$

The resultant force, shown in Fig. 2–20c, has a *magnitude* of

$$F_R = \sqrt{(-383.2)^2 + (296.8)^2}$$
$$= 485 \text{ N} \qquad\qquad Ans.$$

From the vector addition in Fig. 2–20c, the direction angle θ is

$$\theta = \tan^{-1}\left(\frac{296.8}{383.2}\right) = 37.8° \qquad\qquad Ans.$$

Realize that the single force $\mathbf{F}_R$ shown in Fig. 2–20c creates the *same* *effect* on the boom as the three forces in Fig. 2–20a.

(c)

Fig. 2–20

PROBLEMS

2–27.   Express $\mathbf{F}_1$ and $\mathbf{F}_2$ as Cartesian vectors.

*2–28.   Determine the magnitude of the resultant force and its orientation measured clockwise from the positive x axis.

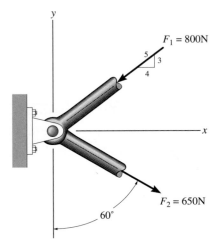

Probs. 2–27/2–28

2–31.   Express $\mathbf{F}_1$, $\mathbf{F}_2$, and $\mathbf{F}_3$ as Cartesian vectors.

*2–32.   Determine the magnitude of the resultant force and its orientation measured counterclockwise from the positive x axis.

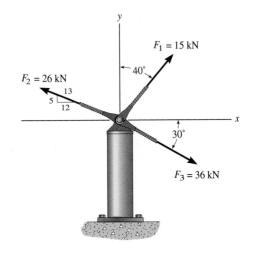

Probs. 2–31/2–32

2–29.   Express $\mathbf{F}_1$ and $\mathbf{F}_2$ as Cartesian vectors.

2–30.   Determine the magnitude of the resultant force and its orientation measured counterclockwise from the positive x axis.

Probs. 2–29/2–30

2–33.   Determine the magnitudes of the x and y components of $\mathbf{F}_1$ and $\mathbf{F}_2$.

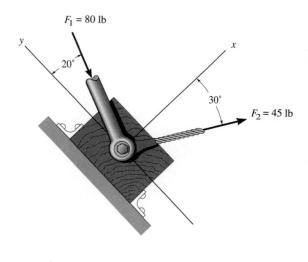

Prob. 2–33

2–34.   Solve Prob. 2–3 by summing the rectangular or x, y components of the forces to obtain the resultant force.

2–35.   Solve Prob. 2–4 by summing the rectangular or x, y components of the forces to obtain the resultant force.

*2–36.   Solve Prob. 2–9 by summing the rectangular or x, y components of the forces to obtain the resultant force.

2–37.   Solve Prob. 2–11 by summing the rectangular or x, y components of the forces.

2–38.   Solve Prob. 2–25 by summing the rectangular or x, y components of the forces to obtain the resultant force.

2–39.   A force of 40 lb is needed to push the car A forward. Determine the force F that the jeep B must exert on the car as a function of θ, which is necessary to maintain the motion. Plot the result, F versus θ for $0 \le \theta \le 90°$.

*2–40.   Express each force acting on the bracket in Cartesian vector form. What is the magnitude of the resultant force?

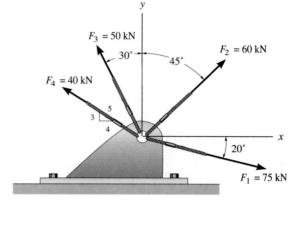

Prob. 2–40

2–41.   Four concurrent forces act on the plate. Determine the magnitude of the resultant force and its orientation measured counterclockwise from the positive x axis.

Prob. 2–39   Prob. 2–41

2–42. The *gusset plate* is subjected to four forces which are concurrent at point O. Determine the magnitude of the resultant force and its orientation measured counterclockwise from the positive x axis.

Prob. 2–42

2–45. The three concurrent forces acting on the screw eye produce a resultant force $\mathbf{F}_R = \mathbf{0}$. If $F_2 = \frac{2}{3}F_1$, and $\mathbf{F}_1$ is to be $90°$ from $\mathbf{F}_2$ as shown, determine the orientation θ and the required magnitude of $\mathbf{F}_3$ expressed in terms of F_1.

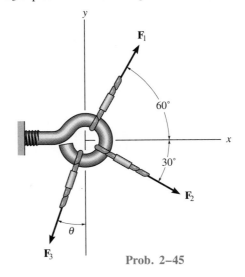

Prob. 2–45

2–43. Three forces act on the bracket. Determine the magnitude and orientation θ of $\mathbf{F}_2$ so that the resultant force is directed along the positive u axis and has a magnitude of 50 lb.

*2–44. If $F_2 = 150$ lb and $\theta = 55°$, determine the magnitude and orientation, measured clockwise from the positive u axis, of the resultant force of the three forces acting on the bracket.

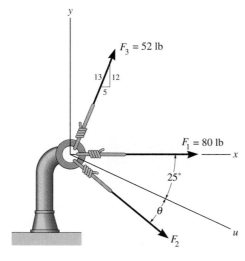

Probs. 2–43/2–44

2–46. Determine the magnitude and orientation θ of $\mathbf{F}_B$ so that the resultant force is directed along the positive y axis and has a magnitude of 1500 N.

2–47. Determine the magnitude and orientation, measured counterclockwise from the positive y axis, of the resultant force acting on the bracket, if $F_B = 600$ N and $\theta = 20°$.

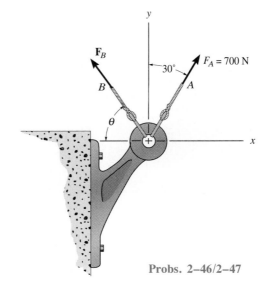

Probs. 2–46/2–47

*2–48. The brace is used to support the wall. When used for this purpose, the pin exerts a horizontal force F_x and a vertical force F_y on the brace at A. If the maximum resultant force that can be developed along the brace is 6 kN, and the ratio $F_x/F_y \leq 0.5$, determine the minimum angle θ of placement for the brace.

Prob. 2–48

2–49. The crate is to be hoisted using two chains. Determine the magnitudes of forces F_A and F_B acting on each chain in order to develop a resultant force of 600 N directed along the positive y axis. Set $\theta = 45°$.

2–50. The crate is to be hoisted using two chains. If the resultant force is to be 600 N, directed along the positive y axis, determine the magnitudes of forces F_A and F_B acting on each chain and the orientation θ of F_B so that the magnitude of F_B is a *minimum*. F_A acts at 30° from the y axis as shown.

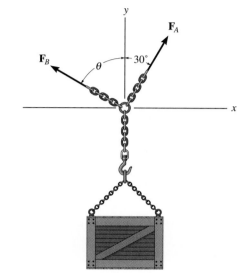

Probs. 2–49/2–50

2–51. Determine the magnitude of force F so that the resultant F_R of the three forces is as small as possible.

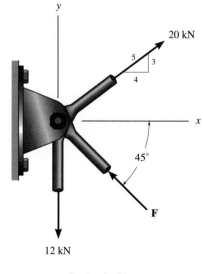

Prob. 2–51

2.5 Cartesian Vectors

The operations of vector algebra, when applied to solving problems in *three dimensions,* are greatly simplified if the vectors are first represented in Cartesian vector form. In this section we will present a general method for doing this, then in Sec. 2.6 we will apply this method to solving problems involving the addition of forces. Similar applications will be illustrated for the position and moment vectors given in later sections of the text.

Right-Handed Coordinate System. A right-handed coordinate system will be used for developing the theory of vector algebra that follows. A rectangular or Cartesian coordinate system is said to be *right-handed* provided the thumb of the right hand points in the direction of the positive z axis when the right-hand fingers are curled about this axis and directed from the positive x toward the positive y axis, Fig. 2–21. Furthermore, according to this rule, the z axis for a two-dimensional problem as in Fig. 2–20 would be directed outward, perpendicular to the page.

Fig. 2–21

Rectangular Components of a Vector. A vector **A** may have one, two, or three rectangular components along the x, y, z coordinate axes, depending on how the vector is oriented relative to the axes. In general, though, when **A** is directed within an octant of the x, y, z frame, Fig. 2–22, then by two successive applications of the parallelogram law, we may resolve the vector into components as $\mathbf{A} = \mathbf{A}' + \mathbf{A}_z$ and then $\mathbf{A}' = \mathbf{A}_x + \mathbf{A}_y$. Combining these equations, **A** is represented by the vector sum of its *three* rectangular components,

$$\mathbf{A} = \mathbf{A}_x + \mathbf{A}_y + \mathbf{A}_z \qquad (2-2)$$

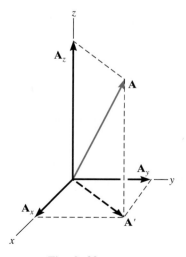

Fig. 2–22

Unit Vector.

In general, a *unit vector* is a vector having a magnitude of 1. If **A** is a vector having a magnitude $A \neq 0$, then a unit vector having the *same direction* as **A** is represented by

$$\mathbf{u}_A = \frac{\mathbf{A}}{A} \qquad (2\text{--}3)$$

Rewriting this equation gives

$$\mathbf{A} = A\mathbf{u}_A \qquad (2\text{--}4)$$

Since vector **A** is of a certain type, e.g., a force vector, it is customary to use the proper set of units for its description. The magnitude A also has this same set of units; hence, from Eq. 2–3, the *unit vector will be dimensionless* since the units will cancel out. Equation 2–4 therefore indicates that vector **A** may be expressed in terms of both its magnitude and direction *separately;* i.e., A (a positive scalar) defines the *magnitude* of **A**, and $\mathbf{u}_A$ (a dimensionless vector) defines the *direction* and sense of **A**, Fig. 2–23.

Fig. 2–23

Cartesian Unit Vectors.

In three dimensions, the set of Cartesian unit vectors, **i, j, k,** is used to designate the directions of the *x, y, z* axes respectively. As stated in Sec. 2–4, the *sense* (or arrowhead) of these vectors will be described analytically by a plus or minus sign, depending on whether they are pointing along the positive or negative *x, y,* or *z* axis. Thus the positive unit vectors are shown in Fig. 2–24.

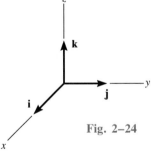

Fig. 2–24

Cartesian Vector Representation.

Using Cartesian unit vectors, the three vector components of Eq. 2–2 may be written in "Cartesian vector form." Since the components act in the positive **i, j,** and **k** directions, Fig. 2–25, we have

$$\boxed{\mathbf{A} = A_x\mathbf{i} + A_y\mathbf{j} + A_z\mathbf{k}} \qquad (2\text{--}5)$$

There is a distinct advantage to writing vectors in terms of their Cartesian components. Since each of these components has the same form as Eq. 2–4, the *magnitude* and *direction* of each *component vector* are *separated,* and it will be shown that this will simplify the operations of vector algebra, particularly in three dimensions.

Fig. 2–25

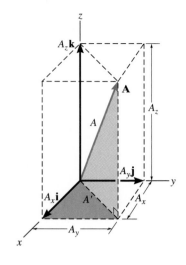

Fig. 2–26

Magnitude of a Cartesian Vector. It is always possible to obtain the magnitude of vector **A** provided the vector is expressed in Cartesian vector form. As shown in Fig. 2–26, from the colored right triangle, $A = \sqrt{A'^2 + A_z^2}$, and from the shaded right triangle, $A' = \sqrt{A_x^2 + A_y^2}$. Combining these equations yields

$$A = \sqrt{A_x^2 + A_y^2 + A_z^2} \qquad (2\text{–}6)$$

Hence, the magnitude of **A** *is equal to the positive square root of the sum of the squares of its components.*

Direction of a Cartesian Vector. The *orientation* of vector **A** is defined by the *coordinate direction angles* α (alpha), β (beta), and γ (gamma), measured between the *tail* of **A** and the *positive x, y, z* axes located at the tail of **A**, Fig. 2–27. Note that regardless of where **A** is directed, each of these angles will be between 0° and 180°. To determine α, β, and γ, consider the projection of **A** onto the *x, y, z* axes, Fig. 2–28. Referring to the colored right triangles shown in each figure, we have

$$\cos\alpha = \frac{A_x}{A} \qquad \cos\beta = \frac{A_y}{A} \qquad \cos\gamma = \frac{A_z}{A} \qquad (2\text{–}7)$$

These numbers are known as the *direction cosines* of **A.** Once they have been obtained, the coordinate direction angles α, β, γ can then be determined from the inverse cosines.

An easy way of obtaining the direction cosines of **A** is to form a unit vector in the direction of **A**, Eq. 2–3. Provided **A** is expressed in Cartesian

Fig. 2–27

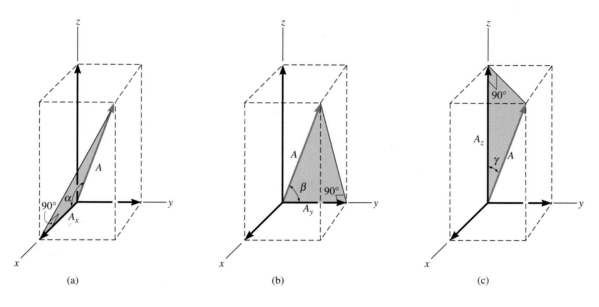

(a)   (b)   (c)

Fig. 2–28

vector form as $\mathbf{A} = A_x\mathbf{i} + A_y\mathbf{j} + A_z\mathbf{k}$ (Eq. 2–5), we have

$$\mathbf{u}_A = \frac{\mathbf{A}}{A} = \frac{A_x}{A}\mathbf{i} + \frac{A_y}{A}\mathbf{j} + \frac{A_z}{A}\mathbf{k} \qquad (2\text{–}8)$$

where $A = \sqrt{(A_x)^2 + (A_y)^2 + (A_z)^2}$ (Eq. 2–6). By comparison with Eqs. 2–7, it is seen that *the* **i, j,** *and* **k** *components of* $\mathbf{u}_A$ *represent the direction cosines of* **A,** i.e.,

$$\mathbf{u}_A = \cos\alpha\mathbf{i} + \cos\beta\mathbf{j} + \cos\gamma\mathbf{k} \qquad (2\text{–}9)$$

Since the magnitude of a vector is equal to the positive square root of the sum of the squares of the magnitudes of its components, and $\mathbf{u}_A$ has a magnitude of 1, then from Eq. 2–9 an important relation between the direction cosines can be formulated as

$$\cos^2\alpha + \cos^2\beta + \cos^2\gamma = 1 \qquad (2\text{–}10)$$

Provided vector **A** lies in a known octant, this equation can be used to determine one of the coordinate direction angles if the other two are known. (See Example 2–10.)

Finally, if the magnitude and coordinate direction angles of **A** are given, **A** may be expressed in Cartesian vector form as

$$\begin{aligned}\mathbf{A} &= A\mathbf{u}_A \\ &= A\cos\alpha\mathbf{i} + A\cos\beta\mathbf{j} + A\cos\gamma\mathbf{k} \qquad (2\text{–}11) \\ &= A_x\mathbf{i} + A_y\mathbf{j} + A_z\mathbf{k}\end{aligned}$$

2.6 Addition and Subtraction of Cartesian Vectors

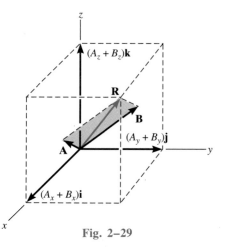

Fig. 2–29

The vector operations of addition and subtraction of two or more vectors are greatly simplified if the vectors are expressed in terms of their Cartesian components. For example, consider the two vectors **A** and **B,** both of which are directed within the positive octant of the x, y, z frame, Fig. 2–29. If $\mathbf{A} = A_x\mathbf{i} + A_y\mathbf{j} + A_z\mathbf{k}$ and $\mathbf{B} = B_x\mathbf{i} + B_y\mathbf{j} + B_z\mathbf{k}$, then the resultant vector, **R,** has components which represent the scalar sums of the **i, j,** and **k** components of **A** and **B,** i.e.,

$$\mathbf{R} = \mathbf{A} + \mathbf{B} = (A_x + B_x)\mathbf{i} + (A_y + B_y)\mathbf{j} + (A_z + B_z)\mathbf{k}$$

Vector subtraction, being a special case of vector addition, simply requires a scalar subtraction of the respective **i, j,** and **k** components of either **A** or **B.** For example,

$$\mathbf{R}' = \mathbf{A} - \mathbf{B} = (A_x - B_x)\mathbf{i} + (A_y - B_y)\mathbf{j} + (A_z - B_z)\mathbf{k}$$

Concurrent Force Systems. In particular, the above concept of vector addition may be generalized and applied to a system of several concurrent forces. In this case, the force resultant is the vector sum of all the forces in the system and can be written as

$$\mathbf{F}_R = \Sigma\mathbf{F} = \Sigma F_x\mathbf{i} + \Sigma F_y\mathbf{j} + \Sigma F_z\mathbf{k} \tag{2–12}$$

Here ΣF_x, ΣF_y, and ΣF_z represent the algebraic sums of the respective x, y, z, or **i, j, k** components of each force in the system.

The following examples illustrate numerically the methods used to apply the above theory to the solution of problems involving force as a vector quantity.

Example 2–9

Determine the magnitude and the coordinate direction angles of the resultant force acting on the ring in Fig. 2–30a.

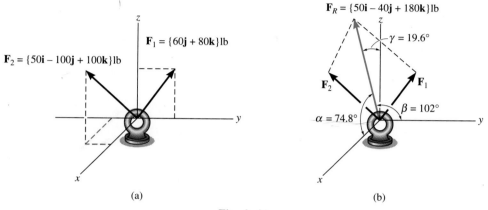

(a)   (b)

Fig. 2–30

SOLUTION

Since each force is represented in Cartesian vector form, the resultant force, shown in Fig. 2–30b, is

$$\mathbf{F}_R = \Sigma \mathbf{F} = \mathbf{F}_1 + \mathbf{F}_2 = (60\mathbf{j} + 80\mathbf{k}) + (50\mathbf{i} - 100\mathbf{j} + 100\mathbf{k})$$
$$= \{50\mathbf{i} - 40\mathbf{j} + 180\mathbf{k}\}\ \text{lb}$$

The magnitude of $\mathbf{F}_R$ is found from Eq. 2–6, i.e.,

$$F_R = \sqrt{(50)^2 + (-40)^2 + (180)^2}$$
$$= 191.0\ \text{lb} \hspace{2cm} \textit{Ans.}$$

The coordinate direction angles α, β, γ are determined from the components of the unit vector acting in the direction of $\mathbf{F}_R$.

$$\mathbf{u}_{F_R} = \frac{\mathbf{F}_R}{F_R} = \frac{50}{191.0}\mathbf{i} - \frac{40}{191.0}\mathbf{j} + \frac{180}{191.0}\mathbf{k}$$
$$= 0.2617\mathbf{i} - 0.2094\mathbf{j} + 0.9422\mathbf{k}$$

so that

$$\cos \alpha = 0.2617 \hspace{1cm} \alpha = 74.8° \hspace{1.5cm} \textit{Ans.}$$
$$\cos \beta = -0.2094 \hspace{1cm} \beta = 102° \hspace{1.5cm} \textit{Ans.}$$
$$\cos \gamma = 0.9422 \hspace{1cm} \gamma = 19.6° \hspace{1.5cm} \textit{Ans.}$$

These angles are shown in Fig. 2–30b. In particular, note that $\beta > 90°$ since the **j** component of $\mathbf{u}_{F_R}$ is negative.

Example 2–10

Express the force **F** shown in Fig. 2–31 as a Cartesian vector.

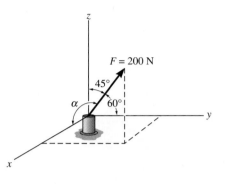

Fig. 2–31

SOLUTION

Since only two coordinate direction angles are specified, the third angle α is determined from Eq. 2–10; i.e.,

$$\cos^2 \alpha + \cos^2 \beta + \cos^2 \gamma = 1$$
$$\cos^2 \alpha + \cos^2 60° + \cos^2 45° = 1$$
$$\cos \alpha = \sqrt{1 - (0.707)^2 - (0.5)^2} = \pm 0.5$$

Hence,

$$\alpha = \cos^{-1}(0.5) = 60° \qquad \text{or} \qquad \alpha = \cos^{-1}(-0.5) = 120°$$

By inspection of Fig. 2–31, however, it is necessary that $\alpha = 60°$, since $\mathbf{F}_x$ is in the $+x$ direction.

Using Eq. 2–11, with $F = 200$ N, we have

$$\mathbf{F} = F \cos \alpha \mathbf{i} + F \cos \beta \mathbf{j} + F \cos \gamma \mathbf{k}$$
$$= 200 \cos 60° \mathbf{i} + 200 \cos 60° \mathbf{j} + 200 \cos 45° \mathbf{k}$$
$$= \{100.0\mathbf{i} + 100.0\mathbf{j} + 141.4\mathbf{k}\} \text{ N} \qquad\qquad Ans.$$

By applying Eq. 2–6, note that indeed the magnitude of $F = 200$ N.

$$F = \sqrt{F_x^2 + F_y^2 + F_z^2}$$
$$= \sqrt{(100.0)^2 + (100.0)^2 + (141.4)^2} = 200 \text{ N}$$

Example 2–11

Express the force **F** shown acting on the hook in Fig. 2–32*a* as a Cartesian vector.

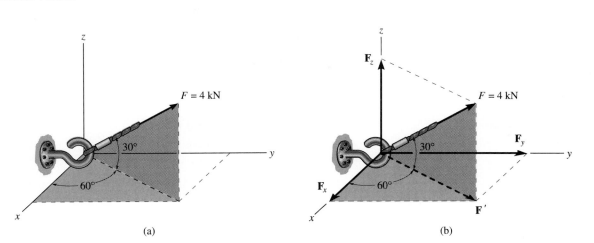

Fig. 2–32

SOLUTION

In this case the angles 60° and 30° defining the direction of **F** are *not* coordinate direction angles. Why? By two successive applications of the parallelogram law, however, **F** can be resolved into its *x, y, z* components as shown in Fig. 2–32*b*. First, from the colored triangle,

$$F' = 4 \cos 30° = 3.46 \text{ kN}$$
$$F_z = 4 \sin 30° = 2.00 \text{ kN}$$

Next, using **F**′ and the shaded triangle,

$$F_x = 3.46 \cos 60° = 1.73 \text{ kN}$$
$$F_y = 3.46 \sin 60° = 3.00 \text{ kN}$$

Thus,

$$\mathbf{F} = \{1.73\mathbf{i} + 3.00\mathbf{j} + 2.00\mathbf{k}\} \text{ kN} \qquad \textit{Ans.}$$

As an exercise, show that the magnitude of **F** is indeed 4 kN, and that the coordinate direction angle $\alpha = 64.3°$.

Example 2–12

Express force **F** shown in Fig. 2–33a as a Cartesian vector.

SOLUTION

As in Example 2–11, the angles of 60° and 45° defining the direction of **F** are *not* coordinate direction angles. The two successive applications of the parallelogram law needed to resolve **F** into its x, y, z components are shown in Fig. 2–33b. By trigonometry, the magnitudes of the components are

$$F_z = 100 \sin 60° = 86.6 \text{ lb}$$
$$F' = 100 \cos 60° = 50 \text{ lb}$$
$$F_x = 50 \cos 45° = 35.4 \text{ lb}$$
$$F_y = 50 \sin 45° = 35.4 \text{ lb}$$

Realizing that $\mathbf{F}_y$ has a direction defined by $-\mathbf{j}$, we have

$$\mathbf{F} = \mathbf{F}_x + \mathbf{F}_y + \mathbf{F}_z$$
$$\mathbf{F} = \{35.4\mathbf{i} - 35.4\mathbf{j} + 86.6\mathbf{k}\} \text{ lb} \qquad \textit{Ans.}$$

To show that the magnitude of this vector is indeed 100 lb, apply Eq. 2–6,

$$F = \sqrt{F_x^2 + F_y^2 + F_z^2}$$
$$= \sqrt{(35.4)^2 + (-35.4)^2 + (86.6)^2} = 100 \text{ lb}$$

If needed, the coordinate direction angles of **F** can be determined from the components of the unit vector acting in the direction of **F**. Hence,

$$\mathbf{u} = \frac{\mathbf{F}}{F} = \frac{F_x}{F}\mathbf{i} + \frac{F_y}{F}\mathbf{j} + \frac{F_z}{F}\mathbf{k}$$

$$= \frac{35.4}{100}\mathbf{i} - \frac{35.4}{100}\mathbf{j} + \frac{86.6}{100}\mathbf{k}$$

$$= 0.354\mathbf{i} - 0.354\mathbf{j} + 0.866\mathbf{k}$$

so that

$$\alpha = \cos^{-1}(0.354) = 69.3°$$
$$\beta = \cos^{-1}(-0.354) = 111°$$
$$\gamma = \cos^{-1}(0.866) = 30.0°$$

These results are shown in Fig. 2–33c.

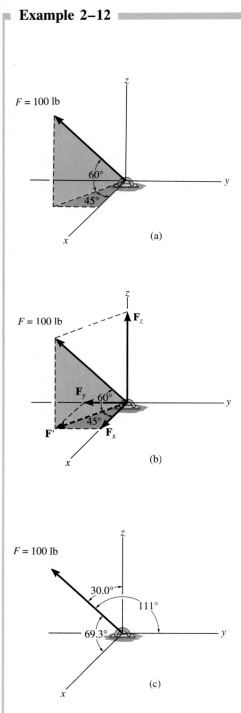

Fig. 2–33

Example 2–13

Two forces act on the hook shown in Fig. 2–34a. Specify the coordinate direction angles of $\mathbf{F}_2$ so that the resultant force $\mathbf{F}_R$ acts along the positive y axis and has a magnitude of 800 N.

SOLUTION

To solve this problem, the resultant force and its two components, $\mathbf{F}_1$ and $\mathbf{F}_2$, will each be expressed in Cartesian vector form. Then, as shown in Fig. 2–34b, it is necessary that $\mathbf{F}_R = \mathbf{F}_1 + \mathbf{F}_2$.

Applying Eq. 2–11,

$$\mathbf{F}_1 = F_1\mathbf{u}_{F_1} = F_1 \cos \alpha_1\mathbf{i} + F_1 \cos \beta_1\mathbf{j} + F_1 \cos \gamma_1\mathbf{k}$$
$$= 300 \cos 45°\mathbf{i} + 300 \cos 60°\mathbf{j} + 300 \cos 120°\mathbf{k}$$
$$= \{212.1\mathbf{i} + 150\mathbf{j} - 150\mathbf{k}\} \text{ N}$$
$$\mathbf{F}_2 = F_2\mathbf{u}_{F_2} = F_{2x}\mathbf{i} + F_{2y}\mathbf{j} + F_{2z}\mathbf{k}$$

According to the problem statement, the resultant force $\mathbf{F}_R$ has a magnitude of 800 N and acts in the $+\mathbf{j}$ direction. Hence,

$$\mathbf{F}_R = (800 \text{ N})(+\mathbf{j}) = \{800\mathbf{j}\} \text{ N}$$

We require

$$\mathbf{F}_R = \mathbf{F}_1 + \mathbf{F}_2$$
$$800\mathbf{j} = 212.1\mathbf{i} + 150\mathbf{j} - 150\mathbf{k} + F_{2x}\mathbf{i} + F_{2y}\mathbf{j} + F_{2z}\mathbf{k}$$
$$800\mathbf{j} = (212.1 + F_{2x})\mathbf{i} + (150 + F_{2y})\mathbf{j} + (-150 + F_{2z})\mathbf{k}$$

To satisfy this equation, the corresponding $\mathbf{i}$, $\mathbf{j}$, and $\mathbf{k}$ components on the left and right sides must be equal. This is equivalent to stating that the x, y, z components of $\mathbf{F}_R$ be equal to the corresponding x, y, z components of $(\mathbf{F}_1 + \mathbf{F}_2)$. Hence,

$$0 = 212.1 + F_{2x} \qquad F_{2x} = -212.1 \text{ N}$$
$$800 = 150 + F_{2y} \qquad F_{2y} = 650 \text{ N}$$
$$0 = -150 + F_{2z} \qquad F_{2z} = 150 \text{ N}$$

Since the magnitudes of $\mathbf{F}_2$ and its components are known, we can use Eq. 2–11 to determine α, β, γ.

$$-212.1 = 700 \cos \alpha_2; \qquad \alpha_2 = \cos^{-1}\left(\frac{-212.1}{700}\right) = 108° \quad Ans.$$

$$650 = 700 \cos \beta_2; \qquad \beta_2 = \cos^{-1}\left(\frac{650}{700}\right) = 21.8° \quad Ans.$$

$$150 = 700 \cos \gamma_2; \qquad \gamma_2 = \cos^{-1}\left(\frac{150}{700}\right) = 77.6° \quad Ans.$$

These results are shown in Fig. 2–34b.

(a)

(b)

Fig. 2–34

PROBLEMS

*2–52. Express each force as a Cartesian vector and then determine the resultant force $\mathbf{F}_R$. Find the magnitude and coordinate direction angles of the resultant force and sketch this vector on the coordinate system.

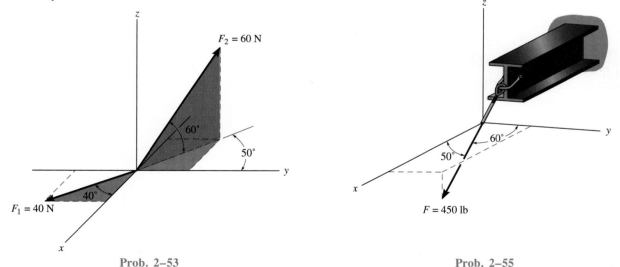

Prob. 2–52

2–53. Express each force as a Cartesian vector and then determine the resultant force $\mathbf{F}_R$. Find the magnitude and coordinate direction angles of the resultant force and sketch this vector on the coordinate system.

Prob. 2–53

2–54. Express each force as a Cartesian vector and then determine the resultant force $\mathbf{F}_R$. Find the magnitude and coordinate direction angles of the resultant force and sketch this vector on the coordinate system.

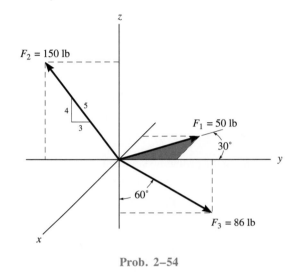

Prob. 2–54

2–55. The cable at the end of the beam exerts a force of 450 lb on the beam as shown. Express $\mathbf{F}$ as a Cartesian vector.

Prob. 2–55

*2–56. The bracket is subjected to the two forces shown. Express each force in Cartesian vector form and then determine the resultant force $\mathbf{F}_R$. Find the magnitude and coordinate direction angles of the resultant force.

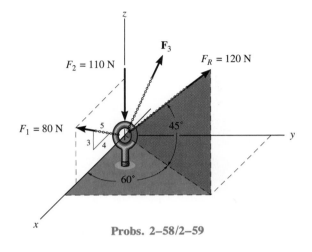

2–58. Three forces act on the screw eye. If the resultant force $\mathbf{F}_R$ has a magnitude and orientation as shown, determine the magnitude and coordinate direction angles of force $\mathbf{F}_3$.

2–59. Determine the coordinate direction angles of $\mathbf{F}_1$ and $\mathbf{F}_R$.

Prob. 2–56

Probs. 2–58/2–59

*2–60. The two forces $\mathbf{F}_1$ and $\mathbf{F}_2$ acting at the end of the pipe have a resultant force of $\mathbf{F}_R = \{120\mathbf{i}\}$ N. Determine the magnitude and coordinate direction angles of $\mathbf{F}_2$.

2–57. Three forces act on the ring. Determine the magnitude and coordinate direction angles of the resultant force.

2–61. Determine the coordinate direction angles of the force $\mathbf{F}_1$ and indicate them on the figure.

Prob. 2–57

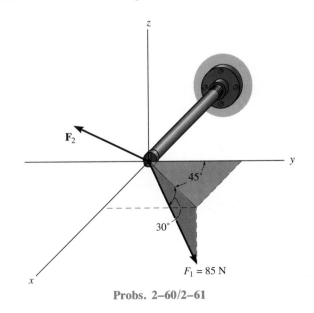

Probs. 2–60/2–61

2–62. The force **F** acting on the stake has a component of 40 N acting in the x–y plane as shown. Express **F** as a Cartesian vector.

2–63. Determine the magnitude and coordinate direction angles of the force **F** acting on the stake.

2–66. The pipe is subjected to the force **F,** which has components acting along the x, y, z axes as shown. If the magnitude of **F** is 12 kN, and $\alpha = 120°$ and $\gamma = 45°$, determine the magnitudes of its three components.

2–67. The pipe is subjected to the force **F** which has components $F_x = 1.5$ kN and $F_z = 1.25$ kN. If $\beta = 75°$, determine the magnitudes of **F** and **F**$_y$.

Probs. 2–62/2–63

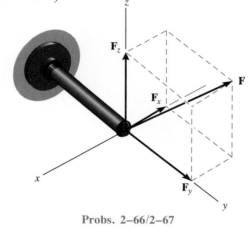

Probs. 2–66/2–67

*2–64. The force **F** has a magnitude of 80 lb and acts within the octant shown such that $\alpha = 60°$ and $\beta = 45°$. Express **F** as a Cartesian vector.

2–65. If $F_x = 200$ lb, express **F** as a Cartesian vector.

*2–68. Determine the magnitude and coordinate direction angles of **F**$_2$ so that the resultant of the two forces acts along the positive x axis and has a magnitude of 350 N.

2–69. Determine the magnitude and coordinate direction angles of **F**$_2$ so that the resultant of the two forces is zero.

Probs. 2–64/2–65

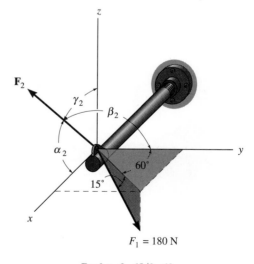

Probs. 2–68/2–69

2.7 Position Vectors

In this section we will introduce the concept of a position vector. It will be shown in Sec. 2.8 that this vector is of importance in formulating a Cartesian force vector directed between any two points in space, and later, in Chapter 4, we will use it for finding the moment of a force.

x, y, z **Coordinates.**   Throughout this text we will use a *right-handed* coordinate system to reference the location of points in space. Furthermore, we will use the convention followed in many technical books, and that is to require the positive *z* axis to be directed *upward* (the zenith direction) so that it measures the height of an object or the altitude of a point. The *x, y* axes then lie in the horizontal plane, Fig. 2–35. Points in space are located relative to the origin of coordinates, *O*, by successive measurements along the *x, y, z* axes. For example, in Fig. 2–35 the coordinates of point *A* are obtained by starting at *O* and measuring $x_A = +4$ m along the *x* axis, $y_A = +2$ m along the *y* axis, and $z_A = -6$ m along the *z* axis. Thus, $A(4, 2, -6)$. In a similar manner, measurements along the *x, y, z* axes from *O* to *B* yield the coordinates of *B*, i.e., $B(0, 2, 0)$. Also notice that $C(6, -1, 4)$.

Fig. 2–35

Position Vector.   The *position vector* **r** is defined as a fixed vector which locates a point in space relative to another point. For example, if **r** extends from the origin of coordinates, *O*, to point $P(x, y, z)$, Fig. 2–36*a*, then **r** can be expressed in Cartesian vector form as

$$\mathbf{r} = x\mathbf{i} + y\mathbf{j} + z\mathbf{k}$$

In particular, note how the head-to-tail vector addition of the three components yields vector **r**, Fig. 2–36*b*. Starting at the origin *O*, one travels *x* in the $+\mathbf{i}$ direction, then *y* in the $+\mathbf{j}$ direction, and finally *z* in the $+\mathbf{k}$ direction to arrive at point $P(x, y, z)$.

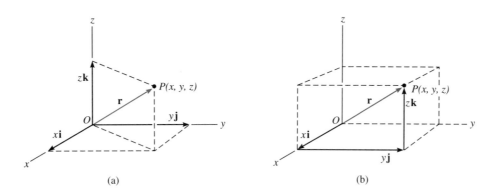

(a)

(b)

Fig. 2–36

In the more general case, the position vector may be directed from point A to point B in space, Fig. 2–37a. As noted, this vector is also designated by the symbol **r**. As a matter of convention, however, we will *sometimes* refer to this vector with *two subscripts* to indicate from and to the point where it is directed, thus, **r** can also be designated as $\mathbf{r}_{AB}$. Also, note that $\mathbf{r}_A$ and $\mathbf{r}_B$ in Fig. 2–37a are referenced with only one subscript since they extend from the origin of coordinates.

From Fig. 2–37a, by the head-to-tail vector addition, we require

$$\mathbf{r}_A + \mathbf{r} = \mathbf{r}_B$$

Solving for **r** and expressing $\mathbf{r}_A$ and $\mathbf{r}_B$ in Cartesian vector form yields

$$\mathbf{r} = \mathbf{r}_B - \mathbf{r}_A = (x_B\mathbf{i} + y_B\mathbf{j} + z_B\mathbf{k}) - (x_A\mathbf{i} + y_A\mathbf{j} + z_A\mathbf{k})$$

or

$$\mathbf{r} = (x_B - x_A)\mathbf{i} + (y_B - y_A)\mathbf{j} + (z_B - z_A)\mathbf{k} \qquad (2\text{–}13)$$

Thus, the **i, j, k** *components of the position vector* **r** *may be formed by taking the coordinates of the tail of the vector,* $A(x_A, y_A, z_A)$, *and subtracting them from the corresponding coordinates of the head,* $B(x_B, y_B, z_B)$. Again note how the head-to-tail addition of these three components yields **r,** i.e., going from A to B, Fig. 2–37b, one first travels $(x_B - x_A)$ in the $+\mathbf{i}$ direction, then $(y_B - y_A)$ in the $+\mathbf{j}$ direction, and finally $(z_B - z_A)$ in the $+\mathbf{k}$ direction.

(a)

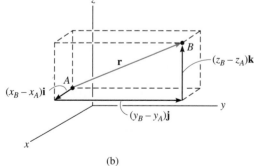

(b)

Fig. 2–37

Example 2–14

Determine the magnitude and direction of the position vector extending from A to B in Fig. 2–38a.

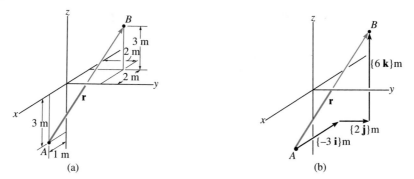

(a) (b)

SOLUTION

In accordance with Eq. 2–13, the coordinates of the tail $A(1, 0, -3)$ are subtracted from the coordinates of the head $B(-2, 2, 3)$, which yields

$$\mathbf{r} = (-2 - 1)\mathbf{i} + (2 - 0)\mathbf{j} + [3 - (-3)]\mathbf{k}$$
$$= \{-3\mathbf{i} + 2\mathbf{j} + 6\mathbf{k}\} \text{ m}$$

As shown in Fig. 2–38b, the three components of $\mathbf{r}$ can also be obtained in a more *direct manner* by realizing that in going from A to B one must move along the x axis $\{-3\mathbf{i}\}$ m, along the y axis $\{2\mathbf{j}\}$ m, and finally along the z axis $\{6\mathbf{k}\}$ m.

The magnitude of $\mathbf{r}$ is thus

$$r = \sqrt{(-3)^2 + (2)^2 + (6)^2} = 7 \text{ m} \qquad Ans.$$

Formulating a unit vector in the direction of $\mathbf{r}$, we have

$$\mathbf{u} = \frac{\mathbf{r}}{r} = \frac{-3}{7}\mathbf{i} + \frac{2}{7}\mathbf{j} + \frac{6}{7}\mathbf{k}$$

The components of this unit vector yield the coordinate direction angles

$$\alpha = \cos^{-1}\left(\frac{-3}{7}\right) = 115° \qquad Ans.$$

$$\beta = \cos^{-1}\left(\frac{2}{7}\right) = 73.4° \qquad Ans.$$

$$\gamma = \cos^{-1}\left(\frac{6}{7}\right) = 31.0° \qquad Ans.$$

These angles are measured from the *positive axes* of a localized coordinate system placed at the tail of $\mathbf{r}$ as shown in Fig. 2–38c.

(c)

Fig. 2–38

2.8 Force Vector Directed Along a Line

Quite often in three-dimensional statics problems, the direction of a force is specified by two points through which its line of action passes. Such a situation is shown in Fig. 2–39, where the force $\mathbf{F}$ is directed along the cord AB. We can formulate $\mathbf{F}$ as a Cartesian vector by realizing that it has the *same direction* and *sense* as the position vector $\mathbf{r}$ directed from point A to point B on the cord. This common direction is specified by the *unit vector* $\mathbf{u} = \mathbf{r}/r$. Hence,

$$\mathbf{F} = F\mathbf{u} = F\left(\frac{\mathbf{r}}{r}\right)$$

Although we have represented $\mathbf{F}$ symbolically in Fig. 2–39, note that it has units of force, and unlike $\mathbf{r}$, or coordinates x, y, z, which have units of length, $\mathbf{F}$ cannot be scaled along the coordinate axes.

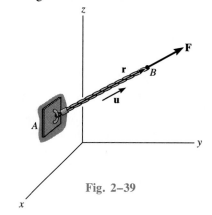

Fig. 2–39

PROCEDURE FOR ANALYSIS

When $\mathbf{F}$ is directed along a line which extends from point A to point B, then $\mathbf{F}$ can be expressed in Cartesian vector form as follows:

Position Vector. Determine the position vector $\mathbf{r}$ directed from A to B, and compute its magnitude r.

Unit Vector. Determine the unit vector $\mathbf{u} = \mathbf{r}/r$ which defines the *direction* and *sense* of *both* $\mathbf{r}$ and $\mathbf{F}$.

Force Vector. Determine $\mathbf{F}$ by combining its magnitude F and direction $\mathbf{u}$, i.e., $\mathbf{F} = F\mathbf{u}$.

This procedure is illustrated numerically in the following example problems.

Example 2–15

The man shown in Fig. 2–40*a* pulls on the cord with a force of 70 lb. Represent this force, acting on the support *A*, as a Cartesian vector and determine its direction.

SOLUTION

Force **F** is shown in Fig. 2–40*b*. The *direction* of this vector, **u,** is determined from the position vector **r,** which extends from *A* to *B*, Fig. 2–40*b*. To formulate **F** as a Cartesian vector we use the following procedure.

Position Vector. The coordinates of the end points of the cord are $A(0, 0, 30)$ and $B(12, -8, 6)$. Forming the position vector by subtracting the corresponding x, y, and z coordinates of A from those of B, we have

$$\mathbf{r} = (12 - 0)\mathbf{i} + (-8 - 0)\mathbf{j} + (6 - 30)\mathbf{k}$$
$$= \{12\mathbf{i} - 8\mathbf{j} - 24\mathbf{k}\} \text{ ft}$$

Show on Fig. 2–40*a* how one can write **r** *directly* by going from A $\{12\mathbf{i}\}$ ft, then $\{-8\mathbf{j}\}$ ft, and finally $\{-24\mathbf{k}\}$ ft to get to B.

The magnitude of **r,** which represents the *length* of cord AB, is

$$r = \sqrt{(12)^2 + (-8)^2 + (-24)^2} = 28 \text{ ft}$$

Unit Vector. Forming the unit vector that defines the direction and sense of both **r** and **F** yields

$$\mathbf{u} = \frac{\mathbf{r}}{r} = \frac{12}{28}\mathbf{i} - \frac{8}{28}\mathbf{j} - \frac{24}{28}\mathbf{k}$$

Force Vector. Since **F** has a *magnitude* of 70 lb and a *direction* specified by **u,** then

$$\mathbf{F} = F\mathbf{u} = 70 \text{ lb} \left(\frac{12}{28}\mathbf{i} - \frac{8}{28}\mathbf{j} - \frac{24}{28}\mathbf{k} \right)$$
$$= \{30\mathbf{i} - 20\mathbf{j} - 60\mathbf{k}\} \text{ lb} \qquad \textit{Ans.}$$

As shown in Fig. 2–40*b*, the coordinate direction angles are measured between **r** (or **F**) and the *positive axes* of a localized coordinate system with origin placed at *A*. From the components of the unit vector:

$$\alpha = \cos^{-1}\left(\frac{12}{28}\right) = 64.6° \qquad \textit{Ans.}$$

$$\beta = \cos^{-1}\left(\frac{-8}{28}\right) = 107° \qquad \textit{Ans.}$$

$$\gamma = \cos^{-1}\left(\frac{-24}{28}\right) = 149° \qquad \textit{Ans.}$$

(a)

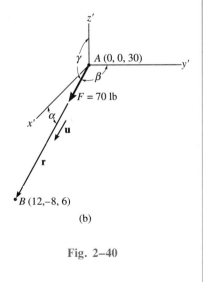

(b)

Fig. 2–40

Example 2–16

(a)

The circular plate shown in Fig. 2–41a is partially supported by the cable AB. If the force of the cable on the hook at A is $F = 500$ N, express **F** as a Cartesian vector.

SOLUTION

As shown in Fig. 2–41b, **F** has the same direction and sense as the position vector **r,** which extends from A to B.

Position Vector. The coordinates of the end points of the cable are $A(0, 0, 2)$ and $B(1.707, 0.707, 0)$, as indicated in the figure. Thus,

$$\mathbf{r} = (1.707 - 0)\mathbf{i} + (0.707 - 0)\mathbf{j} + (0 - 2)\mathbf{k}$$
$$= \{1.707\mathbf{i} + 0.707\mathbf{j} - 2\mathbf{k}\} \text{ m}$$

Note how one can calculate these components *directly* by going from A, $\{-2\mathbf{k}\}$ m along the z axis, then $\{1.707\mathbf{i}\}$ m along the x axis, and finally $\{0.707\mathbf{j}\}$ m along the y axis to get to B.

The magnitude of **r** is

$$r = \sqrt{(1.707)^2 + (0.707)^2 + (-2)^2} = 2.72 \text{ m}$$

Unit Vector. Thus,

$$\mathbf{u} = \frac{\mathbf{r}}{r} = \frac{1.707}{2.72}\mathbf{i} + \frac{0.707}{2.72}\mathbf{j} - \frac{2}{2.72}\mathbf{k}$$
$$= 0.627\mathbf{i} + 0.260\mathbf{j} - 0.735\mathbf{k}$$

Force Vector. Since $F = 500$ N and **F** has the direction **u,** we have

$$\mathbf{F} = F\mathbf{u} = 500 \text{ N}(0.627\mathbf{i} + 0.260\mathbf{j} - 0.735\mathbf{k})$$
$$= \{314\mathbf{i} + 130\mathbf{j} - 368\mathbf{k}\} \text{ N} \qquad \textit{Ans.}$$

Using these components, notice that indeed the magnitude of **F** is 500 N; i.e.,

$$F = \sqrt{(314)^2 + (130)^2 + (-368)^2} = 500 \text{ N}$$

Show that the coordinate direction angle $\gamma = 137°$, and indicate this angle on the figure.

Fig. 2–41

(b)

Example 2–17

The cables exert forces $F_{AB} = 100$ N and $F_{AC} = 120$ N on the ring at A as shown in Fig. 2–42a. Determine the magnitude of the resultant force acting at A.

SOLUTION

The resultant force $\mathbf{F}_R$ is shown graphically in Fig. 2–42b. We can express this force as a Cartesian vector by first formulating $\mathbf{F}_{AB}$ and $\mathbf{F}_{AC}$ as Cartesian vectors and then adding their components. The directions of $\mathbf{F}_{AB}$ and $\mathbf{F}_{AC}$ are specified by forming unit vectors $\mathbf{u}_{AB}$ and $\mathbf{u}_{AC}$ along the cables. These unit vectors are obtained from the associated position vectors $\mathbf{r}_{AB}$ and $\mathbf{r}_{AC}$. With reference to Fig. 2–42b for $\mathbf{F}_{AB}$ we have

$$\mathbf{r}_{AB} = (4 - 0)\mathbf{i} + (0 - 0)\mathbf{j} + (0 - 4)\mathbf{k}$$
$$= \{4\mathbf{i} - 4\mathbf{k}\} \text{ m}$$
$$r_{AB} = \sqrt{(4)^2 + (-4)^2} = 5.66 \text{ m}$$
$$\mathbf{F}_{AB} = 100 \text{ N}\left(\frac{\mathbf{r}_{AB}}{r_{AB}}\right) = 100 \text{ N}\left(\frac{4}{5.66}\mathbf{i} - \frac{4}{5.66}\mathbf{k}\right)$$
$$\mathbf{F}_{AB} = \{70.7\mathbf{i} - 70.7\mathbf{k}\} \text{ N}$$

For $\mathbf{F}_{AC}$ we have

$$\mathbf{r}_{AC} = (4 - 0)\mathbf{i} + (2 - 0)\mathbf{j} + (0 - 4)\mathbf{k}$$
$$= \{4\mathbf{i} + 2\mathbf{j} - 4\mathbf{k}\} \text{ m}$$
$$r_{AC} = \sqrt{(4)^2 + (2)^2 + (-4)^2} = 6 \text{ m}$$
$$\mathbf{F}_{AC} = 120 \text{ N}\left(\frac{\mathbf{r}_{AC}}{r_{AC}}\right) = 120 \text{ N}\left(\frac{4}{6}\mathbf{i} + \frac{2}{6}\mathbf{j} - \frac{4}{6}\mathbf{k}\right)$$
$$= \{80\mathbf{i} + 40\mathbf{j} - 80\mathbf{k}\} \text{ N}$$

The resultant force is therefore

$$\mathbf{F}_R = \mathbf{F}_{AB} + \mathbf{F}_{AC} = (70.7\mathbf{i} - 70.7\mathbf{k}) + (80\mathbf{i} + 40\mathbf{j} - 80\mathbf{k})$$
$$= \{150.7\mathbf{i} + 40\mathbf{j} - 150.7\mathbf{k}\} \text{ N}$$

The magnitude of $\mathbf{F}_R$ is thus

$$F_R = \sqrt{(150.7)^2 + (40)^2 + (-150.7)^2}$$
$$= 217 \text{ N} \qquad\qquad Ans.$$

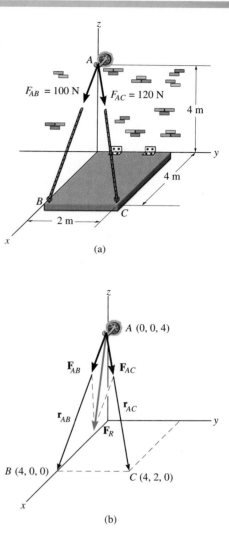

(a)

(b)

Fig. 2–42

PROBLEMS

2–70. A position vector extends from the origin to the point (2 m, 3 m, 6 m). Determine the coordinate direction angles α, β, and γ which the tail of the vector makes with the x, y, and z axes, respectively.

2–71. Represent the position vector $\mathbf{r}$ acting from point A(3 m, 5 m, 6 m) to point B(5 m, −2 m, 1 m) in Cartesian vector form. Calculate its coordinate direction angles, and find the distance between points A and B.

*2–72. Given the three position vectors

$$\mathbf{r}_1 = \{3\mathbf{i} - 4\mathbf{j} + 3\mathbf{k}\} \text{ m}$$
$$\mathbf{r}_2 = \{4\mathbf{i} - 5\mathbf{k}\} \text{ m}$$
$$\mathbf{r}_3 = \{3\mathbf{i} - 2\mathbf{j} + 5\mathbf{k}\} \text{ m}$$

determine the magnitude and coordinate direction angles of $\mathbf{r} = 2\mathbf{r}_1 - \mathbf{r}_2 + 3\mathbf{r}_3$.

2–73. Given the three position vectors

$$\mathbf{r}_1 = \{2\mathbf{i} + 5\mathbf{j} + 4\mathbf{k}\} \text{ m}$$
$$\mathbf{r}_2 = \{3\mathbf{i} + 2\mathbf{k}\} \text{ m}$$
$$\mathbf{r}_3 = \{-2\mathbf{i} + 4\mathbf{j}\} \text{ m}$$

determine the magnitude and coordinate direction angles of $\mathbf{r} = \mathbf{r}_1 - \mathbf{r}_2 + \frac{1}{2}\mathbf{r}_3$.

2–74. Express the position vector $\mathbf{r}$ in Cartesian vector form; then determine its magnitude and coordinate direction angles.

2–75. Express the position vector $\mathbf{r}$ in Cartesian vector form; then determine its magnitude and coordinate direction angles.

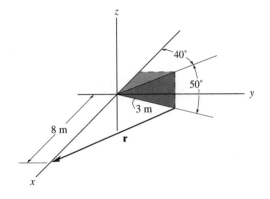

Prob. 2–75

*2–76. Express the position vector $\mathbf{r}$ in Cartesian vector form; then determine its magnitude and coordinate direction angles.

Prob. 2–74

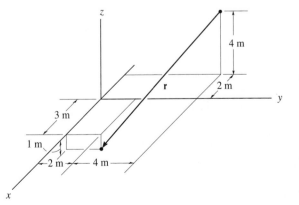

Prob. 2–76

2–77. Determine the distance between the end points A and B on the pipe. *Hint:* Formulate a position vector directed from A to B and then determine its magnitude.

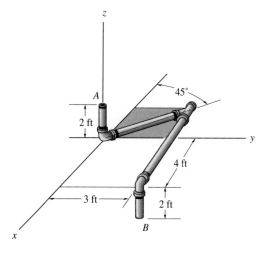

Prob. 2–77

2–79. The 80-ft-long cable is attached to the boom at A. If x = 60 ft, determine the coordinate location (y, z) to the end of the cable B. Take z = 3y.

*2–80. The cable is attached to the boom at A and to point B, which has coordinates of x = 80 ft and y = 40 ft. If z = 40 ft, determine the length of the cable and the angle between the positive z axis and the cable.

Probs. 2–79/2–80

2–78. Determine the length of the connecting rod AB by first formulating a Cartesian position vector from A to B and then determining its magnitude.

2–81. The positions of point A on the flagpole and point B on the light have been measured relative to the electronic distance meter (EDM) at O. Determine the distance between A and B. *Hint:* Formulate a position vector directed from A to B, then determine its magnitude.

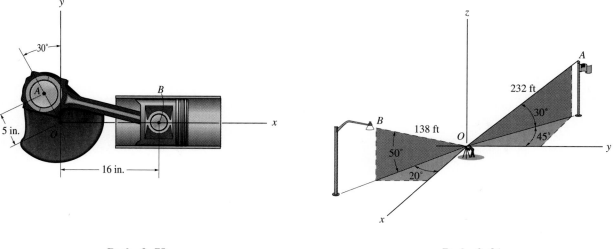

Prob. 2–78   Prob. 2–81

2–82.   Determine the lengths of wires AD, BD, and CD. The ring at D is midway between A and B.

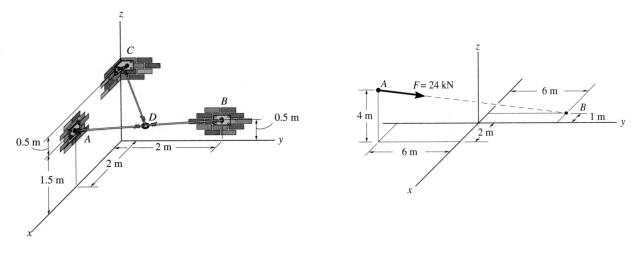

Prob. 2–82

2–83.   Express the force as a Cartesian vector and then determine its coordinate direction angles.

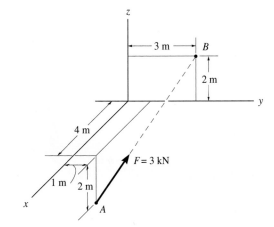

Prob. 2–83

*2–84.   Express the force as a Cartesian vector; then determine its coordinate direction angles.

Prob. 2–84

2–85.   Express the force as a Cartesian vector and then determine its coordinate direction angles.

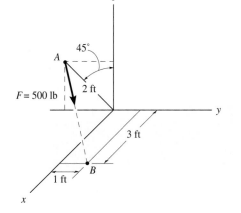

Prob. 2–85

2–86. Express each of the two forces in Cartesian vector form.

*2–88. The cable, attached to the shear-leg derrick, exerts a force on the derrick of 350 lb. Express this force as a Cartesian vector.

Prob. 2–86

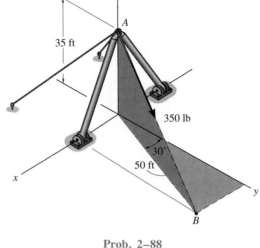

Prob. 2–88

2–87. A force **F** is applied at the top of the tower at *A*. If it acts in the direction shown such that its vertical component has a magnitude of 80 lb, express **F** as a Cartesian vector and then determine its coordinate direction angles.

2–89. Two tractors pull on the tree with the forces shown. Represent each force as a Cartesian vector and then determine the magnitude and coordinate direction angles of the resultant force.

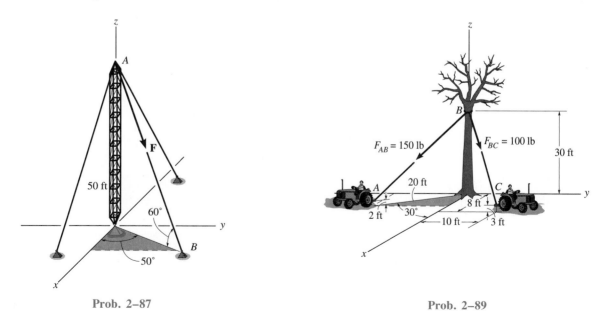

Prob. 2–87

Prob. 2–89

2–90. The window is held open by chain AB. Determine the length of the chain, and express the 50-lb force acting at A along the chain as a Cartesian vector and determine its coordinate direction angles.

Prob. 2–90

2–91. Express each of the two forces in Cartesian vector form and then determine the magnitude and coordinate direction angles of the resultant force.

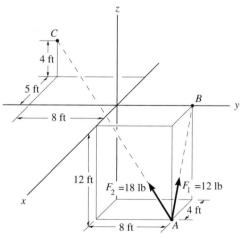

Prob. 2–91

*2–92. The cable OA exerts a force on point O of $\mathbf{F} = \{-4\mathbf{i} + 3\mathbf{j} + 10\mathbf{k}\}$ lb, directed from O to A. If the length of the cable is 5 ft, what are the coordinates $(-x, y, z)$ of point A?

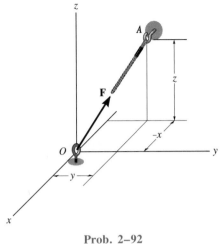

Prob. 2–92

2–93. The three supporting cables exert the forces shown on the sign. Represent each force as a Cartesian vector.

Prob. 2–93

2–94. The cylindrical vessel is supported by three cables which are concurrent at point D. Express each force which the cables exert on the vessel as a Cartesian vector, and determine the magnitude and coordinate direction angles of the resultant force.

2–95. The pole is held in place by three cables. If the force of each cable acting on the pole is shown, determine the position $(x, y, 0)$ for fixing cable BA so that the resultant force exerted on the pole is directed along its axis, from B toward O. Also, what is the magnitude of the resultant force?

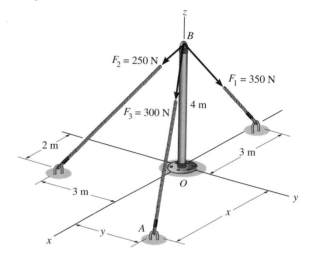

Prob. 2–95

*2–96.** The pole is held in place by three cables. If the force of each cable acting on the pole is shown, determine the magnitude and coordinate direction angles α, β, γ of the resultant force. Set $x = 4$ m, $y = 2$ m.

Prob. 2–94

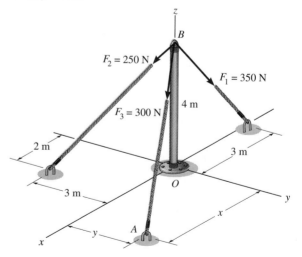

Prob. 2–96

2.9 Dot Product

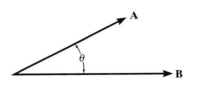

Fig. 2–43

Occasionally in statics one has to find the angle between two lines or the components of a force parallel and perpendicular to a line. In two dimensions, these problems can readily be solved by trigonometry since the geometry is easy to visualize. In three dimensions, however, this is often difficult, and consequently vector methods should be employed for the solution. The dot product defines a particular method for "multiplying" two vectors and is used to solve the above-mentioned problems.

The *dot product* of vectors **A** and **B,** written **A · B**, and read "**A** dot **B,**" is defined as the product of the magnitudes of **A** and **B** and the cosine of the angle θ between their tails, Fig. 2–43. Expressed in equation form,

$$\boxed{\mathbf{A} \cdot \mathbf{B} = AB \cos \theta} \tag{2–14}$$

where $0° \leq \theta \leq 180°$. The dot product is often referred to as the *scalar product* of vectors, since the result is a *scalar* and not a vector.

Laws of Operation

1. Commutative law:

$$\mathbf{A} \cdot \mathbf{B} = \mathbf{B} \cdot \mathbf{A}$$

2. Multiplication by a scalar:

$$a(\mathbf{A} \cdot \mathbf{B}) = (a\mathbf{A}) \cdot \mathbf{B} = \mathbf{A} \cdot (a\mathbf{B}) = (\mathbf{A} \cdot \mathbf{B})a$$

3. Distributive law:

$$\mathbf{A} \cdot (\mathbf{B} + \mathbf{D}) = (\mathbf{A} \cdot \mathbf{B}) + (\mathbf{A} \cdot \mathbf{D})$$

It is easy to prove the first and second laws by using Eq. 2–14. The proof of the distributive law is left as an exercise (see Prob. 2–97).

Cartesian Vector Formulation.

Equation 2–14 may be used to find the dot product for each of the Cartesian unit vectors. For example, $\mathbf{i} \cdot \mathbf{i} = (1)(1) \cos 0° = 1$ and $\mathbf{i} \cdot \mathbf{j} = (1)(1) \cos 90° = 0$. In a similar manner,

$$\mathbf{i} \cdot \mathbf{i} = 1 \qquad \mathbf{j} \cdot \mathbf{j} = 1 \qquad \mathbf{k} \cdot \mathbf{k} = 1$$
$$\mathbf{i} \cdot \mathbf{j} = 0 \qquad \mathbf{i} \cdot \mathbf{k} = 0 \qquad \mathbf{k} \cdot \mathbf{j} = 0$$

These results should not be memorized; rather, it should be clearly understood how each is obtained.

Consider now the dot product of two general vectors **A** and **B** which are expressed in Cartesian vector form. We have

$$\mathbf{A} \cdot \mathbf{B} = (A_x\mathbf{i} + A_y\mathbf{j} + A_z\mathbf{k}) \cdot (B_x\mathbf{i} + B_y\mathbf{j} + B_z\mathbf{k})$$
$$= A_xB_x(\mathbf{i} \cdot \mathbf{i}) + A_xB_y(\mathbf{i} \cdot \mathbf{j}) + A_xB_z(\mathbf{i} \cdot \mathbf{k})$$
$$+ A_yB_x(\mathbf{j} \cdot \mathbf{i}) + A_yB_y(\mathbf{j} \cdot \mathbf{j}) + A_yB_z(\mathbf{j} \cdot \mathbf{k})$$
$$+ A_zB_x(\mathbf{k} \cdot \mathbf{i}) + A_zB_y(\mathbf{k} \cdot \mathbf{j}) + A_zB_z(\mathbf{k} \cdot \mathbf{k})$$

Carrying out the dot-product operations, the final result becomes

$$\mathbf{A} \cdot \mathbf{B} = A_x B_x + A_y B_y + A_z B_z \qquad (2\text{--}15)$$

Thus, to determine the dot product of two Cartesian vectors, multiply their corresponding x, y, z components and sum their products algebraically. Since the result is a scalar, be careful *not* to include any unit vectors in the final result.

Applications. The dot product has two important applications in mechanics.

1. *The angle formed between two vectors or intersecting lines.* The angle θ between the tails of vectors **A** and **B** in Fig. 2–43 can be determined from Eq. 2–14 and written as

$$\theta = \cos^{-1}\left(\frac{\mathbf{A} \cdot \mathbf{B}}{AB}\right) \qquad 0° \le \theta \le 180°$$

Here $\mathbf{A} \cdot \mathbf{B}$ is computed from Eq. 2–15. In particular, notice that if $\mathbf{A} \cdot \mathbf{B} = 0$, $\theta = \cos^{-1} 0 = 90°$, so that **A** will be *perpendicular* to **B.**

2. *The components of a vector parallel and perpendicular to a line.* The component of vector **A** parallel to or collinear with the line aa' in Fig. 2–44 is defined by $\mathbf{A}_\parallel$, where $A_\parallel = A \cos\theta$. This component is sometimes referred to as the *projection* of **A** onto the line, since a right angle is formed in the construction. If the *direction* of the line is specified by the unit vector **u,** then, since $u = 1$, we can determine $A_\parallel$ directly from the dot product (Eq. 2–14); i.e.,

$$A_\parallel = A \cos\theta = \mathbf{A} \cdot \mathbf{u}$$

Hence, the scalar projection of **A** *along a line is determined from the dot product of* **A** *and the unit vector* **u** *which defines the direction of the line.* Notice that if this result is positive, then **A** has a directional sense which is the same as **u,** whereas if $A_\parallel$ is a negative scalar, then **A** has the opposite sense of direction to **u.** The component $\mathbf{A}_\parallel$ represented as a *vector* is therefore

$$\mathbf{A}_\parallel = A \cos\theta\, \mathbf{u} = (\mathbf{A} \cdot \mathbf{u})\mathbf{u}$$

Note that the component of **A** which is *perpendicular* to line aa' can also be obtained, Fig. 2–44. Since $\mathbf{A} = \mathbf{A}_\parallel + \mathbf{A}_\perp$, then $\mathbf{A}_\perp = \mathbf{A} - \mathbf{A}_\parallel$. Two possible ways of obtaining $A_\perp$ would be first to determine θ from the dot product, $\theta = \cos^{-1}(\mathbf{A} \cdot \mathbf{u}/A)$, then $A_\perp = A \sin\theta$. Alternatively, if $A_\parallel$ is known, then by the Pythagorean theorem we can also write $A_\perp = \sqrt{A^2 - A_\parallel^2}$.

The above two applications are illustrated numerically in the following example problems.

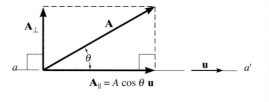

$$\mathbf{A}_\parallel = A \cos\theta\, \mathbf{u}$$

Fig. 2–44

Example 2–18

The frame shown in Fig. 2–45a is subjected to a horizontal force $\mathbf{F} = \{300\mathbf{j}\}$ N acting at its corner. Determine the magnitude of the components of this force parallel and perpendicular to member AB.

Fig. 2–45

SOLUTION

The magnitude of the component of $\mathbf{F}$ along AB is equal to the dot product of $\mathbf{F}$ and the unit vector $\mathbf{u}_B$ which defines the direction of AB, Fig. 2–45b. Since

$$\mathbf{u}_B = \frac{\mathbf{r}_B}{r_B} = \frac{2\mathbf{i} + 6\mathbf{j} + 3\mathbf{k}}{\sqrt{(2)^2 + (6)^2 + (3)^2}} = 0.286\mathbf{i} + 0.857\mathbf{j} + 0.429\mathbf{k}$$

Then

$$F_{AB} = F \cos \theta = \mathbf{F} \cdot \mathbf{u}_B = (300\mathbf{j}) \cdot (0.286\mathbf{i} + 0.857\mathbf{j} + 0.429\mathbf{k})$$
$$= (0)(0.286) + (300)(0.857) + (0)(0.429)$$
$$= 257.1 \text{ N}$$

Since the result is a positive scalar, $\mathbf{F}_{AB}$ has the same sense of direction as $\mathbf{u}_B$, Fig. 2–45b.

Expressing $\mathbf{F}_{AB}$ in Cartesian vector form, we have

$$\mathbf{F}_{AB} = F_{AB}\mathbf{u}_B = 257.1 \text{ N}(0.286\mathbf{i} + 0.857\mathbf{j} + 0.429\mathbf{k})$$
$$= \{73.5\mathbf{i} + 220\mathbf{j} + 110\mathbf{k}\} \text{ N} \qquad Ans.$$

The perpendicular component, Fig. 2–45b, is therefore

$$\mathbf{F}_{\perp} = \mathbf{F} - \mathbf{F}_{AB} = 300\mathbf{j} - (73.5\mathbf{i} + 220\mathbf{j} + 110\mathbf{k})$$
$$= \{-73.5\mathbf{i} + 80\mathbf{j} - 110\mathbf{k}\} \text{ N}$$

Its magnitude can be determined either from this vector or from the Pythagorean theorem, Fig. 2–45b:

$$F_{\perp} = \sqrt{F^2 - F_{AB}^2}$$
$$= \sqrt{(300)^2 - (257.1)^2}$$
$$= 155 \text{ N} \qquad Ans.$$

Example 2–19

The pipe in Fig. 2–46a is subjected to the force $F = 80$ lb at its end B. Determine the angle θ between $\mathbf{F}$ and the pipe segment BA, and the magnitudes of the components of $\mathbf{F}$, which are parallel and perpendicular to BA.

(a)

SOLUTION

Angle θ. Position vectors along BA and BC are first determined.

$$\mathbf{r}_{BA} = \{-2\mathbf{i} - 2\mathbf{j} + 1\mathbf{k}\} \text{ ft}$$
$$\mathbf{r}_{BC} = \{-3\mathbf{j} + 1\mathbf{k}\} \text{ ft}$$

Thus, the angle θ between their tails is

$$\cos \theta = \frac{\mathbf{r}_{BA} \cdot \mathbf{r}_{BC}}{r_{BA}r_{BC}} = \frac{(-2)(0) + (-2)(-3) + (1)(1)}{3\sqrt{10}}$$
$$= 0.7379$$
$$\theta = 42.5° \hspace{3cm} Ans.$$

Components of $\mathbf{F}$. We must first formulate the unit vector along BA and force $\mathbf{F}$ as Cartesian vectors.

$$\mathbf{u}_{BA} = \frac{\mathbf{r}_{BA}}{r_{BA}} = \frac{-2\mathbf{i} - 2\mathbf{j} + 1\mathbf{k}}{3} = -\frac{2}{3}\mathbf{i} - \frac{2}{3}\mathbf{j} + \frac{1}{3}\mathbf{k}$$
$$F = 80 \text{ lb}\left(\frac{\mathbf{r}_{BC}}{r_{BC}}\right) = 80\left(\frac{-3\mathbf{j} + 1\mathbf{k}}{\sqrt{10}}\right) = -75.89\mathbf{j} + 25.30\mathbf{k}$$

Thus,

$$F_{BA} = \mathbf{F} \cdot \mathbf{u}_{BA} = (-75.89\mathbf{j} + 25.30\mathbf{k}) \cdot \left(-\frac{2}{3}\mathbf{i} - \frac{2}{3}\mathbf{j} + \frac{1}{3}\mathbf{k}\right)$$
$$= 0 + 50.60 + 8.43$$
$$= 59.0 \text{ lb} \hspace{3cm} Ans.$$

Since θ was calculated in Fig. 2–46b, this same result can also be obtained directly from trigonometry.

$$F_{BA} = 80 \cos 42.5° = 59.0 \text{ lb} \hspace{2cm} Ans.$$

Realize that both solutions are valid since, by the dot product,

$$F_{BA} = \mathbf{F} \cdot \mathbf{u}_{BA} = F(1) \cos \theta$$

Also,

$$F_{\perp} = F \sin \theta$$
$$= 80 \sin 42.5°$$
$$= 54.0 \text{ lb} \hspace{3cm} Ans.$$

(b) Fig. 2–46

PROBLEMS

2–97. Given the three vectors **A**, **B**, and **D**, show that $\mathbf{A} \cdot (\mathbf{B} + \mathbf{D}) = (\mathbf{A} \cdot \mathbf{B}) + (\mathbf{A} \cdot \mathbf{D})$.

2–98. Determine the angle θ between the two position vectors.

2–99. Determine the projected component of vector $\mathbf{r}_1$ acting in the direction of $\mathbf{r}_2$. Express the result as a Cartesian vector.

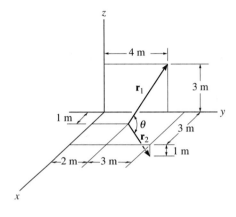

Probs. 2–98/2–99

2–102. Determine the magnitude of the projected component of the position vector **r** along the Oa axis.

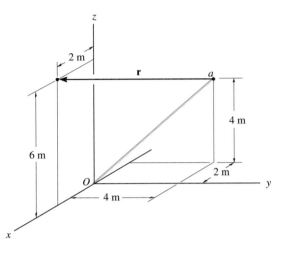

Prob. 2–102

*2–100. Determine the angle θ between the two position vectors.

2–101. Determine the projected component of vector $\mathbf{r}_1$ acting in the direction of $\mathbf{r}_2$. Express the result as a Cartesian vector.

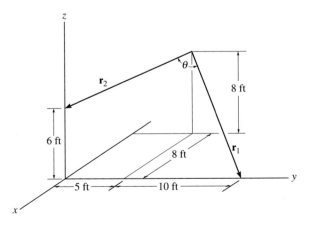

Probs. 2–100/2–101

2–103. Determine the magnitude of the projected component of the position vector **r** along the Oa axis.

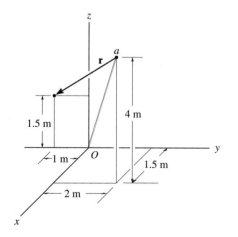

Prob. 2–103

*2–104.   A force of $F = 80$ N is applied to the handle of the wrench. Determine the angle θ between the tail of the force and the handle AB.

2–105.   A force of $F = 80$ N is applied to the handle of the wrench. Determine the magnitudes of the components of the force acting along the axis AB of the wrench handle and perpendicular to it.

F = 80 N

B

θ

30°

45°

A

300 mm

500 mm

z

y

x

Probs. 2–104/2–105

2–107.   Determine the magnitudes of the projected components of the force $\mathbf{F} = \{-80\mathbf{i} + 30\mathbf{j} + 20\mathbf{k}\}$ lb in the direction of the cables AB and AC.

z

F

A

8 ft

3 ft

C

4 ft

y

5 ft

B

4 ft

x

Prob. 2–107

2–106.   A force of $\mathbf{F} = \{-40\mathbf{k}\}$ lb acts at the end of the pipe. Determine the magnitudes of the components $\mathbf{F}_1$ and $\mathbf{F}_2$ which act along the pipe's axis and perpendicular to it.

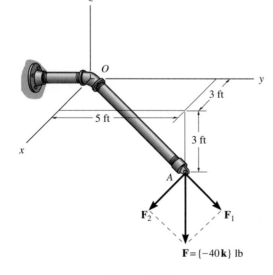

z

O

y

3 ft

5 ft

3 ft

x

A

$\mathbf{F}_2$

$\mathbf{F}_1$

$\mathbf{F} = \{-40\mathbf{k}\}$ lb

Prob. 2–106

*2–108.   Determine the magnitude of the projected component of the 150-lb force acting along the axis BC of the pipe.

2–109.   Determine the angle θ between pipe segments BA and BC.

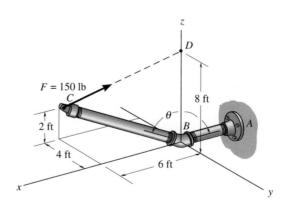

z

D

F = 150 lb

C

8 ft

2 ft

θ

B

A

4 ft

6 ft

x

y

Probs. 2–108/2–109

2–110. Determine the angle θ between cables AB and AC.

2–111. If F has a magnitude of 55 lb, determine the magnitude of its projected component acting along the x axis and along cable AC.

Probs. 2–110/2–111

2–114. Two cables exert forces on the pipe. Determine the magnitude of the projected component of F_1 along the line of action of F_2.

2–115. Determine the angle θ between the two cables attached to the pipe.

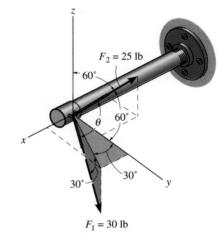

Probs. 2–114/2–115

*2–112. Determine the angles θ and ϕ between the axis OA of the pole and each cable, AB and AC.

2–113. The two cables exert the forces shown on the pole. Determine the magnitude of the projected component of each force acting along the axis OA of the pole.

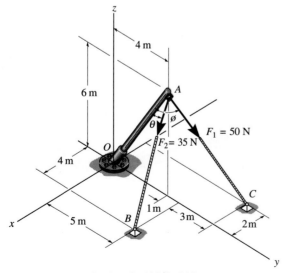

Probs. 2–112/2–113

*2–116. The force $F = \{25i - 50j + 10k\}$ N acts at the end A of the pipe assembly. Determine the magnitudes of the components F_1 and F_2 which act along the axis of AB and perpendicular to it.

Prob. 2–116

REVIEW PROBLEMS

2–117. The door is held open by means of two chains. If the tension in *AB* and *CD* is $F_A = 300$ N and $F_C = 250$ N, respectively, express each of these forces in Cartesian vector form.

Prob. 2–117

2–118. Express each force acting on the screw eye as a Cartesian vector, and then determine the magnitude and orientation of the resultant force.

2–119. Use the parallelogram law to determine the magnitude of the resultant force acting on the screw eye. Determine its orientation, measured counterclockwise from the positive *x* axis.

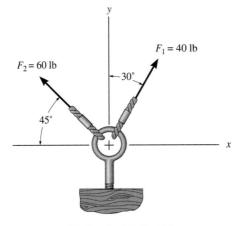

Probs. 2–118/2–119

*2–120. The backbone exerts a vertical force of 110 lb on the lumbosacral joint of a person who is standing erect. Determine the magnitudes of the components of this force directed perpendicular to the surface of the sacrum, v axis, and parallel to it, u axis.

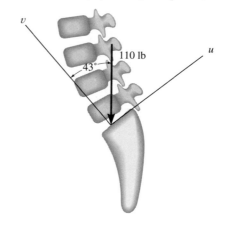

Prob. 2–120

2–121. The cord *AB* exerts a force of $\mathbf{F} = \{12\mathbf{i} + 9\mathbf{j} - 8\mathbf{k}\}$ lb on the hook. If the cord is 8 ft long, determine the location x, y of the point of attachment *B*, and the height z of the hook.

2–122. The cord exerts a force of $F = 30$ lb on the hook. If the cord is 8 ft long, $z = 4$ ft, and the *x* component of the force is $F_x = 25$ lb, determine the location x, y of the point of attachment *B* of the cord to the ground.

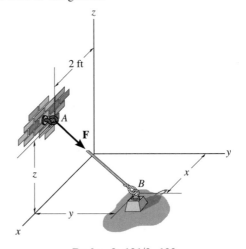

Probs. 2–121/2–122

2–123. Determine the magnitude of the projected component of force **F** along the *Oa* axis.

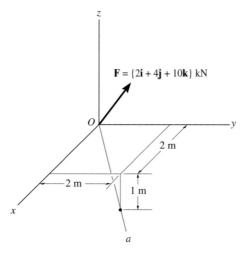

Prob. 2–123

*2–124. Express each of the three forces acting on the column in Cartesian vector form and determine the magnitude of the resultant force.

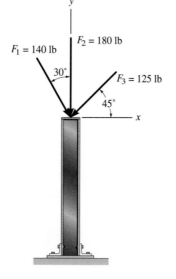

Prob. 2–124

2–125. Determine the angle θ ($0° \leq \theta \leq 90°$) for cable *AB* so that the 500-N vertical force has a component of 200 N directed along cable *AB* from *A* toward *B*. What is the corresponding component of force acting along cable *CB*? Set $\phi = 60°$.

2–126. Determine the angle ϕ ($0° \leq \phi \leq 90°$) between cables *BA* and *BC* so that the 500-N vertical force has a component of 250 N which acts along cable *CB*, from *C* to *B*. Set $\theta = 60°$.

Probs. 2–125/2–126

3 Equilibrium of a Particle

In this chapter the methods for resolving a force into components and expressing a force as a Cartesian vector will be used to solve problems involving the equilibrium of a particle. To simplify the discussion, particle equilibrium for a concurrent coplanar force system will be considered first. Then, in the last part of the chapter, equilibrium problems involving concurrent three-dimensional force systems will be considered.

3.1 Condition for the Equilibrium of a Particle

A particle is in *equilibrium* provided it is at rest if originally at rest, or has a constant velocity if originally in motion. Most often, however, the term "equilibrium" or, more specifically, "static equilibrium" is used to describe an object at rest. To maintain a state of equilibrium, it is *necessary* to satisfy Newton's first law of motion, which states that if the *resultant force* acting on a particle is *zero,* then the particle is in equilibrium. This condition may be stated mathematically as

$$\Sigma \mathbf{F} = \mathbf{0} \qquad (3-1)$$

where $\Sigma \mathbf{F}$ is the vector *sum of all the forces* acting on the particle.

Not only is Eq. 3–1 a necessary condition for equilibrium, it is also a *sufficient* condition. This follows from Newton's second law of motion, which can be written as $\Sigma \mathbf{F} = m\mathbf{a}.$ Since the force system satisfies Eq. 3–1, then $m\mathbf{a} = \mathbf{0},$ and therefore the particle's acceleration $\mathbf{a} = \mathbf{0},$ and consequently the particle indeed moves with constant velocity or remains at rest.

3.2 The Free-Body Diagram

To apply the equation of equilibrium correctly, we must account for *all* the known and unknown forces ($\Sigma\mathbf{F}$) which act *on* the particle. The best way to do this is to draw the particle's *free-body diagram*. This diagram is a sketch of the particle which represents it as being isolated or "free" from its surroundings. On this sketch it is necessary to show *all* the forces that act *on* the particle. Once this diagram is drawn, it will then be easy to apply Eq. 3–1.

Before presenting a formal procedure as to how to draw a free-body diagram, we will first discuss two types of connections often encountered in particle equilibrium problems.

Springs. If a *linear elastic spring* is used for support, the length of the spring will change in direct proportion to the force acting on it. A characteristic that defines the "elasticity" of a spring is the *spring constant* or *stiffness k*. Specifically, the magnitude of force developed by a linear elastic spring which has a stiffness k, and is deformed (elongated or compressed) a distance s measured from its unloaded position, is

$$F = ks \tag{3–2}$$

Note that s is determined from the difference in the spring's deformed length l and its undeformed length l_o, i.e., $s = l - l_o$. Thus, if s is positive, $\mathbf{F}$ "pulls" on the spring; whereas if s is negative, $\mathbf{F}$ must "push" on it. For example, the spring shown in Fig. 3–1 has an undeformed length $l_o = 0.4$ m and stiffness $k = 500$ N/m. To stretch it so that $l = 0.6$ m, a force $F = ks = (500$ N/m$)(0.6$ m $- 0.4$ m$) = 100$ N is needed. Likewise, to compress it to a length $l = 0.2$ m, a force $F = ks = (500$ N/m$)(0.2$ m $- 0.4$ m$) = -100$ N is required, Fig. 3–1.

Fig. 3–1

Cables and Pulleys.   Throughout this book, except in Sec. 7.4, all cables (or cords) are assumed to have negligible weight and they cannot be stretched. A cable can support *only* a tension or "pulling" force, and this force always acts in the direction of the cable. In Chapter 5 it will be shown that the tension force developed in a *continuous cable* which passes over a frictionless pulley must have a *constant* magnitude to keep the cable in equilibrium. Hence, for any angle θ, shown in Fig. 3–2, the cable is subjected to a constant tension T throughout its length.

Fig. 3–2

PROCEDURE FOR DRAWING
A FREE-BODY DIAGRAM

Since we must account for all the forces acting on the particle, the importance of drawing a free-body diagram before applying the equation of equilibrium to the solution of a problem cannot be overemphasized. To construct a free-body diagram, the following three steps are necessary.

Step 1.   Imagine the particle to be *isolated* or cut "free" from its surroundings. Draw or sketch its outlined shape.

Step 2.   Indicate on this sketch *all* the forces that act *on the particle*. These forces can be *active forces,* which tend to set the particle in motion, such as those caused by attached cables, weight, or magnetic and electrostatic interaction. Also, *reactive forces* will occur, such as those caused by the constraints or supports that tend to prevent motion. To account for all these forces, it may help to trace around the particle's boundary, carefully noting each force acting on it.

Step 3.   The forces that are *known* should be labeled with their proper magnitudes and directions. Letters are used to represent the magnitudes and directions of forces that are unknown. In particular, if a force has a known line of action but unknown magnitude, the "arrowhead," which defines the sense of the force, can be *assumed*. The correct sense will become apparent after solving for the unknown magnitude. By definition, the *magnitude* of a force is *always positive* so that, if the solution yields a "negative" scalar, the *minus sign* indicates that the arrowhead or sense of the force is opposite to that which was originally assumed.

Application of the above steps is illustrated in the following two examples.

Example 3–1

The crate in Fig. 3–3a has a weight of 20 lb. Draw a free-body diagram of the crate, the cord BD, and the ring at B.

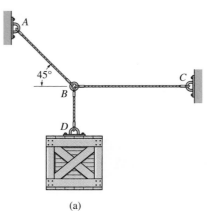

(a)

SOLUTION

If we imagine the crate to be isolated from its surroundings, then by inspection there are only two forces acting on it, namely, the gravitational force or weight of 20 lb, and the force of the cord BD. Thus the free-body diagram is shown in Fig. 3–3b.

If the cord BD is isolated, then there are only two forces acting on it, Fig. 3–3c, namely, the force of the connection on the top of the crate, $\mathbf{F}_D$, and the force $\mathbf{F}_B$ at B which is caused by the ring. Since these forces tend to *stretch* the cord, we can state that the cord is in *tension*. (This must be the case, since compressive forces would cause the cord to collapse.)

When the ring at B is isolated from its surroundings, it is realized that three forces act on it. All these forces are caused by the attached cords, Fig. 3–3d. Notice that $\mathbf{F}_B$ shown here is equal but opposite to that shown in Fig. 3–3c, a consequence of Newton's third law.

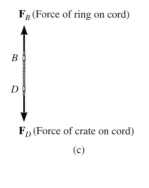

$\mathbf{F}_D$ (Force of cord on crate)

20 lb (Weight of crate)

(b)

$\mathbf{F}_B$ (Force of ring on cord)

$\mathbf{F}_D$ (Force of crate on cord)

(c)

Fig. 3–3

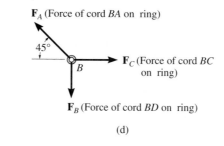

$\mathbf{F}_A$ (Force of cord BA on ring)

$\mathbf{F}_C$ (Force of cord BC on ring)

$\mathbf{F}_B$ (Force of cord BD on ring)

(d)

Example 3–2

The sphere in Fig. 3–4a has a mass of 6 kg and is supported as shown. Draw a free-body diagram of the sphere and the knot at C.

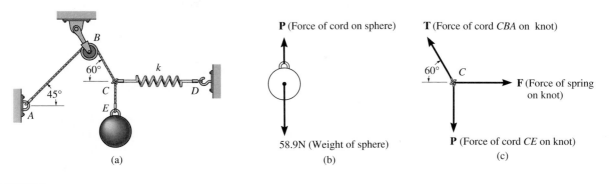

(a)

P (Force of cord on sphere)

58.9N (Weight of sphere)

(b)

T (Force of cord CBA on knot)

F (Force of spring on knot)

P (Force of cord CE on knot)

(c)

SOLUTION

There are two forces acting on the sphere, namely, its weight and the force **P** of the cord CE. The sphere has a weight of (6 kg)(9.81 m/s^2) = 58.9 N. Its free-body diagram is shown in Fig. 3–4b.

By inspection, three forces act on the knot at C. They are caused by cords CBA and CE, and the spring CD. Thus, the free-body diagram of the knot is shown in Fig. 3–4c.

Fig. 3–4

3.3 Coplanar Force Systems

Many particle equilibrium problems involve a coplanar force system. If the forces lie in the x–y plane, they can each be resolved into their respective **i** and **j** components and Eq. 3–1 can be written as

$$\Sigma \mathbf{F} = \mathbf{0}$$
$$\Sigma F_x \mathbf{i} + \Sigma F_y \mathbf{j} = \mathbf{0}$$

For this vector equation to be satisfied, both the x and y components must equal zero, otherwise $\Sigma \mathbf{F} \neq \mathbf{0}$. Hence, we require

$$\boxed{\begin{aligned} \Sigma F_x &= 0 \\ \Sigma F_y &= 0 \end{aligned}}$$

(3–3)

These scalar equilibrium equations state that the algebraic sum of the x and y components of all the forces acting on the particle be equal to zero. As a result, Eqs. 3–3 can be solved for at most two unknowns, generally represented as angles and magnitudes of forces shown on the particle's free-body diagram.

Scalar Notation. Since each of the two equilibrium equations requires the resolution of vector components along a specified axis (x or y), we will use scalar notation to represent the components when applying these equations. By doing this, the sense of direction for each component is accounted for by an *algebraic sign* which corresponds to the arrowhead direction of the component as indicated graphically on the free-body diagram. In particular, if a force component has an unknown magnitude, then the arrowhead sense of the force on the free-body diagram can be *assumed*. Since the magnitude of a force is *always positive,* then if the *solution* yields a *negative scalar,* it indicates that the sense of the force as shown on the free-body diagram is opposite to that which was assumed.

For example, consider the free-body diagram of the particle subjected to the two forces as shown in Fig. 3–5. For the sake of discussion, we have *assumed* that the *unknown force* **F** acts to the right to maintain equilibrium. Application of the equation of equilibrium along the x axis yields

Fig. 3–5

$$\xrightarrow{+}\Sigma F_x = 0; \qquad\qquad +F + 10 \text{ N} = 0$$

Both terms are "positive" since both forces act in the positive x direction, as indicated by the arrow placed alongside the equation. When this equation is solved, $F = -10$ N. Here the *negative sign* refers to the fact that **F** in Fig. 3–5 is shown in the opposite sense of the actual direction. In other words, **F** must act to the left to hold the particle in equilibrium. Notice that if the $+x$ axis in Fig. 3–5 was directed to the left both terms in the above equation would be negative, but again $F = -10$ N, indicating F would be directed to the left.

PROCEDURE FOR ANALYSIS

The following procedure provides a method for solving coplanar force problems involving particle equilibrium:

Free-Body Diagram. Draw a free-body diagram of the particle. As outlined in Sec. 3.2, this requires that all the known and unknown force magnitudes and angles be labeled on the diagram. The sense of a force having an unknown magnitude can be assumed.

Equations of Equilibrium. Establish the x, y axes in *any* suitable direction and apply the two equations of equilibrium, $\Sigma F_x = 0$, and $\Sigma F_y = 0$. For application, components are positive if they are directed along the positive axes, and negative if they are directed along the negative axes. If more than two unknowns exist and the problem involves a spring, apply $F = ks$ (Eq. 3–2) to relate the spring force to the deformation s of the spring.

The following example problems illustrate this solution procedure numerically.

Example 3-3

Determine the tension in cords AB and AD for equilibrium of the 10-kg crate shown in Fig. 3–6a.

SOLUTION

The free-body diagram of the crate is shown in Fig. 3–6b. Here the weight of the crate is $(10 \text{ kg})(9.81 \text{ m/s}^2) = 98.1 \text{ N}$. As a result, the force of cord CA on the crate must also be 98.1 N in order to hold the crate in equilibrium. By Newton's third law, this force acts in an equal but opposite manner on the cord CA, Fig. 3–6c, and so the cord is held in equilibrium by the 98.1-N force of the ring. The force in cords AB and AD can now be obtained by investigating the equilibrium of the ring at A, because this "particle" is subjected to the force of both of these cords.

Free-Body Diagram. As shown in Fig. 3–6d, there are three concurrent forces *acting on the ring*. The cord tension forces $\mathbf{T}_B$ and $\mathbf{T}_D$ have unknown magnitudes but known directions. Cord AC exerts a downward force on A equal to 98.1 N. Why?

(d)

Equations of Equilibrium. Since the equilibrium equations require a summation of the x and y components of each force, $\mathbf{T}_B$ must be resolved into x and y components. These components, shown dashed on the free-body diagram, have magnitudes of $T_B \cos 30°$ and $T_B \sin 30°$, respectively. Applying the equations of equilibrium, we have

$$\xrightarrow{+} \Sigma F_x = 0; \qquad T_B \cos 30° - T_D = 0 \qquad (1)$$
$$+\uparrow \Sigma F_y = 0; \qquad T_B \sin 30° - 98.1 = 0 \qquad (2)$$

Solving Eq. (2) for T_B and substituting into Eq. (1) to obtain T_D yields

$$T_B = 196 \text{ N} \qquad \qquad Ans.$$
$$T_D = 170 \text{ N} \qquad \qquad Ans.$$

The accuracy of these results, of course, depends on the accuracy of the data, i.e., measurements of geometry and loads. For most engineering work involving a problem such as this, the data as measured to three significant figures would be sufficient. Also, note that here we have neglected the weights of the cords, a reasonable assumption since they would be small in comparison with the weight of the crate.

(a)

98.1N (Force of cord on crate)

98.1N (Weight of crate)

(b)

98.1N (Force of ring A on cord)

98.1N (Force of crate on cord)

(c)

Fig. 3–6

Example 3–4

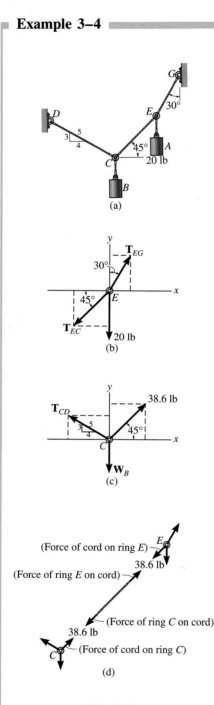

(a)

(b)

(c)

(Force of cord on ring E)

(Force of ring E on cord)

(Force of ring C on cord)

38.6 lb

(Force of cord on ring C)

(d)

Fig. 3–7

If the cylinder at A in Fig. 3–7a has a weight of 20 lb, determine the weight of the cylinder at B and the force in each cord needed to hold the system in the equilibrium position shown.

SOLUTION

Since the weight of A is known, the unknown tension in cords EG and EC can be determined by investigating the equilibrium of the ring at E. Why?

Free-Body Diagram. There are three forces acting on E, as shown in Fig. 3–7b.

Equations of Equilibrium. Resolving each force into its x and y components using trigonometry, and applying the equations of equilibrium, we have

$$\xrightarrow{+}\Sigma F_x = 0; \qquad T_{EG} \sin 30° - T_{EC} \cos 45° = 0 \qquad (1)$$
$$+\uparrow \Sigma F_y = 0; \qquad T_{EG} \cos 30° - T_{EC} \sin 45° -. 20 = 0 \qquad (2)$$

Solving Eq. 1 for T_{EG} in terms of T_{EC} and substituting the result into Eq. 2 allows a solution for T_{EC}. One then obtains T_{EG} from Eq. 1. The results are

$$T_{EC} = 38.6 \text{ lb} \qquad\qquad Ans.$$
$$T_{EG} = 54.6 \text{ lb} \qquad\qquad Ans.$$

Using the calculated result for T_{EC}, the equilibrium of the ring at C can now be investigated to determine the tension in CD and the weight of B.

Free-Body Diagram. As shown in Fig. 3–7c, $T_{EC} = 38.6$ lb "pulls" on C. The reason for this becomes clear when one draws the free-body diagram of cord CE and applies both equilibrium and the principle of action, equal but opposite force reaction (Newton's third law), Fig. 3–7d.

Equations of Equilibrium. Noting the components of $\mathbf{T}_{CD}$ are proportional to the slope of the cord as defined by the 3–4–5 triangle, we have

$$\xrightarrow{+}\Sigma F_x = 0; \qquad 38.6 \cos 45° - (\tfrac{4}{5})T_{CD} = 0 \qquad (3)$$
$$+\uparrow \Sigma F_y = 0; \qquad (\tfrac{3}{5})T_{CD} + 38.6 \sin 45° - W_B = 0 \qquad (4)$$

Solving Eq. 3 and substituting the result into Eq. 4 yields

$$T_{CD} = 34.1 \text{ lb} \qquad\qquad Ans.$$
$$W_B = 47.8 \text{ lb} \qquad\qquad Ans.$$

Example 3–5

Determine the required length of cord AC in Fig. 3–8a so that the 8-kg lamp is suspended in the position shown. The *undeformed* length of the spring AB is $l'_{AB} = 0.4$ m, and the spring has a stiffness of $k_{AB} = 300$ N/m.

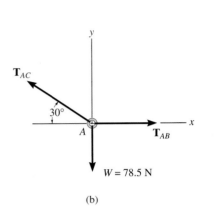

(a)

(b)

Fig. 3–8

SOLUTION

If the force in spring AB is known, the stretch of the spring can be found ($F = ks$). Using the problem geometry, it is then possible to calculate the required length of AC.

Free-Body Diagram.   The lamp has a weight $W = 8(9.81) = 78.5$ N. The free-body diagram of the ring at A is shown in Fig. 3–8b.

Equations of Equilibrium

$$\xrightarrow{+} \Sigma F_x = 0; \qquad T_{AB} - T_{AC} \cos 30° = 0$$
$$+ \uparrow \Sigma F_y = 0; \qquad T_{AC} \sin 30° - 78.5 = 0$$

Solving, we obtain

$$T_{AC} = 157.0 \text{ N}$$
$$T_{AB} = 136.0 \text{ N}$$

The stretch of spring AB is therefore

$$T_{AB} = k_{AB} s_{AB}; \qquad 136.0 \text{ N} = 300 \text{ N/m}(s_{AB})$$
$$s_{AB} = 0.453 \text{ m}$$

so the stretched length is

$$l_{AB} = l'_{AB} + s_{AB}$$
$$l_{AB} = 0.4 \text{ m} + 0.453 \text{ m} = 0.853 \text{ m}$$

The horizontal distance from C to B, Fig. 3–8a, requires

$$2 \text{ m} = l_{AC} \cos 30° + 0.853 \text{ m}$$
$$l_{AC} = 1.32 \text{ m} \qquad\qquad\qquad \textit{Ans.}$$

PROBLEMS

3–1.   Determine the magnitudes of F_1 and F_2 so that the particle is in equilibrium.

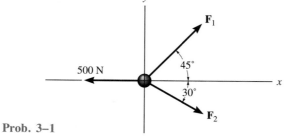

Prob. 3–1

3–2.   Determine the magnitude and orientation θ of F so that the particle is in equilibrium.

Prob. 3–2

3–3.   Determine the magnitude and orientation θ of F so that the particle is in equilibrium.

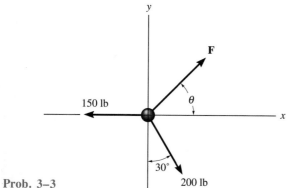

Prob. 3–3

*3–4.   The particle is subjected to the action of four *coplanar* forces. Determine the magnitude of F and the orientation θ of the 750-N force for equilibrium.

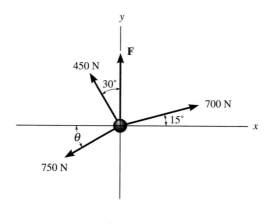

Prob. 3–4

3–5.   The members of a truss are pin-connected at joint O as shown. Determine the magnitudes of F_1 and F_2 for equilibrium.

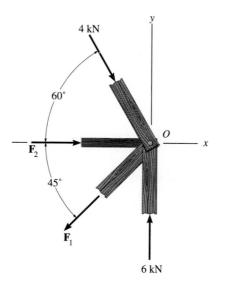

Prob. 3–5

3–6. The joint of a light metal truss is formed by riveting four angles to the *gusset plate*. Knowing the force in members A and C, determine the forces $\mathbf{F}_B$ and $\mathbf{F}_D$ acting on members B and D that are required for equilibrium. The force system is concurrent at point O.

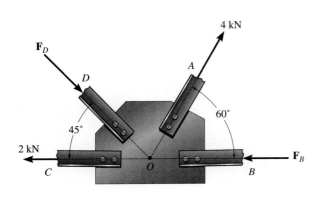

Prob. 3–6

3–7. Determine the maximum load that can be suspended without exceeding a cable tension of 780 lb in either cable AB or AC.

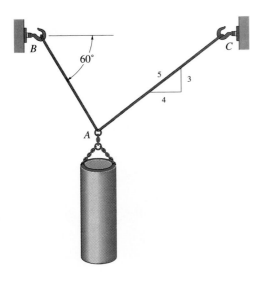

Prob. 3–7

*3–8. Determine the force in cables AB and AC necessary to support the 15-kg light fixture.

Prob. 3–8

3–9. The cables act to brace the structural framework. If cable AB exerts a force of $F_{AB} = 60$ lb on the ring at A, determine the force in cables AC and AD.

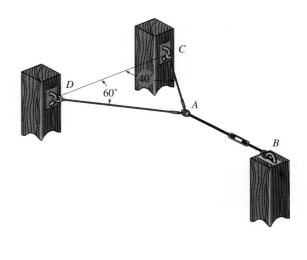

Prob. 3–9

3–10. The sling is used to support a beam having a weight of 900 lb. Determine the force in cables AB and AC for equilibrium.

Prob. 3–10

3–11. Determine the forces in cables AC and AB needed to hold the 20-kg cylinder D in equilibrium. Set $F = 300$ N and $d = 1$ m.

*3–12. The cylinder D has a mass of 20 kg. If a force of $F = 100$ N is applied horizontally to the ring at A, determine the largest dimension d so that the force in cable AC is zero.

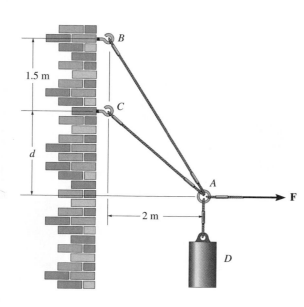

Probs. 3–11/3–12

3–13. The 200-lb tank is suspended by means of a 6-ft-long cable, which is attached to the sides of the tank and passes over the small pulley located at O. If the cable can be attached at either points A and B, or C and D, determine which attachment produces the least amount of tension in the cable. What is this tension?

Prob. 3–13

3–14. Determine the maximum weight of the engine that can be supported without exceeding a cable tension of 450 lb in cable AB and 480 lb in cable AC.

Prob. 3–14

3–15. The pan and contents on the scale have a weight of 10 lb. If each spring has an unstretched length of 4 ft and a stiffness of $k = 5$ lb/ft, determine the angle θ for equilibrium.

3–17. Determine the stretch of each spring for equilibrium of the 20-kg block. The springs are shown in their equilibrium position.

Prob. 3–15

Prob. 3–17

*3–16. The elastic cord (or spring) *ABC* has a stiffness of 500 N/m and an unstretched length of 6 m. Determine the horizontal force **F** applied to the cord, which is attached to the *small* pulley *B*, so that the displacement of the pulley from the supports is $d = 1.5$ m.

3–18. Three weights are connected by the light inextensible cords. If the cords pass over small frictionless pulleys, determine the distance *s* for equilibrium.

Prob. 3–16

Prob. 3–18

3–19. Determine the force in each cord for equilibrium needed to support the bucket, which has a mass of 60 kg.

Prob. 3–19

3–21. The two spheres A and B have an equal weight and are electrostatically charged, such that the repulsive force acting between them has a magnitude of 0.006 lb and is directed along the dashed line. Determine the angle θ, the tension in the cords, and the weight W of each sphere.

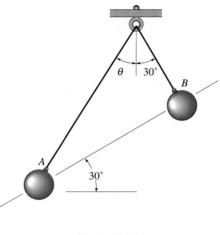

Prob. 3–21

*3–20. The pipe has a weight of 150 lb and is supported by the six cords tied together as shown. Determine the tension in each cord and the angle θ for equilibrium of DC. Cord BC is horizontal. *Hint:* Analyze the equilibrium of the knots in the following sequence: A, B, C, using the results previously calculated.

3–22. Prove *Lami's theorem*, which states that if three concurrent forces are in equilibrium, each is proportional to the sine of the angle of the other two; that is, $P/\sin \alpha = Q/\sin \beta = R/\sin \gamma$.

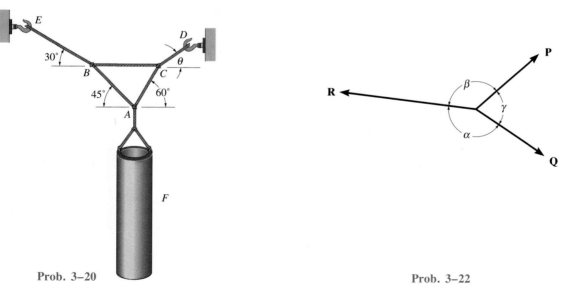

Prob. 3–20

Prob. 3–22

3–23. Determine the tension T in chains AB and AC as a function of the angle θ. The lamp has a weight W. Plot the results as a curve representing T (vertical axis) versus θ (horizontal axis) for $0° \le \theta \le 90°$.

Prob. 3–23

3–25. The springs on the rope assembly are originally unstretched when $\theta = 0°$. Determine the tension in each rope when $F = 90$ lb. Neglect the size of the pulleys at B and D.

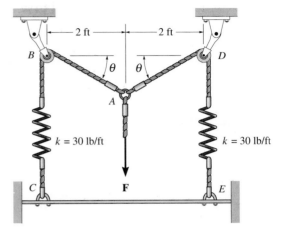

Prob. 3–25

*3–24. The ring at A is attached to the end of the rope. If the rope passes over the beam and then through the ring, determine the angle θ of the rope if a 25-lb load is suspended from its opposite end.

25 lb

Prob. 3–24

3–26. The springs on the rope assembly are originally unstretched when $\theta = 0°$. Determine the vertical force F required to displace the ring at A so that $\theta = 30°$. Neglect the size of the pulleys at B and D.

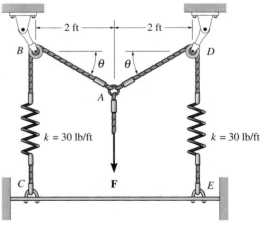

Prob. 3–26

3–27. The spring has a stiffness of $k = 80$ N/m and an unstretched length of 2 m. Determine the force in cables BC and BD when the spring is held in the position shown.

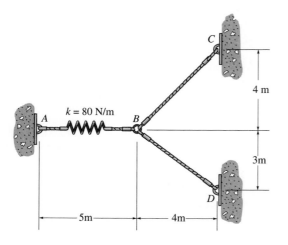

Prob. 3–27

3–29. The ends of the 6-m-long cable AB are attached to the fixed walls. If a bucket and its contents have a mass of 10 kg and are suspended from the cable by means of a *small* pulley at C, determine the location x of the pulley for equilibrium.

Prob. 3–29

3–30. The man attempts to pull the log at C by using the three ropes. Determine the direction θ in which he should pull on his rope with a force of 80 lb, so that he exerts a maximum force on the log. Calculate the force on the log for this case. Also, determine the direction in which he should pull in order to maximize the force in the rope attached to B. What is this maximum force?

*3–28. Three blocks are supported using the cords and two pulleys as shown. If the blocks have weights of $W_A = W$, $W_B = 0.25W$, and $W_C = W$, determine the angles θ and ϕ for equilibrium.

Prob. 3–28

Prob. 3–30

3.4 Three-Dimensional Force Systems

It was shown in Sec. 3.1 that particle equilibrium requires

$$\Sigma \mathbf{F} = \mathbf{0} \qquad\qquad (3\text{--}4)$$

If the forces acting on the particle are resolved into their respective $\mathbf{i}, \mathbf{j}, \mathbf{k}$ components, we can then write

$$\Sigma F_x \mathbf{i} + \Sigma F_y \mathbf{j} + \Sigma F_z \mathbf{k} = \mathbf{0}$$

To ensure that Eq. 3–4 is satisfied, we must therefore require that the following three scalar component equations be satisfied:

$$\boxed{\begin{aligned} \Sigma F_x &= 0 \\ \Sigma F_y &= 0 \\ \Sigma F_z &= 0 \end{aligned}} \qquad\qquad (3\text{--}5)$$

These equations represent the *algebraic sums* of the x, y, z force components acting on the particle. Using them we can solve for at most three unknowns, generally represented as angles or magnitudes of forces shown on the particle's free-body diagram.

PROCEDURE FOR ANALYSIS

The following procedure provides a method for solving three-dimensional force equilibrium problems.

Free-Body Diagram.   Draw a free-body diagram of the particle and label all the known and unknown forces on this diagram.

Equations of Equilibrium.   Establish the x, y, z coordinate axes with origin located at the particle and apply the equations of equilibrium. Use the three scalar Eqs. 3–5 in cases where it is easy to resolve each force acting on the particle into its x, y, z components. If this appears difficult, first express each force acting on the particle in Cartesian vector form, and then substitute these vectors into Eq. 3–4. By setting the respective $\mathbf{i}, \mathbf{j}, \mathbf{k}$ components equal to zero, the three scalar Eqs. 3–5 can be generated. If more than three unknowns exist and the problem involves a spring, consider using $F = ks$ to relate the spring force to the deformation s of the spring.

The following example problems numerically illustrate this solution procedure.

Example 3–6

The 90-lb cylinder shown in Fig. 3–9a is supported by two cables and a spring having a stiffness $k = 500$ lb/ft. Determine the force in the cables and the stretch of the spring for equilibrium. Cable AD lies in the x–y plane and cable AC lies in the x–z plane.

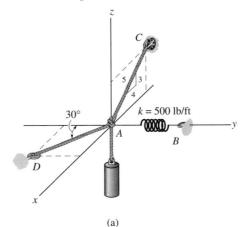

(a)

SOLUTION

The stretch of the spring can be determined once the force in the spring is determined.

Free-Body Diagram. The connection at A is chosen for the analysis since the cable forces are concurrent at this point. Fig. 3–9b.

Equations of Equilibrium. By inspection, each force can easily be resolved into its x, y, z components, and therefore the three scalar equations of equilibrium can be directly applied. Considering components directed along the positive axes as "positive," we have

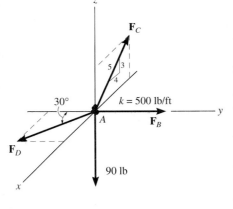

(b)

Fig. 3–9

$$\Sigma F_x = 0; \qquad F_D \sin 30° - \tfrac{4}{5}F_C = 0 \qquad (1)$$

$$\Sigma F_y = 0; \qquad -F_D \cos 30° + F_B = 0 \qquad (2)$$

$$\Sigma F_z = 0; \qquad \tfrac{3}{5}F_C - 90 \text{ lb} = 0 \qquad (3)$$

Solving Eq. 3 for F_C, then Eq. 1 for F_D, and finally Eq. 2 for F_B, we get

$$F_C = 150 \text{ lb} \qquad \qquad \textit{Ans.}$$
$$F_D = 240 \text{ lb} \qquad \qquad \textit{Ans.}$$
$$F_B = 208 \text{ lb} \qquad \qquad \textit{Ans.}$$

The stretch of the spring is therefore

$$F_B = ks_{AB}$$
$$208 \text{ lb} = 500 \text{ lb/ft } (s_{AB})$$
$$s_{AB} = 0.416 \text{ ft} \qquad \qquad \textit{Ans.}$$

Example 3–7

Determine the magnitude and coordinate direction angles of force **F** in Fig. 3–10*a* that are required for equilibrium of particle *O*.

SOLUTION

Free-Body Diagram. Four forces act on particle *O*, Fig. 3–10*b*.

Equations of Equilibrium. A Cartesian vector analysis will be used for the solution in order to formulate the components of **F**$_3$. Hence,

$$\Sigma \mathbf{F} = \mathbf{0}; \qquad \mathbf{F}_1 + \mathbf{F}_2 + \mathbf{F}_3 + \mathbf{F} = \mathbf{0} \qquad (1)$$

Expressing each of these forces in Cartesian vector form, noting that the coordinates of *B* are $B(-2, -3, 6)$, we have

$$\mathbf{F}_1 = \{400\mathbf{j}\} \text{ N}$$

$$\mathbf{F}_2 = \{-800\mathbf{k}\} \text{ N}$$

$$\mathbf{F}_3 = F_3 \mathbf{u}_B = F_3 \left(\frac{\mathbf{r}_B}{r_B} \right) = 700 \text{ N} \left[\frac{-2\mathbf{i} - 3\mathbf{j} + 6\mathbf{k}}{\sqrt{(-2)^2 + (-3)^2 + (6)^2}} \right]$$

$$= \{-200\mathbf{i} - 300\mathbf{j} + 600\mathbf{k}\} \text{ N}$$

$$\mathbf{F} = F_x\mathbf{i} + F_y\mathbf{j} + F_z\mathbf{k}$$

Substituting into Eq. 1 yields

$$400\mathbf{j} - 800\mathbf{k} - 200\mathbf{i} - 300\mathbf{j} + 600\mathbf{k} + F_x\mathbf{i} + F_y\mathbf{j} + F_z\mathbf{k} = \mathbf{0}$$

Equating the respective **i**, **j**, **k** components to zero, we have

$$\Sigma F_x = 0; \qquad -200 + F_x = 0 \qquad F_x = 200 \text{ N}$$

$$\Sigma F_y = 0; \qquad 400 - 300 + F_y = 0 \qquad F_y = -100 \text{ N}$$

$$\Sigma F_z = 0; \qquad -800 + 600 + F_z = 0 \qquad F_z = 200 \text{ N}$$

Thus,

$$\mathbf{F} = \{200\mathbf{i} - 100\mathbf{j} + 200\mathbf{k}\} \text{ N}$$

$$F = \sqrt{(200)^2 + (-100)^2 + (200)^2} = 300 \text{ N} \qquad \textit{Ans.}$$

$$\mathbf{u}_F = \frac{\mathbf{F}}{F} = \frac{200}{300}\mathbf{i} - \frac{100}{300}\mathbf{j} + \frac{200}{300}\mathbf{k}$$

$$\alpha = \cos^{-1}\left(\frac{200}{300}\right) = 48.2° \qquad \textit{Ans.}$$

$$\beta = \cos^{-1}\left(\frac{-100}{300}\right) = 110° \qquad \textit{Ans.}$$

$$\gamma = \cos^{-1}\left(\frac{200}{300}\right) = 48.2° \qquad \textit{Ans.}$$

The magnitude and correct direction of **F** are shown in Fig. 3–10*c*.

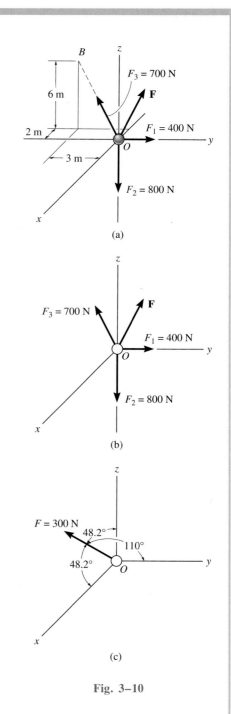

(a)

(b)

(c)

Fig. 3–10

Example 3–8

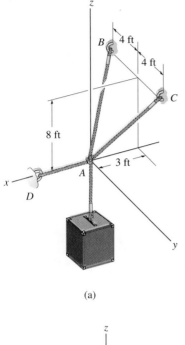

(a)

Determine the force developed in each cable used to support the 40-lb crate shown in Fig. 3–11a.

SOLUTION

Free-Body Diagram.   As shown in Fig. 3–11b, the free-body diagram of point A is considered in order to "expose" the three unknown forces in the cables and thereby obtain their magnitudes.

Equations of Equilibrium

$$\Sigma \mathbf{F} = \mathbf{0}; \qquad \mathbf{F}_B + \mathbf{F}_C + \mathbf{F}_D + \mathbf{W} = \mathbf{0} \qquad (1)$$

Since the coordinates of points B and C are $B(-3, -4, 8)$ and $C(-3, 4, 8)$, we have

$$\mathbf{F}_B = F_B\left[\frac{-3\mathbf{i} - 4\mathbf{j} + 8\mathbf{k}}{\sqrt{(-3)^2 + (-4)^2 + (8)^2}}\right]$$

$$= -0.318F_B\mathbf{i} - 0.424F_B\mathbf{j} + 0.848F_B\mathbf{k}$$

$$\mathbf{F}_C = F_C\left[\frac{-3\mathbf{i} + 4\mathbf{j} + 8\mathbf{k}}{\sqrt{(-3)^2 + (4)^2 + (8)^2}}\right]$$

$$= -0.318F_C\mathbf{i} + 0.424F_C\mathbf{j} + 0.848F_C\mathbf{k}$$

$$\mathbf{F}_D = F_D\mathbf{i}$$

$$\mathbf{W} = -40\mathbf{k}$$

Substituting these forces into Eq. 1 gives

$$-0.318F_B\mathbf{i} - 0.424F_B\mathbf{j} + 0.848F_B\mathbf{k} - 0.318F_C\mathbf{i} + 0.424F_C\mathbf{j}$$
$$+ 0.848F_C\mathbf{k} + F_D\mathbf{i} - 40\mathbf{k} = \mathbf{0}$$

Equating the respective **i, j, k** components to zero yields

$$\Sigma F_x = 0; \qquad -0.318F_B - 0.318F_C + F_D = 0 \qquad (2)$$
$$\Sigma F_y = 0; \qquad -0.424F_B + 0.424F_C = 0 \qquad (3)$$
$$\Sigma F_z = 0; \qquad 0.848F_B + 0.848F_C - 40 = 0 \qquad (4)$$

Equation 3 states that $F_B = F_C$. Thus, solving Eq. 4 for F_B and F_C and substituting the result into Eq. 2 to obtain F_D, we have

$$F_B = F_C = 23.6 \text{ lb} \qquad \textit{Ans.}$$
$$F_D = 15.0 \text{ lb} \qquad \textit{Ans.}$$

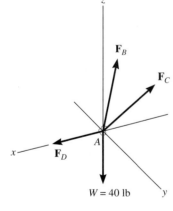

W = 40 lb

(b)

Fig. 3–11

Example 3–9

The 100-kg cylinder shown in Fig. 3–12a is supported by three cords, one of which is connected to a spring. Determine the tension in each cord and the stretch of the spring.

(a)

SOLUTION

Free-Body Diagram. The force in each cord can be determined by investigating the free-body diagram of point A, Fig. 3–12b. Why? The weight of the cylinder is $W = 100(9.81) = 981$ N.

Equations of Equilibrium

$$\Sigma \mathbf{F} = 0; \qquad \mathbf{F}_B + \mathbf{F}_C + \mathbf{F}_D + \mathbf{W} = 0 \qquad (1)$$

Each of these vectors can be expressed in Cartesian vector form. Using Eq. 2–11 for $\mathbf{F}_C$, and noting point $D(-1, 2, 2)$ for $\mathbf{F}_D$, we have

$$\mathbf{F}_B = F_B \mathbf{i}$$
$$\mathbf{F}_C = F_C \cos 120°\mathbf{i} + F_C \cos 135°\mathbf{j} + F_C \cos 60°\mathbf{k}$$
$$= -0.5F_C\mathbf{i} - 0.707F_C\mathbf{j} + 0.5F_C\mathbf{k}$$
$$\mathbf{F}_D = F_D\left[\frac{-1\mathbf{i} + 2\mathbf{j} + 2\mathbf{k}}{\sqrt{(-1)^2 + (2)^2 + (2)^2}}\right]$$
$$= -0.333F_D\mathbf{i} + 0.667F_D\mathbf{j} + 0.667F_D\mathbf{k}$$
$$\mathbf{W} = -981\mathbf{k}$$

Substituting these forces into Eq. 1 yields

$$F_B\mathbf{i} - 0.5F_C\mathbf{i} - 0.707F_C\mathbf{j} + 0.5F_C\mathbf{k} - 0.333F_D\mathbf{i} + 0.667F_D\mathbf{j}$$
$$+ 0.667F_D\mathbf{k} - 981\mathbf{k} = 0$$

Equating the respective **i, j, k** components to zero,

$$\Sigma F_x = 0; \qquad F_B - 0.5F_C - 0.333F_D = 0 \qquad (2)$$
$$\Sigma F_y = 0; \qquad -0.707F_C + 0.667F_D = 0 \qquad (3)$$
$$\Sigma F_z = 0; \qquad 0.5F_C + 0.667F_D - 981 = 0 \qquad (4)$$

(b)

Fig. 3–12

Solving Eq. 3 for F_D in terms of F_C and substituting into Eq. 4 yields F_C. F_D is determined from Eq. 3. Finally, substituting the results into Eq. 2 yields F_B. Hence,

$$F_C = 813 \text{ N} \qquad \qquad Ans.$$
$$F_D = 862 \text{ N} \qquad \qquad Ans.$$
$$F_B = 693.7 \text{ N} \qquad \qquad Ans.$$

The stretch of the spring is therefore

$$F = ks; \qquad 693.7 = 1500s$$
$$s = 0.462 \text{ m} \qquad \qquad Ans.$$

PROBLEMS

3–31. Determine the magnitudes of forces F_1, F_2, and F_3 for equilibrium of the particle. Force F_3 is located in the positive x, y, z octant.

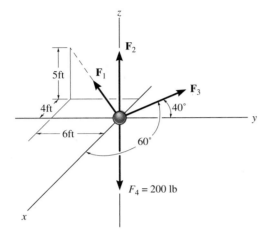

Prob. 3–31

3–33. Determine the magnitudes of F_1, F_2, and F_3 for equilibrium of the particle.

Prob. 3–33

*3–32. Determine the magnitude and coordinate direction angles of the force **P** required to keep the concurrent force system in equilibrium.

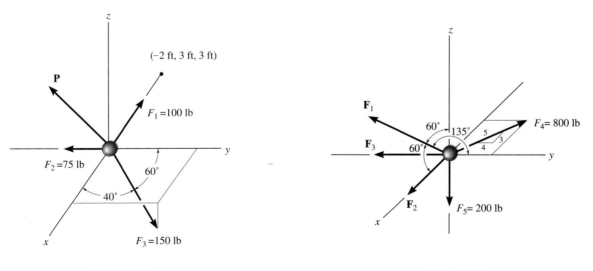

Prob. 3–32

3–34. Determine the magnitudes of F_1, F_2, and F_3 for equilibrium of the particle.

Prob. 3–34

3–35. Determine the magnitudes of F_1, F_2, and F_3 for equilibrium of the particle.

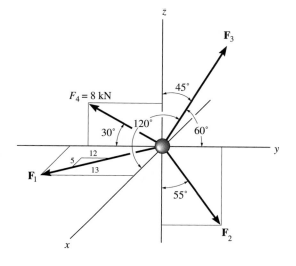

Prob. 3–35

3–37. The joint of a space frame is subjected to four member forces. Member *OA* lies in the *x–y* plane and member *OB* lies in the *y–z* plane. Determine the forces acting in each of the members required for equilibrium of the joint.

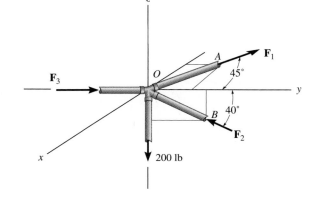

Prob. 3–37

*3–36. Determine the magnitude of **P** and the orientation of the 200-lb force required to keep the concurrent force system in equilibrium.

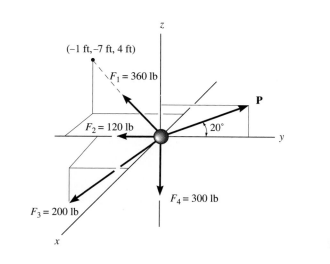

Prob. 3–36

3–38. The three cables are used to support the 8-kg lamp. Determine the force developed in each cable for equilibrium.

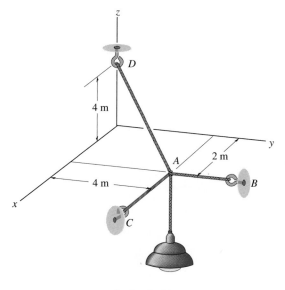

Prob. 3–38

3–39.   Determine the force in each cable needed to support the 500-lb load.

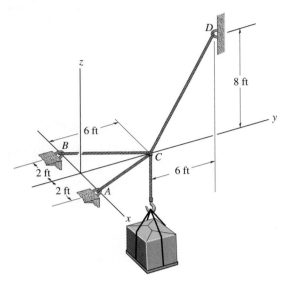

Prob. 3–39

3–41.   The 25-kg pot is supported at *A* by the three cables. Determine the force in each cable for equilibrium.

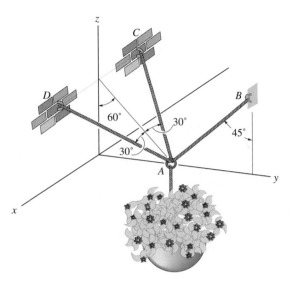

Prob. 3–41

*3–40.   If the bucket and its contents have a total weight of 20 lb, determine the force in the supporting cables *DA*, *DB*, and *DC*.

3–42.   Determine the force in each of the three cables needed to lift the tractor which has a mass of 8 Mg.

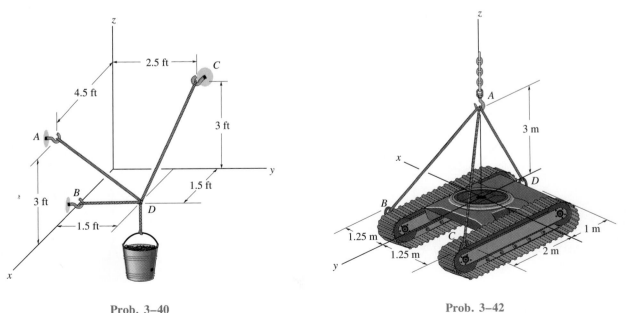

Prob. 3–40

Prob. 3–42

3–43. If cable *AB* is subjected to a tension of 700 N, determine the tension in cables *AC* and *AD* and the magnitude of the vertical force **F**.

*3–44. Three 10-lb balls and one 15-lb ball are suspended from the pulley-and-cable system. If the pulleys are frictionless and the centers of all of them lie in the same horizontal plane, determine the sag *s* for equilibrium of the system.

Prob. 3–43

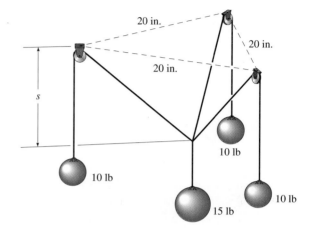

Prob. 3–44

REVIEW PROBLEMS

3–45. The crate is hoisted from the hold of a ship using the cable arrangement shown. If a vertical force of 2500 N is applied to the hook at *A*, determine the tension in each of the three cables for equilibrium. Set *d* = 1 m.

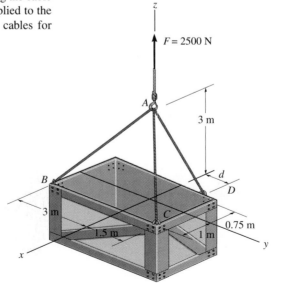

Prob. 3–45

3–46. The crate is hoisted from the hold of a ship using the cable arrangement shown. If a vertical force of 2500 N is applied to the hook at A, determine the tension in each of the three cables for equilibrium. Set $d = 1.5$ m.

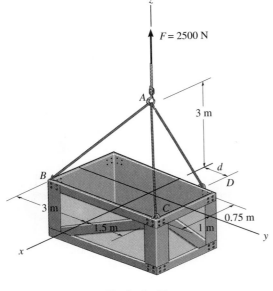

$F = 2500$ N

3 m

d

D

B

3 m

C

0.75 m

1.5 m

1 m

y

x

Prob. 3–46

3–47. Three electrostatically charged pith balls, each having a weight of $0.44(10^{-3})$ lb, are suspended from a common point. If the electrostatic repulsive force acting between any two balls is $1(10^{-3})$ lb and is directed along a line of action joining these two balls, determine the tension developed in each cord for equilibrium.

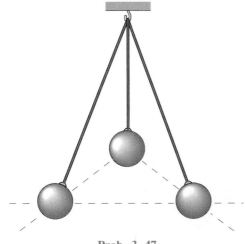

Prob. 3–47

*3–48. A force of 100 lb holds the 400-lb crate in equilibrium. Determine the coordinates $(0, y, z)$ of point A if the tension in cords AC and AB is 700 lb each.

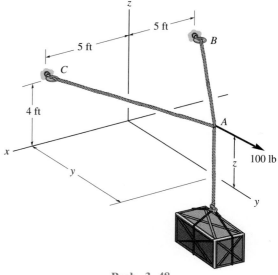

5 ft

5 ft

C

B

4 ft

A

x

y

z

y

100 lb

Prob. 3–48

3–49. The 800-lb cylinder is supported by three chains as shown. Determine the force in each chain for equilibrium of point A. Take $d = 2.4$ ft.

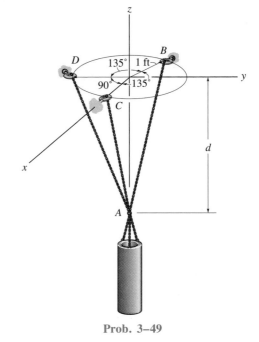

z

D

B

135°   1 ft

90°   135°

C

y

x

d

A

Prob. 3–49

3–50. The 800-lb cylinder is supported by three chains as shown. Determine the force in each chain for equilibrium of point *A*. Take *d* = 1 ft.

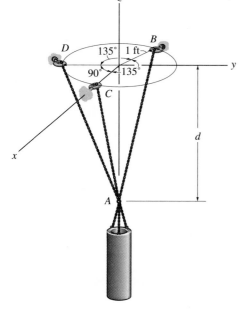

Probs. 3–50

***3–52.** The cords suspend the two buckets in the equilibrium position shown. Determine the weight of bucket *A* if bucket *B* has a weight of 20 lb.

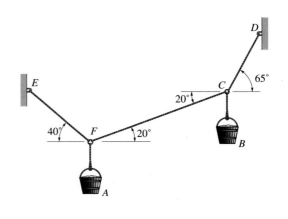

Prob. 3–52

3–51. The cords suspend the two buckets in the equilibrium position shown. Determine the weight of bucket *B*. Bucket *A* has a weight of 60 lb.

3–53. The bucket has a weight of 20 lb. Determine the tension developed in each cord for equilibrium.

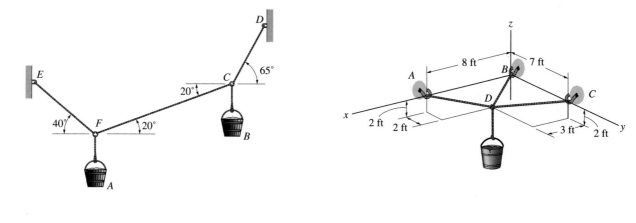

Prob. 3–51   Prob. 3–53

■3–54.   The 5-ft-long cord AB is attached to the end B of a spring having an unstretched length of 5 ft. The other end of the spring is attached to a roller C so that the spring remains horizontal as it stretches. If a 10-lb weight is suspended from B, determine the angle θ of cord AB for equilibrium.

*3–56.   The members of a truss are pin-connected at joint O. Determine the magnitude of $\mathbf{F}_1$ and its angle θ for equilibrium. Set $F_2 = 6$ kN.

Prob. 3–54

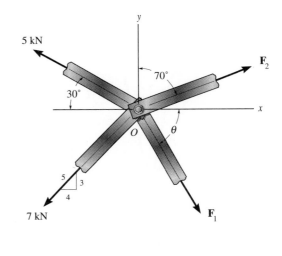

Prob. 3–56

3–55.   The members of a truss are pin-connected at joint O. Determine the magnitudes of $\mathbf{F}_1$ and $\mathbf{F}_2$ for equilibrium. Set $\theta = 60°$.

3–57.   Two electrically charged pith balls, each having a mass of 0.15 g, are suspended from light threads of equal length. Determine the magnitude of the resultant horizontal force of repulsion, $\mathbf{F}$, acting on each ball if the measured distance between them is $r = 200$ mm.

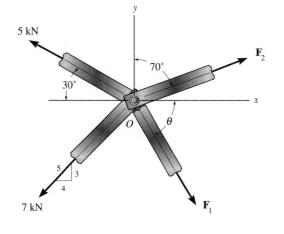

Prob. 3–55

Prob. 3–57

4 Force System Resultants

In Chapter 3 it was shown that the condition for the equilibrium of a particle or a concurrent force system simply requires that the resultant of the force system be equal to zero. In Chapter 5 it will be shown that such a restriction is necessary but not sufficient for the equilibrium of a rigid body. A further restriction must be made with regard to the nonconcurrency of the applied force system, giving rise to the concept of *moment*. In this chapter a formal definition of a moment will be presented and ways of finding the moment of a force about a point or axis will be discussed. We will also present methods for determining the resultants of nonconcurrent force systems. This development is important since application of the equations for force-system simplification is similar to applying the equations of equilibrium for a rigid-body. Furthermore, the resultants of a force system will influence the state of equilibrium or motion of a rigid body in the same way as the force system, and therefore we can study rigid-body behavior in a simpler manner by using the resultants.

4.1 Cross Product

The moment of a force will be formulated using Cartesian vectors in the next section. Before doing this, however, it is first necessary to expand our knowledge of vector algebra and introduce the cross-product method of vector multiplication.

The *cross product* of two vectors **A** and **B** yields the vector **C**, which is written

$$\mathbf{C} = \mathbf{A} \times \mathbf{B}$$

and is read "**C** equals **A** cross **B**."

Magnitude. The *magnitude* of **C** is defined as the product of the magnitudes of **A** and **B** and the sine of the angle θ between their tails ($0° \leq \theta \leq 180°$). Thus, $C = AB \sin \theta$.

Direction. Vector **C** has a *direction* that is perpendicular to the plane containing **A** and **B** such that the direction of **C** is specified by the right-hand rule; i.e., curling the fingers of the right hand from vector **A** (cross) to vector **B,** the thumb then points in the direction of **C,** as shown in Fig. 4–1.

Knowing both the magnitude and direction of **C,** we can write

$$\mathbf{C} = \mathbf{A} \times \mathbf{B} = (AB \sin \theta)\mathbf{u}_C \qquad (4–1)$$

where the scalar $AB \sin \theta$ defines the *magnitude* of **C** and the unit vector $\mathbf{u}_C$ defines the *direction* of **C**. The terms of Eq. 4–1 are illustrated graphically in Fig. 4–2.

Fig. 4–1

Fig. 4–2

Laws of Operation

1. The commutative law is *not* valid; i.e.,

$$\mathbf{A} \times \mathbf{B} \neq \mathbf{B} \times \mathbf{A}$$

Rather,

$$\mathbf{A} \times \mathbf{B} = -\mathbf{B} \times \mathbf{A}$$

This is shown in Fig. 4–3 by using the right-hand rule. The cross product $\mathbf{B} \times \mathbf{A}$ yields a vector that acts in the opposite direction to $\mathbf{C}$; i.e., $\mathbf{B} \times \mathbf{A} = -\mathbf{C}$.

2. Multiplication by a scalar:

$$a(\mathbf{A} \times \mathbf{B}) = (a\mathbf{A}) \times \mathbf{B} = \mathbf{A} \times (a\mathbf{B}) = (\mathbf{A} \times \mathbf{B})a$$

This property is easily shown, since the magnitude of the resultant vector ($|a|AB \sin \theta$) and its direction are the same in each case.

3. The distributive law:

$$\mathbf{A} \times (\mathbf{B} + \mathbf{D}) = (\mathbf{A} \times \mathbf{B}) + (\mathbf{A} \times \mathbf{D})$$

The proof of this identity is left as an exercise (see Prob. 4–1). It is important to note that *proper order* of the cross products must be maintained, since they are not commutative.

Cartesian Vector Formulation. Equation 4–1 may be used to find the cross product of a pair of Cartesian unit vectors. For example, to find $\mathbf{i} \times \mathbf{j}$, the *magnitude* of the resultant vector is $(i)(j)(\sin 90°) = (1)(1)(1) = 1$, and its *direction* is determined using the right-hand rule. As shown in Fig. 4–4, the resultant vector points in the $+\mathbf{k}$ direction. Thus, $\mathbf{i} \times \mathbf{j} = (1)\mathbf{k}$. In a similar manner,

$$
\begin{array}{lll}
\mathbf{i} \times \mathbf{j} = \mathbf{k} & \mathbf{i} \times \mathbf{k} = -\mathbf{j} & \mathbf{i} \times \mathbf{i} = 0 \\
\mathbf{j} \times \mathbf{k} = \mathbf{i} & \mathbf{j} \times \mathbf{i} = -\mathbf{k} & \mathbf{j} \times \mathbf{j} = 0 \\
\mathbf{k} \times \mathbf{i} = \mathbf{j} & \mathbf{k} \times \mathbf{j} = -\mathbf{i} & \mathbf{k} \times \mathbf{k} = 0
\end{array}
$$

These results should *not* be memorized; rather, it should be clearly understood how each is obtained by using the right-hand rule and the definition of the cross product. A simple scheme shown in Fig. 4–5 is helpful for obtaining the same results when the need arises. If the circle is constructed as shown, then "crossing" two unit vectors in a *counterclockwise* fashion around the circle yields the *positive* third unit vector; e.g., $\mathbf{k} \times \mathbf{i} = \mathbf{j}$. Moving *clockwise,* a *negative* unit vector is obtained; e.g., $\mathbf{i} \times \mathbf{k} = -\mathbf{j}$.

Fig. 4–3

Fig. 4–4

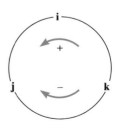

Fig. 4–5

Consider now the cross product of two general vectors **A** and **B** which are expressed in Cartesian vector form. We have

$$\mathbf{A} \times \mathbf{B} = (A_x\mathbf{i} + A_y\mathbf{j} + A_z\mathbf{k}) \times (B_x\mathbf{i} + B_y\mathbf{j} + B_z\mathbf{k})$$
$$= A_xB_x(\mathbf{i} \times \mathbf{i}) + A_xB_y(\mathbf{i} \times \mathbf{j}) + A_xB_z(\mathbf{i} \times \mathbf{k})$$
$$+ A_yB_x(\mathbf{j} \times \mathbf{i}) + A_yB_y(\mathbf{j} \times \mathbf{j}) + A_yB_z(\mathbf{j} \times \mathbf{k})$$
$$+ A_zB_x(\mathbf{k} \times \mathbf{i}) + A_zB_y(\mathbf{k} \times \mathbf{j}) + A_zB_z(\mathbf{k} \times \mathbf{k})$$

Carrying out the cross-product operations and combining terms yields

$$\mathbf{A} \times \mathbf{B} = (A_yB_z - A_zB_y)\mathbf{i} - (A_xB_z - A_zB_x)\mathbf{j} + (A_xB_y - A_yB_x)\mathbf{k} \qquad (4-2)$$

This equation may also be written in a more compact determinant form as

$$\mathbf{A} \times \mathbf{B} = \begin{vmatrix} \mathbf{i} & \mathbf{j} & \mathbf{k} \\ A_x & A_y & A_z \\ B_x & B_y & B_z \end{vmatrix} \qquad (4-3)$$

Thus, to find the cross product of any two Cartesian vectors **A** and **B,** it is necessary to expand a determinant whose first row of elements consists of the unit vectors **i, j,** and **k** and whose second and third rows represent the x, y, z components of the two vectors **A** and **B,** respectively.*

*A determinant having three rows and three columns can be expanded using three minors, each of which is multiplied by one of the three terms in the first row. There are four elements in each minor, e.g.,

$$\begin{vmatrix} A_{11} & A_{12} \\ A_{21} & A_{22} \end{vmatrix}$$

By *definition*, this notation represents the terms $(A_{11}A_{22} - A_{12}A_{21})$, which is simply the product of the two elements of the arrow slanting downward to the right $(A_{11}A_{22})$ *minus* the product of the two elements intersected by the arrow slanting downward to the left $(A_{12}A_{21})$. For a 3×3 determinant, such as Eq. 4–3, the three minors can be generated in accordance with the following scheme:

For element **i:**
$$\begin{vmatrix} \mathbf{i} & \mathbf{j} & \mathbf{k} \\ A_x & A_y & A_z \\ B_x & B_y & B_z \end{vmatrix} = \mathbf{i}(A_yB_z - A_zB_y)$$

For element **j:**
$$\begin{vmatrix} \mathbf{i} & \mathbf{j} & \mathbf{k} \\ A_x & A_y & A_z \\ B_x & B_y & B_z \end{vmatrix} = -\mathbf{j}(A_xB_z - A_zB_x)$$

For element **k:**
$$\begin{vmatrix} \mathbf{i} & \mathbf{j} & \mathbf{k} \\ A_x & A_y & A_z \\ B_x & B_y & B_z \end{vmatrix} = \mathbf{k}(A_xB_y - A_yB_x)$$

Adding the results and noting that the **j** element *must include the minus sign* yields the expanded form of **A** × **B** given by Eq. 4–2.

4.2 Moment of a Force—Scalar Formulation

The *moment* of a force about a point or axis provides a measure of the tendency of the force to cause a body to rotate about the point or axis. For example, consider the horizontal force $\mathbf{F}_x$, which acts perpendicular to the handle of the wrench and is located a distance d_y from point O, Fig. 4–6a. It is seen that this force tends to cause the pipe to turn about the z axis and so the force creates a moment $(\mathbf{M}_O)_z$ about the axis. The larger the force or the length d_y, the greater the effect. This tendency for rotation caused by $\mathbf{F}_x$ is sometimes called a *torque*, but most often it is called the *moment of a force* or simply the *moment* $(\mathbf{M}_O)_z$. In particular, note that the moment axis (z) is perpendicular to the shaded plane (x–y) which contains both $\mathbf{F}_x$ and d_y and that this axis intersects the plane at point O. Now consider applying the force $\mathbf{F}_z$ to the wrench, Fig. 4–6b. This force will *not* rotate the pipe about the z axis. Instead, it tends to rotate it about the x axis. Keep in mind that although it may not be possible actually to "rotate" or turn the pipe in this manner, $\mathbf{F}_z$ still creates the *tendency* for rotation and so the moment $(\mathbf{M}_O)_x$ is produced. As before, the force and distance d_y lie in the shaded plane (y–z) which is perpendicular to the moment axis (x). Lastly, if a force $\mathbf{F}_y$ is applied to the wrench, Fig. 4–6c, no moment is produced about point O. This lack of turning effect results, since the line of action of the force passes through O and therefore no tendency for rotation is possible.

(a)

Fig. 4–6

(b)

(c)

For the general case, consider the force $\mathbf{F}$ and point O which lie in a shaded plane as shown in Fig. 4–7a. The moment $\mathbf{M}_O$ about point O, or about an axis passing through O and perpendicular to the plane, is a *vector quantity* since it has a specified magnitude and direction and adds according to the parallelogram law.

Fig. 4-7

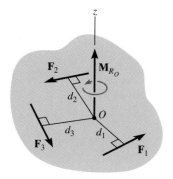

Fig. 4-8

Magnitude.

The magnitude of $\mathbf{M}_O$ is

$$M_O = Fd \qquad (4\text{–}4)$$

where d is referred to as the *moment arm* or perpendicular distance from the axis at point O to the line of action of the force. Units of moment magnitude consist of force times distance, e.g., $N \cdot m$ or $lb \cdot ft$.

Direction.

The direction of $\mathbf{M}_O$ will be specified by using the "right-hand rule." To do this, the fingers of the right hand are curled such that they follow the sense of rotation, which would occur if the force could rotate about point O, Fig. 4–7a. The *thumb* then *points* along the *moment axis* so that it gives the direction and sense of the moment vector, which is *upward* and *perpendicular* to the shaded plane containing $\mathbf{F}$ and d. By this definition, the moment $\mathbf{M}_O$ can be considered as a *sliding vector* and therefore acts at any point along the moment axis.

In three dimensions, $\mathbf{M}_O$ is illustrated by a vector arrow with a curl on it to *distinguish* it from a force vector, Fig. 4–7a. Many problems in mechanics, however, involve coplanar force systems that may be conveniently viewed in two dimensions. For example, a two-dimensional view of Fig. 4–7a is given in Fig. 4–7b. Here $\mathbf{M}_O$ is simply represented by the (counterclockwise) curl, which indicates the action of $\mathbf{F}$. The arrowhead on this curl is used to show the *sense of rotation* caused by $\mathbf{F}$. Using the right-hand rule, however, realize that the direction and sense of the moment vector in Fig. 4–7b are specified by the thumb, which points *out* of the page, since the fingers follow the curl. In particular, notice that *this curl or sense of rotation can always be determined by observing in which direction the force would "orbit" about point O* (counterclockwise in Fig. 4–7b). In two dimensions we will often refer to finding the moment of a force "about a point" (O). Keep in mind, however, that the moment *actually* acts about an axis which is perpendicular to the plane containing $\mathbf{F}$ and d, and this axis intersects the plane at the point (O), Fig. 4–7a.

Resultant Moment of a System of Coplanar Forces.

If a system of forces all lie in the x–y plane, then the moment produced by each force about point O will be directed along the z axis, Fig. 4–8. Consequently, the resultant moment $\mathbf{M}_{R_O}$ of the system can be determined by simply adding the moments of each force *algebraically,* since all the moment vectors are collinear. We can write this vector sum symbolically as

$$\zeta+; \qquad M_{R_O} = \Sigma Fd \qquad (4\text{–}5)$$

Here the counterclockwise curl written alongside the equation indicates that by the scalar sign convention, the moment of any force will be positive if it is directed along the $+z$ axis, whereas a negative moment is directed along the $-z$ axis.

Examples 4–1 through 4–4 illustrate numerical application of Eqs. 4–4 and 4–5.

4.3 Moment of a Force—Vector Formulation

The moment of a force **F** about point O, or actually about an axis passing through O and perpendicular to the plane containing O and **F**, Fig. 4–9a, can also be expressed using the vector cross product, namely,

$$\mathbf{M}_O = \mathbf{r} \times \mathbf{F} \qquad\qquad (4\text{–}6)$$

Here **r** represents a position vector drawn from O to *any point* lying on the line of action of **F**. It will now be shown that indeed the moment $\mathbf{M}_O$, when determined by this cross product, has the proper magnitude and direction.

Magnitude. The magnitude of the above cross product is defined from Eq. 4–1 as $M_O = rF \sin \theta$. Here, the angle θ is measured between the *tails* of **r** and **F**. Hence, **r** must be treated as a sliding vector so that θ can be constructed properly, Fig. 4–9b. As shown, the moment arm $d = r \sin \theta$, so that indeed

$$M_O = rF \sin \theta = F (r \sin \theta) = Fd$$

which agrees with Eq. 4–4.

Direction. The direction and sense of $\mathbf{M}_O$ in Eq. 4–6 are determined by the right-hand rule as it applies to the cross product. Thus, extending **r** to the dashed position and curling the right-hand fingers from **r** toward **F**, "**r** cross **F**," the thumb is directed upward or perpendicular to the plane containing **r** and **F** and this is in the *same direction* as $\mathbf{M}_O$, the moment of the force about point O, Fig. 4–9b. Note that the "curl" of the fingers, like the curl around the moment vector, indicates the sense of rotation caused by the force. Since the cross product is not commutative, it is important that the *proper order* of **r** and **F** be maintained in Eq. 4–6.

(a)

(b)

Fig. 4–9

(a)

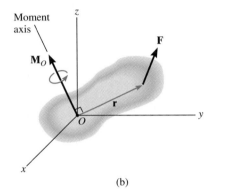

Moment axis

(b)

Fig. 4–10

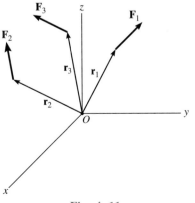

Fig. 4–11

Cartesian Vector Formulation. If the position vector **r** and force **F** are expressed in Cartesian vector form, Fig. 4–10, then from Eq. 4–6 we have

$$M_O = r \times F = \begin{vmatrix} i & j & k \\ r_x & r_y & r_z \\ F_x & F_y & F_z \end{vmatrix} \qquad (4\text{–}7)$$

where r_x, r_y, r_z represent the x, y, z components of the position vector drawn from point O to *any point* on the line of action of the force

F_x, F_y, F_z represent the x, y, z components of the force vector

If the determinant is expanded, then like Eq. 4–2 we have

$$M_O = (r_y F_z - r_z F_y)i - (r_x F_z - r_z F_x)j + (r_x F_y - r_y F_x)k \qquad (4\text{–}8)$$

The physical meaning of these three moment components becomes evident by studying Fig. 4–10*a*. For example, the **i** component of M_O is determined from the moments of F_x, F_y, and F_z about the x axis. In particular, note that F_x does *not* create a moment or tendency to cause turning about the x axis, since this force is *parallel* to the x axis. The line of action of F_y passes through point E, and so the magnitude of the moment of F_y about point A on the x axis is $r_z F_y$. By the right-hand rule this component acts in the negative **i** direction. Likewise, F_z contributes a moment component of $r_y F_z \mathbf{i}$. Thus, $(M_O)_x = (r_y F_y - r_z F_y)$ as shown in Eq. 4–7. As an exercise, establish the **j** and **k** components of M_O in this manner, and show that indeed the expanded form of the determinant, Eq. 4–8, represents the moment of **F** about point O. Once determined, realize that M_O will be *perpendicular* to the shaded plane containing vectors **r** and **F**, Fig. 4–10*b*.

It will be shown in Example 4–5 that the computation of the moment using the cross product has a distinct advantage over the scalar formulation when solving problems in three dimensions. This is because it is generally easier to establish the position vector **r** to the force, rather than determining the moment-arm distance d that must be directed *perpendicular* to the line of action of the force.

Resultant Moment of a System of Forces. The resultant moment of a system of forces about point O can be determined by vector addition resulting from successive applications of Eq. 4–7. This resultant can be written symbolically as

$$M_{R_O} = \Sigma r \times F \qquad (4\text{–}9)$$

and is shown in Fig. 4–11.

Example 4–6 illustrates numerical application of Eqs. 4–7 and 4–9.

Example 4–1

Determine the moment of the 800-N force acting on the frame in Fig. 4–12 about points A, B, C, and D.

SOLUTION (*SCALAR ANALYSIS*)

In general, $M = Fd$, where d is the moment arm or *perpendicular distance* from the *point* on the moment axis to the *line of action* of the force. Hence,

$M_A = 800$ N$(2.5$ m$) = 2000$ N $\cdot$ m $\downarrow$ *Ans.*

$M_B = 800$ N$(1.5$ m$) = 1200$ N $\cdot$ m $\downarrow$ *Ans.*

$M_C = 800$ N$(0) = 0$ (line of action of **F** passes through C)   *Ans.*

$M_D = 800$ N$(0.5$ m$) = 400$ N $\cdot$ m $\uparrow$ *Ans.*

Fig. 4–12

Example 4–2

Determine the location of the point of application P and the direction of a 20-lb force that lies in the plane of the square plate shown in Fig. 4–13a, so that this force creates the greatest counterclockwise moment about point O. What is this moment?

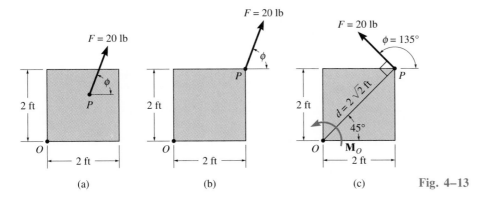

(a)   (b)   (c)   Fig. 4–13

SOLUTION (*SCALAR ANALYSIS*)

Since the maximum moment created by the force is required, the force must act on the plate at a distance *farthest* from point O. As shown in Fig. 4–13b, the point of application must therefore be at the diagonal corner. In order to produce *counterclockwise* rotation of the plate about O, **F** must act at an angle $45° < \phi < 225°$. The greatest moment is produced when the line of action of **F** is *perpendicular* to d, i.e., $\phi = 135°$, Fig. 4–13c. The maximum moment is therefore

$$M_O = Fd = (20 \text{ lb})(2\sqrt{2} \text{ ft}) = 56.6 \text{ lb} \cdot \text{ft} \uparrow \qquad Ans.$$

By the right-hand rule, $\mathbf{M}_O$ is directed out of the page.

Example 4-3

For each case illustrated in Fig. 4–14, determine the moment of the force about point O.

SOLUTION (*SCALAR ANALYSIS*)

The line of action of each force is extended as a dashed line in order to establish the moment arm d. Also, the tendency of rotation caused by the force is shown as a colored curl. Thus,

Fig. 4–14a, $M_O = (100 \text{ N})(2 \text{ m}) = 200 \text{ N} \cdot \text{m}$ ↓   *Ans.*

Fig. 4–14b, $M_O = (50 \text{ N})(0.75 \text{ m}) = 37.5 \text{ N} \cdot \text{m}$ ↓   *Ans.*

Fig. 4–14c, $M_O = (40 \text{ lb})(4 \text{ ft} + 2 \cos 30° \text{ ft}) = 229 \text{ lb} \cdot \text{ft}$ ↓   *Ans.*

Fig. 4–14d, $M_O = (60 \text{ lb})(1 \sin 45° \text{ ft}) = 42.4 \text{ lb} \cdot \text{ft}$ ↑   *Ans.*

Fig. 4–14e, $M_O = (7 \text{ kN})(4 \text{ m} - 1 \text{ m}) = 21.0 \text{ kN} \cdot \text{m}$ ↑   *Ans.*

Fig. 4–14

Example 4-4

Determine the resultant moment of the four forces acting on the member shown in Fig. 4–15 about point O.

SOLUTION

Here it is necessary to apply Eq. 4–5. Assuming that positive moments act in the $+\mathbf{k}$ direction, i.e., counterclockwise, we have

$$\zeta + M_{R_O} = \Sigma Fd;$$

$$M_{R_O} = -50 \text{ N}(2 \text{ m}) + 60 \text{ N}(0) + 20 \text{ N}(3 \sin 30° \text{ m})$$
$$-40 \text{ N}(4 \text{ m} + 3 \cos 30° \text{ m})$$

$$M_{R_O} = -334 \text{ N} \cdot \text{m} = 334 \text{ N} \cdot \text{m} \text{ ↓}$$ *Ans.*

For this calculation, note how the moment-arm distances for the 20-N and 40-N forces are established from the extended (dashed) lines of action of each of these forces.

Fig. 4–15

Example 4–5

The pole in Fig. 4–16a is subjected to a 60-N force that is directed from C to B. Determine the magnitude of the moment created by this force about point A.

SOLUTION *(VECTOR ANALYSIS)*

As shown in Fig. 4–16b, either one of two position vectors can be used for the solution, since $\mathbf{M}_A = \mathbf{r}_B \times \mathbf{F}$ or $\mathbf{M}_A = \mathbf{r}_C \times \mathbf{F}$. The position vectors are represented as

$$\mathbf{r}_B = \{1\mathbf{i} + 3\mathbf{j} + 2\mathbf{k}\} \text{ m} \quad \text{and} \quad \mathbf{r}_C = \{3\mathbf{i} + 4\mathbf{j}\} \text{ m}$$

The force has a magnitude of 60 N and a direction specified by the unit vector $\mathbf{u}_F$, directed from C to B. Thus,

$$\mathbf{F} = (60 \text{ N})\mathbf{u}_F = (60)\left[\frac{(1-3)\mathbf{i} + (3-4)\mathbf{j} + (2-0)\mathbf{k}}{\sqrt{(-2)^2 + (-1)^2 + (2)^2}}\right]$$

$$= \{-40\mathbf{i} - 20\mathbf{j} + 40\mathbf{k}\} \text{ N}$$

Substituting into the determinant formulation, Eq. 4–7, and following the scheme for determinant expansion as stated in the footnote on page 104, we have

$$\mathbf{M}_A = \mathbf{r}_B \times \mathbf{F} = \begin{vmatrix} \mathbf{i} & \mathbf{j} & \mathbf{k} \\ 1 & 3 & 2 \\ -40 & -20 & 40 \end{vmatrix}$$

$$= [3(40) - 2(-20)]\mathbf{i} - [1(40) - 2(-40)]\mathbf{j} + [1(-20) - 3(-40)]\mathbf{k}$$

or

$$\mathbf{M}_A = \mathbf{r}_C \times \mathbf{F} = \begin{vmatrix} \mathbf{i} & \mathbf{j} & \mathbf{k} \\ 3 & 4 & 0 \\ -40 & -20 & 40 \end{vmatrix}$$

$$= [4(40) - 0(-20)]\mathbf{i} - [3(40) - 0(-40)]\mathbf{j} + [3(-20) - 4(-40)]\mathbf{k}$$

In both cases,

$$\mathbf{M}_A = \{160\mathbf{i} - 120\mathbf{j} + 100\mathbf{k}\} \text{ N} \cdot \text{m}$$

The *magnitude* of $\mathbf{M}_A$ is therefore

$$M_A = \sqrt{(160)^2 + (-120)^2 + (100)^2} = 224 \text{ N} \cdot \text{m} \qquad Ans.$$

As expected, $\mathbf{M}_A$ acts perpendicular to the shaded plane containing vectors $\mathbf{F}$, $\mathbf{r}_B$, and $\mathbf{r}_C$, Fig. 4–16c. (How would you find its coordinate direction angles $\alpha = 44.3°$, $\beta = 122°$, $\gamma = 63.4°$?) Had this problem been worked using a scalar approach, where $M_A = Fd$, notice the difficulty that might arise in obtaining the moment arm d.

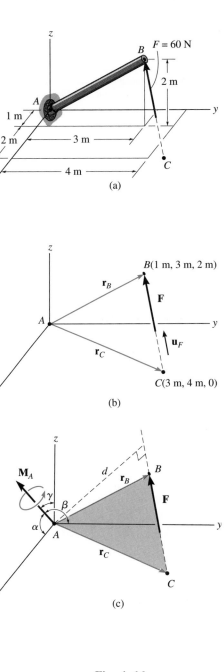

Fig. 4–16

■ Example 4–6 ■

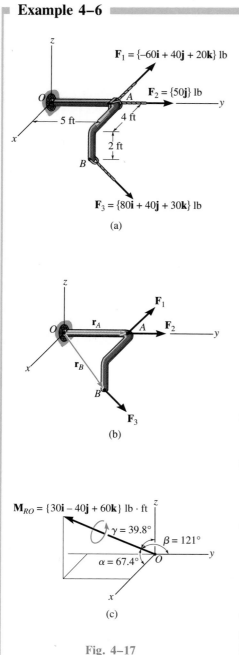

$F_1 = \{-60i + 40j + 20k\}$ lb

$F_2 = \{50j\}$ lb

5 ft

4 ft

2 ft

B

$F_3 = \{80i + 40j + 30k\}$ lb

(a)

r_A

A F_2

r_B

B

F_3

(b)

$M_{RO} = \{30i - 40j + 60k\}$ lb · ft

$\gamma = 39.8°$

$\beta = 121°$

$\alpha = 67.4°$

(c)

Fig. 4–17

Three forces act on the rod shown in Fig. 4–17a. Determine the resultant moment they create about the flange at O, and determine the direction of this moment axis.

SOLUTION

Here we must apply Eq. 4–9. Position vectors are directed from point O to each force as shown in Fig. 4–17b. These vectors are

$$r_A = \{5j\} \text{ ft}$$
$$r_B = \{4i + 5j - 2k\} \text{ ft}$$

Since $F_2 = \{2i\}$ ft, and the Cartesian components of the other forces are given, the resultant moment about O is therefore

$$M_{R_O} = \Sigma r \times F$$
$$= r_A \times F_1 + r_A \times F_2 + r_B \times F_3$$

$$= \begin{vmatrix} i & j & k \\ 0 & 5 & 0 \\ -60 & 40 & 20 \end{vmatrix} + \begin{vmatrix} i & j & k \\ 0 & 5 & 0 \\ 0 & 50 & 0 \end{vmatrix} + \begin{vmatrix} i & j & k \\ 4 & 5 & -2 \\ 80 & 40 & -30 \end{vmatrix}$$

$$= [5(20) - 40(0)]i - [0j] + [0(40) - (-60)(5)]k + [0i + 0j + 0k]$$
$$+ [5(-30) - (40)(-2)]i - [4(-30) - 80(-2)]j + [4(40) - 80(5)]k$$
$$= \{30i - 40j + 60k\} \text{ lb · ft} \qquad Ans.$$

The moment axis is directed along the line of action of M_{R_O}. Since the magnitude of this moment is

$$M_{R_O} = \sqrt{(30)^2 + (-40)^2 + (60)^2} = 78.10 \text{ lb · ft}$$

the unit vector which defines the direction of the moment axis is

$$u = \frac{M_{R_O}}{M_{R_O}} = \frac{30i - 40j + 60k}{78.10} = 0.3841i - 0.5121j + 0.7682k$$

Therefore, the coordinate direction angles of the axis are

$$\cos \alpha = 0.3841; \qquad \alpha = 67.4° \qquad Ans.$$
$$\cos \beta = -0.5121; \qquad \beta = 121° \qquad Ans.$$
$$\cos \gamma = 0.7682; \qquad \gamma = 39.8° \qquad Ans.$$

These results are shown in Fig. 4–17c. Realize that the three forces tend to cause the pipe to rotate about this axis in the manner shown by the curl indicated on the moment vector.

4.4 Transmissibility of a Force and the Principle of Moments

Using the definition of a moment formalized by the cross product, we will now present two important concepts often used throughout our study of mechanics.

Transmissibility of a Force. Consider the force **F** applied at point *A* in Fig. 4–18. The moment created by **F** about point *O* is $\mathbf{M}_O = \mathbf{r}_A \times \mathbf{F}$; however, it was shown that the position vector "**r**" can extend from *O* to *any point* on the line of action of **F**. Consequently, **F** may be applied at point *B* and the same moment $\mathbf{M}_O = \mathbf{r}_B \times \mathbf{F}$ will be computed. As a result, **F** has the properties of a *sliding vector* and can therefore act at *any point along its line of action and still create the same moment about point O*. We refer to **F** in this regard as being "transmissible," and we will discuss this property further in Sec. 4.7.

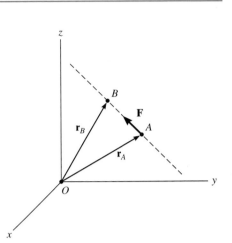

Fig. 4–18

Principle of Moments. A concept often used in mechanics is the *principle of moments,* which is sometimes referred to as *Varignon's theorem* since it was originally developed by the French mathematician Varignon (1654–1722). It states that *the moment of a force about a point is equal to the sum of the moments of the force's components about the point*. The proof follows directly from the distributive law of the vector cross product. To show this, consider the force **F** and two of its components, where $\mathbf{F} = \mathbf{F}_1 + \mathbf{F}_2$, Fig. 4–19. We have

$$\mathbf{M}_O = \mathbf{r} \times \mathbf{F}_1 + \mathbf{r} \times \mathbf{F}_2 = \mathbf{r} \times (\mathbf{F}_1 + \mathbf{F}_2) = \mathbf{r} \times \mathbf{F}$$

This concept has important applications to the solution of problems and proofs of theorems that follow, since it is often easier to determine the moments of a force's components rather than the moment of the force itself.

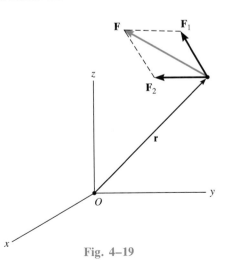

Fig. 4–19

Example 4–7

A 200-N force acts on the bracket shown in Fig. 4–20a. Determine the moment of the force about point A.

(a)

SOLUTION I

The moment arm d can be found by trigonometry, using the construction shown in Fig. 4–16b. From triangle BCD,

$$CB = d = 100 \cos 45° = 70.71 \text{ mm} = 0.070\ 71 \text{ m}$$

Thus,

$$M_A = Fd = 200 \text{ N}(0.070\ 71 \text{ m}) = 14.1 \text{ N} \cdot \text{m} \ ↺$$

According to the right-hand rule, $\mathbf{M}_A$ is directed in the $+\mathbf{k}$ direction since the force tends to rotate *counterclockwise* about point A. Hence, reporting the moment as a Cartesian vector, we have

$$\mathbf{M}_A = \{14.1\mathbf{k}\} \text{ N} \cdot \text{m} \qquad\qquad Ans.$$

(b)

SOLUTION II

The 200-N force may be resolved into x and y components, as shown in Fig. 4–20c. In accordance with the principle of moments, the moment of $\mathbf{F}$ computed about point A is equivalent to the sum of the moments produced by the two force components. Assuming counterclockwise rotation as positive, i.e., in the $+\mathbf{k}$ direction, we can apply Eq. 4–5, in which case

$$↺+M_A = (200 \sin 45°)(0.20) - (200 \cos 45°)(0.10)$$
$$= 14.1 \text{ N} \cdot \text{m} \ ↺$$

Thus

$$\mathbf{M}_A = \{14.1\mathbf{k}\} \text{ N} \cdot \text{m} \qquad\qquad Ans.$$

By comparison, it is seen that Solution II provides a more *convenient method* for analysis than Solution I, since the moment arm for each component force is easier to establish.

(c)

Fig. 4–20

Example 4–8

The force **F** acts at the end of the beam shown in Fig. 4–21a. Determine the moment of the force about point O.

SOLUTION I (SCALAR ANALYSIS)

The force is resolved into its x and y components as shown in Fig. 4–21b and the moments of the components are computed about point O. Taking positive moments as counterclockwise, i.e., in the $+\mathbf{k}$ direction, we have

$$\downarrow + \; M_O = 400 \sin 30°(0.2) - 400 \cos 30°(0.4)$$
$$= -98.6 \text{ N} \cdot \text{m} = 98.6 \text{ N} \cdot \text{m} \downarrow$$

or

$$\mathbf{M}_O = \{-98.6\mathbf{k}\} \text{ N} \cdot \text{m} \qquad\qquad \textit{Ans.}$$

SOLUTION II (VECTOR ANALYSIS)

Using a Cartesian vector approach, the force and position vectors shown in Fig. 4–21c can be represented as

$$\mathbf{r} = \{0.4\mathbf{i} - 0.2\mathbf{j}\} \text{ m}$$
$$\mathbf{F} = \{400 \sin 30°\mathbf{i} - 400 \cos 30°\mathbf{j}\} \text{ N}$$
$$= \{200.0\mathbf{i} - 346.4\mathbf{j}\} \text{ N}$$

The moment is therefore

$$\mathbf{M}_O = \mathbf{r} \times \mathbf{F} = \begin{vmatrix} \mathbf{i} & \mathbf{j} & \mathbf{k} \\ 0.4 & -0.2 & 0 \\ 200.0 & -346.4 & 0 \end{vmatrix}$$
$$= 0\mathbf{i} - 0\mathbf{j} + [0.4(-346.4) - (-0.2)(200.0)]\mathbf{k}$$
$$= \{-98.6\mathbf{k}\} \text{ N} \cdot \text{m} \qquad\qquad \textit{Ans.}$$

Fig. 4–21

By comparison, it is seen that the scalar analysis (Solution I) provides a more *convenient method* for analysis than Solution II, since the direction of the moment and the moment arm for each component force are easy to establish. Hence, this method is generally recommended for solving problems displayed in two dimensions. On the other hand, Cartesian vector analysis is generally recommended only for solving three-dimensional problems, where the moment arms and force components are often more difficult to determine.

PROBLEMS

4.1 If **A, B,** and **D** are given vectors, prove the distributive law for the vector cross product, i.e., **A** × (**B** + **D**) = (**A** × **B**) + (**A** × **D**).

4–2. Prove the triple scalar product identity **A** · **B** × **C** = **A** × **B** · **C.**

4–3. Given the three nonzero vectors **A, B,** and **C,** show that if **A** · (**B** × **C**) = 0, the three vectors *must* lie in the same plane.

***4–4.** Determine the magnitude and directional sense of the moment of the force at *A* about point *O*.

4–5. Determine the magnitude and directional sense of the moment of the force at *A* about point *P*.

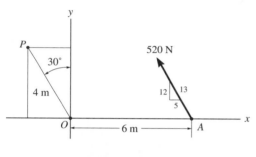

Probs. 4–4/4–5

4–6. Determine the magnitude and directional sense of the resultant moment of the forces at *A* and *B* about point *O*.

4–7. Determine the magnitude and directional sense of the resultant moment of the forces at *A* and *B* about point *P*.

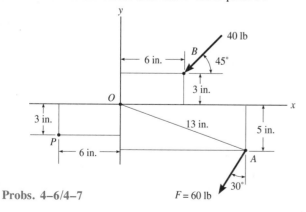

Probs. 4–6/4–7

***4–8.** Determine the magnitude and directional sense of the resultant moment of the forces at *A* and *B* about point *O*.

4–9. Determine the magnitude and directional sense of the resultant moment of the forces at *A* and *B* about point *P*.

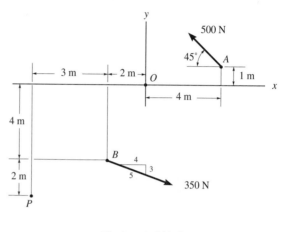

Probs. 4–8/4–9

4–10. A force of 30 lb is applied to the handle of the wrench. Determine the moment of this force about point *O*.

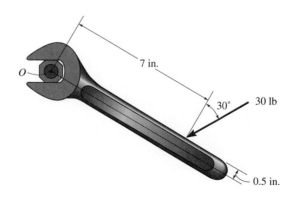

Prob. 4–10

4–11. Using one hand, the man attempts to lift the sledge hammer, which has a head having a weight of 10 lb. Determine the greatest angle θ to which he can hold it if the maximum moment he can develop at his wrist is 18 lb · ft. Neglect the weight of the handle.

Prob. 4–11

4–13. Determine the resultant moment about point A of the three forces acting on the beam.

4–14. Determine the resultant moment about point B of the three forces acting on the beam.

Probs. 4–13/4–14

*4–12.** The crane can be adjusted for any angle $0° \leq \theta \leq 90°$ and any extension $0 \leq x \leq 5$ m. For a suspended mass of 150 kg, determine the moment developed at A as a function of x and θ. What values of both x and θ develop the maximum possible moment at A? Compute this moment. Neglect the size of the pulley at B.

4–15. Determine the magnitude of force $\mathbf{F}_2$ that must be applied perpendicular to the handle so that the resultant moment of both $\mathbf{F}_1$ and $\mathbf{F}_2$ at O is zero.

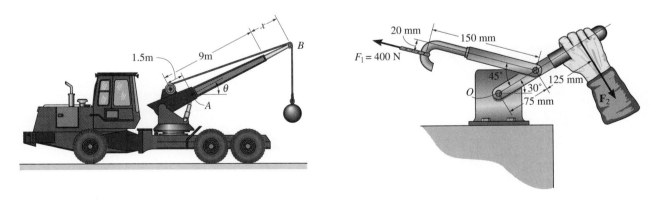

Prob. 4–12

Prob. 4–15

*4–16.   Determine the orientation $\theta(0° \le \theta \le 180°)$ of the 40-lb force **F** so that **F** produces (*a*) the maximum moment about point A and (*b*) no moment about point A. Compute the moment in each case.

4–17.   Determine the moment of the force **F** about point A as a function of θ. Plot the results of M (ordinate) versus θ (abscissa) for $0° \le \theta \le 180°$.

4–19.   The tongs are used to grip the ends of the drilling pipe P. Determine the torque (moment) M_P that the applied force $F = 150$ lb exerts on the pipe as a function of θ. Plot this moment M_P versus θ for $0 \le \theta \le 90°$.

*4–20.   The tongs are used to grip the ends of the drilling pipe P. If a torque (moment) of $M_P = 800$ lb · ft is needed at P to turn the pipe, determine the cable force **F** that must be applied to the tongs. Set $\theta = 30°$.

Probs. 4–16/4–17

Probs. 4–19/4–20

4–18.   The man exerts the two forces on the handle of the shovel. Determine the resultant moment of these forces about the blade at A.

4–21.   Determine the moment of the force **F** at A about point O. Express the result as a Cartesian vector.

4–22.   Determine the moment of the force **F** at A about point P. Express the result as a Cartesian vector.

Prob. 4–18

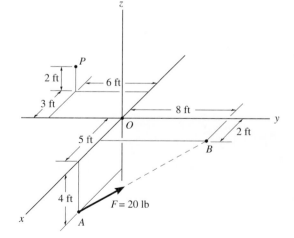

Probs. 4–21/4–22

4–23. Determine the resultant moment of the forces about point O. Express the result as a Cartesian vector.

*4–24. Determine the resultant moment of the forces about point P. Express the result as a Cartesian vector.

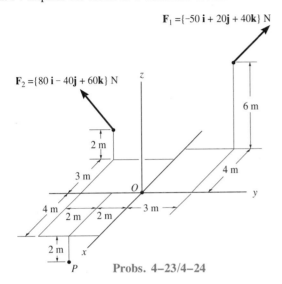

Probs. 4–23/4–24

4–26. Using Cartesian vectors, compute the moment of each of the two forces acting on the pipe assembly about point O. Add these moments and calculate the magnitude and coordinate direction angles of the resultant moment.

4–27. Using Cartesian vectors, determine the moment of each of the two forces acting on the pipe assembly about point A. Add these moments and calculate the magnitude and coordinate direction angles of the resultant moment.

Probs. 4–26/4–27

4–25. The cable exerts a 140-N force on the telephone pole as shown. Determine the moment of this force at the base A of the pole. Solve the problem two ways, i.e., by using a position vector from A to C, then A to B.

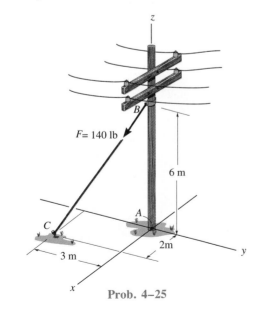

Prob. 4–25

*4–28. The curved pipe has a radius of 5 ft. If a force of 80 lb acts at its end as shown, determine the moment of this force about point C. Solve the problem by using two different position vectors.

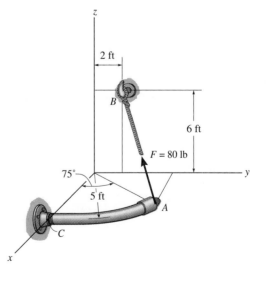

Prob. 4–28

4–29. A force **F** having a magnitude of $F = 100$ N acts along the diagonal of the parallelepiped. Using Cartesian vectors, determine the moment of **F** about point A where $\mathbf{M}_A = \mathbf{r}_B \times \mathbf{F}$ and $\mathbf{M}_A = \mathbf{r}_C \times \mathbf{F}$.

Prob. 4–29

4–30. Find the resultant moment of the two forces acting at the end of the pipe assembly about each of the joints at A, B, and C.

$\mathbf{F}_2 = \{9\mathbf{i} - 4\mathbf{j} - 10\mathbf{k}\}$ lb
$\mathbf{F}_1 = \{4\mathbf{i} + 10\mathbf{j} - 8\mathbf{k}\}$ lb

Prob. 4–30

4–31. If $\mathbf{F} = \{50\mathbf{i} + 60\mathbf{j} + 30\mathbf{k}\}$ lb, determine the magnitude and coordinate direction angles of the moment of **F** about point A.

*4–32. Determine the coordinate direction angles of the force **F** applied at the end of the pipe such that the moment created by **F** about point A is zero.

Probs. 4–31/4–32

4–33. The curved rod has a radius of 4 ft and is supported by the cable AB, which exerts a force of 80 lb on the rod. Determine the moment of this force about point C. Solve the problem by using two different position vectors.

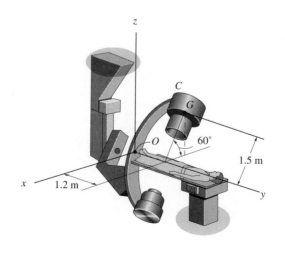

Prob. 4–33

4–34. The x-ray machine is used for medical diagnosis. If the camera and housing at C have a mass of 150 kg and a mass center at G, determine the moment of its weight about point O when it is in the position shown.

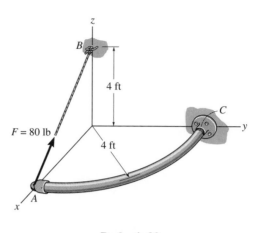

Prob. 4–34

4–35. The pipe assembly is subjected to the 80-N force. Determine the moment of this force about point A.

*4–36. The pipe assembly is subjected to the 80-N force. Determine the moment of this force about point B.

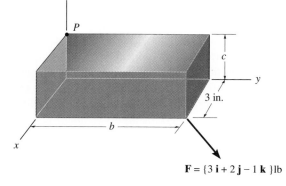

Probs. 4–35/4–36

4–37. The moment at point P created by the applied force $\mathbf{F}$ acting on the box is $\mathbf{M}_P = \{-1\mathbf{i} - 3\mathbf{j} - 9\mathbf{k}\}$ lb · in. Determine the dimensions b and c of the box.

Prob. 4–37

4–38. A force of $\mathbf{F} = \{-5\mathbf{i} + 3\mathbf{j} - 4\mathbf{k}\}$ kN produces a moment of $\mathbf{M}_O = \{-17\mathbf{i} - 7\mathbf{j} + 16\mathbf{k}\}$ kN · m about point O. If the force acts at point P having a y coordinate of $y = 2$ m, determine the x and z coordinates.

4–39. The force $\mathbf{F} = \{-5\mathbf{i} + 3\mathbf{j} - 4\mathbf{k}\}$ kN creates a moment about point O of $\mathbf{M}_O = \{-17\mathbf{i} - 3\mathbf{j} + 19\mathbf{k}\}$ kN · m. If the force passes through point P having a z coordinate of $z = 3$ m, determine the x and y coordinates of the point. Also, realizing that $M_O = Fd$, determine the perpendicular distance d from point O to the line of action of $\mathbf{F}$.

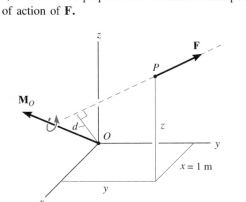

Probs. 4–38/4–39

*4–40. A force of $\mathbf{F} = \{6\mathbf{i} - 2\mathbf{j} + 1\mathbf{k}\}$ kN produces a moment of $\mathbf{M}_O = \{4\mathbf{i} + 5\mathbf{j} - 14\mathbf{k}\}$ kN · m about the origin of coordinates, point O. If the force acts at a point having an x coordinate of $x = 1$ m, determine the y and z coordinates.

4–41. The force $\mathbf{F} = \{6\mathbf{i} + 8\mathbf{j} + 10\mathbf{k}\}$ N creates a moment about point O of $\mathbf{M}_O = \{-14\mathbf{i} + 8\mathbf{j} + 2\mathbf{k}\}$ N · m. If the force passes through a point having an x coordinate of $x = 1$ m, determine the y and z coordinates of the point. Also, realizing that $M_O = Fd$, determine the perpendicular distance d from point O to the line of action of $\mathbf{F}$.

Probs. 4–40/4–41

4.5 Moment of a Force About a Specified Axis

Recall that when the moment of a force is computed about a point, the moment and its axis are *always* perpendicular to the plane containing the force and the moment arm. In some problems it is important to find the *component* of this moment along a specified axis that passes through the point. To solve this problem either a scalar or vector analysis can be used.

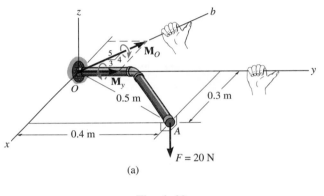

(a)

Fig. 4–22

Scalar Analysis. As a numerical example of this problem, consider the pipe assembly shown in Fig. 4–22a, which lies in the horizontal plane and is subjected to the vertical force of $F = 20$ N applied at point A. The moment of this force about point O has a *magnitude* of $M_O = (20 \text{ N})(0.5 \text{ m}) = 10 \text{ N} \cdot \text{m}$ and a *direction* defined by the right-hand rule, as shown in Fig. 4–22a. This moment tends to turn the pipe about the Ob axis. For practical reasons, however, it may be necessary to determine the *component* of $\mathbf{M}_O$ about the y axis, $\mathbf{M}_y$, since this component tends to unscrew the pipe from the flange at O. From Fig. 4–22a, $\mathbf{M}_y$ has a magnitude of $M_y = \frac{3}{5}(10 \text{ N} \cdot \text{m}) = 6 \text{ N} \cdot \text{m}$ and a sense of direction shown by the vector resolution. Rather than performing this *two-step* process of first finding the moment of the force about point O and then resolving the moment along the y axis, it is also possible to solve this problem *directly*. To do so, it is necessary to determine the perpendicular or moment-arm distance from the line of action of $\mathbf{F}$ to the y axis. From Fig. 4–22a this distance is 0.3 m. Thus the *magnitude* of the moment of the force about the y axis is again $M_y = 0.3(20 \text{ N}) = 6 \text{ N} \cdot \text{m}$, and the *direction* is determined by the right-hand rule as shown.

In general, then, *if the line of action of a force* **F** *is perpendicular to an axis aa*, the magnitude of the moment of **F** about the axis can be determined from the equation

$$M_a = Fd_a$$
(4–10)

Here d_a is the *perpendicular or shortest distance* from the force line of action to the axis. The direction is determined from the thumb of the right hand when the fingers are curled in accordance with the direction of rotation as produced by the force. In particular, realize that a *force will not contribute a moment about an axis if its line of action is parallel to the axis or its line of action passes through the axis.*

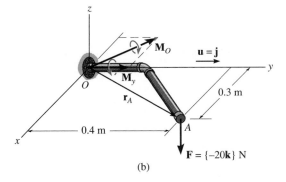

(b)

Vector Analysis.

The previous two-step solution of first finding the moment of the force about a point on the axis and then finding the projected component of the moment about the axis can also be performed using a vector analysis, Fig. 4–22b. Here the moment about point O is first determined from $\mathbf{M}_O = \mathbf{r}_A \times \mathbf{F} = (0.3\mathbf{i} + 0.4\mathbf{j}) \times (-20\mathbf{k}) = \{-8\mathbf{i} + 6\mathbf{j}\} \text{ N} \cdot \text{m}$. The component or projection of this moment along the y axis is then determined from the dot product (Sec. 2.9). Since the unit vector for the axis (or line) is $\mathbf{u} = \mathbf{j}$, then $M_y = \mathbf{M}_O \cdot \mathbf{u} = (-8\mathbf{i} + 6\mathbf{j}) \cdot \mathbf{j} = 6 \text{ N} \cdot \text{m}$. This result, of course, is to be expected, since it represents the $\mathbf{j}$ component of $\mathbf{M}_O$.

A vector analysis such as this is particularly advantageous for finding the moment of a force about an axis when the force components or the appropriate moment arms are difficult to determine. For this reason, the above two-step process will now be generalized and applied to a body of arbitrary shape. To do so, consider the body in Fig. 4–23, which is subjected to the force **F** acting at point A. Here we wish to determine the effect of **F** in tending to rotate the body about the aa' axis. This tendency for rotation is measured by the moment component $\mathbf{M}_a$. To determine $\mathbf{M}_a$ we first compute the moment of **F**

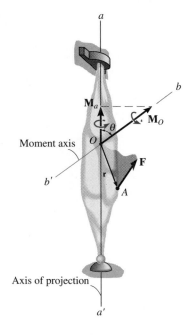

Fig. 4–23

about an *arbitrary point O* that lies on the axis. In this case, $\mathbf{M}_O$ is expressed by the cross product $\mathbf{M}_O = \mathbf{r} \times \mathbf{F}$, where $\mathbf{r}$ is directed from O to A. Since $\mathbf{M}_O$ acts along the moment axis bb', which is perpendicular to the plane containing $\mathbf{r}$ and $\mathbf{F}$, the component or projection of $\mathbf{M}_O$ onto the aa' axis is then represented by $\mathbf{M}_a$. The *magnitude* of $\mathbf{M}_a$ is determined by the dot product, $M_a = M_O \cos \theta = \mathbf{M}_O \cdot \mathbf{u}_a$, where $\mathbf{u}_a$ is a unit vector used to define the direction of the aa' axis. Combining these two steps as a general expression, we have $M_a = (\mathbf{r} \times \mathbf{F}) \cdot \mathbf{u}_a$. Since the dot product is commutative, we can also write

$$M_a = \mathbf{u}_a \cdot (\mathbf{r} \times \mathbf{F})$$

In vector algebra, this combination of dot and cross product yielding the scalar M_a is termed the *triple scalar product*. Provided the Cartesian components of each of the vectors are known, the mixed triple product may be written in determinant form as

$$M_a = (u_{a_x}\mathbf{i} + u_{a_y}\mathbf{j} + u_{a_z}\mathbf{k}) \cdot \begin{vmatrix} \mathbf{i} & \mathbf{j} & \mathbf{k} \\ r_x & r_y & r_z \\ F_x & F_y & F_z \end{vmatrix}$$

or simply

$$M_a = \mathbf{u}_a \cdot (\mathbf{r} \times \mathbf{F}) = \begin{vmatrix} u_{a_x} & u_{a_y} & u_{a_z} \\ r_x & r_y & r_z \\ F_x & F_y & F_z \end{vmatrix} \qquad (4\text{--}11)$$

where $u_{a_x}, u_{a_y}, u_{a_z}$ represent the x, y, z components of the unit vector defining the direction of the aa' axis

r_x, r_y, r_z represent the x, y, z components of the position vector drawn from any point O on the aa' axis to any point A on the line of action of the force

F_x, F_y, F_z represent the x, y, z components of the force vector.

When M_a is evaluated from Eq. 4–11, it will yield a positive or negative scalar. The sign of this scalar indicates the sense of direction of $\mathbf{M}_a$ along the aa' axis. If it is positive, then $\mathbf{M}_a$ will have the same sense as $\mathbf{u}_a$, whereas if it is negative, then $\mathbf{M}_a$ will act opposite to $\mathbf{u}_a$.

Once M_a is determined, we can then express $\mathbf{M}_a$ as a Cartesian vector, namely,

$$\mathbf{M}_a = M_a\mathbf{u}_a = [\mathbf{u}_a \cdot (\mathbf{r} \times \mathbf{F})]\mathbf{u}_a \qquad (4\text{--}12)$$

Finally, if the resultant moment of a series of forces is to be computed about the axis, then the moment components of each force are added together *algebraically*, since each component lies along the same axis.

The following examples illustrate a numerical application of the above concepts.

Example 4–9

The force $\mathbf{F} = \{-40\mathbf{i} + 20\mathbf{j} + 10\mathbf{k}\}$ N acts at point A shown in Fig. 4–24a. Determine the moments of this force about the x and Oa axes.

$\mathbf{F} = \{-40\mathbf{i} + 20\mathbf{j} + 10\mathbf{k}\}$ N

SOLUTION I (*VECTOR ANALYSIS*)

We can solve this problem by using the position vector $\mathbf{r}_A$. Why? Since $\mathbf{r}_A = \{-3\mathbf{i} + 4\mathbf{j} + 6\mathbf{k}\}$ m, and $\mathbf{u}_x = \mathbf{i}$, then applying Eq. 4–11,

$$M_x = \mathbf{i} \cdot (\mathbf{r}_A \times \mathbf{F}) = \begin{vmatrix} 1 & 0 & 0 \\ -3 & 4 & 6 \\ -40 & 20 & 10 \end{vmatrix}$$

$$= 1[4(10) - 6(20)] - 0[(-3)(10) - 6(-40)] + 0[(-3)(20) - 4(-40)]$$

$$= -80 \text{ N} \cdot \text{m} \qquad \textit{Ans.}$$

The negative sign indicates that the sense of $\mathbf{M}_x$ is opposite to $\mathbf{i}$.

Realize that this result also represents the $\mathbf{i}$ component of the moment of $\mathbf{F}$ about point A, i.e., $\mathbf{M}_A = \mathbf{r}_A \times \mathbf{F}$.

We can compute M_{Oa} using $\mathbf{r}_A$ and $\mathbf{u}_{Oa} = -\frac{3}{5}(\mathbf{i}) + \frac{4}{5}(\mathbf{j})$. Thus,

$$M_{Oa} = \mathbf{u}_{Oa} \cdot (\mathbf{r}_A \times \mathbf{F}) = \begin{vmatrix} -\frac{3}{5} & \frac{4}{5} & 0 \\ -3 & 4 & 6 \\ -40 & 20 & 10 \end{vmatrix}$$

$$= -\tfrac{3}{5}[4(10) - 6(20)] - \tfrac{4}{5}[(-3)(10) - 6(-40)] + 0[(-3)(20) - 4(-40)]$$

$$= -120 \text{ N} \cdot \text{m} \qquad \textit{Ans.}$$

What does the negative sign indicate?

The moment components are shown in Fig. 4–24b.

SOLUTION II (*SCALAR ANALYSIS*)

Since the force components and moment arms are easy to determine for computing M_x a scalar analysis can be used to solve this problem. Referring to Fig. 4–24c, only the 10-N and 20-N forces contribute moments about the x axis. (The line of action of the 40-N force is *parallel* to this axis and hence its moment about the x axis is zero.) Using the right-hand rule, the algebraic sum of the moment components about the x axis is therefore

$$M_x = (10 \text{ N})(4 \text{ m}) - (20 \text{ N})(6 \text{ m}) = -80 \text{ N} \cdot \text{m} \qquad \textit{Ans.}$$

Although not required here, note also that

$$M_y = (10 \text{ N})(3 \text{ m}) - (40 \text{ N})(6 \text{ m}) = -210 \text{ N} \cdot \text{m}$$
$$M_z = (40 \text{ N})(4 \text{ m}) - (20 \text{ N})(3 \text{ m}) = 100 \text{ N} \cdot \text{m}$$

If we were to determine M_{Oa} by this scalar method it would require much more effort, since the force components of 40 N and 20 N are *not perpendicular* to the direction of Oa. The vector analysis yields a more direct solution.

Fig. 4–24

Example 4–10

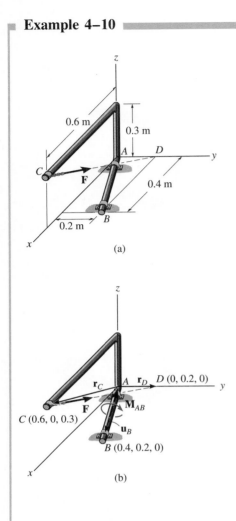

(a)

(b)

Fig. 4–25

The rod shown in Fig. 4–25a is supported by two brackets at A and B. Determine the moment $\mathbf{M}_{AB}$ produced by $\mathbf{F} = \{-600\mathbf{i} + 200\mathbf{j} - 300\mathbf{k}\}$ N, which tends to rotate the rod about the AB axis.

SOLUTION

A vector analysis using Eq. 4–11 will be considered for the solution since the moment arm is difficult to determine. In general, $M_{AB} = \mathbf{u}_B \cdot (\mathbf{r} \times \mathbf{F})$. Each of the terms in this equation will now be identified.

Unit vector $\mathbf{u}_B$ defines the direction of the AB axis of the rod, Fig. 4–25b, where

$$\mathbf{u}_B = \frac{\mathbf{r}_B}{r_B} = \frac{0.4\mathbf{i} + 0.2\mathbf{j}}{\sqrt{(0.4)^2 + (0.2)^2}} = 0.894\mathbf{i} + 0.447\mathbf{j}$$

Vector $\mathbf{r}$ is directed from *any point* on the AB axis to *any point* on the line of action of the force. For example, position vectors $\mathbf{r}_C$ and $\mathbf{r}_D$ are suitable, Fig. 4–25b. (Although not shown, $\mathbf{r}_{BC}$ or $\mathbf{r}_{BD}$ can also be used.) For simplicity, we choose $\mathbf{r}_D$, where

$$\mathbf{r}_D = \{0.2\mathbf{j}\} \text{ m}$$

The force is

$$\mathbf{F} = \{-600\mathbf{i} + 200\mathbf{j} - 300\mathbf{k}\} \text{ N}$$

Substituting these vectors into the determinant form and expanding, we have

$$M_{AB} = \mathbf{u}_B \cdot (\mathbf{r}_D \times \mathbf{F}) = \begin{vmatrix} 0.894 & 0.447 & 0 \\ 0 & 0.2 & 0 \\ -600 & 200 & -300 \end{vmatrix}$$

$$= 0.894[0.2(-300) - 0(200)] - 0.447[0(-300) - 0(-600)] + 0[0(200) - 0.2(-600)]$$

$$= -53.64 \text{ N} \cdot \text{m}$$

The negative sign indicates that the sense of $\mathbf{M}_{AB}$ is opposite to that of $\mathbf{u}_B$. Expressing $\mathbf{M}_{AB}$ as a Cartesian vector yields

$$\mathbf{M}_{AB} = M_{AB}\mathbf{u}_B = (-53.64 \text{ N} \cdot \text{m})(0.894\mathbf{i} + 0.447\mathbf{j})$$
$$= \{-48.0\mathbf{i} - 24.0\mathbf{j}\} \text{ N} \cdot \text{m} \qquad \qquad \textit{Ans.}$$

The result is shown in Fig. 4–25b.

It should be mentioned that if axis AB is defined using a unit vector directed from B toward A, then in the above formulation $-\mathbf{u}_B$ would have to be used. This would lead to $M_{AB} = +53.64 \text{ N} \cdot \text{m}$. Consequently, $\mathbf{M}_{AB} = M_{AB}(-\mathbf{u}_B)$, and the above result would again be determined.

PROBLEMS

4–42. Determine the moment of the force **F** about the *Oa* axis. Express the result as a Cartesian vector.

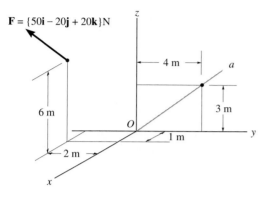

$\mathbf{F} = \{50\mathbf{i} - 20\mathbf{j} + 20\mathbf{k}\}\mathrm{N}$

4 m

3 m

6 m

2 m

1 m

Prob. 4–42

*4–44. Determine the moment of the force **F** about the *aa* axis. Express the result as a Cartesian vector.

$F = 3$ kN

6 m

60°

45°

60°

5 m

2 m

2 m

Prob. 4–44

4–43. Determine the resultant moment of two forces about the *aa* axis. Express the result as a Cartesian vector.

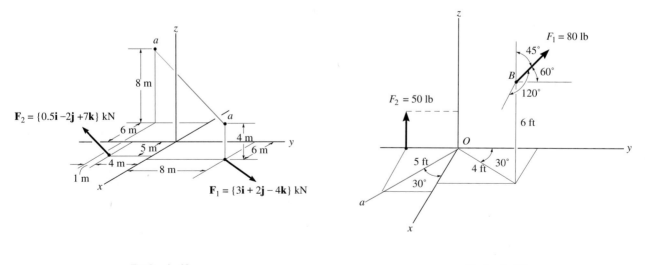

$\mathbf{F}_2 = \{0.5\mathbf{i} - 2\mathbf{j} + 7\mathbf{k}\}$ kN

8 m

6 m

5 m

4 m

4 m

6 m

8 m

1 m

$\mathbf{F}_1 = \{3\mathbf{i} + 2\mathbf{j} - 4\mathbf{k}\}$ kN

Prob. 4–43

4–45. Determine the resultant moment of the two forces about the *Oa* axis. Express the result as a Cartesian vector.

$F_1 = 80$ lb

45°

60°

B

120°

$F_2 = 50$ lb

6 ft

5 ft

4 ft

30°

30°

Prob. 4–45

4–46. Determine the moments of the force **F** about the x, y, and z axes. Solve the problem (a) using a Cartesian vector approach and (b) using a scalar approach.

*4–48. Determine the magnitude of the resultant moment of the three forces about the axis AB. Solve the problem (a) using a Cartesian vector approach and (b) using a scalar approach.

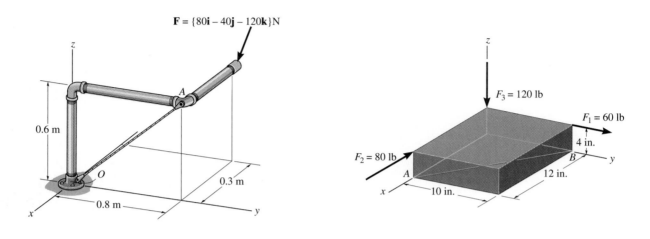

Prob. 4–46

Prob. 4–48

4–47. Determine the moment of the force **F** about an axis extending between O and A. Express the result as a Cartesian vector.

4–49. A force of $\mathbf{F} = \{8\mathbf{i} - 1\mathbf{j} + 1\mathbf{k}\}$ lb is applied to the handle of the box wrench. Determine the component of the moment of this force about the z axis which is effective in tightening the bolt.

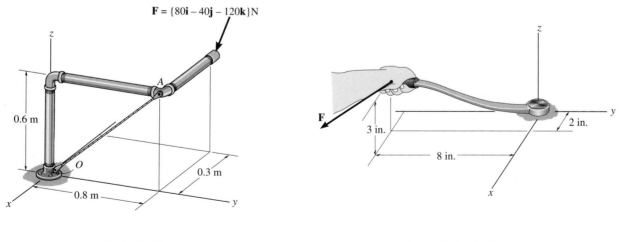

Prob. 4–47

Prob. 4–49

4–50. The chain AB exerts a force of 20 lb on the door at B. Determine the magnitude of the moment of this force along the hinged axis x of the door.

Prob. 4–50

4–51. The lug and box wrenches are used in combination to remove the lug nut from the wheel hub. If the applied force on the end of the box wrench is $\mathbf{F} = \{4\mathbf{i} - 12\mathbf{j} + 2\mathbf{k}\}$ N, determine the magnitude of the moment of this force about the x axis which is effective in unscrewing the lug nut.

Prob. 4–51

*4–52. The 50-lb force acts on the gear in the direction shown. Determine the moment of this force about the y axis.

Prob. 4–52

4–53. Determine the moment that the force $\mathbf{F}$ exerts about the y axis of the shaft. Solve the problem using a Cartesian vector approach and using a scalar approach. Express the result as a Cartesian vector.

Prob. 4–53

4–54.   The bracket is acted upon by a 600-N force at A. Determine the moment of this force about the y axis.

Prob. 4–54

4–55.   A force of 50 N is applied to the handle of the door. Determine the moment of this force about the hinged axis z. Neglect the size of the doorknob. *Suggestion:* Use a scalar analysis.

Prob. 4–55

*4–56.   The force of $F = 80$ lb acts along the edge DB of the tetrahedron. Determine the magnitude of the moment of this force about the edge AC.

4–57.   If the moment of the force $\mathbf{F}$ about the edge AC of the tetrahedron has a magnitude of $M = 200$ lb · ft and is directed from C toward A, determine the magnitude of $\mathbf{F}$.

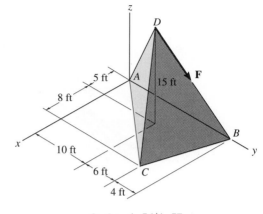

Probs. 4–56/4–57

4–58.   A vertical force of $F = 60$ N is applied to the handle of the pipe wrench. Determine the moment that this force exerts along the axis AB (x axis) of the pipe assembly. Both the wrench and pipe assembly ABC lie in the x–y plane. *Suggestion:* Use a scalar analysis.

4–59.   Determine the magnitude of the vertical force $\mathbf{F}$ acting on the handle of the wrench so that this force produces a component of moment along the AB axis (x axis) of the pipe assembly of $(\mathbf{M}_A)_x = \{-5\mathbf{i}\}$ N · m. Both the pipe assembly ABC and the wrench lie in the x–y plane. *Suggestion:* Use a scalar analysis.

Probs. 4–58/4–59

4.6 Moment of a Couple

A *couple* is defined as two parallel forces that have the same magnitude, opposite directions, and are separated by a perpendicular distance *d*, Fig. 4–26. Since the resultant force of the two forces composing the couple is zero, the only effect of a couple is to produce a rotation or tendency of rotation in a specified direction.

Fig. 4–26

The moment produced by a couple, called a *couple moment*, is equivalent to the sum of the moments of both couple forces, determined about *any* arbitrary point *O* in space. To show this, consider position vectors $\mathbf{r}_A$ and $\mathbf{r}_B$, directed from *O* to points *A* and *B* lying on the line of action of $-\mathbf{F}$ and $\mathbf{F}$, Fig. 4–27. The couple moment computed about *O* is therefore

$$\mathbf{M} = \mathbf{r}_A \times (-\mathbf{F}) + \mathbf{r}_B \times (\mathbf{F})$$
$$= (\mathbf{r}_B - \mathbf{r}_A) \times \mathbf{F}$$

By the triangle law of vector addition, $\mathbf{r}_A + \mathbf{r} = \mathbf{r}_B$ or $\mathbf{r} = \mathbf{r}_B - \mathbf{r}_A$, so that

$$\mathbf{M} = \mathbf{r} \times \mathbf{F} \qquad (4\text{–}13)$$

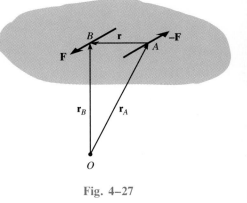

Fig. 4–27

This result indicates that a couple moment is a *free vector*, i.e., it can act at *any point*, since $\mathbf{M}$ depends *only* upon the position vector directed *between* the forces and *not* the position vectors $\mathbf{r}_A$ and $\mathbf{r}_B$, directed from point *O* to the forces. This concept is therefore unlike the moment of a force, which requires a definite point (or axis) about which moments are determined.

Scalar Formulation. The moment of a couple, $\mathbf{M}$, Fig. 4–28, is defined as having a *magnitude* of

$$\boxed{M = Fd} \qquad (4\text{–}14)$$

where *F* is the magnitude of one of the forces and *d* is the perpendicular distance or moment arm between the forces. The *direction* and sense of the couple moment are determined by the right-hand rule, where the thumb indicates the direction when the fingers are curled with the sense of rotation caused by the two forces. In all cases, $\mathbf{M}$ acts perpendicular to the plane containing these forces.

Fig. 4–28

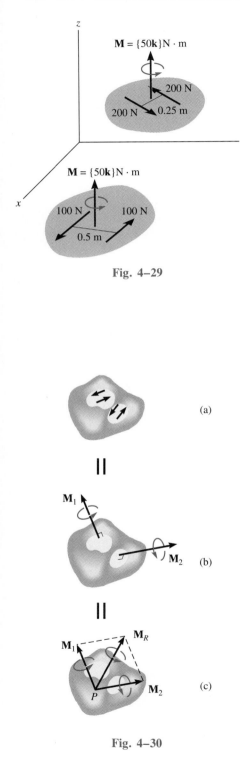

Fig. 4–29

Fig. 4–30

Vector Formulation.

The moment of a couple can also be expressed by the vector cross product using Eq. 4–13,

$$M = r \times F \qquad (4\text{–}15)$$

Application of this equation is easily remembered if one thinks of taking the moments of both forces about a point lying on the line of action of one of the forces. For example, if moments are taken about point A in Fig. 4–27 the moment of $-F$ is zero about this point and the moment of F is defined from Eq. 4–15. Therefore, in the formulation r is crossed with the force F to which it is directed.

Equivalent Couples.

Two couples are said to be equivalent if they produce the same moment. Since the moment produced by a couple is always perpendicular to the plane containing the couple forces, it is therefore necessary that the forces of equal couples lie either in the same plane or in planes that are *parallel* to one another. In this way, the direction of each couple moment will be the same, that is, perpendicular to the parallel planes. For example, the two couples shown in Fig. 4–29 are equivalent. One couple is produced by a pair of 100-N forces separated by a distance of $d = 0.5$ m, and the other is produced by a pair of 200-N forces separated by a distance of 0.25 m. Since the planes in which the forces act are parallel to the x–y plane, the moment produced by each of the couples may be expressed as $M = \{50k\}$ N $\cdot$ m.

Resultant Couple Moment.

Since couple moments are free vectors, they may be applied at any point P on a body and added vectorially. For example, the two couples acting on different planes of the rigid body in Fig. 4–30a may be replaced by their corresponding couple moments M_1 and M_2, Fig. 4–30b, and then these free vectors may be moved to the *arbitrary point P* and added to obtain the resultant couple moment $M_R = M_1 + M_2$, shown in Fig. 4–30c.

If more than two couple moments act on the body, we may generalize this concept and write the vector resultant as

$$M_R = \Sigma r \times F \qquad (4\text{–}16)$$

where each couple moment is computed in accordance with Eq. 4–15.

The following examples illustrate these concepts numerically. In general, problems projected in two dimensions should be solved using a scalar analysis, since the moment arms and force components are easy to compute.

Example 4–11

A couple acts at the end of the beam shown in Fig. 4–31a. Replace it by an equivalent couple having a pair of forces that act through (a) points A and B, and (b) points D and E.

Fig. 4–31

SOLUTION (*SCALAR ANALYSIS*)

The couple has a magnitude of $M = Fd = 400(0.2) = 80 \text{ N} \cdot \text{m}$ and a direction that is into the page since the forces tend to rotate clockwise. **M** is a free vector so that it can be placed at any point on the beam, Fig. 4–31b.

Part (a).   To preserve the direction of **M,** *horizontal* forces acting through points A and B must be directed as shown in Fig. 4–31c. The magnitude of each force is

$$M = Fd$$
$$80 = F(0.25)$$
$$F = 320 \text{ N} \qquad\qquad Ans.$$

Part (b).   To generate the required clockwise rotation, forces acting through points D and E must be *vertical* and directed as shown in Fig. 4–31d. The magnitude of each force is

$$M = Pd$$
$$80 = P(0.1)$$
$$P = 800 \text{ N} \qquad\qquad Ans.$$

Example 4–12

Determine the moment of the couple acting on the beam shown in Fig. 4–32a.

(a)

SOLUTION (*SCALAR ANALYSIS*)

Here it is somewhat difficult to determine the perpendicular distance between the forces and compute the couple moment as $M = Fd$. Instead, we can resolve each force into its horizontal and vertical components, $F_x = \frac{4}{5}(150) = 120$ lb and $F_y = \frac{3}{5}(150) = 90$ lb, Fig. 4–32b, and then use the principle of moments. The couple moment can be determined about *any point*. For example, if point D is chosen, we have for all four forces,

$$\zeta+M = 120 \text{ lb}(0 \text{ ft}) - 90 \text{ lb}(2 \text{ ft}) + 90 \text{ lb}(5 \text{ ft}) + 120 \text{ lb}(1 \text{ ft})$$
$$= 390 \text{ lb} \cdot \text{ft} \uparrow \qquad\qquad\qquad\qquad \textit{Ans.}$$

It is easier, however, to determine the moments about point A or B in order to *eliminate* the moment of the forces acting at the moment point. For point A, Fig. 4–28b, we have

$$\zeta+M = 90 \text{ lb}(3 \text{ ft}) + 120 \text{ lb}(1 \text{ ft})$$
$$= 390 \text{ lb} \cdot \text{ft} \uparrow \qquad\qquad\qquad\qquad \textit{Ans.}$$

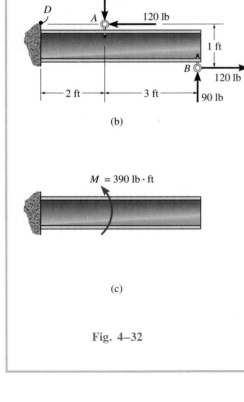

(b)

$M = 390$ lb · ft

(c)

Fig. 4–32

Show that one obtains this same result if moments are summed about point B. Notice also that the couple in Fig. 4–32a has been replaced by *two* couples in Fig. 4–32b. Using $M = Fd$, one couple has a moment of $M_1 = 90$ lb(3 ft) $= 270$ lb · ft and the other has a moment of $M_2 = 120$ lb(1 ft) $= 120$ lb · ft. By the right-hand rule, both couple moments are counterclockwise and are therefore directed out of the page. Since these couples are free vectors, they can be moved to any point and added, which yields $M = 270$ lb · ft $+ 120$ lb · ft $= 390$ lb · ft $\uparrow$, the same result determined above. **M** is a free vector and can therefore act at any point on the beam, Fig. 4–32c.

Example 4–13

Determine the couple moment acting on the pipe shown in Fig. 4–33a. Segment AB is directed $30°$ below the x–y plane.

(a)

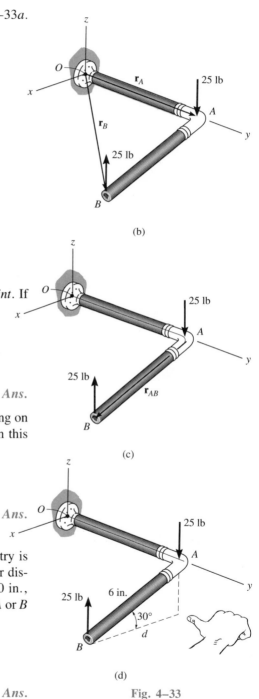

(b)

(c)

(d)

Fig. 4–33

SOLUTION I *(VECTOR ANALYSIS)*

The moment of the two couple forces can be found about *any point*. If point O is considered, Fig. 4–33b, we have

$$\mathbf{M} = \mathbf{r}_A \times (-25\mathbf{k}) + \mathbf{r}_B \times (25\mathbf{k})$$
$$= (8\mathbf{j}) \times (-25\mathbf{k}) + (6 \cos 30°\mathbf{i} + 8\mathbf{j} - 6 \sin 30°\mathbf{k}) \times (25\mathbf{k})$$
$$= -200\mathbf{i} - 129.9\mathbf{j} + 200\mathbf{i}$$
$$= \{-130\mathbf{j}\} \text{ lb} \cdot \text{in.} \qquad\qquad Ans.$$

It is *easier* to take moments of the couple forces about a point lying on the line of action of one of the forces, e.g., point A, Fig. 4–33c. In this case the moment of the force at A is zero, so that

$$\mathbf{M} = \mathbf{r}_{AB} \times (25\mathbf{k})$$
$$= (6 \cos 30°\mathbf{i} - 6 \sin 30°\mathbf{k}) \times (25\mathbf{k})$$
$$= \{-130\mathbf{j}\} \text{ lb} \cdot \text{in.} \qquad\qquad Ans.$$

SOLUTION II *(SCALAR ANALYSIS)*

Although this problem is shown in three dimensions, the geometry is simple enough to use the scalar equation $M = Fd$. The perpendicular distance between the lines of action of the forces is $d = 6 \cos 30° = 5.20$ in., Fig. 4–33d. Hence, taking moments of the forces about either point A or B yields

$$M = Fd = 25(5.20) = 129.9 \text{ lb} \cdot \text{in.}$$

Applying the right-hand rule, $\mathbf{M}$ acts in the $-\mathbf{j}$ direction. Thus,

$$\mathbf{M} = \{-130\mathbf{j}\} \text{ lb} \cdot \text{in.} \qquad\qquad Ans.$$

Example 4–14

Replace the two couples acting on the triangular block in Fig. 4–34a by a resultant couple moment.

Fig. 4–34

(a)   (b)   (c)

SOLUTION (*VECTOR ANALYSIS*)

The couple moment $\mathbf{M}_1$, developed by the forces at A and B, can easily be determined from a scalar formulation.

$$M_1 = Fd = 200(0.3) = 60 \text{ N} \cdot \text{m}$$

By the right-hand rule, $\mathbf{M}_1$ acts in the $+\mathbf{i}$ direction, Fig. 4–34b. Hence,

$$\mathbf{M}_1 = \{60\mathbf{i}\} \text{ N} \cdot \text{m}$$

Vector analysis will be used to determine $\mathbf{M}_2$, caused by forces at C and D. If moments are computed about point D, Fig. 4–34a, $\mathbf{M}_2 = \mathbf{r}_{DC} \times \mathbf{F}_C$, then

$$\mathbf{M}_2 = \mathbf{r}_{DC} \times \mathbf{F}_C = (0.2\mathbf{i}) \times [100(\tfrac{4}{5})\mathbf{j} - 100(\tfrac{3}{5})\mathbf{k}]$$
$$= (0.2\mathbf{i}) \times [80\mathbf{j} - 60\mathbf{k}] = 16(\mathbf{i} \times \mathbf{j}) - 12(\mathbf{i} \times \mathbf{k})$$
$$= \{12\mathbf{j} + 16\mathbf{k}\} \text{ N} \cdot \text{m}$$

Try to establish $\mathbf{M}_2$ by using a scalar formulation, Fig. 4–34b.

Since $\mathbf{M}_1$ and $\mathbf{M}_2$ are free vectors, they may be moved to some arbitrary point P on the block and added vectorially, Fig. 4–34c. The resultant couple moment becomes

$$\mathbf{M}_R = \mathbf{M}_1 + \mathbf{M}_2 = \{60\mathbf{i} + 12\mathbf{j} + 16\mathbf{k}\} \text{ N} \cdot \text{m} \qquad Ans.$$

PROBLEMS

*4–60. Determine the magnitude and direction of the couple moment.

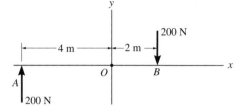

Prob. 4–60

4–61. Determine the magnitude and direction of the couple moment.

Prob. 4–61

4–62. Determine the magnitude and direction of the couple moment.

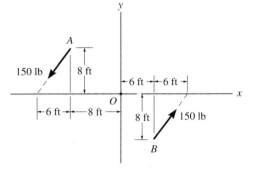

Prob. 4–62

4–63. The hanger H is used to support the end of the floor beam B. If the loading on the hanger consists of two couples, determine the magnitude of the horizontal forces **F** and $-$**F** so that the resultant couple moment on the hanger is zero.

Prob. 4–63

*4–64. The road reacts with a torque of $M_A = 400$ N $\cdot$ m and $M_B = 200$ N $\cdot$ m on the brushes of the road sweeper. Determine the magnitude of the couple forces that are developed by the road on the rear wheels of the sweeper, so that the resultant couple moment on the sweeper is zero. What is the magnitude of these forces if the brush at B is turned off?

Prob. 4–64

4–65.   A clockwise couple moment having a magnitude of $M = 40$ lb · ft is resisted by the shaft of an electric motor. Determine the magnitudes of the reactive forces **R** and −**R,** which act at the supports *A* and *B*, so that the resultant of the two couples is zero.

4–66.   Determine the magnitude of the couple moment **M** acting on the electric motor if the reactive forces **R** and −**R** each have a magnitude of 20 lb and the resultant of the two couples is zero.

Probs. 4–65/4–66

4–67.   The resultant couple moment created by the two couples acting on the disk is $\{10\mathbf{k}\}$ kip · in. Determine the magnitude of force **T.**

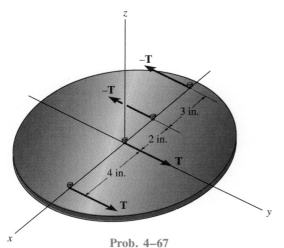

Prob. 4–67

*4–68.**   The cord passing over the two small pegs *A* and *B* of the board is subjected to a tension of 10 lb. Determine the *minimum* tension *P* and the orientation θ of the cord passing over the pegs *C* and *D*, so that the resultant couple moment produced by the two cords is 20 lb · in., clockwise.

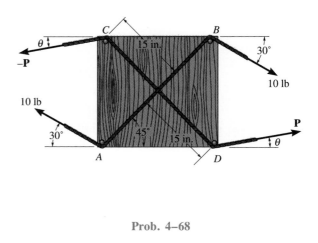

Prob. 4–68

4–69.   Two couples act on the beam. Determine the magnitude of **F** so that the resultant couple moment is 450 lb · ft, counterclockwise. Where on the beam does the resultant couple moment act?

Prob. 4–69

4–70. The ends of the triangular plate are subjected to three couples. Determine the magnitude of the force **F** so that the resultant couple moment is 400 N · m clockwise.

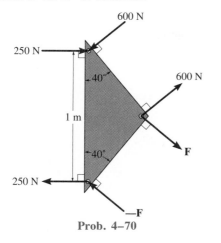

Prob. 4–70

4–71. Two couples act on the frame. If the resultant couple moment is to be zero, determine the distance d between the 100-lb couple forces.

*4–72. Two couples act on the frame. If $d = 4$ ft, determine the resultant couple moment. Compute the result by resolving each force into x and y components and (*a*) finding the moment of each couple (Eq. 4–14) and (*b*) summing the moments of all the force components about point A.

4–73. Two couples act on the frame. If $d = 6$ ft, determine the resultant couple moment. Compute the result by resolving each force into x and y components and (*a*) finding the moment of each couple (Eq. 4–14) and (*b*) summing the moments of all the force components about point B.

Probs. 4–71/4–72/4–73

4–74. Determine the resultant couple moment acting on the beam. Solve the problem two ways: (*a*) sum moments about point O; and (*b*) sum moments about point A.

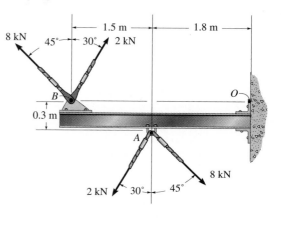

Prob. 4–74

4–75. Determine the couple moment. Express the result as a Cartesian vector.

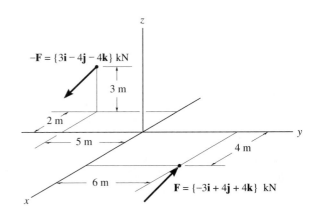

Prob. 4–75

*4–76.   Determine the couple moment. Express the result as a Cartesian vector.

Prob. 4–76

4–77.   The gear reducer is subjected to the couple moments shown. Determine the resultant couple moment and specify its magnitude and coordinate direction angles.

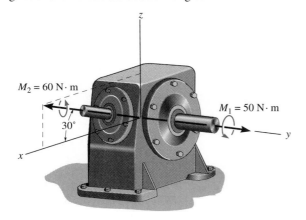

Prob. 4–77

4–78.   Express the moment of the couple acting on the pipe assembly in Cartesian vector form. What is the magnitude of the couple moment?

Prob. 4–78

4–79.   Determine the resultant couple moment of the two couples that act on the pipe assembly. The distance from A to B is d = 400 mm. Express the result as a Cartesian vector.

*4–80.   Determine the distance d between A and B so that the resultant couple moment has a magnitude of $M_R = 20$ N · m.

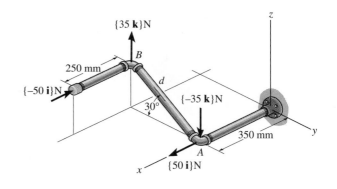

Probs. 4–79/4–80

4–81.   If the resultant couple moment of the three couples acting on the triangular block is to be zero, determine the magnitude of forces **F** and **P**.

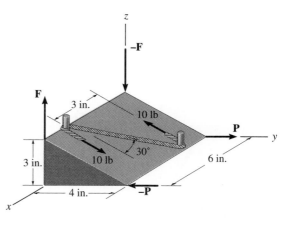

Prob. 4–81

4.7 Movement of a Force on a Rigid Body

Many problems in statics, including the reduction of a force system to its simplest possible form, require moving a force from one point to another on a rigid body. Since a force tends to both *translate* and *rotate* a body, it is important that these two "external" effects remain the same if the force is moved from one point to another on the body. Two cases for the location of the point O to which the force is moved will now be considered.

1. *Point O is on the line of action of the force*. Consider the rigid body shown in Fig. 4–35a, which is subjected to the force **F** applied to point A. In order to move the force to point O we will first apply equal but opposite forces **F** and −**F** at O, as shown in Fig. 4–35b. This step in no way alters the external effects on the body; however, the two forces indicated by the slash across them can be canceled, leaving the force at point O as required, Fig. 4–35c. By using this construction procedure, an *equivalent system* has been maintained between each of the diagrams, as indicated by the equal sign. Note, however, that the force has simply been "transmitted" along its line of action, from point A, Fig. 4–35a, to point O, Fig. 4–35c. In other words, the force can be considered as a *sliding vector* since it can act at any point O along its line of action. In Sec. 4.4 we referred to this concept as the *principle of transmissibility*. It can be formally stated as follows: *The external effects on a rigid body remain unchanged when a force, acting at a given point on the body, is applied to another point lying on the line of action of the force*. It is important to realize that only the *external effects* caused by the force remain *unchanged* after moving it. Certainly the *internal effects* depend on where **F** is located. For example, when **F** acts at A, the internal forces in the body have a high intensity around A; whereas if **F** acts at O, the effect of **F** on generating internal forces at A will be less.

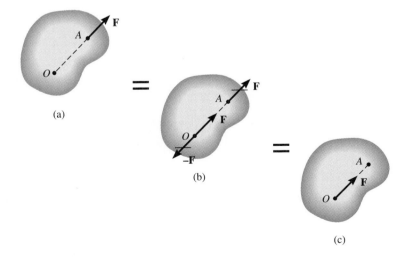

(a)

(b)

(c)

Fig. 4–35

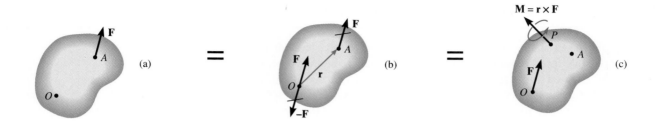

Fig. 4–36

2. *Point O is not on the line of action of the force.* This case is shown in Fig. 4–36a. Following the same procedure as above, we first apply equal but opposite forces **F** and −**F** at point O, Fig. 4–36b. Here the two forces indicated by a slash across them form a couple which has a moment defined by the cross product **M** = **r** × **F.** Since the couple moment is *a free vector,* it may be applied at any point P on the body as shown in Fig. 4–36c. In addition to this couple moment, **F** now acts at point O as required.

As a physical illustration of these two statements, consider the effect on the hand when holding the end of a stick of negligible weight. If a vertical force **F** is applied at the other end and the stick is held in the vertical position, Fig. 4–37a, then, by the principle of transmissibility, the same force is felt at the grip, Fig. 4–37b, regardless of where the force is applied along its line of action. When the stick is held in the horizontal position, Fig. 4–37c, the force has the effect of producing *both* a downward force at the grip and a clockwise twist, Fig. 4–37d. These same effects are felt at the grip if the force **F** and the couple moment $M = Fd$ are applied to the stick at the grip, Fig. 4–37c.

Fig. 4–37

4.8 Resultants of a Force and Couple System

When a rigid body is subjected to a system of forces and couple moments, it is often simpler to study the external effects on the body by using the force and couple moment resultants, rather than the force and couple moment system. To show how to simplify a system of forces and couple moments to their resultants, consider the rigid body in Fig. 4–38a. The force and couple moment system acting on it will be simplified by moving the forces and couple moments to the arbitrary point O. In this regard, the couple moment $\mathbf{M}$ is simply moved to O, since it is a free vector. Forces $\mathbf{F}_1$ and $\mathbf{F}_2$ are sliding vectors, and since O does not lie on the line of action of these forces, each must be moved to O in accordance with the procedure stated in Sec. 4.7. For example, when $\mathbf{F}_1$ is applied at O, a corresponding couple moment $\mathbf{M}_1 = \mathbf{r}_1 \times \mathbf{F}_1$ must also be applied to the body, Fig. 4–38b. By vector addition, the force and couple moment system shown in Fig. 4–38b can now be reduced to an *equivalent* resultant force $\mathbf{F}_R = \mathbf{F}_1 + \mathbf{F}_2$ and resultant couple moment $\mathbf{M}_{R_O} = \mathbf{M} + \mathbf{M}_1 + \mathbf{M}_2$ as shown in Fig. 4–38c. Note that both the magnitude and direction of $\mathbf{F}_R$ are independent of the location of point O; however, $\mathbf{M}_{R_O}$ depends upon this location, since the moments $\mathbf{M}_1$ and $\mathbf{M}_2$ are computed using the position vectors $\mathbf{r}_1$ and $\mathbf{r}_2$. Realize also that $\mathbf{M}_{R_O}$ is a free vector and can act at *any point* on the body, although point O is generally chosen as its point of application.

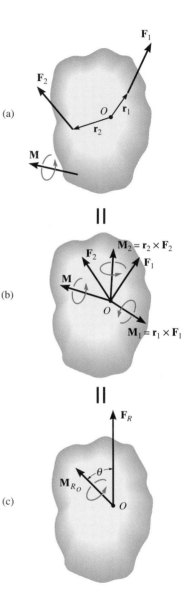

Fig. 4–38

PROCEDURE FOR ANALYSIS

The above method for simplifying any force and couple moment system to a resultant force acting at point O and a resultant couple moment will now be stated in general terms. To apply this method it is first necessary to establish an x, y, z coordinate system.

Three-Dimensional Systems.   A Cartesian vector analysis is generally used to solve problems involving three-dimensional force and couple systems for which the force components and moment arms are difficult to determine.

Force Summation.   The resultant force is equivalent to the vector sum of all the forces in the system; i.e.,

$$\boxed{\mathbf{F}_R = \Sigma \mathbf{F}} \qquad (4\text{–}17)$$

Moment Summation.   The resultant couple moment is equivalent to the vector sum of all the couple moments in the system plus the moments about point O of all the forces in the system; i.e.,

$$\boxed{\mathbf{M}_{R_O} = \Sigma \mathbf{M}_O} \qquad (4\text{–}18)$$

Coplanar Force Systems.   Since force components and moment arms are easy to determine in two dimensions, a scalar analysis provides the most convenient solution to problems involving coplanar force systems. Assuming that the forces lie in the x–y plane and any couple moments are perpendicular to this plane (along the z axis), then the resultants are determined as follows:

Force Summation.   The *resultant force* $\mathbf{F}_R$ is equivalent to the vector sum of its two components $\mathbf{F}_{R_x}$ and $\mathbf{F}_{R_y}$. Each component is found from the scalar (algebraic) sum of the components of all the forces in the system that act in the same direction; i.e.,

$$
\begin{aligned}
F_{R_x} &= \Sigma F_x \\
F_{R_y} &= \Sigma F_y
\end{aligned}
\qquad (4\text{--}19)
$$

Moment Summation.   The *resultant couple moment* $\mathbf{M}_{R_O}$ is perpendicular to the plane containing the forces and is equivalent to the scalar (algebraic) sum of all the couple moments in the system *plus* the moments about point O of all the forces in the system; i.e.,

$$
\mathbf{M}_{R_O} = \Sigma M_O
\qquad (4\text{--}20)
$$

When determining the moments of the forces about O, it is generally advantageous to use the *principle of moments;* i.e., determine the moments of the *components* of each force rather than the moment of the force itself.

It is important to remember that, when applying any of these equations, attention should be paid to the sense of direction of the force components and the moments of the forces. If they are along the positive coordinate axes, they represent positive scalars; whereas if these components have a directional sense along the negative coordinate axes, they are negative scalars. By following this convention, a positive result, for example, indicates that the resultant vector has a sense of direction along the positive coordinate axis.

The following two examples illustrate these procedures numerically.

Example 4–15

Replace the forces acting on the pipe shown in Fig. 4–39a by an equivalent resultant force and couple moment acting at point A.

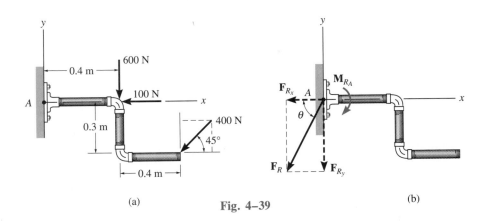

(a)

Fig. 4–39

(b)

SOLUTION (*SCALAR ANALYSIS*)

The principle of moments will be applied to the 400-N force, whereby the moments of its two rectangular components will be considered.

Force Summation.   The resultant force has x and y components of

$$\xrightarrow{+} F_{R_x} = \Sigma F_x; \quad F_{R_x} = -100 - 400 \cos 45° = -382.8 \text{ N} = 382.8 \text{ N} \leftarrow$$

$$+ \uparrow F_{R_y} = \Sigma F_y; \quad F_{R_y} = -600 - 400 \sin 45° = -882.8 \text{ N} = 882.8 \text{ N} \downarrow$$

As shown in Fig. 4–39b, $\mathbf{F}_R$ has a magnitude of

$$F_R = \sqrt{(F_{R_x})^2 + (F_{R_y})^2} = \sqrt{(382.8)^2 + (882.8)^2} = 962 \text{ N} \quad \textit{Ans.}$$

and a direction defined from the vector sketch of

$$\theta = \tan^{-1}\left(\frac{F_{R_y}}{F_{R_x}}\right) = \tan^{-1}\left(\frac{882.8}{382.8}\right) = 66.6° \quad {}_{\theta}\!\nearrow \quad \textit{Ans.}$$

Moment Summation.   The resultant couple moment $\mathbf{M}_{R_A}$ is determined by summing the moments of the forces about point A. Assuming that positive moments act counterclockwise, i.e., in the $+\mathbf{k}$ direction, we have

$$\zeta + M_{R_A} = \Sigma M_A;$$

$$M_{R_A} = 100(0) - 600(0.4) - (400 \sin 45°)(0.8) - (400 \cos 45°)(0.3)$$

$$= -551 \text{ N} \cdot \text{m} = 551 \text{ N} \cdot \text{m} \downarrow \quad \textit{Ans.}$$

In conclusion, when $\mathbf{M}_{R_A}$ and $\mathbf{F}_R$ act on the pipe at point A, Fig. 4–39b, they will produce the *same* external effect or reaction at the support A as that produced by the force system in Fig. 4–39a.

Example 4–16

(a)

A beam is subjected to a couple moment **M** and forces $\mathbf{F}_1$ and $\mathbf{F}_2$ as shown in Fig. 4–40a. Replace this system by an equivalent resultant force and couple moment acting at point O.

SOLUTION *(VECTOR ANALYSIS)*

Expressing the forces and couple moment as Cartesian vectors, we have

$$\mathbf{F}_1 = \{-800\mathbf{k}\} \text{ N}$$

$$\mathbf{F}_2 = (300 \text{ N})\mathbf{u}_{CB} = (300 \text{ N})\left(\frac{\mathbf{r}_{CB}}{r_{CB}}\right)$$

$$= 300\left[\frac{-0.15\mathbf{i} + 0.1\mathbf{j}}{\sqrt{(-0.15)^2 + (0.1)^2}}\right] = \{-249.6\mathbf{i} + 166.4\mathbf{j}\} \text{ N}$$

$$\mathbf{M} = -500(\tfrac{4}{5})\mathbf{j} + 500(\tfrac{3}{5})\mathbf{k} = \{-400\mathbf{j} + 300\mathbf{k}\} \text{ N} \cdot \text{m}$$

Force Summation

$$\mathbf{F}_R = \Sigma\mathbf{F}; \qquad \mathbf{F}_R = \mathbf{F}_1 + \mathbf{F}_2 = -800\mathbf{k} - 249.6\mathbf{i} + 166.4\mathbf{j}$$
$$= \{-249.6\mathbf{i} + 166.4\mathbf{j} - 800\mathbf{k}\} \text{ N} \qquad \textit{Ans.}$$

Moment Summation

$$\mathbf{M}_{R_O} = \Sigma\mathbf{M}_O;$$
$$\mathbf{M}_{R_O} = \mathbf{M} + \mathbf{r}_C \times \mathbf{F}_1 + \mathbf{r}_B \times \mathbf{F}_2$$

$$= (-400\mathbf{j} + 300\mathbf{k}) + (1\mathbf{k}) \times (-800\mathbf{k}) + \begin{vmatrix} \mathbf{i} & \mathbf{j} & \mathbf{k} \\ -0.15 & 0.1 & 1 \\ -249.6 & 166.4 & 0 \end{vmatrix}$$

$$= (-400\mathbf{j} + 300\mathbf{k}) + (\mathbf{0}) + (-166.4\mathbf{i} - 249.6\mathbf{j})$$

$$= \{-166.4\mathbf{i} - 649.6\mathbf{j} + 300\mathbf{k}\} \text{ N} \cdot \text{m} \qquad \textit{Ans.}$$

The results are shown in Fig. 4–40b.

(b)

Fig. 4–40

4.9 Further Reduction of a Force and Couple System

Simplification to a Single Resultant Force. Consider now a special case for which the system of forces and couple moments acting on a rigid body, Fig. 4–41a, reduces at point O to a resultant force $\mathbf{F}_R = \Sigma\mathbf{F}$ and resultant couple moment $\mathbf{M}_{R_O} = \Sigma\mathbf{M}_O$, which are *perpendicular* to one another, Fig. 4–41b. Whenever this occurs, we can further simplify the force and couple moment system by moving $\mathbf{F}_R$ to another point P, located either on or off the body so that no resultant couple moment has to be applied to the body, Fig. 4–41c. In other words, if the force and couple moment system in Fig. 4–41a is reduced to a resultant system at point P, only the force resultant will have to be applied to the body, Fig. 4–41c.

The location of point P, measured from point O, can always be determined provided $\mathbf{F}_R$ and $\mathbf{M}_{R_O}$ are known, Fig. 4–41b. As shown in Fig. 4–41c, P must lie on the bb axis, which is perpendicular to the line of action of $\mathbf{F}_R$ and the aa axis. This point is chosen such that the distance d satisfies the scalar equation $M_{R_O} = F_R d$ or $d = M_{R_O}/F_R$. With $\mathbf{F}_R$ so located, it will produce the same external effects on the body as the force and couple moment system in Fig. 4–41a, or the force and couple moment resultants in Fig. 4–41b. We refer to the force and couple moment system in Fig. 4–41a as being *equivalent or equipollent* to the single force "system" in Fig. 4–41c, because each system produces the *same* resultant force and resultant moment when replaced at point O.

If a system of forces is either concurrent, coplanar, or parallel, it can always be reduced, as in the above case, to a single resultant force $\mathbf{F}_R$ acting through a unique point P. This is because in each of these cases $\mathbf{F}_R$ and $\mathbf{M}_{R_O}$ will always be perpendicular to each other when the force system is simplified at *any* point O.

Fig. 4–41

Fig. 4–42

Concurrent force systems have been treated in detail in Chapter 2. Obviously, all the forces act at a point for which there is no resultant couple moment, so the point P is automatically specified, Fig. 4–42.

Coplanar force systems, which may include couple moments directed perpendicular to the plane of the forces, as shown in Fig. 4–43a, can be reduced to a single resultant force, because when each force in the system is moved to any point O in the x–y plane, it produces a couple moment that is *perpendicular* to the plane, i.e., in the $\pm\mathbf{k}$ direction. The resultant moment $\mathbf{M}_{R_O} = \Sigma\mathbf{M} + \Sigma\mathbf{r} \times \mathbf{F}$ is thus perpendicular to the resultant force $\mathbf{F}_R$, Fig. 4–43b.

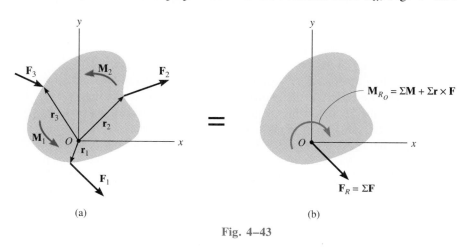

(a)　　　　　　　　　(b)

Fig. 4–43

Parallel force systems, which can include couple moments which are perpendicular to the forces, as shown in Fig. 4–44a, can be reduced to a single resultant force, because when each force is moved to any point O in the x–y plane, it produces a couple moment that has components only about the x and y axes. The resultant moment $\mathbf{M}_{R_O} = \Sigma\mathbf{M} + \Sigma\mathbf{r} \times \mathbf{F}$ is thus perpendicular to the resultant force $\mathbf{F}_R$, Fig. 4–44b.

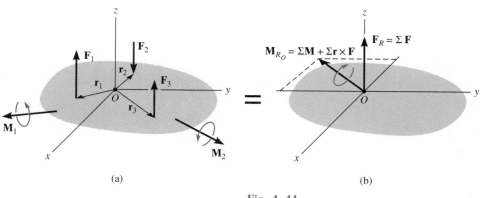

(a)　　　　　　　　　(b)

Fig. 4–44

PROCEDURE FOR ANALYSIS

The technique used to reduce a coplanar or parallel force system to a single resultant force follows the general procedure outlined in the previous section. First establish an x, y, z coordinate system. Then simplification requires the following two steps:

Force Summation.   The resultant force $\mathbf{F}_R$ equals the sum of all the forces of the system, Fig. 4–41a and 4–41c; i.e.,

$$\boxed{\mathbf{F}_R = \Sigma\mathbf{F}}$$

Moment Summation.   The distance d from the arbitrary point O to the line of action of $\mathbf{F}_R$ is determined by equating the moment of $\mathbf{F}_R$ about O, $\mathbf{M}_{R_O}$, Fig. 4–41c, to the sum of the moments about point O of all the couple moments and forces in the system, $\Sigma\mathbf{M}_O$, Fig. 4–41a; i.e.,

$$\boxed{\mathbf{M}_{R_O} = \Sigma\mathbf{M}_O}$$

Most often a scalar analysis can be used to apply these equations, since the force components and the moment arms are easily determined for either coplanar or parallel force systems.

Reduction to a Wrench.

In the general case, the force and couple moment system acting on a body, Fig. 4–38a, will reduce to a single resultant force $\mathbf{F}_R$ and couple moment $\mathbf{M}_{R_O}$ at O which are *not* perpendicular. Instead, $\mathbf{F}_R$ will act at an angle θ from $\mathbf{M}_{R_O}$, Fig. 4–38c. As shown in Fig. 4–45a, however, $\mathbf{M}_{R_O}$ may be resolved into two components: one perpendicular, $\mathbf{M}_\perp$, and the other parallel, $\mathbf{M}_\parallel$, to the line of action of $\mathbf{F}_R$. As in the previous discussion, the perpendicular component $\mathbf{M}_\perp$ may be *eliminated* by moving $\mathbf{F}_R$ to point P, as shown in Fig. 4–45b. This point lies on axis bb, which is perpendicular to both $\mathbf{M}_{R_O}$ and $\mathbf{F}_R$. In order to maintain an equivalency of loading, the distance from O to P is $d = M_\perp / F_R$. Furthermore, when $\mathbf{F}_R$ is applied at P, the moment of $\mathbf{F}_R$ tending to cause rotation of the body *about O* is in the *same direction* as $\mathbf{M}_\perp$, Fig. 4–45a. Finally, since $\mathbf{M}_\parallel$ is a free vector, it may be moved to P so that it is collinear with $\mathbf{F}_R$, Fig. 4–45c. This combination of a collinear force and couple moment is called a *wrench* or *screw*. The *axis of the wrench* has the same line of action as the force. Hence, the wrench tends to cause both a translation along and a rotation about this axis. Comparing Fig. 4–45a to Fig. 4–45c, it is seen that a general force and couple moment system acting on a body can be reduced to a wrench. The axis of the wrench and a point through which this axis passes are unique and can always be determined.

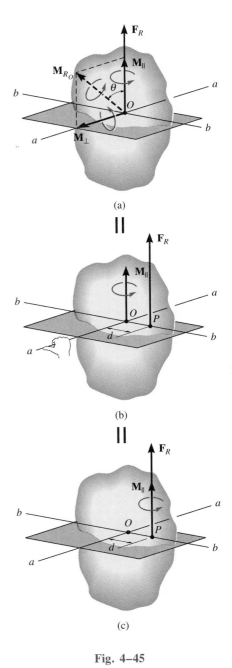

(a)

(b)

(c)

Fig. 4–45

Example 4–17

Replace the system of forces acting on the beam shown in Fig. 4–46a by an equivalent resultant force. Specify the distance the force acts from point A.

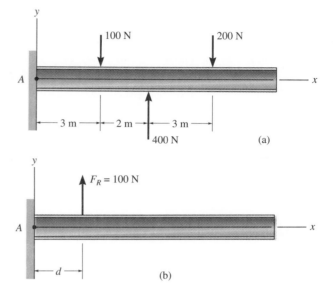

Fig. 4–46

SOLUTION

Here the force system is parallel and coplanar. We will use the established x, y, z axes to solve this problem.

Force Summation.   From Fig. 4–46a the force resultant $\mathbf{F}_R$ is

$$+\uparrow F_R = \Sigma F; \; F_R = -100 \text{ N} + 400 \text{ N} - 200 \text{ N} = 100 \text{ N} \uparrow \qquad Ans.$$

Moment Summation.   Moments will be summed about point A. Considering counterclockwise rotations as positive, i.e., positive moment vectors act in the $+\mathbf{k}$ direction, then from Figs. 4–46a and 4–46b we require the moment of $\mathbf{F}_R$ about A to equal the moments of the force system about A; i.e.,

$$\zeta + M_{R_A} = \Sigma M_A;$$
$$100 \text{ N}(d) = -(100 \text{ N})(3 \text{ m}) + (400 \text{ N})(5 \text{ m}) - (200 \text{ N})(8 \text{ m})$$
$$(100)(d) = 100$$
$$d = 1 \text{ m} \qquad Ans.$$

Note that using a clockwise sign convention would yield the same result. Since d is *positive*, $\mathbf{F}_R$ acts to the right of A as shown in Fig. 4–46b.

Example 4–18

The beam AE in Fig. 4–47a is subjected to a system of coplanar forces. Determine the magnitude, direction, and location on the beam of a resultant force which is equivalent to the given system of forces.

(a)

(b)

Fig. 4–47

SOLUTION

The origin of coordinates is arbitrarily located at point E as shown in Fig. 4–47a.

Force Summation.   Resolving the 500-N force into x and y components, and summing the force components, yields

$$\xrightarrow{+}F_{R_x} = \Sigma F_x; \qquad F_{R_x} = 500 \cos 60° + 100 = 350.0 \text{ N} \rightarrow$$
$$+\uparrow F_{R_y} = \Sigma F_y; \qquad F_{R_y} = -500 \sin 60° + 200 = -233.0 \text{ N}$$
$$= 233.0 \text{ N} \downarrow$$

The magnitude and direction of the resultant force are established from the vector addition shown in Fig. 4–47b. We have

$$F_R = \sqrt{(350.0)^2 + (233.0)^2} = 420.5 \text{ N} \qquad\qquad Ans.$$

$$\theta = \tan^{-1}\left(\frac{233.0}{350.0}\right) = 33.7° \quad \searrow_\theta \qquad\qquad Ans.$$

Moment Summation.   Moments will be summed about point E. Hence, from Figs. 4–47a and 4–47b, we require the moments of the components of $\mathbf{F}_R$ (or the moment of $\mathbf{F}_R$) about point E to equal the moments of the force system about E; i.e.,

$$\zeta + M_{R_E} = \Sigma M_E;$$
$$233.0(d) + 350.0(0) = (500 \sin 60°)(4) + (500 \cos 60°)(0) - (100)(0.5) - (200)(2.5)$$

$$d = \frac{1182.1}{233.0} = 5.07 \text{ m} \qquad\qquad Ans.$$

Example 4–19

(a)

(b)

Fig. 4–48

The frame shown in Fig. 4–48a is subjected to three coplanar forces. Replace this loading by an equivalent resultant force and specify where the resultant's line of action intersects members AB and BC.

SOLUTION

Force Summation. Resolving the 75-lb force into x and y components and summing the force components yields

$$\xrightarrow{+}F_{R_x} = \Sigma F_x; \quad F_{R_x} = -75(\tfrac{3}{5}) - 15 = -60 \text{ lb} = 60 \text{ lb} \leftarrow$$
$$+\uparrow F_{R_y} = \Sigma F_y; \; F_{R_y} = -75(\tfrac{4}{5}) - 90 = -150 \text{ lb} = 150 \text{ lb} \downarrow$$

As shown by the vector addition in Fig. 4–48b,

$$F_R = \sqrt{(60)^2 + (150)^2} = 161.6 \text{ lb} \qquad Ans.$$

$$\theta = \tan^{-1}\left(\frac{150}{60}\right) = 68.2° \quad {}^{\theta}\!\!\nearrow \qquad Ans.$$

Moment Summation. Moments will be summed about point A. If the line of action of $\mathbf{F}_R$ intersects AB, Fig. 4–48b, we require the moment of the components of $\mathbf{F}_R$ in Fig. 4–48b about A to equal the moments of the force system in Fig. 4–48a about A; i.e.,

$$\zeta + M_{R_A} = \Sigma M_A; \; 0(150) + y(60) = 4(15) + 0(90) + 7(75)(\tfrac{3}{5}) - 2(75)(\tfrac{4}{5})$$
$$y = 4.25 \text{ ft} \qquad Ans.$$

By the principle of transmissibility, $\mathbf{F}_R$ can also intersect BC, Fig. 4–48b, in which case we have

$$\zeta + M_{R_A} = \Sigma M_A; \; 7(60) - x(150) = 4(15) + 0(90) + 7(75)(\tfrac{3}{5}) - 2(75)(\tfrac{4}{5})$$
$$x = 1.10 \text{ ft} \qquad Ans.$$

We can also solve for these positions by assuming $\mathbf{F}_R$ acts at the arbitrary point (x, y) on its line of action, Fig. 4–48b. Summing moments about point A yields

$$\zeta + M_{R_A} = \Sigma M_A; \; y(60) - x(150) = 4(15) + 0(90) + 7(75)(\tfrac{3}{5}) - 2(75)(\tfrac{4}{5})$$
$$60y - 150x = 255$$

which is the equation of the colored dashed line in Fig. 4–48b. To find the points of intersection with the frame, set $x = 0$, then $y = 4.25$ ft, and set $y = 7$ ft, then $x = 1.10$ ft.

Example 4–20

The slab in Fig. 4–49a is subjected to four parallel forces. Determine the magnitude and direction of a resultant force equivalent to the given force system, and locate its point of application on the slab.

(a)   Fig. 4–49   (b)

SOLUTION (*SCALAR ANALYSIS*)

Force Summation.   From Fig. 4–49a, the resultant force is

$$+\uparrow F_R = \Sigma F; \qquad F_R = -600 + 100 - 400 - 500$$
$$= -1400 \text{ N} = 1400 \text{ N} \downarrow \qquad\qquad Ans.$$

Moment Summation.   We require the moment about the x axis of the resultant force, Fig. 4–49b, to be equal to the sum of the moments about the x axis of all the forces in the system, Fig. 4–49a. The moment arms are determined from the y coordinates since these coordinates represent the *perpendicular distances* from the x axis to the lines of action of the forces. Using the right-hand rule, where positive moments act in the $+\mathbf{i}$ direction, we have

$$M_{R_x} = \Sigma M_x;$$
$$-1400y = 600(0) + 100(5) - 400(10) + 500(0)$$
$$-1400y = -3500 \qquad y = 2.50 \text{ m} \qquad\qquad Ans.$$

In a similar manner, assuming that positive moments act in the $+\mathbf{j}$ direction, a moment equation can be written about the y axis using moment arms defined by the x coordinates of each force.

$$M_{R_y} = \Sigma M_y;$$
$$1400x = 600(8) - 100(6) + 400(0) + 500(0)$$
$$1400x = 4200 \qquad x = 3.00 \text{ m} \qquad\qquad Ans.$$

Hence, a force of $F_R = 1400$ N placed at point $P(3.00 \text{ m}, 2.50 \text{ m})$ on the slab, Fig. 4–49b, is equivalent to the parallel force system acting on the slab in Fig. 4–49a.

■ Example 4–21

Three parallel forces act on the rim of the circular plate in Fig. 4–50a. Determine the magnitude and direction of a resultant force equivalent to the given force system and locate its point of application, P, on the plate.

(a)

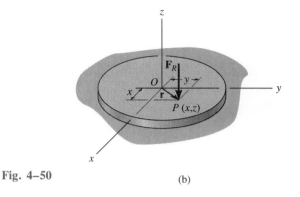

(b)

Fig. 4–50

SOLUTION (*VECTOR ANALYSIS*)

Force Summation. From Fig. 4–50a, the force resultant $\mathbf{F}_R$ is

$$\mathbf{F}_R = \Sigma\mathbf{F}; \qquad \mathbf{F}_R = -300\mathbf{k} - 200\mathbf{k} - 150\mathbf{k}$$
$$= \{-650\mathbf{k}\} \text{ lb} \qquad\qquad \textit{Ans.}$$

Moment Summation. Choosing point O as a reference for computing moments and assuming that $\mathbf{F}_R$ acts at a point $P(x, y)$, Fig. 4–50b, we require

$$\mathbf{M}_{R_O} = \Sigma\mathbf{M}_O; \quad \mathbf{r} \times \mathbf{F}_R = \mathbf{r}_A \times (-300\mathbf{k}) + \mathbf{r}_B \times (-200\mathbf{k}) + \mathbf{r}_C \times (-150\mathbf{k})$$
$$(x\mathbf{i} + y\mathbf{j}) \times (-650\mathbf{k}) = (8\mathbf{i}) \times (-300\mathbf{k}) + (-8\mathbf{j}) \times (-200\mathbf{k})$$
$$+ (-8 \sin 45°\mathbf{i} + 8 \cos 45°\mathbf{j}) \times (-150\mathbf{k})$$
$$650x\mathbf{j} - 650y\mathbf{i} = 2400\mathbf{j} + 1600\mathbf{i} - 848.5\mathbf{j} - 848.5\mathbf{i}$$

Equating the corresponding $\mathbf{j}$ and $\mathbf{i}$ components yields

$$650x = 2400 - 848.5 \qquad\qquad (1)$$
$$-650y = 1600 - 848.5 \qquad\qquad (2)$$

Solving these equations, we obtain the coordinates of point P,

$$x = 2.39 \text{ ft} \qquad y = -1.16 \text{ ft} \qquad\qquad \textit{Ans.}$$

The negative sign indicates that it was wrong to have assumed a $+y$ position for $\mathbf{F}_R$ as shown in Fig. 4–50b.

As a review, try to establish Eqs. 1 and 2 by using a scalar analysis; i.e., apply the sum of moments about the x and y axes, respectively.

PROBLEMS

4–82. Replace the force at *A* by an equivalent force and couple moment at *P*.

*4–84. Replace the force and couple moment system by an equivalent force and couple moment acting at point *P*.

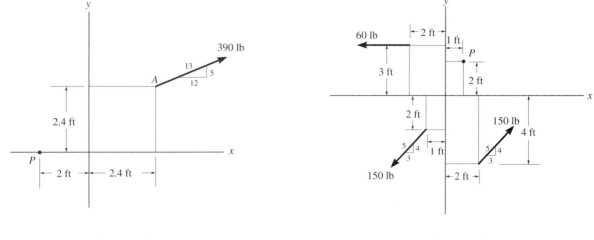

Prob. 4–82

Prob. 4–84

4–83. Replace the force at *A* by an equivalent force and couple moment at *P*.

4–85. Replace the force and couple moment system by an equivalent force and couple moment acting at point *P*.

Prob. 4–83

Prob. 4–85

4–86.   Replace the force and couple moment system by an equivalent force and couple moment acting at point O.

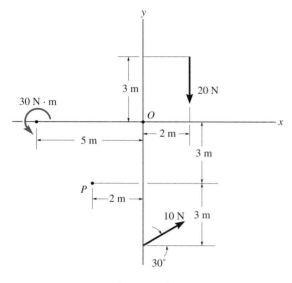

Prob. 4–86

4–88.   Replace the force system by an equivalent resultant force and specify its point of application measured along the x axis from point O.

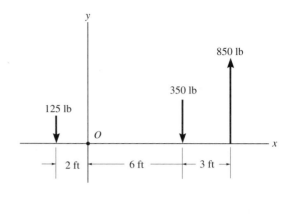

Prob. 4–88

4–89.   Replace the coplanar force system by an equivalent resultant force and specify where its line of action intersects the x axis.

4–90.   Replace the coplanar force system by an equivalent resultant force and specify where its line of action intersects the y axis.

4–87.   Replace the force and couple moment system by an equivalent force and couple moment acting at point P.

Prob. 4–87

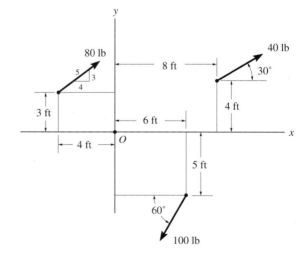

Probs. 4–89/4–90

4–91. Replace the loading system acting on the beam by an equivalent resultant force and specify its point of application on the beam measured from point O.

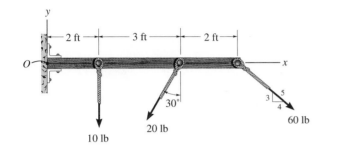

Prob. 4–91

4–93. Determine the magnitude of the vertical force **F** acting tangent to the gear so that **F** and **M** can be replaced by an equivalent resultant force at point A.

Prob. 4–93

*4–92. The gear is subjected to the two forces shown. Replace these forces by an equivalent resultant force and couple moment acting at point O.

Prob. 4–92

4–94. Replace the two forces by an equivalent resultant force and couple moment at point O. Set $F = 20$ lb.

4–95. Replace the two forces by an equivalent resultant force and couple moment at point O. Set $F = 15$ lb.

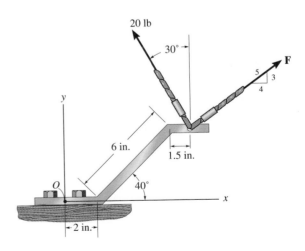

Probs. 4–94/4–95

*4–96.   Replace the force system acting on the beam by an equivalent resultant force and couple moment at point A.

4–97.   Replace the force system acting on the beam by an equivalent resultant force and couple moment at point O.

*4–100.   Determine the magnitudes of F_1 and F_2 and the angle of F_1 so that the loading creates a zero resultant force and couple moment on the handle.

Probs. 4–96/4–97

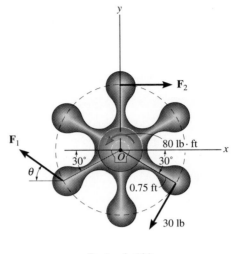

Prob. 4–100

4–98.   Replace the force system acting on the beam by an equivalent resultant force and couple moment at point B.

4–99.   Replace the force system acting on the beam by an equivalent resultant force and couple moment at point A.

4–101.   Replace the force system acting on the frame by an equivalent resultant force, and specify where the resultant's line of action intersects member AB measured from point A. Neglect the thickness of the members.

4–102.   Replace the force system acting on the frame by an equivalent resultant force and couple moment acting at point C. Neglect the thickness of the members.

Probs. 4–98/4–99

Probs. 4–101/4–102

4–103. Determine the magnitude and orientation θ of force **F** and its placement d on the beam so that the loading system is equivalent to a resultant force of 15 kN acting vertically downward at point O and a clockwise couple moment of 60 kN · m.

*4–104. Determine the magnitude and orientation θ of force **F** and its placement d on the beam so that the loading system is equivalent to a resultant force of 20 kN acting vertically downward at point O, and a clockwise couple moment of 80 kN · m.

4–107. Replace the force system acting on the frame by an equivalent resultant force and couple moment acting at point A.

*4–108. Replace the force system acting on the frame by an equivalent resultant force and specify where the resultant's line of action intersects member AB measured from point A.

*4–109. Replace the force system acting on the frame by an equivalent resultant force and specify where the resultant's line of action intersects member BC measured from point B.

Probs. 4–107/4–108/4–109

Probs. 4–103/4–104

4–105. The frame is subjected to the coplanar system of loads. Replace this system by an equivalent resultant force and couple moment acting at point A.

4–106. The frame is subjected to the coplanar system of loads. Replace this system by an equivalent resultant force and couple moment acting at point B.

4–110. The system of parallel forces acts on the top cord of the truss. Determine the equivalent resultant force of the system and specify its location measured from point A.

Probs. 4–105/4–106

Prob. 4–110

4–111. Replace the force at *A* by an equivalent resultant force and couple moment at point *O*. Express the results in Cartesian vector form.

*4–112. Replace the force at *A* by an equivalent resultant force and couple moment at point *P*. Express the results in Cartesian vector form.

Probs. 4–111/4–112

4–115. Replace the force at *A* by an equivalent resultant force and couple moment at point *P*. Express the results in Cartesian vector form.

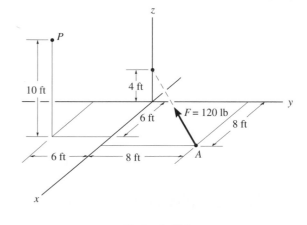

Prob. 4–115

4–113. Replace the force and couple moment system by an equivalent resultant force and couple moment at point *O*. Express the results in Cartesian vector form.

4–114. Replace the force and couple moment system by an equivalent resultant force and couple moment at point *P*. Express the results in Cartesian vector form.

*4–116. Replace the force **F** having a magnitude of *F* = 70 lb and acting at point *A* by an equivalent resultant force and couple moment at point *O*.

Probs. 4–113/4–114

Prob. 4–116

4–117. The tube supports the four parallel forces. Determine the magnitudes of forces $\mathbf{F}_C$ and $\mathbf{F}_D$ acting at C and D so that the equivalent resultant force of the system acts through the midpoint O of the tube.

4–119. A force and couple act on the pipe assembly. Replace this system by an equivalent resultant force. Specify the location of the resultant force along the x axis, measured from A. The pipe lies in the x–y plane and the three forces are vertical.

Prob. 4–119

Prob. 4–117

4–118. The building slab is subjected to four parallel column loadings. Determine the equivalent resultant force and specify its location (x, y) on the slab.

*4–120. The crate is to be hoisted using the three slings shown. Replace the system of forces acting on the slings by an equivalent resultant force and couple moment at point O. The force $\mathbf{F}_1$ is vertical.

Prob. 4–118

Prob. 4–120

4–121.   Replace the three forces acting on the plate by a wrench. Specify the magnitude of the force and couple moment for the wrench and the point $P(y, z)$ where its line of action intersects the plate.

4–123.   Replace the two wrenches and the force, acting on the pipe assembly, by an equivalent resultant force and couple moment at point O.

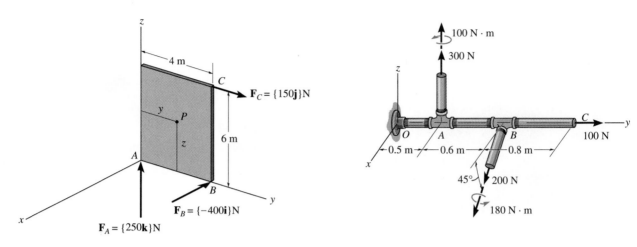

Prob. 4–121

Prob. 4–123

4–122.   Replace the three forces acting on the plate by a wrench. Specify the force and couple moment for the wrench and the point $P(x, y)$ where its line of action intersects the plate.

*4–124.   The three forces acting on the block each have a magnitude of 10 lb. Replace this system by a wrench and specify the point where the wrench intersects the z axis measured from point O.

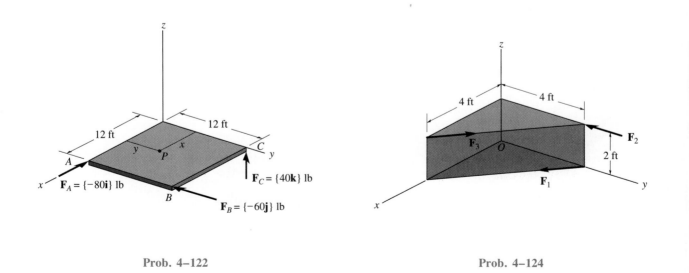

Prob. 4–122

Prob. 4–124

4.10 Reduction of a Simple Distributed Loading

In many situations a very large surface area of a body may be subjected to *distributed loadings* such as those caused by wind, fluids, or simply the weight of material supported over the body's surface. The *intensity* of these loadings at each point on the surface is defined as the *pressure p* (force per unit area), which can be measured in units of lb/ft^2 or pascals (Pa), where 1 Pa = 1 N/m^2.

   In this section we will consider the most common case of a distributed pressure loading, which is *uniform* along one axis of a flat rectangular body upon which the loading is applied.* An example of such a loading is shown in Fig. 4–51a. The direction of the intensity of the pressure load is indicated by arrows shown on the *load-intensity diagram*. The entire loading on the plate is therefore a system of parallel forces, infinite in number and each acting on a separate differential area of the plate. Here the *loading function, p = p(x)* Pa, is only a function of x since the pressure is uniform along the y axis. If we multiply $p = p(x)$ by the *width a* m of the plate, we obtain $w = [p(x) \text{ N/m}^2]a$ m $= w(x)$ N/m. This loading function, shown in Fig. 4–51b, is a measure of load distribution along the line $y = 0$ which is in the plane of symmetry of the loading, Fig. 4–51a. As noted, it is measured as a force per unit length, rather than a force per unit area. Consequently, the load-intensity diagram for $w = w(x)$ can be represented by a system of *coplanar* parallel forces, shown in two dimensions in Fig. 4–51b. Using the methods of Sec. 4.9, this system of forces can be simplified to a single resultant force $\mathbf{F}_R$ and its location $\bar{x}$ specified, Fig. 4–51c.

| (a) | (b) | (c) |

*The more general case of a nonuniform surface loading acting on a body is considered in Sec. 9.5.

Fig. 4–51

Fig. 4–51

Magnitude of Resultant Force. From Eq. 4–19 ($F_R = \Sigma F$), the magnitude of $\mathbf{F}_R$ is equivalent to the sum of all the forces in the system. In this case integration must be used, since there is an infinite number of parallel forces $d\mathbf{F}$ acting along the plate, Fig. 4–51b. Since $d\mathbf{F}$ is acting on an element of length dx, and $w(x)$ is a force per unit length, then at the location x, $dF = w(x)\ dx = dA$. In other words, the magnitude of $d\mathbf{F}$ is determined from the colored differential *area dA* under the loading curve. For the entire plate length,

$$+\downarrow F_R = \Sigma F; \qquad \boxed{F_R = \int_L w(x)\ dx = \int_A dA = A} \qquad (4\text{--}21)$$

Hence, the magnitude of the resultant force is equal to the total area under the loading diagram $w = w(x)$.

Location of Resultant Force. Applying Eq. 4–20 ($M_{R_O} = \Sigma M_O$), the location $\bar{x}$ of the line of action of $\mathbf{F}_R$ can be determined by equating the moments of the force resultant and the force distribution about point O (the y axis). Since $d\mathbf{F}$ produces a moment of $x\ dF = x\ w(x)\ dx$ about O, Fig. 4–51b, then for the entire plate, Fig. 4–51c,

$$\gamma + M_{R_O} = \Sigma M_O; \qquad \bar{x}F_R = \int_L x\ w(x)\ dx$$

Solving for $\bar{x}$, using Eq. 4–21, we can write

$$\boxed{\bar{x} = \frac{\displaystyle\int_L x\ w(x)\ dx}{\displaystyle\int_L w(x)\ dx} = \frac{\displaystyle\int_A x\ dA}{\displaystyle\int_A dA}} \qquad (4\text{--}22)$$

This equation represents the x coordinate for the geometric center or *centroid* of the *area* under the distributed-loading diagram $w(x)$. *Therefore, the resultant force has a line of action which passes through the centroid C (geometric center) of the area defined by the distributed-loading diagram* $w(x)$, Fig. 4–51c.

Once $\bar{x}$ is determined, $\mathbf{F}_R$ by symmetry passes through point ($\bar{x}$, 0) on the surface of the plate, Fig. 4–51d. If we now consider the three-dimensional pressure loading $p(x)$, Fig. 4–51a, we can therefore conclude that *the resultant force has a magnitude equal to the volume under the distributed-loading curve p = p(x) and a line of action which passes through the centroid (geometric center) of this volume.* Detailed treatment of the integration techniques for computing the centroids of volumes or areas is given in Chapter 9. In many cases, however, the distributed-loading diagram is in the shape of a rectangle, triangle, or other simple geometric form. The centroids for such common shapes do not have to be determined from Eq. 4–22; rather, they can be obtained directly from the tabulation given on the inside back cover.

Example 4–22

In each case, determine the magnitude and location of the resultant of the distributed load acting on the beams in Fig. 4–52.

(a)

(c)

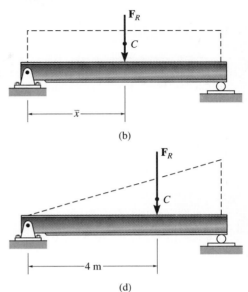

(b)

(d)

Fig. 4–52

SOLUTION

Uniform Loading.   As indicated $w = 400$ lb/ft, which is constant over the entire beam, Fig. 4–52a. This loading forms a rectangle, the area of which is equal to the resultant force, Fig. 4–52b; i.e.,

$$F_R = (400 \text{ lb/ft})(10 \text{ ft}) = 4000 \text{ lb} \qquad \textit{Ans.}$$

The location of $\mathbf{F}_R$ passes through the geometric center or centroid C of this rectangular area, so

$$\bar{x} = 5 \text{ ft} \qquad \textit{Ans.}$$

Triangular Loading.   Here the loading varies uniformly in intensity from 0 to 600 N/m, Fig. 4–52c. These values can be verified by substitution of $x = 0$ and $x = 6$ m into the loading function $w = 100x$ N/m. The area of this triangular loading is equal to $\mathbf{F}_R$, Fig. 4–52d. From the table on the inside back cover, $A = \frac{1}{2}bh$, so that

$$F_R = \tfrac{1}{2}(6 \text{ m})(600 \text{ N/m}) = 1800 \text{ N} \qquad \textit{Ans.}$$

The line of action of $\mathbf{F}_R$ passes through the centroid C of the triangle. Using the table on the inside back cover, this point lies at a distance of one third the length of the beam, measured from the right side. Hence,

$$\bar{x} = 6 \text{ m} - \tfrac{1}{3}(6 \text{ m}) = 4 \text{ m} \qquad \textit{Ans.}$$

Example 4–23

Determine the magnitude and location of the resultant of the distributed load acting on the beam shown in Fig. 4–53a.

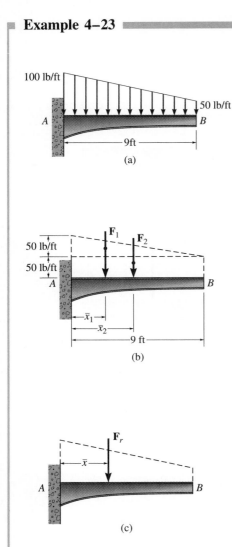

SOLUTION

The area of the loading diagram is a *trapezoid*, and therefore the solution can be obtained directly from the area and centroid formulas for a trapezoid listed on the inside back cover. Since these formulas are not easily remembered, instead we will solve this problem by using "composite" areas. In this regard, we can divide the trapezoidal loading into a rectangular and triangular loading as shown in Fig. 4–53b. The magnitude of the force represented by each of these loadings is equal to its associated *area,*

$$F_1 = \tfrac{1}{2}(9 \text{ ft})(50 \text{ lb/ft}) = 225 \text{ lb}$$
$$F_2 = (9 \text{ ft})(50 \text{ lb/ft}) = 450 \text{ lb}$$

The lines of action of these parallel forces act through the *centroid* of their associated areas and therefore intersect the beam at

$$\bar{x}_1 = \tfrac{1}{3}(9 \text{ ft}) = 3 \text{ ft}$$
$$\bar{x}_2 = \tfrac{1}{2}(9 \text{ ft}) = 4.5 \text{ ft}$$

The two parallel forces F_1 and F_2 can be reduced to a single resultant F_R. The magnitude of F_R is

$$+\downarrow F_R = \Sigma F; \qquad F_R = 225 + 450 = 675 \text{ lb} \qquad \textit{Ans.}$$

With reference to point A, Fig. 4–53b and c, we can define the location of F_R. We require

$$\curvearrowright + M_{R_A} = \Sigma M_A; \qquad \bar{x}(675) = 3(225) + 4.5(450)$$
$$\bar{x} = 4 \text{ ft} \qquad \textit{Ans.}$$

Note: The trapezoidal area in Fig. 4–53a can also be divided into two triangular areas as shown in Fig. 4–53d. In this case

$$F_1 = \tfrac{1}{2}(9 \text{ ft})(100 \text{ lb/ft}) = 450 \text{ lb}$$
$$F_2 = \tfrac{1}{2}(9 \text{ ft})(50 \text{ lb/ft}) = 225 \text{ lb}$$

and

$$\bar{x}_1 = \tfrac{1}{3}(9 \text{ ft}) = 3 \text{ ft}$$
$$\bar{x}_2 = \tfrac{1}{3}(9 \text{ ft}) = 3 \text{ ft}$$

Fig. 4–53

Using these results, show that again $F_R = 675$ lb and $\bar{x} = 4$ ft.

Example 4–24

Determine the magnitude and location of the resultant force acting on the beam in Fig. 4–54a.

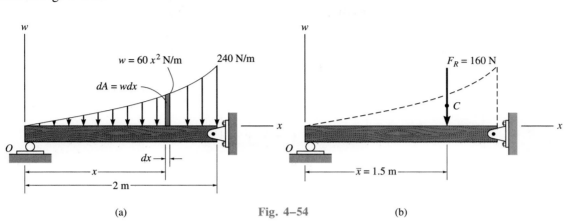

(a)   Fig. 4–54   (b)

SOLUTION

Since $w = w(x)$ is given, this problem will be solved by integration. The colored differential area element $dA = w\,dx = 60x^2\,dx$. Applying Eq. 4–21, by summing these elements from $x = 0$ to $x = 2$ m, we obtain the resultant force $\mathbf{F}_R$,

$F_R = \Sigma F$;

$$F_R = \int_A dA = \int_0^2 60x^2\,dx = 60\left[\frac{x^3}{3}\right]_0^2 = 60\left[\frac{2^3}{3} - \frac{0^3}{3}\right]$$

$$= 160\text{ N} \qquad\qquad Ans.$$

Since the element of area dA is located an arbitrary distance x from O, the location $\bar{x}$ of $\mathbf{F}_R$ *measured from* O, Fig. 4–54b, is determined from Eq. 4–22.

$$\bar{x} = \frac{\displaystyle\int_A x\,dA}{\displaystyle\int_A dA} = \frac{\displaystyle\int_0^2 x(60x^2)\,dx}{160} = \frac{60\left[\dfrac{x^4}{4}\right]_0^2}{160} = \frac{60\left[\dfrac{2^4}{4} - \dfrac{0^4}{4}\right]}{160}$$

$$= 1.5\text{ m} \qquad\qquad Ans.$$

These results may be checked by using the table on the inside back cover, where it is shown that for an exparabolic area of length a, height b, and shape shown in Fig. 4–54a,

$$A = \frac{ab}{3} = \frac{2(240)}{3} = 160\text{ N} \quad\text{and}\quad \bar{x} = \frac{3}{4}a = \frac{3}{4}(2) = 1.5\text{ m}$$

Example 4–25

A distributed loading of $p = 800x$ Pa acts over the top surface of the beam shown in Fig. 4–55a. Determine the magnitude and location of the resultant force.

(a)   (b)   (c)

Fig. 4–55

SOLUTION

The loading function $p = 800x$ Pa indicates that the load intensity varies uniformly from $p = 0$ at $x = 0$ to $p = 7200$ Pa at $x = 9$ m. Since the intensity is uniform along the width of the beam (the y axis), the loading may be viewed in two dimensions as shown in Fig. 4–55b. Here

$$w = (800x \text{ N/m}^2)(0.2 \text{ m})$$
$$= (160x) \text{ N/m}$$

At $x = 9$ m, note that $w = 1440$ N/m. Although we may again apply Eqs. 4–21 and 4–22 as in Example 4–24, it is simpler to use the table on the inside back cover.

The magnitude of the resultant force is

$$F_R = \tfrac{1}{2}(9 \text{ m})(1440 \text{ N/m}) = 6480 \text{ N} = 6.48 \text{ kN} \qquad \textit{Ans.}$$

The line of action of $\mathbf{F}_R$ passes through the *centroid C* of the triangle. Hence,

$$\bar{x} = 9 \text{ m} - \tfrac{1}{3}(9 \text{ m}) = 6 \text{ m} \qquad \textit{Ans.}$$

The results are shown in Fig. 4–55c.

We may also view the resultant $\mathbf{F}_R$ as *acting* through the *centroid* of the *volume* of the loading diagram $p = p(x)$ in Fig. 4–55a. Hence $\mathbf{F}_R$ intersects the x–y plane at the point (6 m, 0). Furthermore, the *magnitude* of $\mathbf{F}_R$ is equal to the *volume* under the loading diagram; i.e.,

$$F_R = V = \tfrac{1}{2}(7200 \text{ N/m}^2)(9 \text{ m})(0.2 \text{ m}) = 6.48 \text{ kN} \qquad \textit{Ans.}$$

PROBLEMS

4–125. Determine the resultant moment of both the 100-lb force and the triangular distributed load about point O.

Prob. 4–125

4–126. Replace the loading by an equivalent resultant force and specify the location of the force on the beam, measured from point B.

Prob. 4–126

4–127. The loading on the book shelf is distributed as shown. Determine the magnitude of the equivalent resultant force and specify its location measured from point O.

*4–128. The loading on the book shelf is distributed as shown. Determine the magnitude of the equivalent resultant force and specify its location measured from point A.

Probs. 4–127/4–128

4–129. Determine the magnitude of the equivalent resultant force of the distributed loading and specify its location on the beam, measured from point A.

Prob. 4–129

4–130. The column is used to support the floor which exerts a force of 3000 lb on the top of the column. The effect of soil pressure along the side of the column is distributed as shown. Replace this loading by an equivalent resultant force and specify where it acts along the column measured from its base A.

Prob. 4–130

*4–132. Replace the loading by an equivalent resultant force and couple moment acting at point O.

Prob. 4–132

4–131. The distribution of soil loading on the bottom of a building slab is shown. Replace this loading by an equivalent resultant force and specify its location measured from point O.

4–133. Replace the distributed loading by an equivalent resultant force, and specify its location on the beam measured from the pin at C.

Prob. 4–131

Prob. 4–133

4–134. Replace the loading by an equivalent resultant force, and specify its location measured from point A.

Prob. 4–134

4–135. The beam is subjected to the distributed loading. Determine the length b of the uniform load and its position a on the beam such that the resultant force and couple moment acting on the beam are zero.

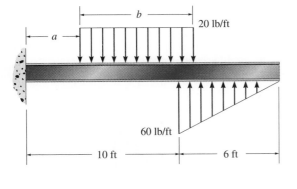

Prob. 4–135

*4–136. Determine the length b of the triangular load and its position a on the beam such that the resultant force is zero and the resultant couple moment acting on the beam is 8 kN · m clockwise.

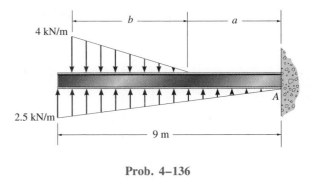

Prob. 4–136

4–137. Replace the loading by an equivalent resultant force and specify its location on the beam, measured from point O.

Prob. 4–137

4–138. Replace the loading by an equivalent resultant force and couple moment acting on the beam at point O.

Prob. 4–138

4–139. Wet concrete exerts a pressure distribution along the wall of the form. Determine the resultant force of this distribution and specify the height h where the bracing strut should be placed so that it lies through the line of action of the resultant force. The wall has a width of 5 m.

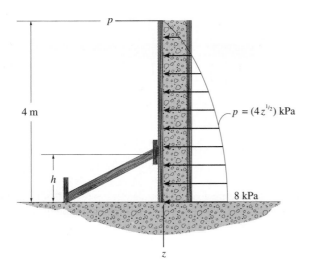

$$p = (4z^{1/2})\ kPa$$

Prob. 4–139

*4–140. Determine the resultant force of the loading acting on the beam.

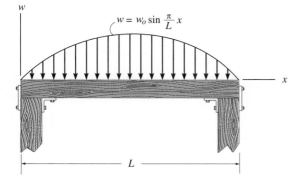

$$w = w_0 \sin \frac{\pi}{L} x$$

Prob. 4–140

4–141. Determine the magnitude of the resultant force of the loading acting on the beam and specify where it acts from point O.

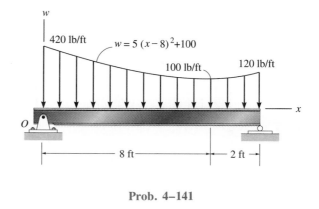

$$w = 5\,(x-8)^2 + 100$$

Prob. 4–141

4–142. Determine the resultant force of the loading acting on the beam, and specify its location measured from end A.

$$w = 10\sqrt{5x - x^2}\ \text{lb/ft}$$

Prob. 4–142

4–143. The wing of the jet aircraft is subjected to the distributed loading shown. Determine the resultant force of this loading and specify its location measured from A.

$$w = 0.5\sqrt{10x + (16 + x^2)^{1/2}}$$

Prob. 4–143

REVIEW PROBLEMS

*4–144. Determine the moment about point A of each of the three forces acting on the beam. What is the resultant moment of all the forces about point A?

4–145. Determine the moment about point B of each of the three forces acting on the beam. What is the resultant moment of all the forces about point B?

*4–148. The main beam along the wing of an airplane is swept back at an angle of 25°. From load calculations it is determined that the beam is subjected to couple moments $M_x = 17$ kip · ft and $M_y = 25$ kip · ft. Determine the resultant couple moments created about the x' and y' axes. The axes all lie in the same horizontal plane.

Probs. 4–144/4–145

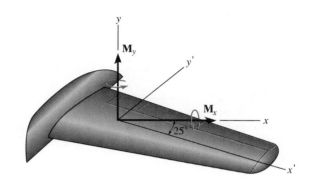

Prob. 4–148

4–146. Determine the moment of the force $\mathbf{F}$ which is directed along ED of the block about the axis BA. Solve the problem by using two different position vectors $\mathbf{r}$. Express the result as a Cartesian vector.

4–147. Determine the moment of the force $\mathbf{F}$ which is directed along ED of the block about the axis OC. Solve the problem by using two different position vectors $\mathbf{r}$. Express the result as a Cartesian vector.

4–149. The curved rod lies in the x–y plane and has a radius of 3 m. If a force of $F = 80$ N acts at its end as shown, determine the moment of this force about point O.

4–150. The curved rod lies in the x–y plane and has a radius of 3 m. If a force of $F = 80$ N acts at its end as shown, determine the moment of this force about point B.

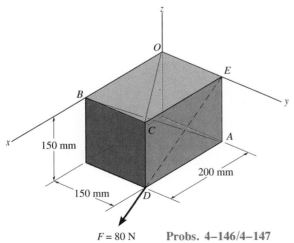

$F = 80$ N   Probs. 4–146/4–147

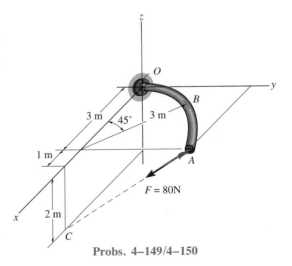

Probs. 4–149/4–150

4–151. The weights of the various components of the truck are shown. Replace this system of forces by an equivalent resultant force and couple moment acting at point A.

***4–152.** The weights of the various components of the truck are shown. Replace this system of forces by an equivalent resultant force and specify its location measured from point A.

4–153. The belt passing over the pulley is subjected to forces $\mathbf{F}_1$ and $\mathbf{F}_2$, each having a magnitude of 40 N. $\mathbf{F}_1$ acts in the $-\mathbf{k}$ direction. Replace these forces by an equivalent resultant force and couple moment at point A. Express the results in Cartesian vector form. Set $\theta = 0°$ so that $\mathbf{F}_2$ acts in the $-\mathbf{j}$ direction.

4–154. The belt passing over the pulley is subjected to two forces $\mathbf{F}_1$ and $\mathbf{F}_2$, each having a magnitude of 40 N. $\mathbf{F}_1$ acts in the $-\mathbf{k}$ direction. Replace these forces by an equivalent force and couple moment at point A. Express the results in Cartesian vector form. Take $\theta = 45°$.

Probs. 4–151/4–152

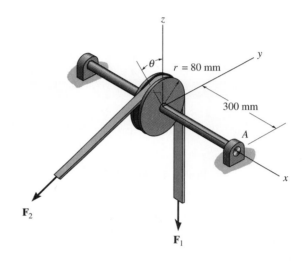

Probs. 4–153/4–154

5 Equilibrium of a Rigid Body

In this chapter the fundamental concepts of rigid-body equilibrium will be discussed. It will be shown that equilibrium requires both a *balance of forces,* to prevent the body from translating with accelerated motion, and a *balance of moments,* to prevent the body from rotating.

Many types of engineering problems involve symmetric loadings and can be solved by projecting all the forces acting on a body onto a single plane. Hence, in the first part of this chapter, the equilibrium of a body subjected to a *coplanar* or *two-dimensional force system* will be considered. Ordinarily the geometry of such problems is not very complex, so a scalar solution is suitable for analysis. The more general discussion of rigid bodies subjected to *three-dimensional force systems* is given in the second part of this chapter. It will be seen that many of these types of problems can best be solved by using vector analysis.

5.1 Conditions for Rigid-Body Equilibrium

In Chapter 3 it was stated that a particle is in equilibrium if it remains at rest or moves with constant velocity. For this to be the case, it is both necessary and sufficient to require the resultant force acting on the particle to be equal to zero. Using this fact, we will now develop the conditions required to maintain equilibrium for a rigid body. To do this, consider the rigid body in Fig. 5–1a, which is fixed in the *x*, *y*, *z* reference and is either at rest or moves with the

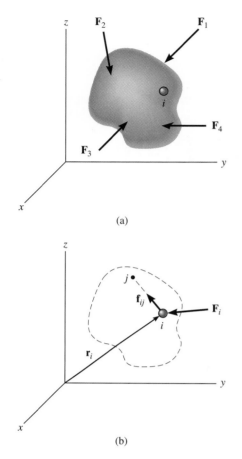

(a)

(b)

Fig. 5–1

reference at constant velocity. A free-body diagram of the arbitrary ith particle of the body is shown in Fig. 5–1b. There are two types of forces which act
on it. The *internal forces,* represented symbolically as

$$\sum_{\substack{j=1 \\ (j \neq 1)}}^{n} \mathbf{f}_{ij} = \mathbf{f}_i$$

are forces which all the other particles exert on the ith particle and produce the
resultant $\mathbf{f}_i$. Although only one of these particles is shown in Fig. 5–1b, the
summation extends over all n particles composing the body. For this summation note that it is meaningless for $i = j$ since the ith particle cannot exert a
force on itself. The resultant *external force* $\mathbf{F}_i$ represents, for example, the
effects of gravitational, electrical, magnetic, or contact forces between the ith
particle and adjacent bodies or particles *not* included within the body. If the
particle is in equilibrium, then applying Newton's first law we have

$$\mathbf{F}_i + \mathbf{f}_i = \mathbf{0}$$

When the equation of equilibrium is applied to each of the other particles
of the body, similar equations will result. If all these equations are added
together *vectorially,* we obtain

$$\Sigma \mathbf{F}_i + \Sigma \mathbf{f}_i = \mathbf{0}$$

The summation of the internal forces if carried out will equal zero since the
internal forces between particles within the body will occur in equal but opposite collinear pairs, Newton's third law. Consequently, only the sum of the
external forces will remain; and therefore, letting $\Sigma \mathbf{F}_i = \Sigma \mathbf{F}$, the above equation can be written as

$$\Sigma \mathbf{F} = \mathbf{0}$$

Let us now consider the moments of the forces acting on the ith particle
about the arbitrary point O, Fig. 5–1b. Using the particle equilibrium equation and the distributive law of the vector cross product yields

$$\mathbf{r}_i \times (\mathbf{F}_i + \mathbf{f}_i) = \mathbf{r}_i \times \mathbf{F}_i + \mathbf{r}_i \times \mathbf{f}_i = \mathbf{0}$$

Similar equations can be written for the other particles of the body, and
adding them together vectorially, we obtain

$$\Sigma \mathbf{r}_i \times \mathbf{F}_i + \Sigma \mathbf{r}_i \times \mathbf{f}_i = \mathbf{0}$$

The second term is zero since, as stated above, the internal forces occur in
equal but opposite collinear pairs, and by the transmissibility of a force, as
discussed in Sec. 4.4, the moment of each pair of forces about point O is
therefore zero. Hence, using the notation $\Sigma \mathbf{M}_O = \Sigma \mathbf{r}_i \times \mathbf{F}_i$, we can write the
previous equation as

$$\Sigma \mathbf{M}_O = \mathbf{0}$$

Hence the *equations of equilibrium* for a rigid body can be summarized as follows:

$$\boxed{\begin{aligned} \Sigma \mathbf{F} &= \mathbf{0} \\ \Sigma \mathbf{M}_O &= \mathbf{0} \end{aligned}} \qquad (5\text{–}1)$$

These equations require that a rigid body will be in equilibrium provided the sum of all the *external forces* acting on the body is equal to zero and the sum of the moments of the external forces about a point is equal to zero. The fact that these conditions are *necessary* for equilibrium has now been proven. They are also *sufficient* conditions. To show this, let us assume that the body is *not* in equilibrium, and yet the force system acting on it satisfies Eqs. 5–1. Suppose that an *additional force* $\mathbf{F}'$ is required to hold the body in equilibrium. As a result, the equilibrium equations become

$$\Sigma \mathbf{F} + \mathbf{F}' = \mathbf{0}$$
$$\Sigma \mathbf{M}_O + \mathbf{M}_O' = \mathbf{0}$$

where $\mathbf{M}_O'$ is the moment of $\mathbf{F}'$ about O. Since $\Sigma \mathbf{F} = \mathbf{0}$ and $\Sigma \mathbf{M}_O = \mathbf{0}$, then we require $\mathbf{F}' = \mathbf{0}$ (also $\mathbf{M}_O' = \mathbf{0}$). Consequently, the additional force $\mathbf{F}'$ is not required for holding the body, and indeed Eqs. 5–1 are also sufficient conditions for equilibrium.

Equilibrium in Two Dimensions

5.2 Free-Body Diagrams

Successful application of the equations of equilibrium requires a complete specification of *all* the known and unknown external forces that act *on the* body. The best way to account for these forces is to draw the body's free-body diagram. This diagram is a sketch of the outlined shape of the body, which represents it as being *isolated* or "free" from its surroundings. On this sketch it is necessary to show *all* the forces and couple moments that the surroundings exert *on the body*. By using this diagram the effects of all the applied forces and couple moments acting on the body can be accounted for when the equations of equilibrium are applied. For this reason, *a thorough understanding of how to draw a free-body diagram is of primary importance for solving problems in mechanics.*

Table 5–1   **Supports for Rigid Bodies Subjected to Two-Dimensional Force Systems**

Types of Connection	Reaction	Number of Unknowns

(1) cable

One unknown. The reaction is a tension force which acts away from the member in the direction of the cable.

(2) weightless link

One unknown. The reaction is a force which acts along the axis of the link.

(3) roller

One unknown. The reaction is a force which acts perpendicular to the surface at the point of contact.

(4) roller or pin in confined smooth slot

One unknown. The reaction is a force which acts perpendicular to the slot.

(5) rocker

One unknown. The reaction is a force which acts perpendicular to the surface at the point of contact.

(6) smooth contacting surface

One unknown. The reaction is a force which acts perpendicular to the surface at the point of contact.

(7) member pin connected to collar on smooth rod

One unknown. The reaction is a force which acts perpendicular to the rod.

Table 5-1 (Contd.)

Types of Connection	Reaction	Number of Unknowns
(8) smooth pin or hinge		Two unknowns. The reactions are two components of force, or the magnitude and direction ϕ of the resultant force. Note that ϕ and θ are not necessarily equal [usually not, unless the rod shown is a link as in (2)].
(9) member fixed connected to collar on smooth rod		Two unknowns. The reactions are the couple moment and the force which acts perpendicular to the rod.
(10) fixed support		Three unknowns. The reactions are the couple moment and the two force components, or the couple moment and the magnitude and direction ϕ of the resultant force.

Support Reactions. Before presenting a formal procedure as to how to draw a free-body diagram, we will first consider the various types of reactions that occur at supports and points of support between bodies subjected to coplanar force systems.

The principles involved for determining these reactions can be illustrated by considering three ways in which a horizontal member, such as a beam, is commonly supported at its end. The first method of support consists of a *roller* or cylinder, Fig. 5–2a. Since this type of support only prevents the beam from translating in the vertical direction, it is necessary that the roller exerts a force on the beam in this direction, Fig. 5–2b.

The beam can be supported in a more restrictive manner by using a *pin* as shown in Fig. 5–3a. The pin passes through holes in the beam and two leaves which are fixed to the ground. Here the pin will prevent translation of the beam in *any direction* ϕ, Fig. 5–3b, and so it must exert a force $\mathbf{F}$ on the beam in this direction. For purposes of analysis, it is generally easier to represent this effect by its two components $\mathbf{F}_x$ and $\mathbf{F}_y$, Fig. 5–3c.

Fig. 5–2

Fig. 5–3

(a)

(b)

Fig. 5–4

The most restrictive way to support the beam would be to use a *fixed support* as shown in Fig. 5–4*a*. This support will prevent both translation and rotation of the beam, and so to do this a force and couple moment must be developed on the beam at its point of connection, Fig. 5–4*b*. As in the case of the pin, the force is usually represented by its components F_x and F_y.

From this discussion, we can now make a general statement as to how to determine the types of reactions that are developed on a contacting member by *any kind* of support. In this regard, *if a support prevents translation in a given direction, then a force is developed on the contacting member in that direction. Likewise, if rotation is prevented, a couple moment is exerted on the member.* Table 5–1 lists some common types of supports used to support bodies subjected to coplanar force systems. (In all cases the angle θ is assumed to be known.) Carefully study each of the symbols used to represent these supports and the types of reactions they exert on their contacting members. Although concentrated forces and couple moments are shown in this table, they actually represent the *resultants* of *distributed surface loads* that exist between each support and its contacting member. It is these *resultants* which are determined in practice, and it is generally not important to determine the actual distribution of the load, since the surface area over which it acts is considerably *smaller* than the *total surface area* of the connected member.

External and Internal Forces. Since a rigid body is a composition of particles, both *external* and *internal* loadings may act on it. It is important to realize, however, that if the free-body diagram for the body is drawn, the forces that are *internal* to the body are *not represented* on the free-body diagram. As discussed in Sec. 5.1, these forces always occur in equal but opposite collinear pairs, and therefore their *net effect* on the body is zero.

In some problems, a free-body diagram for a "system" of connected bodies may be used for an analysis. An example would be the free-body diagram of an entire automobile (system) composed of its many parts. Obviously, the connecting forces between its parts would represent *internal forces* which would *not* be included on the free-body diagram of the automobile. To summarize, then, internal forces act between particles which are located *within* a specified system which is contained within the boundary of the free-body diagram. Particles or bodies outside this boundary exert external forces on the system, and these alone must be shown on the free-body diagram.

Weight and the Center of Gravity. When a body is subjected to a gravitational field, each of its particles has a specified weight as defined by Newton's law of gravitation, $F = Gm_1m_2/r^2$, Eq. 1–2. If we assume the size of the body to be "small" in relation to the size of the earth, then it is appropriate to consider these gravitational forces to be represented as a *system of parallel forces* acting on the particles contained within the boundary of the body. It was shown in Sec. 4.9 that such a system can be reduced to a single resultant force acting through a specified point. We refer to this force resultant as the *weight* **W** of the body, and to the location of its point of application as the *center of gravity G*. The methods used for its calculation will be developed in Chapter 9.

In the examples and problems that follow, if the weight of the body is important for the analysis, this force will then be reported in the problem statement. Also, when the body is *uniform* or made of homogeneous material, the center of gravity will be located at the body's *geometric center* or *centroid;* however, if the body is nonhomogeneous or has an unusual shape, then its center of gravity will be given.

PROCEDURE FOR DRAWING
A FREE-BODY DIAGRAM

To construct a free-body diagram for a rigid body or group of bodies considered as a single system, the following steps should be performed:

Step 1. Imagine the body to be *isolated* or cut "free" from its constraints and connections, and draw (sketch) its outlined shape.

Step 2. Identify all the external forces and couple moments that act on the body. Those generally encountered are due to (1) applied loadings, (2) reactions occurring at the supports or at points of contact with other bodies (see Table 5–1), and (3) the weight of the body. To account for all these effects, it may help to trace over the boundary, carefully noting each force or couple moment acting on it.

Step 3. Indicate the dimensions of the body necessary for computing the moments of forces. The forces and couple moments that are known should be labeled with their proper magnitudes and directions. Letters are used to represent the magnitudes and direction angles of forces and couple moments that are *unknown*. In particular, if a force or couple moment has a known line of action but unknown magnitude, the arrowhead which defines the sense of the vector can be assumed. The correctness of the assumed sense will become apparent after solving the equilibrium equations for the unknown magnitude. By definition, the *magnitude* of a vector is *always positive,* so that if the solution yields a "negative" scalar, the *minus sign* indicates that the vector's sense is *opposite* to that which was originally assumed.

Before proceeding, review this section; then carefully study the following examples. Afterward, attempt to draw the free-body diagrams for the objects in Fig. 5–5 through 5–9 without "looking" at the solutions. Further practice in drawing free-body diagrams should be gained by solving *all* the problems given at the end of this section.

Example 5–1

Draw the free-body diagram of the uniform beam shown in Fig. 5–5a. The beam has a mass of 100 kg.

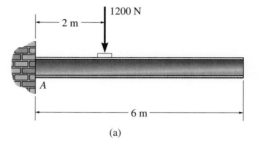

(a)

SOLUTION

The free-body diagram of the beam is shown in Fig. 5–5b. Since the support at A is a fixed wall, there are three reactions acting *on the beam* at A, denoted as $\mathbf{A}_x$, $\mathbf{A}_y$, and $\mathbf{M}_A$. The magnitudes of these vectors are *unknown,* and their sense has been *assumed.* (How does one obtain the *correct* sense of these vectors?) The weight of the beam, $W = 100(9.81) = 981$ N, acts through the beam's center of gravity G, 3 m from A since the beam is uniform.

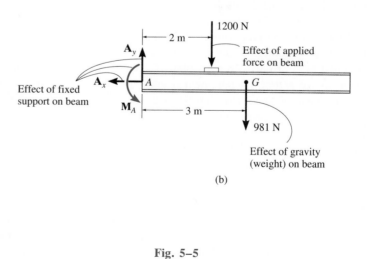

(b)

Fig. 5–5

Example 5–2

Draw the free-body diagram for the bell crank *ABC* shown in Fig. 5–6*a*.

SOLUTION

The free-body diagram is shown in Fig. 5–6*b*. The pin support at *B* exerts force components $\mathbf{B}_x$ and $\mathbf{B}_y$ *on the crank,* each having a known line of action but unknown magnitude. The link at *C* exerts a force $\mathbf{F}_C$ acting in the direction of the link and having an unknown magnitude. The dimensions of the crank are also labeled on the free-body diagram, since this information will be useful in computing the moments of the forces. As usual, the sense of the three unknown forces has been assumed. The correct sense will become apparent after solving the equilibrium equations.

Although not part of this problem, three-dimensional views of the free-body diagrams of the pin and two pin leaves at *B* are shown in Fig. 5–6*c*. Since the leaves are *fixed-connected* to the wall, there are three unknowns that the wall exerts on each leaf; namely, $\mathbf{B}_x''$, $\mathbf{B}_y''$, $\mathbf{M}_B''$. These reactions are shown to be equal in magnitude and direction on each leaf due to the symmetry of the loading and geometry. Note carefully how the principle of action—equal but opposite collinear reaction is used when applying the forces $\mathbf{B}_x'$ and $\mathbf{B}_y'$ to each leaf and the pin. All of these unknowns can be obtained from the equations of equilibrium once $\mathbf{B}_x$ and $\mathbf{B}_y$ are obtained.

(a)

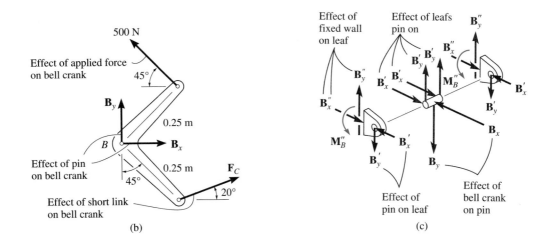

(b)

(c)

Fig. 5–6

■ Example 5–3

Two smooth tubes A and B, each having a mass of 2 kg, rest between the inclined planes shown in Fig. 5–7a. Draw the free-body diagrams for tube A, tube B, and tubes A and B together.

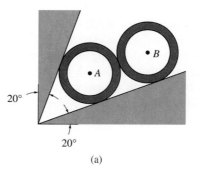

(a)

SOLUTION

The free-body diagram for tube A is shown in Fig. 5–7b. Its weight is $W = 2(9.81) = 19.62$ N. Since all contacting surfaces are *smooth,* the reactive forces **T, F, R** act in a direction *normal* to the tangent at their surfaces of contact.

The free-body diagram of tube B is shown in Fig. 5–7c. Can you identify each of the three forces acting *on the tube?* In particular, note that **R,** representing the force of tube A on tube B, Fig. 5–7c, is equal and opposite to **R** representing the force of tube B on tube A, Fig. 5–7b. This is a consequence of Newton's third law of motion.

The free-body diagram of both tubes combined ("system") is shown in Fig. 5–7d. Here the contact force **R,** which acts between A and B, is considered as an *internal* force and hence is not shown on the free-body diagram. That is, it represents a pair of equal but opposite collinear forces which cancel each other.

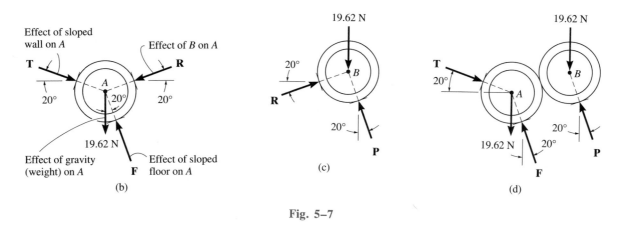

Fig. 5–7

Example 5–4

The free-body diagram of each object in Fig. 5–8 is drawn and the forces acting on the object are identified. The weight of the objects is neglected except where indicated.

SOLUTION

Effect of applied force on member

Effect of pin on member

Effect of smooth support on member

(a)

Effect of applied couple moment on member

Effect of short link on member

Effect of pin on member

(b)

Effect of smooth rod on member

Effect of applied couple moment on member

Effect of smooth support on member

(c)

Effect of applied forces on truss

Effect of pin on truss

Effect of roller on truss

Note: Internal forces of one member on another are equal but opposite collinear forces which are not to be included here.

(d)

Fig. 5–8

Example 5–5

The highway sign shown in Fig. 5–9a has a mass of 100 kg with a center of gravity at G. It is supported by pins at C and D and a cable AB. Draw a free-body diagram of the sign and the supporting frame. Neglect the weight of the frame.

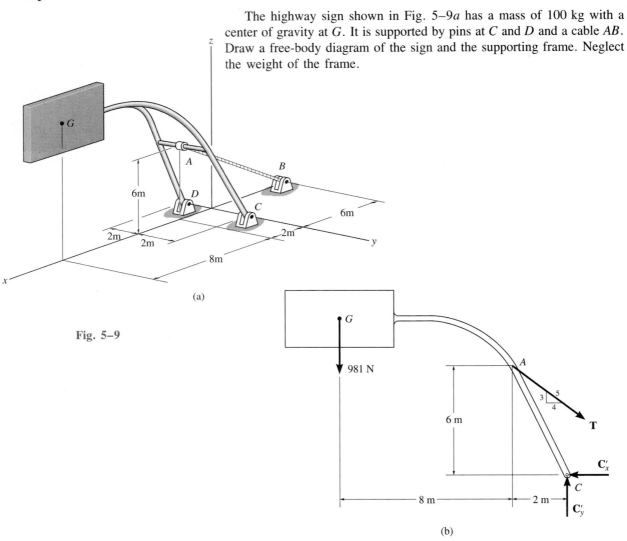

Fig. 5–9

(a)

(b)

SOLUTION

By observation, the frame, sign, and the loading are all symmetrical about the vertical x–y plane, hence the problem may be analyzed using a system of *coplanar forces*. The free-body diagram is shown in Fig. 5–9b. Note that the force **T** that the cable exerts on the frame has a known line of action indicated by the 3–4–5 slope triangle. The force components $\mathbf{C}'_x$ and $\mathbf{C}'_y$ represent the horizontal and vertical reactions of *both* pins C and D, respectively. Consequently, after the solution for these reactions is obtained, *half* their magnitude is applied at C and half at D.

PROBLEMS

5–1. Draw the free-body diagram of the 50-kg pipe which is supported by the smooth contacts at A and B. Explain the significance of each force on the diagram. (See Fig. 5–2b.)

Prob. 5–1

5–2. Draw the free-body diagram of the beam. The support at B is smooth. Explain the significance of each force on the diagram.

Prob. 5–2

5–3. Draw the free-body diagram of the automobile, which has a mass of 5 Mg and center of mass at G. The tires are free to roll, so rolling resistance can be neglected. Explain the significance of each force on the diagram.

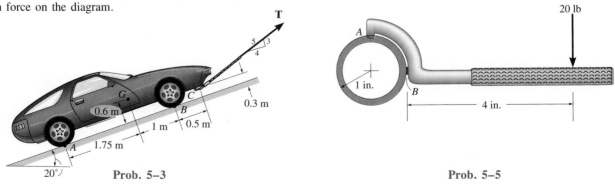

Prob. 5–3

*5–4. Draw the free-body diagram of the beam which supports the 80-kg load and is supported by the pin at A and a cable which wraps around the pulley at D. Explain the significance of each force on the diagram. (See Fig. 5–5b.)

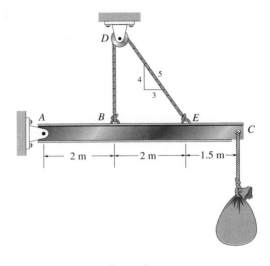

Prob. 5–4

5–5. Draw the free-body diagram of the spanner wrench which is subjected to a force of 20 lb at the grip. The wrench is pinned at A, and the surface of contact at B is smooth. Explain the significance of each force on the diagram. (See Fig. 5–2b.)

Prob. 5–5

5–6. Draw a free-body diagram of the crane boom ABC, which has a mass of 45 kg, center of gravity at G, and supports a load of 30 kg. The boom is pin-connected to the frame at B and to a vertical chain CD. The chain supporting the load is attached to the boom at A.

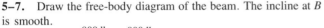

Prob. 5–6

5–7. Draw the free-body diagram of the beam. The incline at B is smooth.

Prob. 5–7

*5–8. Draw the free-body diagram of the winch, which consists of a drum of radius 4 in. It is pin-connected at its center C, and at its outer rim is a ratchet gear having a mean radius of 6 in. The pawl AB serves as a two-force member (short link) and holds the drum from rotating.

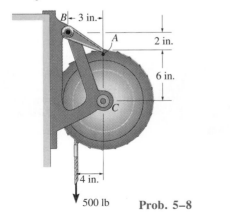

Prob. 5–8

5–9. Draw the free-body diagram of the smooth bar, which has points of contact at A, B, and C.

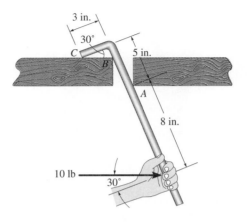

Prob. 5–9

5–10. Draw the free-body diagram of the uniform beam, which has a mass of 100 kg and is supported by making contact with the smooth surfaces at A, B, and C.

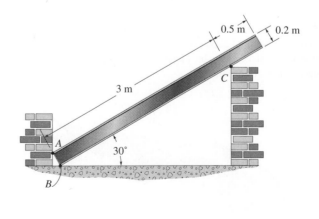

Prob. 5–10

5.3 Equations of Equilibrium

In Sec. 5.1 we developed the two equations which are both necessary and sufficient for the equilibrium of a rigid body, namely, $\Sigma \mathbf{F} = \mathbf{0}$ and $\Sigma \mathbf{M}_O = \mathbf{0}$. When the body is subjected to a system of forces, which all lie in the x–y plane, then the forces can be resolved into their x and y components. Consequently, the conditions for equilibrium in two dimensions are

$$
\begin{array}{|c|}
\hline
\Sigma F_x = 0 \\
\Sigma F_y = 0 \\
\Sigma M_O = 0 \\
\hline
\end{array}
\qquad (5\text{--}2)
$$

Here ΣF_x and ΣF_y represent, respectively, the algebraic sums of the x and y components of all the forces acting on the body, and ΣM_O represents the algebraic sum of the couple moments and the moments of all the force components about an axis perpendicular to the x–y plane and passing through the arbitrary point O, which may lie either on or off the body.

Alternative Sets of Equilibrium Equations. Although Eqs. 5–2 are *most often* used for solving equilibrium problems involving coplanar force systems, two *alternative* sets of three independent equilibrium equations may also be used. One such set is

$$
\begin{aligned}
\Sigma F_a &= 0 \\
\Sigma M_A &= 0 \qquad (5\text{--}3) \\
\Sigma M_B &= 0
\end{aligned}
$$

When using these equations it is required that the moment points A and B do *not* lie on a line that is *perpendicular* to the a axis. To prove that Eqs. 5–3 provide the *conditions* for equilibrium, consider the free-body diagram of an arbitrarily shaped body shown in Fig. 5–10a. Using the methods of Sec. 4.8, the loading acting on the body may be replaced by a single resultant force $\mathbf{F}_R = \Sigma \mathbf{F}$, acting at point A, and a resultant couple moment $\mathbf{M}_{R_A} = \Sigma \mathbf{M}_A$, Fig. 5–10$b$. For equilibrium it is required that $\mathbf{F}_R = \mathbf{0}$ and $\mathbf{M}_{R_A} = \mathbf{0}$. Hence, if $\Sigma M_A = 0$ is satisfied, it is necessary that $\mathbf{M}_{R_A} = \mathbf{0}$. Furthermore, in order that $\mathbf{F}_R$ satisfy $\Sigma F_a = 0$, it must have no component along the a axis, and therefore its line of action must be perpendicular to the a axis, Fig. 5–10c. Finally, if it is required that $\Sigma M_B = 0$, where B does not lie on the line of action of $\mathbf{F}_R$, then $\mathbf{F}_R = \mathbf{0}$, and indeed the body shown in Fig. 5–10a must be in equilibrium.

A second alternative set of equilibrium equations is

$$
\begin{aligned}
\Sigma M_A &= 0 \\
\Sigma M_B &= 0 \qquad (5\text{--}4) \\
\Sigma M_C &= 0
\end{aligned}
$$

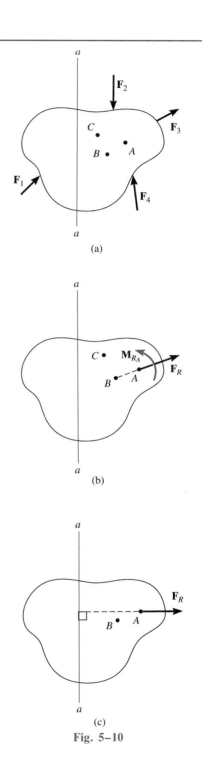

(a)

(b)

(c)

Fig. 5–10

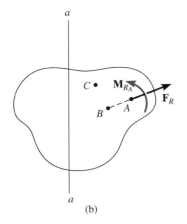

Fig. 5–10

Here it is necessary that points A, B, and C do not lie on the same line. To prove that these equations when satisfied ensure equilibrium, consider again the free-body diagram in Fig. 5–10b. If $\Sigma M_A = 0$ is to be satisfied, then the resultant couple moment $\mathbf{M}_{R_A} = \mathbf{0}$. $\Sigma M_B = 0$ is satisfied if the line of action of $\mathbf{F}_R$ passes through point B as shown, and finally, if we require $\Sigma M_C = 0$, where C does not lie on line AB, it is necessary that $\mathbf{F}_R = \mathbf{0}$, and the body in Fig. 5–10a must then be in equilibrium.

PROCEDURE FOR ANALYSIS

The following procedure provides a method for solving coplanar force equilibrium problems:

Free-Body Diagram.   Draw a free-body diagram of the body as discussed in Sec. 5.2. Briefly, this requires showing all the external forces and couple moments acting *on the body*. The magnitudes of these vectors must be labeled and their directions specified. Dimensions of the body, necessary for computing the moments of forces, are also included on the free-body diagram. Identify the unknowns. The sense of a force or couple moment having an *unknown* magnitude but known line of action can be *assumed*.

Equations of Equilibrium.   Establish the x, y, z axes and apply the equations of equilibrium: $\Sigma F_x = 0$, $\Sigma F_y = 0$, $\Sigma M_O = 0$ (or the alternative sets of Eqs. 5–3 or 5–4). To *avoid* having to solve simultaneous equations, apply the moment equation $\Sigma M_O = 0$ about a point (O) *that lies at the intersection of the lines of action of two unknown forces*. In this way, the moments of these unknowns are *zero* about O, and one can obtain a *direct solution* for the third unknown. When applying the force equations $\Sigma F_x = 0$ and $\Sigma F_y = 0$, orient the x and y axes along lines that will provide the simplest resolution of the forces into their x and y components. If the solution of the equilibrium equations yields a *negative* scalar for an unknown force or couple moment, it indicates that the sense is *opposite* to that which was assumed on the free-body diagram.

The following example problems illustrate this procedure numerically.

Example 5–6

Determine the horizontal and vertical components of reaction for the beam loaded as shown in Fig. 5–11a. Neglect the weight and thickness of the beam in the calculations.

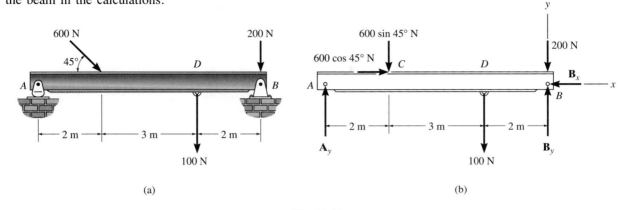

(a)

(b)

Fig. 5–11

SOLUTION

Free-Body Diagram.   Can you identify each of the forces shown on the free-body diagram of the beam, Fig. 5–11b? For simplicity in applying the equilibrium equations, the 600-N force is represented by its x and y components as shown. Also, note that a 200-N force acts on the beam at B, and is independent of the force components $\mathbf{B}_x$ and $\mathbf{B}_y$ which represent the effect of the pin on the beam.

Equations of Equilibrium.   Summing forces in the x direction yields

$$\xrightarrow{+}\Sigma F_x = 0; \qquad 600 \cos 45° \text{ N} - B_x = 0$$
$$B_x = 424 \text{ N} \qquad\qquad Ans.$$

A direct solution for $\mathbf{A}_y$ can be obtained by applying the moment equation $\Sigma M_B = 0$ about point B. For the calculation, it should be apparent that forces 200 N, 600 cos 45° N, $\mathbf{B}_x$, and $\mathbf{B}_y$ all create zero moment about B. Assuming counterclockwise rotation about B to be positive (in the $+\mathbf{k}$ direction), Fig. 5–11b, we have

$$\zeta + \Sigma M_B = 0; \quad 100 \text{ N}(2 \text{ m}) + (600 \sin 45° \text{ N})(5 \text{ m}) - A_y(7 \text{ m}) = 0$$
$$A_y = 332 \text{ N} \qquad\qquad Ans.$$

Summing forces in the y direction, using the result $A_y = 332$ N, gives

$$+\uparrow \Sigma F_y = 0; \quad 332 \text{ N} - 600 \sin 45° \text{ N} - 100 \text{ N} - 200 \text{ N} + B_y = 0$$
$$B_y = 393 \text{ N} \qquad\qquad Ans.$$

Example 5–7

The cord shown in Fig. 5–12a supports a force of 100 lb and wraps over the frictionless pulley. Determine the tension in the cord at C and the horizontal and vertical components of reaction at pin A.

(a)

SOLUTION

Free-Body Diagrams.   The free-body diagrams of the cord and pulley are shown in Fig. 5–12b. Note that the principle of action, equal but opposite reaction, must be carefully observed when drawing each of these diagrams: the cord exerts an unknown load distribution p along part of the pulley's surface, whereas the pulley exerts an equal but opposite effect on the cord. For the solution, however, it is simpler to *combine* the free-body diagrams of the pulley and the contacting portion of the cord, so that the distributed load becomes *internal* to the system and is therefore eliminated from the analysis, Fig. 5–12c.

Equations of Equilibrium.   Summing moments about point A to eliminate A_x and A_y, Fig. 5–12c, we have

$$\zeta+\Sigma M_A = 0; \qquad 100 \text{ lb}(0.5 \text{ ft}) - T(0.5 \text{ ft}) = 0$$

$$T = 100 \text{ lb} \qquad \qquad \textit{Ans.}$$

It is seen that the tension remains *constant* as the cord passes over the pulley. (This of course is true for *any angle* θ at which the cord is directed and for *any radius* r of the pulley.) Using the result for T, a force summation is applied to determine the components of reaction at pin A.

$$\xrightarrow{+}\Sigma F_x = 0; \qquad -A_x + 100 \sin 30° \text{ lb} = 0$$

$$A_x = 50.0 \text{ lb} \qquad \qquad \textit{Ans.}$$

$$+\uparrow \Sigma F_y = 0; \qquad A_y - 100 \text{ lb} - 100 \cos 30° \text{ lb} = 0$$

$$A_y = 187 \text{ lb} \qquad \qquad \textit{Ans.}$$

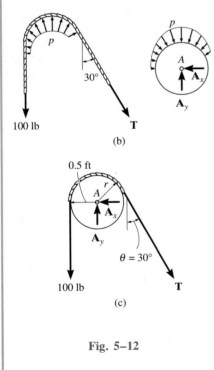

(b)

(c)

Fig. 5–12

Example 5–8

Determine the reactions at the fixed support A for the loaded frame shown in Fig. 5–13a.

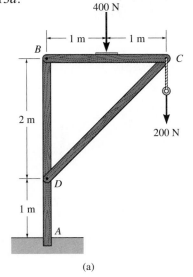

(a)

SOLUTION

Free-Body Diagram. The free-body diagram for the *entire frame* is shown in Fig. 5–13b. There are three unknowns at the fixed support, represented by the magnitudes of $\mathbf{A}_x$, $\mathbf{A}_y$, and $\mathbf{M}_A$.

Equations of Equilibrium

$$\xrightarrow{+}\Sigma F_x = 0; \qquad\qquad A_x = 0 \qquad\qquad\qquad Ans.$$
$$+\uparrow \Sigma F_y = 0; \qquad A_y - 400 \text{ N} - 200 \text{ N} = 0$$
$$A_y = 600 \text{ N} \qquad\qquad\qquad Ans.$$
$$\zeta+\Sigma M_A = 0; \qquad M_A - 400 \text{ N}(1 \text{ m}) - 200 \text{ N}(2 \text{ m}) = 0$$
$$M_A = 800 \text{ N} \cdot \text{m} \qquad\qquad Ans.$$

Point A was chosen for summing moments since the lines of action of the *unknown* forces $\mathbf{A}_x$ and $\mathbf{A}_y$ pass through this point, and therefore these forces were not included in the moment summation. Realize, however, that $\mathbf{M}_A$ must be *included* in the moment summation. This couple moment is a free vector and represents the effect of the fixed support on the frame.

Although only *three* independent equilibrium equations can be written for a rigid body, it is a good practice to *check* the calculations using a fourth equilibrium equation. For example, the above computations may be verified by summing moments about point C:

$$\zeta+\Sigma M_C = 0; \quad 400 \text{ N}(1 \text{ m}) - 600 \text{ N}(2 \text{ m}) + 800 \text{ N} \cdot \text{m} \equiv 0$$
$$400 \text{ N} \cdot \text{m} - 1200 \text{ N} \cdot \text{m} + 800 \text{ N} \cdot \text{m} \equiv 0$$

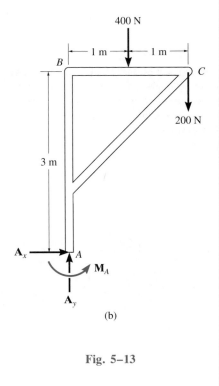

(b)

Fig. 5–13

Example 5–9

The uniform rod shown in Fig. 5–14a is subjected to a force and couple moment. If the rod is supported at A by a smooth wall and at B and C either at the top or bottom by smooth contacts, determine the reactions at these supports. Neglect the weight of the rod.

(a)

SOLUTION

Free-Body Diagram.   As shown in Fig. 5–14b, all the support reactions act normal to the surface of contact since the contacting surfaces are smooth. The reactions at B and C are shown acting in the positive y' direction. This assumes that only the contacts located on the bottom of the rod are used for support.

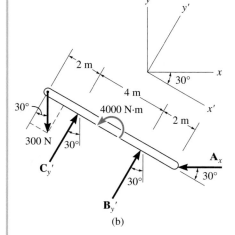

(b)

Fig. 5–14

Equations of Equilibrium.   Using the x, y coordinate system in Fig. 5–14b, we have

$$\xrightarrow{+}\Sigma F_x = 0; \qquad C_{y'} \sin 30° + B_{y'} \sin 30° - A_x = 0 \qquad (1)$$

$$+\uparrow \Sigma F_y = 0; \quad -300\text{ N} + C_{y'} \cos 30° + B_{y'} \cos 30° = 0 \qquad (2)$$

$$\zeta+\Sigma M_A = 0; \qquad -B_{y'}(2\text{ m}) + 4000\text{ N} \cdot \text{m} - C_{y'}(6\text{ m})$$
$$+ (300 \cos 30°\text{ N})(8\text{ m}) = 0 \qquad (3)$$

When writing the moment equation, it should be noticed that the line of action of the force component $300 \sin 30°$ N passes through point A, and therefore this force is not included in the moment equation.

Solving Eqs. 2 and 3 simultaneously, we obtain

$$B_{y'} = -1000.0\text{ N} \qquad\qquad Ans.$$
$$C_{y'} = 1346.4\text{ N} \qquad\qquad Ans.$$

Since $B_{y'}$ is a negative scalar, the sense of $\mathbf{B}_{y'}$ is opposite to that shown on the free-body diagram in Fig. 5–14b. Therefore, the top contact at B serves as the support rather than the bottom one. Retaining the negative sign for $B_{y'}$ (Why?) and substituting the results into Eq. 1, we obtain

$$1346.4 \sin 30°\text{ N} - 1000.0 \sin 30°\text{ N} - A_x = 0$$
$$A_x = 173.2\text{ N} \qquad\qquad Ans.$$

Example 5–10

The beam shown in Fig. 5–15a is pin-connected at A and rests against a smooth support at B. Compute the horizontal and vertical components of reaction at the pin A.

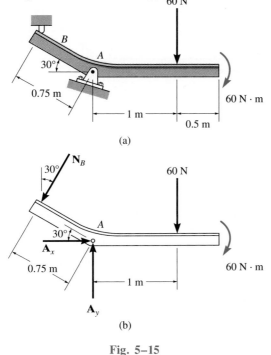

Fig. 5–15

SOLUTION

Free-Body Diagram.   As shown in Fig. 5–15b, the reaction $\mathbf{N}_B$ is perpendicular to the beam at B. Also, horizontal and vertical components of reaction are represented at A, even though the base of the pin support is tilted.

Equations of Equilibrium.   Summing moments about A, we obtain a direct solution for N_B,

$$\zeta + \Sigma M_A = 0; \quad -60 \text{ N} \cdot \text{m} - 60 \text{ N}(1 \text{ m}) + N_B(0.75 \text{ m}) = 0$$

$$N_B = 160 \text{ N}$$

Using this result,

$$\xrightarrow{+} \Sigma F_x = 0; \qquad A_x - 160 \sin 30° \text{ N} = 0$$

$$A_x = 80.0 \text{ N} \qquad \qquad Ans.$$

$$+ \uparrow \Sigma F_y = 0; \qquad A_y - 160 \cos 30° \text{ N} - 60 \text{ N} = 0$$

$$A_y = 199 \text{ N} \qquad \qquad Ans.$$

Example 5–11

A force of 150 lb acts on the end of the beam shown in Fig. 5–16a. Determine the magnitude and direction of the reaction at the pin A and the tension in the cable.

(a)

(b)

Fig. 5–16

SOLUTION

Free-Body Diagram. The forces acting on the beam are shown in Fig. 5–16b.

Equations of Equilibrium. Summing moments about point A to obtain a direct solution for the cable tension yields.

$$\zeta+\Sigma M_A = 0; \quad -(\tfrac{3}{5}T)(2 \text{ ft}) - (\tfrac{4}{5}T)(3 \text{ ft}) + 150 \text{ lb}(10 \text{ ft}) = 0$$
$$-3.6T + 150 \text{ lb}(10 \text{ ft}) = 0 \tag{1}$$
$$T = 416.7 \text{ lb} \qquad \textit{Ans.}$$

Using the principle of transmissibility it is also possible to locate $\mathbf{T}$ at C, even though this point is not on the beam, Fig. 5–16b. In this case, the vertical component of $\mathbf{T}$ creates *zero moment* about A and the moment arm of the horizontal component $(\tfrac{4}{5}T)$ becomes 4.5 ft. Hence, $\Sigma M_A = 0$ yields Eq. 1 directly since $(\tfrac{4}{5}T)(4.5) = 3.6T$.

Summing forces to obtain A_x and A_y, using the result for T, we have

$$\xrightarrow{+}\Sigma F_x = 0; \qquad -A_x + (\tfrac{4}{5})(416.7 \text{ lb}) = 0$$
$$A_x = 333.3 \text{ lb} \rightarrow$$
$$+\uparrow \Sigma F_y = 0; \qquad (\tfrac{3}{5})416.7 \text{ lb} - 150 \text{ lb} - A_y = 0$$
$$A_y = 100 \text{ lb} \downarrow$$

Thus,

$$F_A = \sqrt{(333.3 \text{ lb})^2 + (100 \text{ lb})^2}$$
$$= 348.0 \text{ lb} \qquad \textit{Ans.}$$

$$\theta = \tan^{-1}\frac{100 \text{ lb}}{333.3 \text{ lb}} = 16.7° \qquad \textit{Ans.}$$

Example 5–12

The 100-kg uniform beam *AB* shown in Fig. 5–17*a* is supported at *A* by a pin and at *B* and *C* by a continuous cable which wraps around a friction-less pulley located at *D*. If a maximum tension force of 800 N can be developed in the cable before it breaks, determine the greatest distance *d* at which the 6-kN force can be placed on the beam. What are the horizontal and vertical components of reaction at *A* just before the cable breaks? Neglect the thickness of the beam in the calculation.

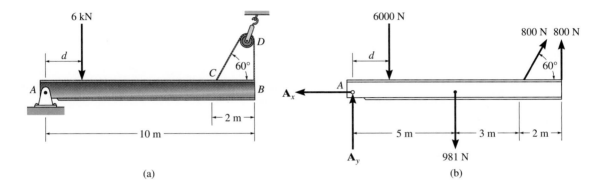

(a)   (b)

Fig. 5–17

SOLUTION

Free-Body Diagram.   Since the cable is continuous and passes over a frictionless pulley, the entire cable is subjected to its maximum tension of 800 N when the 6-kN force acts at *d*. Hence, the cable exerts an 800-N force at points *C* and *B* on the beam in the direction of the cable, Fig. 5–17*b*.

Equations of Equilibrium.   By summing moments of the force system about point *A* it is possible to obtain a direct solution for the dimension *d*. Why?

$\zeta+\Sigma M_A = 0;$
$$-(6000 \text{ N})(d) - 981 \text{ N}(5 \text{ m}) + (800 \sin 60° \text{ N})(8 \text{ m}) + 800 \text{ N}(10 \text{ m}) = 0$$
$$d = 1.44 \text{ m} \qquad \qquad Ans.$$

Summing forces in the *x* and *y* directions, we have

$\xrightarrow{+}\Sigma F_x = 0; \qquad \qquad -A_x + 800 \cos 60° \text{ N} = 0$
$$A_x = 400 \text{ N} \qquad \qquad Ans.$$

$+ \uparrow \Sigma F_y = 0; \quad A_y - 6000 \text{ N} - 981 \text{ N} + 800 \sin 60° \text{ N} + 800 \text{ N} = 0$
$$A_y = 5.49 \text{ kN} \qquad \qquad Ans.$$

5.4 Two- and Three-Force Members

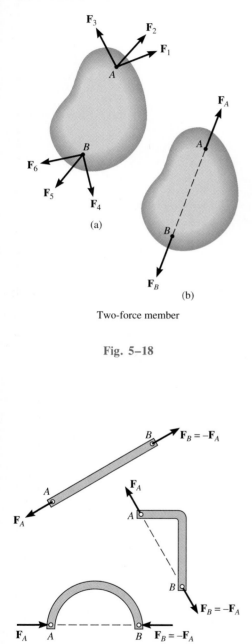

(a)

(b)

Two-force member

Fig. 5–18

The solution to some equilibrium problems can be simplified if one is able to recognize members that are subjected to only two or three forces.

Two-Force Members. When a member is subject to *no couple moments* and forces are applied at only two points on a member, the member is called a *two-force member*. An example of this situation is shown in Fig. 5–18*a*. The forces at A and B are first summed to obtain their respective *resultants* $\mathbf{F}_A$ and $\mathbf{F}_B$, Fig. 5–18*b*. These two forces will maintain *translational or force equilibrium* ($\Sigma\mathbf{F} = \mathbf{0}$) provided $\mathbf{F}_A$ is of equal magnitude and opposite direction to $\mathbf{F}_B$. Furthermore, *rotational or moment equilibrium* ($\Sigma\mathbf{M}_O = \mathbf{0}$) is satisfied if $\mathbf{F}_A$ is *collinear* with $\mathbf{F}_B$. As a result, the line of action of both forces is known, since it always passes through A and B. Hence, only the force magnitude must be determined or stated. Other examples of two-force members held in equilibrium are shown in Fig. 5–19.

Three-Force Members. If a member is subjected to three coplanar forces, then it is necessary that the forces be either *concurrent* or *parallel* if the member is to be in equilibrium. To show this, consider the body in Fig. 5–20 and suppose that any two of the three forces acting on the body have lines of action that intersect at point O. To satisfy moment equilibrium about O, i.e., $\Sigma\mathbf{M}_O = 0$, the third force must also pass through O, which then makes the force system *concurrent*. If two of the three forces are parallel, the point of concurrency, O, is considered to be at "infinity" and the third force must be parallel to the other two forces to intersect at this "point." Provided the three forces are concurrent at a point, only force equilibrium ($\Sigma\mathbf{F} = \mathbf{0}$) must be satisfied.

Two-force members

Fig. 5–19

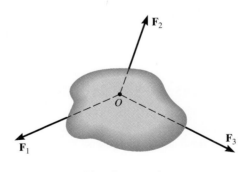

Three-force member

Fig. 5–20

Example 5–13

The lever *ABC* is pin-supported at *A* and connected to a short link *BD* as shown in Fig. 5–21*a*. If the weight of the members is negligible, determine the force of the pin on the lever at *A*.

SOLUTION

Free-Body Diagrams.   As shown by the free-body diagram, Fig. 5–21*b*, the short link *BD* is a *two-force member,* so the *resultant forces* at pins *D* and *B* must be equal, opposite, and collinear. Although the magnitude of the force is unknown, the line of action is known, since it passes through *B* and *D*.

Lever *ABC* is a *three-force member,* and therefore, in order to satisfy moment equilibrium, the three nonparallel forces acting on it must be concurrent at *O*, Fig. 5–21*c*. In particular, note that the force **F** on the lever is equal but opposite to **F** acting at *B* on the link. Why? The distance *CO* must be 0.5 m, since the lines of action of **F** and the 400-N force are known.

Equations of Equilibrium.   By requiring the force system to be concurrent at *O*, so that $\Sigma M_O = 0$, the angle θ which defines the line of action of $\mathbf{F}_A$ can be determined from trigonometry,

$$\theta = \tan^{-1}\left(\frac{0.7}{0.4}\right) = 60.3° \quad \measuredangle^\theta \qquad \textit{Ans.}$$

Applying the force equilibrium equations, we can obtain F and F_A.

$\xrightarrow{+} \Sigma F_x = 0;$ $F_A \cos 60.3° - F \cos 45° + 400 \text{ N} = 0$

$+\uparrow \Sigma F_y = 0;$ $F_A \sin 60.3° - F \sin 45° = 0$

Solving, we get

$$F_A = 1075 \text{ N} \qquad \textit{Ans.}$$
$$F = 1320 \text{ N}$$

Note: We can also solve this problem by representing the force at *A* by its two components $\mathbf{A}_x$ and $\mathbf{A}_y$ and applying $\Sigma M_A = 0$, $\Sigma F_x = 0$, $\Sigma F_y = 0$. Once A_x and A_y are determined, how would you find F_A and θ?

Fig. 5–21

PROBLEMS

5–11. Determine the reactions at the supports of the beam in Prob. 5–2. The support at B is smooth.

*5–12. Determine the force **T** in the towing cable at C and the reaction on *each* of the four tires of the automobile in Prob. 5–3.

5–13. Determine the tension in the cable and the horizontal and vertical components of reaction at the pin A of the beam in Prob. 5–4. The pulley at D is frictionless. Neglect the thickness of the beam.

5–14. Determine the reaction at the pin support at A and the force at the smooth surface of contact at B of the spanner wrench in Prob. 5–5.

5–15. Determine the reactions at the supports of the beam in Prob. 5–7.

*5–16. Determine the horizontal and vertical components of reaction at the pin C of the winch in Prob. 5–8.

5–17. Determine the reactions at points A, B, and C of the bar in Prob. 5–9, needed to support the horizontal force of 10 lb. Neglect the bar's thickness in the calculation.

5–18. Determine the reactions at the supports.

Prob. 5–18

5–19. Determine the reactions at the supports.

Prob. 5–19

*5–20. Determine the reactions at the supports.

Prob. 5–20

5–21. Determine the reactions at the pins A and B. The spring has an unstretched length of 80 mm.

Prob. 5–21

5–22. Determine the reactions at the fixed support A.

Prob. 5–22

5–23. Determine the reactions at the roller A and the pin at B.

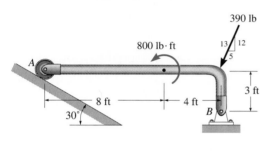

Prob. 5–23

*****5–24.** The uniform rod AB has a weight of 15 lb. Determine the force in the cable when the rod is in the position shown.

Prob. 5–24

5–25. The wheel is subjected to a torque of 500 N · m. Determine the force **F** that acts along the axis AB of the toggle needed to prevent motion. Also, determine the magnitude of the resultant force that acts at the pinned axis C.

Prob. 5–25

5–26. Determine the horizontal and vertical components of reaction at the pin A and the force in the cable BC.

Prob. 5–26

5–27. Determine the horizontal and vertical components of reaction at the pin A and the force in the short link BD.

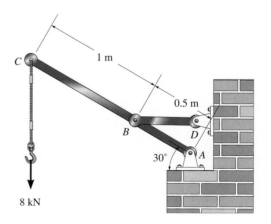

Prob. 5–27

*5–28. Determine the reactions at A and B acting on the 50-kg roll of paper, which has a center of mass at G and rests on the smooth blade of the paper hauler.

Prob. 5–28

5–29. Determine the support reactions on the beam.

Prob. 5–29

5–30. Determine the support reactions on the beam.

Prob. 5–30

5–31. The V-block clamp is used to hold the metal tube in place. If the force which the screw exerts on the tube is $F = 85$ N as shown, determine the normal forces which the block exerts on the tube at A and B. Draw a free-body diagram of the ring and calculate the vertical reactions at C and D. Assume that all points of contact are smooth.

Prob. 5–31

*5–32. The cutter is subjected to a horizontal force of 580 lb and a normal force of 350 lb. Determine the horizontal and vertical components of force acting on the pin A and the force along the hydraulic cylinder BC (a two-force member).

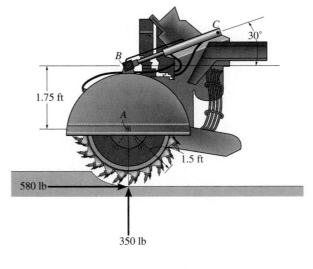

Prob. 5–32

5–33. The crane consists of three parts, which have weights of $W_A = 3500$ lb, $W_B = 900$ lb, $W_C = 1500$ lb and centers of gravity at A, B, and C, respectively. Neglecting the weight of the boom, determine (*a*) the reactions on each of the four tires if the load is hoisted at constant velocity and has a weight of 800 lb, and (*b*), with the boom held in the position shown, the maximum load the crane can lift without tipping over.

Prob. 5–33

5–35. The crane lifts a 400-kg load L. If the primary boom AB has a mass of 1.20 Mg and a center of mass at G_1, whereas the secondary boom BC has a mass of 0.6 Mg and a center of mass at G_2, determine the tension in the cable BD and the horizontal and vertical components of reaction at the pin A.

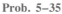

Prob. 5–35

5–34. Determine the resultant normal force acting on *each set of* wheels of the airplane. There is a set of wheels in the front, A, and a set of wheels under each wing, B. Both wings have a total weight of 50 kip and center of gravity at G_w, the fuselage has a weight of 180 kip and center of gravity at G_f, and both engines (one on each side) have a weight of 22 kip and center of gravity at G_e.

*5–36. The picnic table has a weight of 50 lb and a center of gravity at G. If a man weighing 225 lb has a center of gravity at G_M and sits down in the centered position shown, determine the vertical reaction at each of the two legs at B. Neglect the thickness of the legs. What can you conclude from the results?

Prob. 5–34

Prob. 5–36

5–37.   The uniform door has a weight of 100 lb and a center of gravity at G. Determine the reactions at the hinges if the hinge at A supports only a horizontal reaction on the door; whereas the hinge at B exerts both horizontal and vertical reactions.

5–39.   The 25-kg lawn roller is to be lifted over the 100-mm-high step. Compare the magnitudes of force P required to (*a*) push it and (*b*) pull it over the step if in each case the force is directed at $\theta = 30°$ along member AB as shown.

*5–40.   Determine the minimum magnitude of force P and its associated angle θ required to pull the 25-kg roller over the 100-mm step.

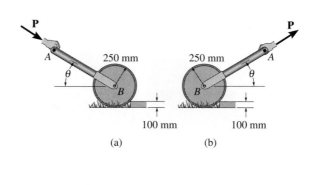

Probs. 5–39/5–40

Prob. 5–37

5–38.   The hold-down clamp exerts a compressive force of 400 N on the wood block at C. If the clamp is loosely bolted to the bench using two symmetrically placed bolts B (one of which is shown), determine the force along the axis of each bolt and the vertical reaction of the bench on the clamp at A.

5–41.   Calculate the tension in the cable at B required to support the load of 200 kg. The cable passes over a frictionless pulley located at C. Neglect the size of this pulley and the weight of the boom.

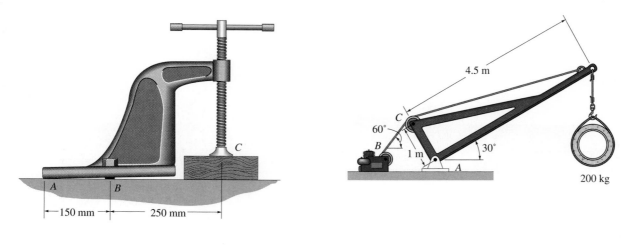

Prob. 5–38

Prob. 5–41

5–42. The dumpster D of the truck has a weight of 5000 lb and a center of gravity at G. Determine the horizontal and vertical components of reaction at the pin A and along the hydraulic cylinder BC (two-force member).

Prob. 5–42

5–43. The boom supports the two vertical loads. Neglect the size of the collars at D and B and the thickness of the boom, and determine the horizontal and vertical components of force at the pin A and the force in cable CB.

Prob. 5–43

*5–44. The davit is used to suspend the lifeboat over the side of a ship. If the boat exerts a force of 800 lb on the cable at D, determine the force acting along the hydraulic cylinder BC (two-force member) and the horizontal and vertical components of reaction at the pin A.

Prob. 5–44

5–45. The chain rests on top of the smooth inclined planes, which intersect at a right angle. If the chain has a weight/length w, prove that it will not slide on the planes.

Prob. 5–45

5–46. Determine the horizontal and vertical components of reaction at the pin A and the reaction at the roller B needed to support the semicircular arch.

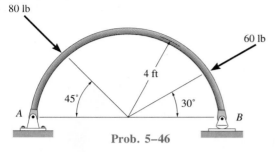

Prob. 5–46

■5–47. The motor has a weight of 850 lb. Determine the force that each of the chains exerts on the supporting hooks at A, B, and C. Neglect the size of the hooks and the thickness of the beam.

Prob. 5–47

***5–48.** The man has a weight W and stands at the center of the plank. If the planes at A and B are smooth, determine the tension in the cord in terms of W and θ.

Prob. 5–48

5–49. A uniform glass rod having a length L is placed in the smooth hemispherical bowl having a radius r. Determine the angle of inclination θ for equilibrium.

Prob. 5–49

5–50. Determine the intensity w_1 and w_2 of the trapezoidal loading if the supports at A and B exert forces of 4 kN and 4.5 kN, respectively, on the beam.

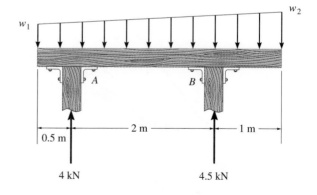

Prob. 5–50

5–51. The rigid metal strip of negligible weight is used as part of an electromagnetic switch. If the stiffness of the springs at A and B is $k = 5$ N/m, and the strip is originally horizontal when the springs are unstretched, determine the smallest vertical force needed to close the contact gap at C.

*5–52. The rigid metal strip of negligible weight is used as part of an electromagnetic switch. Determine the maximum stiffness k of the springs at A and B so that the contact at C closes when the vertical force developed there is 0.5 N. Originally the strip is horizontal as shown.

Probs. 5–51/5–52

5–53. Determine the magnitude of force at the pin A and in the cable BC needed to support the 500-lb load. Neglect the weight of the boom AB.

Prob. 5–53

5–54. Determine the angle θ at which the link ABC is held in equilibrium if member BD moves 2 in. to the right. The springs are originally unstretched when $\theta = 0°$. Each spring has the stiffness shown. The springs remain horizontal since they are attached to roller guides.

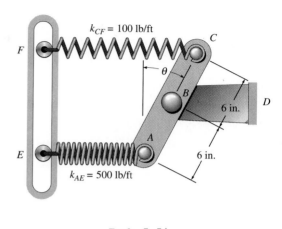

Prob. 5–54

5–55. The exercise machine consists of a rigid bar pin-connected to the frame at A. As the operator pushes up on the bar, it stretches the rubber band BC. If the stiffness of the rubber band is $k = 500$ lb/ft, determine the applied vertical force **F** needed to hold the bar in the position $\theta = 30°$. The rubber band is unstretched when $\theta = 0°$.

Prob. 5–55

*5–56. Determine the tension in the supporting cable BC and the horizontal and vertical components of reaction at pin A necessary to support the bucket of weight W. Express the results in terms of W and the angles θ and ϕ.

Prob. 5–56

5–57. Three uniform books, each having a weight W and length a, are stacked as shown. Determine the maximum distance d that the top book can extend out from the bottom one so the stack does not topple over.

Prob. 5–57

5–58. The uniform rod of length L and weight W is supported on the smooth planes oriented at angles ψ and ϕ. Determine the rod's position θ for equilibrium. Neglect the thickness of the rod.

Prob. 5–58

5–59. A horizontal force $\mathbf{P}$ is applied to the end of the uniform bar of weight W. When the bar is in equilibrium, the force is displaced a distance d as shown. Determine the magnitude of the reaction at the pin A in terms of W, d, and l.

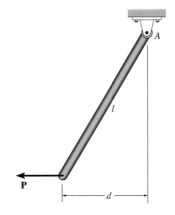

Prob. 5–59

*5–60. The rigid beam of negligible weight is supported horizontally by two springs and a pin. If the springs are unstretched, when the load is removed, determine the force in each spring when the distributed load is applied. Also, compute the vertical deflection of end C. Assume that the spring stiffness k is large enough so that only small deflections occur.

Prob. 5–60

5–61. The uniform rod AB has a weight of 15 lb and the spring is unstretched when $\theta = 0°$. If $\theta = 30°$, determine the stiffness k of the spring.

Prob. 5–61

■**5–62.**   The disk has a mass of 10 kg. If it is suspended from a spring having an unstretched length of 400 mm, determine angle θ for equilibrium.

$k = 600$ N/m

θ

200 mm

Prob. 5–62

5–63.   The light rod of length $2l$ is suspended from a fixed point O using two cords OA and OB each having a length $1.5l$. If weights of $3W$ and W are suspended from the ends A and B, respectively, determine the angle θ of inclination.

$1.5\,l$ O

B

$1.5\,l$

$2\,l$

θ

A

W

$3W$

Prob. 5–63

Equilibrium in Three Dimensions

5.5 Free-Body Diagrams

The first step in solving three-dimensional equilibrium problems, as in the case of two dimensions, is to draw a free-body diagram of the body (or group of bodies considered as a system). Before we show this, however, it is necessary to discuss the types of reactions that can occur at the supports.

Support Reactions.   The reactive forces and couple moments acting at various types of supports and connections, when the members are viewed in three dimensions, are listed in Table 5–2. It is important to recognize the symbols used to represent each of these supports and to understand clearly how the forces and couple moments are developed by each support. As in the two-dimensional case, *a force is developed by a support that restricts the translation of the attached member, whereas a couple moment is developed when rotation of the attached member is prevented*. For example, in Table 5–2, the ball-and-socket joint (4) prevents any translation of the connecting member; therefore, a force must act on the member at the point of connection. This force has three components having unknown magnitudes, F_x, F_y, F_z. Provided these components are known, one can obtain the magnitude of force, $F = \sqrt{F_x^2 + F_y^2 + F_z^2}$, and the force's orientation defined by the coordinate direction angles α, β, γ, Eqs. 2–7.* Since the connecting member is

---

*The three unknowns may also be represented as an unknown force magnitude F and two unknown coordinate direction angles. The third direction angle is obtained using the identity $\cos^2 \alpha + \cos^2 \beta + \cos^2 \gamma = 1$, Eq. 2–10.

allowed to rotate freely about *any* axis, no couple moment is resisted by a ball-and-socket joint.

It should be noted that the *single* bearing supports (5) and (7), the *single* pin (8), and the *single* hinge (9) are shown to support both force and couple-moment components. If, however, these supports are used in conjunction with *other* bearings, pins, or hinges to hold the body in equilibrium, and provided the physical body maintains its *rigidity* when loaded and the supports are *properly aligned* when connected to the body, then the *force reactions* at these supports may *alone* be adequate for supporting the body. In other words, the couple moments become redundant and may be neglected on the free-body diagram. The reason for this will be clear after studying the examples which follow, but essentially the couple moments will not be developed at these supports since the rotation of the body is prevented by the reactions developed at the other supports and not by the supporting couple moments.

Table 5–2 Supports for Rigid Bodies Subjected to Three-Dimensional Force Systems

Types of Connection	Reaction	Number of Unknowns
(1) cable	F	One unknown. The reaction is a force which acts away from the member in the direction of the cable.
(2) smooth surface support	F	One unknown. The reaction is a force which acts perpendicular to the surface at the point of contact.
(3) roller	F	One unknown. The reaction is a force which acts perpendicular to the surface at the point of contact.
(4) ball and socket	F_z F_y F_x	Three unknowns. The reactions are three rectangular force components.

Table 5–2 (Contd.)

Types of Connection	Reaction	Number of Unknowns
(5) single journal bearing		Four unknowns. The reactions are two force and two couple-moment components which act perpendicular to the shaft.
(6) single journal bearing with square shaft		Five unknowns. The reactions are two force and three couple-moment components.
(7) single thrust bearing		Five unknowns. The reactions are three force and two couple-moment components.
(8) single smooth pin		Five unknowns. The reactions are three force and two couple-moment components.
(9) single hinge		Five unknowns. The reactions are three force and two couple-moment components.
(10) fixed support		Six unknowns. The reactions are three force and three couple-moment components.

Free-Body Diagrams. The general procedure for establishing the free-body diagram of a rigid body has been outlined in Sec. 5.2. Essentially it requires first "isolating" the body by drawing its outlined shape. This is followed by a careful *labeling* of *all* the forces and couple moments in reference to an established x, y, z coordinate system. As a general rule, *components of reaction* having an *unknown magnitude* are shown acting on the free-body diagram in the *positive sense*. In this way, if any negative values are obtained, they will indicate that the components act in the negative coordinate directions.

Example 5–14

Several examples of objects along with their associated free-body diagrams are shown in Fig. 5–22. In all cases, the x, y, z axes are established and the unknown reaction components are indicated in the positive sense. The weight of the objects is neglected.

Properly aligned journal bearings at A, B, C.

The force reactions developed by the bearings are sufficient for force and moment equilibrium since they prevent the shaft from rotating about each of the coordinate axes.

(a)

Pin at A and cable BC.

Moment components are developed by the pin on the rod to prevent rotation about the x and z axes.

Fig. 5–22

(b)

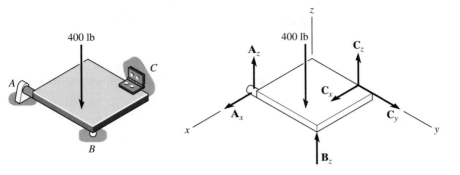

Properly aligned journal bearing at *A* and hinge at *C*. Roller at *B*.

Only force reactions are developed by the bearing and hinge on the plate to prevent rotation about each coordinate axis.

(c)

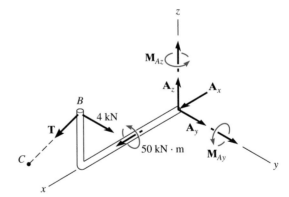

Thrust bearing at *A* and cable *BC*

Moment components are developed by the bearing on the rod in order to prevent rotation about the *y* and *z* axes.

(d)

5.6 Equations of Equilibrium

As stated in Sec. 5.1, the conditions for equilibrium of a rigid body subjected to a three-dimensional force system require that both the *resultant* force and *resultant* couple moment acting on the body be equal to *zero*.

Vector Equations of Equilibrium. The two conditions for equilibrium of a rigid body may be expressed mathematically in vector form as

$$\Sigma \mathbf{F} = \mathbf{0}$$
$$\Sigma \mathbf{M}_O = \mathbf{0}$$

(5–5)

where $\Sigma \mathbf{F}$ is the vector sum of all the external forces acting on the body and $\Sigma \mathbf{M}_O$ is the sum of the couple moments and the moments of all the forces about any point O located either on or off the body.

Scalar Equations of Equilibrium. If all the applied external forces and couple moments are expressed in Cartesian vector form and substituted into Eqs. 5–5, we have

$$\Sigma \mathbf{F} = \Sigma F_x \mathbf{i} + \Sigma F_y \mathbf{j} + \Sigma F_z \mathbf{k} = \mathbf{0}$$
$$\Sigma \mathbf{M}_O = \Sigma M_x \mathbf{i} + \Sigma M_y \mathbf{j} + \Sigma M_z \mathbf{k} = \mathbf{0}$$

Since the $\mathbf{i, j,}$ and $\mathbf{k}$ components are independent from one another, the above equations are satisfied provided

$$\Sigma F_x = 0$$
$$\Sigma F_y = 0$$
$$\Sigma F_z = 0$$

(5–6a)

and

$$\Sigma M_x = 0$$
$$\Sigma M_y = 0$$
$$\Sigma M_z = 0$$

(5–6b)

These *six scalar equilibrium equations* may be used to solve for at most six unknowns shown on the free-body diagram. Equations 5–6a express the fact that the sum of the external force components acting in the x, y, and z directions must be zero, and Eqs. 5–6b require the sum of the moment components about the x, y, and z axes to be zero.

PROCEDURE FOR ANALYSIS

The following procedure provides a method for solving three-dimensional equilibrium problems.

Free-Body Diagram. Construct the free-body diagram for the body. Be sure to include *all* the forces and couple moments that act *on* the body. These interactions are commonly caused by the externally applied loadings, contact forces exerted by adjacent bodies, support reactions, and the weight of the body if it is significant compared to the magnitudes of the other applied forces. Establish the origin of the x, y, z axes at a convenient point, and orient the axes so that they are parallel to as many of the external forces and moments as possible. Identify the unknowns, and in general show all the unknown components having a positive sense if the sense cannot be determined. Dimensions of the body, necessary for calculating the moments of forces, are also included on the free-body diagram.

Equations of Equilibrium. Apply the equations of equilibrium. In many cases, problems can be solved by *direct application* of the six scalar equations $\Sigma F_x = 0$, $\Sigma F_y = 0$, $\Sigma F_z = 0$, $\Sigma M_x = 0$, $\Sigma M_y = 0$, $\Sigma M_z = 0$, Eqs. 5–6; however, if the force components or moment arms seem difficult to determine, it is recommended that the solution be obtained by using vector equations: $\Sigma \mathbf{F} = \mathbf{0}$, $\Sigma \mathbf{M}_O = \mathbf{0}$, Eqs. 5–5. In any case, it is *not necessary* that the set of axes chosen for force summation *coincide* with the set of axes chosen for moment summation. Instead, it is recommended that one *choose the direction of an axis for moment summation such that it intersects the lines of action of as many unknown forces as possible.* The moments of forces passing through points on this axis or forces which are parallel to the axis will then be zero. Furthermore, *any set of three nonorthogonal axes* may be chosen for either the force or moment summations. By the proper choice of axes, it may be possible to solve directly for an unknown quantity, or at least reduce the need for solving a large number of simultaneous equations for the unknowns.

5.7 Constraints for a Rigid Body

To ensure the equilibrium of a rigid body, it is not only necessary to satisfy the equations of equilibrium, but the body must also be properly held or constrained by its supports. Some bodies may have more supports than are necessary for equilibrium, whereas others may not have enough or the supports may be arranged in a particular manner that could cause the body to collapse. Each of these cases will now be discussed.

Redundant Constraints. When a body has redundant supports, that is, more supports than are necessary to hold it in equilibrium, it becomes statically indeterminant. *Statically indeterminate* means that there will be more unknown loadings on the body than equations of equilibrium available for their solution. For example, the two-dimensional problem, Fig. 5–23*a*, and the three-dimensional problem, Fig. 5–23*b*, shown together with their free-body diagrams, are both statically indeterminate because of additional support reactions. In the two-dimensional case, there are five unknowns, that is, M_A, A_x, A_y, B_y, and C_y, for which only three equilibrium equations can be written ($\Sigma F_x = 0$, $\Sigma F_y = 0$, and $\Sigma M_O = 0$, Eqs. 5–2). The three-dimensional problem has eight unknowns, for which only six equilibrium equations can be written, Eqs. 5–6. The additional equations needed to solve indeterminate problems of the type shown in Fig. 5–23 are generally obtained from the deformation conditions at the points of support. These equations involve the physical properties of the body which are studied in subjects dealing with the mechanics of deformation, such as "mechanics of materials."*

Fig. 5–23

(a)

(b)

*See R. C. Hibbeler, *Mechanics of Materials* (New York: Macmillan, 1991).

Improper Constraints. In some cases, there may be as many unknown forces on the body as there are equations of equilibrium; however, *instability* of the body can develop because of *improper constraining* by the supports. In the case of three-dimensional problems, the body is improperly constrained if the support reactions *all intersect a common axis*. For two-dimensional problems, this axis is *perpendicular* to the plane of the forces and therefore appears as a point. Hence, when all the reactive forces are *concurrent* at this point, the body is improperly constrained. Examples of both cases are given in Fig. 5–24. From the free-body diagrams it is seen that the summation of moments about the x axis, Fig. 5–24a, or point O, Fig. 5–24b, will *not* be equal to zero; thus rotation about the x axis or point O will take place.* Furthermore, in both cases, it becomes *impossible* to solve *completely* for all the unknowns, since one can write a moment equation that *does not* involve any of the unknown support reactions, and as a result, this reduces the number of available equilibrium equations by one.

(a)

(b)

*For the three-dimensional problem, $\Sigma M_x = (400 \text{ N})(0.6 \text{ m}) \neq 0$, and for the two-dimensional problem, $\Sigma M_O = (100 \text{ N})(0.2 \text{ m}) \neq 0$.

Fig. 5–24

Fig. 5–25

(a)

(b)

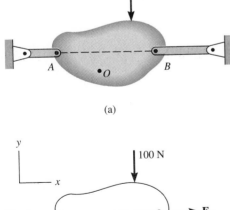

(a)

(b)

Fig. 5–26

Another way in which improper constraining leads to instability occurs when the *reactive forces* are all *parallel*. Three- and two-dimensional examples of this are shown in Fig. 5–25. In both cases, the summation of forces along the x axis will not equal zero.

In some cases, a body may have *fewer* reactive forces than equations of equilibrium that must be satisfied. The body then becomes only *partially constrained*. For example, consider the body shown in Fig. 5–26a with its corresponding free-body diagram in Fig. 5–26b. If O is a point not located on the line AB, the equations $\Sigma F_x = 0$ and $\Sigma M_O = 0$ will be satisfied by proper choice of the reactions $\mathbf{F}_A$ and $\mathbf{F}_B$. The equation $\Sigma F_y = 0$, however, will not be satisfied for the loading conditions and therefore equilibrium will not be maintained.

Proper constraining therefore requires that (1) the lines of action of the reactive forces do not intersect points on a common axis, and (2) the reactive forces must not all be parallel to one another. When the number of reactive forces needed to properly constrain the body in question is a *minimum*, the problem will be statically determinate, and therefore the equations of equilibrium can be used to determine *all* the reactive forces.

Example 5–15

The homogeneous plate shown in Fig. 5–27a has a mass of 100 kg and is subjected to a force and couple moment along its edges. If it is supported in the horizontal plane by means of a roller at A, a ball-and-socket joint at B, and a cord at C, determine the components of reaction at the supports.

SOLUTION (SCALAR ANALYSIS)
Free-Body Diagram. There are five unknown reactions acting on the plate, as shown in Fig. 5-27b. Each of these reactions is assumed to act in a positive coordinate direction.

Equations of Equilibrium. Since the three-dimensional geometry is rather simple, a *scalar analysis* provides a *direct solution* to this problem. A force summation along each axis yields

(a)

$$\Sigma F_x = 0; \qquad B_x = 0 \qquad\qquad\qquad\qquad Ans.$$
$$\Sigma F_y = 0; \qquad B_y = 0 \qquad\qquad\qquad\qquad Ans.$$
$$\Sigma F_z = 0; \qquad A_z + B_z + T_C - 300\text{ N} - 981\text{ N} = 0 \qquad (1)$$

Recall that the moment of a force about an axis is equal to the product of the force magnitude and the perpendicular distance (moment arm) from the line of action of the force to the axis. The sense of the moment is determined by the right-hand rule. Hence, summing moments of the forces on the free-body diagram, with positive moments acting along the positive x or y axis, we have

Fig. 5–27

$$\Sigma M_x = 0; \qquad T_C(2\text{ m}) - 981\text{ N}(1\text{ m}) + B_z(2\text{ m}) = 0 \qquad (2)$$
$$\Sigma M_y = 0;$$
$$300\text{ N}(1.5\text{ m}) + 981\text{ N}(1.5\text{ m}) - B_z(3\text{ m}) - A_z(3\text{ m}) - 200\text{ N}\cdot\text{m} = 0 \qquad (3)$$

The components of force at B can be eliminated if the x', y', z' axes are used. We obtain

$$\Sigma M_{x'} = 0; \qquad 981\text{ N}(1\text{ m}) + 300\text{ N}(2\text{ m}) - A_z(2\text{ m}) = 0 \qquad (4)$$
$$\Sigma M_{y'} = 0;$$
$$-300\text{ N}(1.5\text{ m}) - 981\text{ N}(1.5\text{ m}) - 200\text{ N}\cdot\text{m} + T_C(3\text{ m}) = 0 \qquad (5)$$

Solving Eqs. 1 through 3 or the more convenient Eqs. 1, 4, and 5 yields

$$A_z = 790\text{ N} \qquad B_z = -217\text{ N} \qquad T_C = 707\text{ N} \qquad Ans.$$

The negative sign indicates that $\mathbf{B}_z$ acts downward.

Note that the solution of this problem does not require the use of a summation of moments about the z axis. The plate is partially constrained since the supports will not prevent it from turning about the z axis if a force is applied to it in the x–y plane.

Example 5–16

The windlass shown in Fig. 5–28a is supported by a thrust bearing at A and a smooth journal bearing at B, which are properly aligned on the shaft. Determine the magnitude of the vertical force **P** that must be applied to the handle to maintain equilibrium of the 100-kg crate. Also calculate the reactions at the bearings.

Fig. 5–28

SOLUTION (*SCALAR ANALYSIS*)
Free-Body Diagram.   Since the bearings at A and B are aligned correctly, *only* force reactions occur at these supports, Fig. 5–28b. Why are there no moment reactions?

Equations of Equilibrium.   Summing moments about the x axis yields a direct solution for **P**. Why? For a scalar moment summation, it is necessary to determine the moment of each force as the product of the force magnitude and the *perpendicular distance* from the x axis to the line of action of the force. Using the right-hand rule and assuming positive moments act in the +**i** direction, we have

$$\Sigma M_x = 0; \qquad 981 \text{ N}(0.1 \text{ m}) - P(0.3 \cos 20° \text{ m}) = 0$$
$$P = 348.0 \text{ N} \qquad\qquad Ans.$$

Using this result and summing moments about the y and z axes yields

$$\Sigma M_y = 0;$$
$$-981 \text{ N}(0.5 \text{ m}) + A_z(0.8 \text{ m}) + (348.0 \text{ N})(0.45 \text{ m}) = 0$$
$$A_z = 417.4 \text{ N} \qquad\qquad Ans.$$
$$\Sigma M_z = 0; \qquad\qquad -A_y(0.8 \text{ m}) = 0 \qquad A_y = 0 \qquad\qquad Ans.$$

The reactions at B are determined by a force summation, using the results obtained above.

$$\Sigma F_x = 0; \qquad\qquad A_x = 0 \qquad\qquad Ans.$$
$$\Sigma F_y = 0; \qquad\qquad 0 + B_y = 0 \qquad B_y = 0 \qquad\qquad Ans.$$
$$\Sigma F_z = 0; \quad 417.4 - 981 + B_z - 348.0 = 0 \qquad B_z = 911.6 \text{ N} \qquad Ans.$$

Example 5–17

Determine the tension in cables BC and BD and the reactions at the ball-and-socket joint A for the mast shown in Fig. 5–29a.

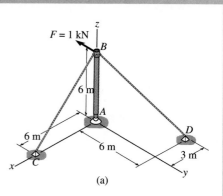

(a)

SOLUTION (*VECTOR ANALYSIS*)
Free-Body Diagram. There are five unknown force magnitudes shown on the free-body diagram, Fig. 5–29b.

Equations of Equilibrium. Expressing each force in Cartesian vector form, we have

$$\mathbf{F} = \{-1000\mathbf{j}\} \text{ N}$$
$$\mathbf{F}_A = A_x\mathbf{i} + A_y\mathbf{j} + A_z\mathbf{k}$$
$$\mathbf{T}_C = 0.707T_C\mathbf{i} - 0.707T_C\mathbf{k}$$
$$\mathbf{T}_D = T_D \left(\frac{\mathbf{r}_{BD}}{r_{BD}}\right) = -0.333T_D\mathbf{i} + 0.667T_D\mathbf{j} - 0.667T_D\mathbf{k}$$

(b)

Applying the force equation of equilibrium gives

$$\Sigma\mathbf{F} = \mathbf{0}; \qquad \mathbf{F} + \mathbf{F}_A + \mathbf{T}_C + \mathbf{T}_D = \mathbf{0}$$
$$(A_x + 0.707T_C - 0.333T_D)\mathbf{i} + (-1000 + A_y + 0.667T_D)\mathbf{j}$$
$$+ (A_z - 0.707T_C - 0.667T_D)\mathbf{k} = \mathbf{0}$$

$$\Sigma F_x = 0; \qquad A_x + 0.707T_C - 0.333T_D = 0 \qquad (1)$$
$$\Sigma F_y = 0; \qquad A_y + 0.667T_D - 1000 = 0 \qquad (2)$$
$$\Sigma F_z = 0; \qquad A_z - 0.707T_C - 0.667T_D = 0 \qquad (3)$$

Fig. 5–29

Summing moments about point A, we have

$$\Sigma\mathbf{M}_A = \mathbf{0}; \qquad \mathbf{r}_B \times (\mathbf{F} + \mathbf{T}_C + \mathbf{T}_D) = \mathbf{0}$$
$$6\mathbf{k} \times (-1000\mathbf{j} + 0.707T_C\mathbf{i} - 0.707T_C\mathbf{k}$$
$$-0.333T_D\mathbf{i} + 0.667T_D\mathbf{j} - 0.667T_D\mathbf{k}) = \mathbf{0}$$

Evaluating the cross product and combining terms yields

$$(-4T_D + 6000)\mathbf{i} + (4.24T_C - 2T_D)\mathbf{j} = \mathbf{0}$$

$$\Sigma M_x = 0; \qquad -4T_D + 6000 = 0 \qquad (4)$$
$$\Sigma M_y = 0; \qquad 4.24T_C - 2T_D = 0 \qquad (5)$$

The moment equation about the z axis, $\Sigma M_z = 0$, is automatically satisfied. Why? Solving Eqs. 1 through 5 we have

$$T_C = 707 \text{ N} \qquad T_D = 1500 \text{ N} \qquad \qquad Ans.$$
$$A_x = 0 \text{ N} \qquad A_y = 0 \text{ N} \qquad A_z = 1500 \text{ N} \qquad Ans.$$

Since the mast is a two-force member, note that the value $A_x = A_y = 0$ could have been determined *by inspection*.

Example 5–18

Rod AB shown in Fig. 5–30a is subjected to the 200-N force. Determine the reactions at the ball-and-socket joint A and the tension in cables BD and BE.

SOLUTION (*VECTOR ANALYSIS*)
Free-Body Diagram.   Fig. 5–30b.

Equations of Equilibrium.   Representing each force on the free-body diagram in Cartesian vector form, we have

$$\mathbf{F}_A = A_x\mathbf{i} + A_y\mathbf{j} + A_z\mathbf{k}$$
$$\mathbf{T}_E = T_E\mathbf{i}$$
$$\mathbf{T}_D = T_D\mathbf{j}$$
$$\mathbf{F} = \{-200\mathbf{k}\} \text{ N}$$

Applying the force equation of equilibrium,

$$\Sigma\mathbf{F} = \mathbf{0}; \qquad \mathbf{F}_A + \mathbf{T}_E + \mathbf{T}_D + \mathbf{F} = \mathbf{0}$$
$$(A_x + T_E)\mathbf{i} + (A_y + T_D)\mathbf{j} + (A_z - 200)\mathbf{k} = \mathbf{0}$$

$$\Sigma F_x = 0; \qquad\qquad A_x + T_E = 0 \tag{1}$$
$$\Sigma F_y = 0; \qquad\qquad A_y + T_D = 0 \tag{2}$$
$$\Sigma F_z = 0; \qquad\qquad A_z - 200 = 0 \tag{3}$$

Summing moments about point A yields

$$\Sigma\mathbf{M}_A = \mathbf{0}; \qquad \mathbf{r}_C \times \mathbf{F} + \mathbf{r}_B \times (\mathbf{T}_E + \mathbf{T}_D) = \mathbf{0}$$

Since $\mathbf{r}_C = \frac{1}{2}\mathbf{r}_B$, then

$$(1\mathbf{i} + 1\mathbf{j} - 0.5\mathbf{k}) \times (-200\mathbf{k}) + (2\mathbf{i} + 2\mathbf{j} - 1\mathbf{k}) \times (T_E\mathbf{i} + T_D\mathbf{j}) = \mathbf{0}$$

Expanding and rearranging terms gives

$$(T_D - 200)\mathbf{i} + (-T_E + 200)\mathbf{j} + (2T_D - 2T_E)\mathbf{k} = \mathbf{0}$$

$$\Sigma M_x = 0; \qquad\qquad T_D - 200 = 0 \tag{4}$$
$$\Sigma M_y = 0; \qquad\qquad -T_E + 200 = 0 \tag{5}$$
$$\Sigma M_z = 0; \qquad\qquad 2T_D - 2T_E = 0 \tag{6}$$

Solving Eqs. 1 through 6, we get

$$A_x = A_y = -200 \text{ N} \qquad\qquad Ans.$$
$$A_z = T_E = T_D = 200 \text{ N} \qquad\qquad Ans.$$

The negative sign indicates that $\mathbf{A}_x$ and $\mathbf{A}_y$ have a sense which is opposite to that shown on the free-body diagram, Fig. 5–30b.

(a)

(b)

Fig. 5–30

Example 5–19

The bent rod in Fig. 5–31*a* is supported at *A* by a journal bearing, at *D* by a ball-and-socket joint, and at *B* by means of cable *BC*. Using only *one equilibrium equation,* obtain a direct solution for the tension in cable *BC*. The bearing at *A* is capable of exerting force components only in the *z* and *y* directions, since it is properly aligned on the shaft.

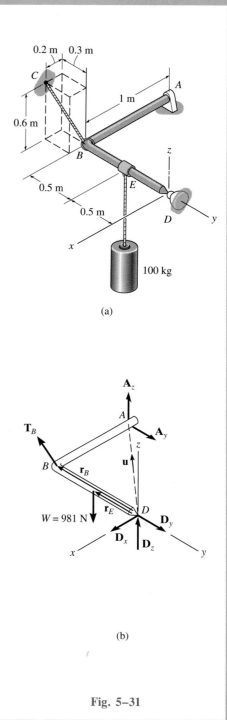

(a)

(b)

Fig. 5–31

SOLUTION (*VECTOR ANALYSIS*)
Free-Body Diagram.   As shown in Fig. 5–31*b*, there are six unknowns: three force components caused by the ball-and-socket joint, two caused by the bearing, and one caused by the cable.

Equations of Equilibrium.   The cable tension $\mathbf{T}_B$ may be obtained *directly* by summing moments about an axis passing through points *D* and *A*. Why? The direction of the axis is defined by the unit vector $\mathbf{u}$, where

$$\mathbf{u} = \frac{\mathbf{r}_{DA}}{r_{DA}} = -\frac{1}{\sqrt{2}}\mathbf{i} - \frac{1}{\sqrt{2}}\mathbf{j}$$

$$= -0.707\mathbf{i} - 0.707\mathbf{j}$$

Hence, the sum of the moments about this axis is zero provided

$$\Sigma M_{DA} = \mathbf{u} \cdot \Sigma(\mathbf{r} \times \mathbf{F}) = 0$$

Here $\mathbf{r}$ represents a position vector drawn from *any point* on the axis *DA* to any point on the line of action of force $\mathbf{F}$ (see Eq. 4–11). With reference to Fig. 5–31*b*, we can therefore write

$$\mathbf{u} \cdot (\mathbf{r}_B \times \mathbf{T}_B + \mathbf{r}_E \times \mathbf{W}) = 0$$

$$(-0.707\mathbf{i} - 0.707\mathbf{j}) \cdot [(-1\mathbf{j}) \times \left(\tfrac{0.2}{0.7}T_B\mathbf{i} - \tfrac{0.3}{0.7}T_B\mathbf{j} + \tfrac{0.6}{0.7}T_B\mathbf{k}\right)$$

$$+ (-0.5\mathbf{j}) \times (-981\mathbf{k})] = 0$$

$$(-0.707\mathbf{i} - 0.707\mathbf{j}) \cdot [(-0.857T_B + 490.5)\mathbf{i} + 0.286T_B\mathbf{k}] = 0$$

$$-0.707(-0.857T_B + 490.5) + 0 + 0 = 0$$

$$T_B = \frac{490.5}{0.857} = 572 \text{ N} \qquad\qquad Ans.$$

The advantage of using Cartesian vectors for this solution should be noted. It would be especially tedious to determine the perpendicular distance from the *DA* axis to the line of action of $\mathbf{T}_B$ using scalar methods.

Note: In Example 5–17, a direct solution for A_z is possible by summing moments about an axis passing through the supports at *C* and *D*, Fig. 5–29*a*. If this is done only the moment of $\mathbf{F}$ and A_z must be considered. Go back to that example and try to apply the above technique to determine the result $A_z = 1500$ N.

PROBLEMS

*5–64.   Determine the x, y, z components of reaction at the fixed wall A.

$$F_2 = \{40i - 30j - 25k\} \ N$$

$$F_1 = \{60i + 50j - 20k\} \ N$$

Prob. 5–64

5–66.   The power drill is subjected to the forces shown acting on the grips. Determine the x, y, z components of force and the y and z components of moment reaction acting on the drill bit at A.

$$F_2 = \{6i - 1j - 1k\} \ lb$$

$$F_1 = \{2i - 3j + 1k\} \ lb$$

Prob. 5–66

5–65.   The nonhomogeneous door of a large pressure vessel has a weight of 125 lb and a center of gravity at G. Determine the magnitudes of the resultant force and resultant couple moment developed at the hinge A, needed to support the door in any open position.

Prob. 5–65

5–67.   Determine the floor reaction on each wheel of the engine stand. The engine weighs 750 lb and has a center of gravity at G.

Prob. 5–67

*5–68. The triangular plate is supported by a ball-and-socket joint at B and rollers at A and C. Determine the x, y, z components of reaction at these supports due to the loading shown.

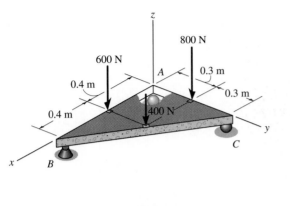

Prob. 5–68

5–70. The plate has a mass of 50 kg/m². Determine the force in each cable if it is suspended in the horizontal plane.

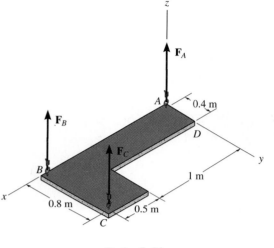

Prob. 5–70

5–69. The concrete slab is being hoisted with constant velocity using the cable arrangement shown. If the slab has a weight of 32,000 lb and a center of gravity at G, determine the tension in the supporting cables AC, BC, and DE.

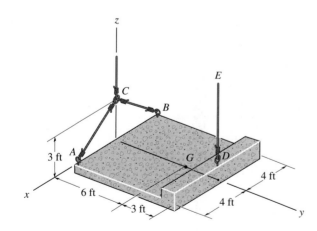

Prob. 5–69

5–71. The boom is supported by a ball-and-socket joint at A and a guy wire at B. If the loads in the cables are each 5 kN and they lie in a plane which is parallel to the x–y plane, determine the x, y, z components of reaction at A and the tension in the cable at B.

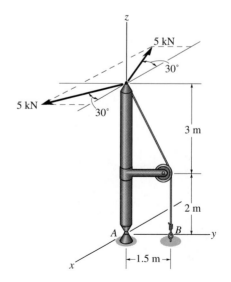

Prob. 5–71

*5–72. The silo has a weight of 3500 lb and a center of gravity at *G*. Determine the vertical component of force that each of the three struts at *A*, *B*, and *C* exerts on the silo if it is subjected to a resultant wind loading of 250 lb which acts in the direction shown.

5–74. The uniform door weighs 50 lb and is supported at *A* and *B* by hinges. The hinge at *A* supports only forces in the horizontal plane, while *B* supports both horizontal and vertical components of force. If the normal force of 35 lb is applied to the handle of the door, determine the horizontal reaction at the smooth door stop *C*, and the *x*, *y*, *z* components of reaction at the hinges *A* and *B*. Neglect the size of the door stop.

Prob. 5–72

Prob. 5–74

5–73. The shaft is supported by a thrust bearing at *A* and a journal bearing at *B*. Determine the *x*, *y*, *z* components of reaction at these supports and the magnitude of force acting on the gear at *C* necessary to hold the shaft in equilibrium. The bearings are in proper alignment and exert only force reactions on the shaft.

5–75. The windlass is subjected to a load of 150 lb. Determine the horizontal force **P** needed to hold the handle in the position shown, and the *x*, *y*, *z* components of reaction at the ball-and-socket *A* and the smooth journal bearing *B*. The bearing is in proper alignment and exerts only force reactions on the shaft.

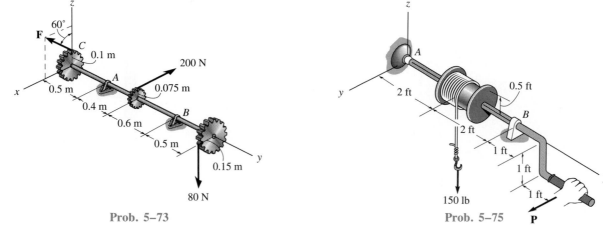

Prob. 5–73   Prob. 5–75   **P**

*5–76. The shaft is supported by journal bearings at A and B. A key is inserted into the bearing at B in order to prevent the shaft from rotating about and translating along its axis. Determine the x, y, z components of reaction at the bearings when the 600-N force is applied to the arm. The bearings are in proper alignment and exert only force reactions on the shaft.

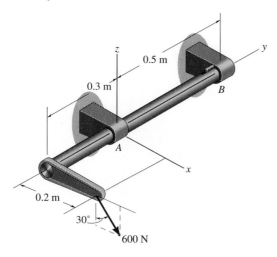

Prob. 5–76

■5–77. The bent rod is supported at A, B, and C by journal bearings. Determine the x, y, z reaction components at the bearings if the rod is subjected to a 200-lb vertical force and a 30-lb·ft couple moment as shown. The bearings are in proper alignment and exert only force reactions on the rod.

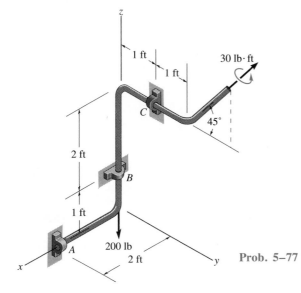

Prob. 5–77

5–78. The rod is supported by journal bearings at A, B, and C. Determine the x, y, z components of reaction at these supports due to the loading shown. The bearings are in proper alignment and exert only force reactions on the rod.

Prob. 5–78

5–79. The forked rod is supported by a collar at A, a thrust bearing at B, and a cable CD. Determine the tension within cable and the x, y, z components of reaction at supports A and B due to the loading shown. The supports at A and B are in proper alignment and exert only force reactions on the rod.

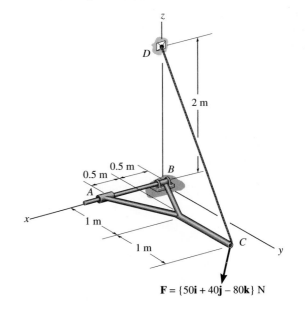

$F = \{50i + 40j - 80k\}$ N

Prob. 5–79

*5–80.   The uniform 20-kg lid on a chest is propped open by the light rod CD. If the hinge at A prevents sliding of the lid along the x axis, whereas the hinge at B does not offer resistance in this direction, calculate the compressive force in CD and the x, y, z components of reaction at the hinges A and B. The hinges are in proper alignment and exert only force reactions on the lid.

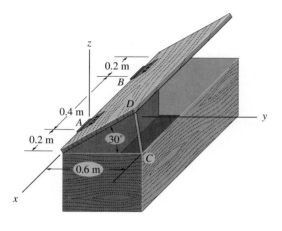

Prob. 5–80

■5–81.   The bent rod is supported at A, B, and C by smooth journal bearings. Determine the x, y, z components of reaction at the bearings if the rod is subjected to a force of $\mathbf{F} = \{-500\mathbf{i} + 400\mathbf{j} + 200\mathbf{k}\}$ N and couple moment of $\mathbf{M} = \{600\mathbf{j}\}$ N · m. The bearings are in proper alignment and exert only force reactions on the rod.

Prob. 5–81

5–82.   The triangular metal plate supports a 150-lb load. Determine the tension in cables BD and CD and the x, y, z components of reaction at the ball-and-socket joint at A.

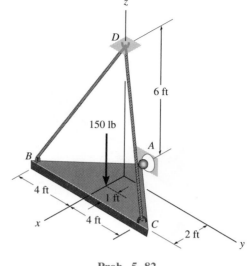

Prob. 5–82

5–83.   Member AB is supported by a cable BC and at A by a smooth fixed *square* rod which fits loosely through the square hole of the collar. If $\mathbf{F} = \{20\mathbf{i} - 40\mathbf{j} - 75\mathbf{k}\}$ lb, determine the x, y, z components of reaction at A and the tension in the cable.

*5–84.   Member AB is supported by a cable BC and at A by a smooth fixed *square* rod which fits loosely through the square hole of the collar. If the force $\mathbf{F} = -\{45\mathbf{k}\}$ lb, determine the tension in cable BC and the x, y, z components of reaction at A.

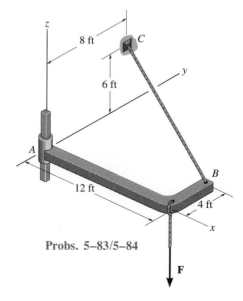

Probs. 5–83/5–84

5–85. The rod is supported by a thrust bearing at A and cable BC. Determine the force in the cable and the x, y, z components of reaction at A.

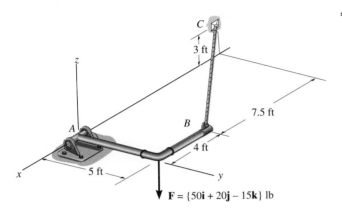

F = {50**i** + 20**j** − 15**k**} lb

Prob. 5–85

5–86. Member AB is supported by a cable BC and at A by a *square* rod which fits loosely through the square hole at the end support of the member as shown. Determine the x, y, z components of reaction at A and the tension in the cable.

5–87. The boom supports a load having a weight of $W = 850$ lb. Determine the x, y, z components of reaction at the ball-and-socket joint A and the tension in cables BC and DE.

*5–88. Cable BC or DE can support a maximum tension of 700 lb before it breaks. Determine the greatest weight W that can be suspended from the end of the boom. Also, determine the x, y, z components of reaction at the ball-and-socket joint A.

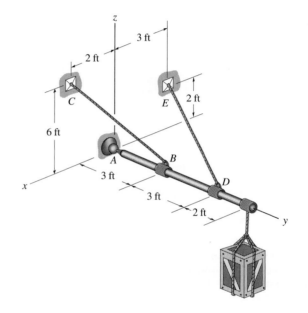

Probs. 5–87/5–88

F = {− 60**i** + 40**j** − 30**k**}lb

Prob. 5–86

REVIEW PROBLEMS

5–89. The horizontal beam is supported by springs at its ends. Each spring has a stiffness of $k = 5$ kN/m and is originally unstretched so that the beam is in the horizontal position. Determine the angle of tilt of the beam if a load of 800 N is applied at point C as shown.

Prob. 5–89

5–91. A vertical force of 80 lb acts on the crankshaft. Determine the horizontal equilibrium force **P** that must be applied to the handle and the x, y, z components of force at the smooth journal bearing A and the thrust bearing B. The bearings are properly aligned and exert only force reactions on the shaft.

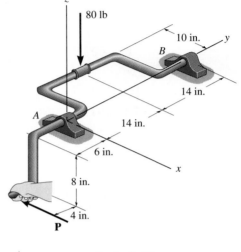

Prob. 5–91

5–90. The horizontal beam is supported by springs at its ends. If the stiffness of the spring at A is $k_A = 5$ kN/m, determine the required stiffness of the spring at B so that if the beam is loaded with the 800-N force it remains in the horizontal position both before and after loading.

Prob. 5–90

*5–92.** The uniform rod has a length l and weight W. It is supported at one end A by a smooth wall and the other end by a cord of length s which is attached to the wall as shown. Show that for equilibrium it is required that $h = [(s^2 - l^2)/3]^{1/2}$.

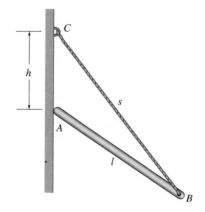

Prob. 5–92

5–93. A uniform beam having a weight of 200 lb supports a vertical load of 800 lb. If the ground pressure varies linearly as shown, determine the load intensities w_1 and w_2, measured in lb/ft, necessary for equilibrium.

Prob. 5–93

5–95. Determine the reactions at the supports A and B of the frame.

Prob. 5–95

5–94. Determine the reactions on the beam.

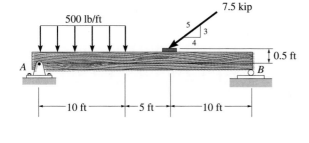

Prob. 5–94

***5–96.** The stiff-leg derrick used on ships is supported by a ball-and-socket joint at D and two cables BA and BC. The cables are attached to a smooth collar ring at B, which allows rotation of the derrick about the z axis. If the derrick supports a crate having a mass of 100 kg, determine the tension in the supporting cables and the x, y, z components of reaction at D.

Prob. 5–96

5–97.   A skeletal diagram of the lower leg is shown in the lower figure. Here it can be noted that this portion of the leg is lifted by the quadriceps muscle attached to the hip at A and to the patella bone at B. This bone slides freely over cartilage at the knee joint. The quadriceps is further extended and attached to the tibia at C. Using the mechanical system shown in the upper figure to model the lower leg, determine the tension **T** in the quadriceps and the magnitude of the resultant force at the femur (pin), D, in order to hold the lower leg in the position shown. The lower leg has a mass of 3.2 kg and a mass center at G_1, the foot has a mass of 1.6 kg and a mass center at G_2.

5–98.   Determine the horizontal and vertical components of reaction at the pin A and the force in the cable BC. Neglect the thickness of the members.

Prob. 5–98

Prob. 5–97

5–99.   Determine the horizontal and vertical components of reaction at the pin and the reaction at the roller support of the beam.

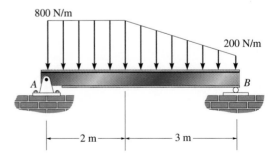

Prob. 5–99

6 Structural Analysis

In this chapter we will use the equations of equilibrium to analyze structures composed of pin-connected members. The analysis is based on the principle that if a structure is in equilibrium, then each of its members is also in equilibrium. By applying the equations of equilibrium to the various parts of a simple truss, frame, or machine, we will be able to determine all the forces acting at the connections.

The topics in this chapter are very important since they provide practice in drawing free-body diagrams, using the principle of action, equal but opposite collinear force reaction, and applying the equations of equilibrium.

6.1 Simple Trusses

A *truss* is a structure composed of slender members joined together at their end points. The members commonly used in construction consist of wooden struts or metal bars. The joint connections are usually formed by bolting or welding the ends of the members to a common plate, called a *gusset plate,* as shown in Fig. 6–1a, or by simply passing a large bolt or pin through each of the members, Fig. 6–1b.

(a)

Fig. 6–1

(b)

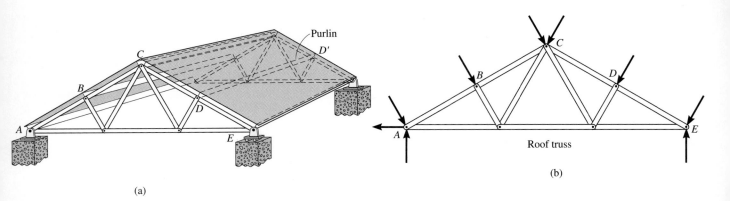

(a)

Roof truss

(b)

Fig. 6–2

Planar Trusses. *Planar* trusses lie in a single plane and are often used to support roofs and bridges. The truss *ABCDE,* shown in Fig. 6–2a, is an example of a typical roof-supporting truss. In this figure, the roof load is transmitted to the truss *at the joints* by means of a series of *purlins,* such as *DD'.* Since the imposed loading acts in the same plane as the truss, Fig. 6–2b, the analysis of the forces developed in the truss members is two-dimensional.

In the case of a bridge, such as shown in Fig. 6–3a, the load on the deck is first transmitted to *stringers,* then to *floor beams,* and finally to the *joints B, C,* and *D* of the two supporting side trusses. Like the roof truss, the bridge truss loading is also coplanar, Fig. 6–3b.

When bridge or roof trusses extend over large distances, a rocker or roller is commonly used for supporting one end, joint *E* in Figs. 6–2a and 6–3a. This type of support allows freedom for expansion or contraction of the members due to temperature or application of loads.

Fig. 6–3

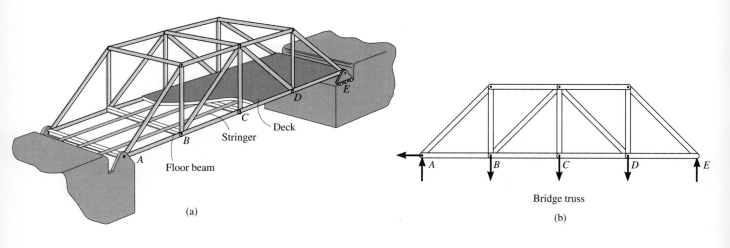

(a)

Bridge truss

(b)

Assumptions for Design. To design both the members and the connections of a truss, it is first necessary to determine the *force* developed in each member when the truss is subjected to a given loading. In this regard, two important assumptions will be made:

1. *All loadings are applied at the joints*. In most situations, such as for bridge and roof trusses, this assumption is true. Frequently in the force analysis the weights of the members are neglected, since the forces supported by the members are usually large in comparison with their weights. If the member's weight is to be included in the analysis, it is generally satisfactory to apply it as a vertical force, half of its magnitude applied at each end of the member.
2. *The members are joined together by smooth pins*. In cases where bolted or welded joint connections are used, this assumption is satisfactory provided the center lines of the joining members are *concurrent*, as in the case of point *A* in Fig. 6–1a.

Because of these two assumptions, *each truss member acts as a two-force member,* and therefore the forces at the ends of the member must be directed along the axis of the member. If the force tends to *elongate* the member, it is a *tensile force* (**T**), Fig. 6–4a; whereas if it tends to *shorten* the member, it is a *compressive force* (**C**), Fig. 6–4b. In the actual design of a truss it is important to state whether the nature of the force is tensile or compressive. Most often, compression members must be made *thicker* than tension members, because of the buckling or column effect that occurs when a member is in compression.

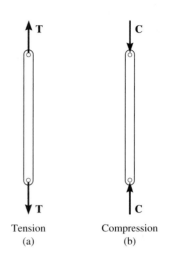

Fig. 6–4

Simple Truss. To prevent collapse, the framework of a truss must be rigid. Obviously, the four-bar frame *ABCD* in Fig. 6–5 will collapse unless a diagonal, such as *AC*, is added for support. The simplest framework which is rigid or stable is a *triangle*. Consequently, a *simple truss* is constructed by *starting* with a basic triangular element, such as *ABC* in Fig. 6–6, and connecting two members (*AD* and *BD*) to form an additional element. Thus it is seen that as each additional element of two members is placed on the truss, the number of joints for a simple truss is increased by one.

Fig. 6–5

Fig. 6–6

6.2 The Method of Joints

If a truss is in equilibrium, then each of its joints must also be in equilibrium. The method of joints is based on this fact, since it consists of satisfying the equilibrium conditions for the forces exerted *on the pin* at each joint of the truss. Because the truss members are all straight two-force members lying in the same plane, the force system acting at each pin is *coplanar and concurrent*. Consequently, rotational or moment equilibrium is automatically satisfied at the joint (or pin), and it is only necessary to satisfy $\Sigma F_x = 0$ and $\Sigma F_y = 0$ to ensure translational or force equilibrium.

When using the method of joints, it is *first* necessary to draw the joint's free-body diagram before applying the equilibrium equations. To do this, recall that the *line of action* of each member force acting on the joint is *specified* from the geometry of the truss, since the force in a member passes along the axis of the member. As an example, consider the pin at joint B of the truss in Fig. 6–7a. Three forces act on the pin, namely, the 500-N force and the forces exerted by members BA and BC. The free-body diagram is shown in Fig. 6–7b. As shown, $\mathbf{F}_{BA}$ is "pulling" on the pin, which means that member BA is in *tension;* whereas $\mathbf{F}_{BC}$ is "pushing" on the pin, and consequently member BC is in *compression*. These effects are clearly demonstrated by isolating the joint with small segments of the member connected to the pin, Fig. 6–7c. Notice that pushing or pulling on these small segments indicates the effect of the member being either in compression or tension.

In all cases, the analysis should start at a joint having at least one known force and at most two unknown forces, as in Fig. 6–7b. In this way, application of $\Sigma F_x = 0$ and $\Sigma F_y = 0$ yields two algebraic equations which can be

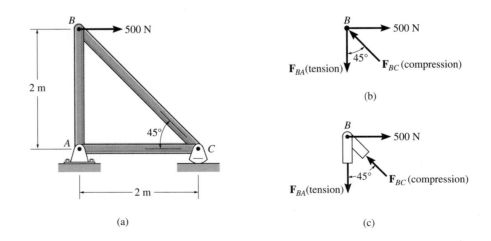

Fig. 6–7

solved for the two unknowns. When applying these equations, the correct sense of an unknown member force can be determined using one of two possible methods.

1. *Always assume* the *unknown member forces* acting on the joint's free-body diagram to be in *tension,* i.e., "pulling" on the pin. If this is done, then numerical solution of the equilibrium equations will yield *positive scalars for members in tension and negative scalars for members in compression.* Once an unknown member force is found, use its *correct* magnitude and sense (T or C) on subsequent joint free-body diagrams.

2. The *correct* sense of direction of an unknown member force can, in many cases, be determined "by inspection." For example, $\mathbf{F}_{BC}$ in Fig. 6–7b must push on the pin (compression) since its horizontal component, $F_{BC} \sin 45°$, must balance the 500-N force ($\Sigma F_x = 0$). Likewise, $\mathbf{F}_{BA}$ is a tensile force since it balances the vertical component, $F_{BC} \cos 45°$ ($\Sigma F_y = 0$). In more complicated cases, the sense of an unknown member force can be *assumed;* then, after applying the equilibrium equations, the assumed sense can be verified from the numerical results. A *positive* answer indicates that the sense is *correct,* whereas a *negative* answer indicates that the sense shown on the free-body diagram must be *reversed.* This is the method we will use in the example problems which follow.

PROCEDURE FOR ANALYSIS

The following procedure provides a typical means for analyzing a truss using the method of joints.

Draw the free-body diagram of a joint having at least one known force and at most two unknown forces. (If this joint is at one of the supports, it generally will be necessary to know the external reactions at the truss support.) Use one of the two methods described above for establishing the sense of an unknown force. Orient the x and y axes such that the forces on the free-body diagram can be easily resolved into their x and y components and then apply the two force equilibrium equations $\Sigma F_x = 0$ and $\Sigma F_y = 0$. Solve for the two unknown member forces and verify their correct sense.

Continue to analyze each of the other joints, where again it is necessary to choose a joint having at most two unknowns and at least one known force. Realize that once the force in a member is found from the analysis of a joint at one of its ends, the result can be used to analyze the forces acting on the joint at its other end. Strict adherence to the principle of action, equal but opposite reaction must, of course, be observed. Remember, a member in *compression* "pushes" on the joint and a member in *tension* "pulls" on the joint.

Once the force analysis of the truss has been completed, the size of the members and their connections can be determined using the theory of mechanics of materials along with information given in engineering design codes.

Example 6-1

(a)

(b)

(c)

(d)

(e)

Fig. 6-8

Determine the force in each member of the truss shown in Fig. 6–8a and indicate whether the members are in tension or compression.

SOLUTION

By inspection of Fig. 6–8a, there are two unknown member forces at joint B, two unknown member forces and an unknown reaction force at joint C, and two unknown member forces and two unknown reaction forces at joint A. Since we must have no more than two unknowns at the joint and at least one known force acting there, we must begin the analysis at joint B.

Joint B. The free-body diagram of the pin at B is shown in Fig. 6–8b. Three forces act on the pin: the external force of 500 N and the *two* unknown forces developed by members BA and BC. Applying the equations of joint equilibrium, we have

$$\xrightarrow{+} \Sigma F_x = 0; \qquad 500 - F_{BC} \sin 45° = 0 \qquad F_{BC} = 707.1 \text{ N} \quad \text{(C)} \quad \textit{Ans.}$$
$$+ \uparrow \Sigma F_y = 0; \qquad F_{BC} \cos 45° - F_{BA} = 0 \qquad F_{BA} = 500 \text{ N} \quad \text{(T)} \qquad \textit{Ans.}$$

Since the force in member BC has been calculated, we can proceed to analyze joint C in order to determine the force in member AC and the support reaction at the rocker.

Joint C. From the free-body diagram of joint C, Fig. 6–8c, we have

$$\xrightarrow{+} \Sigma F_x = 0; \qquad -F_{AC} + 707.1 \cos 45° = 0 \qquad F_{AC} = 500 \text{ N} \quad \text{(T)} \quad \textit{Ans.}$$
$$+ \uparrow \Sigma F_y = 0; \qquad C_y - 707.1 \sin 45° = 0 \qquad C_y = 500 \text{ N} \qquad \textit{Ans.}$$

Joint A. Although not necessary, we can determine the support reactions at joint A using the results of $F_{AC} = 500$ N and $F_{AB} = 500$ N. From the free-body diagram, Fig. 6–8d, we have

$$\xrightarrow{+} \Sigma F_x = 0; \qquad 500 - A_x = 0 \qquad A_x = 500 \text{ N}$$
$$+ \uparrow \Sigma F_y = 0; \qquad 500 - A_y = 0 \qquad A_y = 500 \text{ N}$$

The results of the analysis are summarized in Fig. 6–8e. Note that the free-body diagram of each pin shows the effects of all the connected members and external forces applied to the pin, whereas the free-body diagram of each member shows only the effects of the end pins on the member.

Example 6–2

Determine the forces acting in all the members of the truss shown in Fig. 6–9a.

SOLUTION

By inspection, there are more than two unknowns at each joint. Consequently, the support reactions on the truss must first be determined. Show that they have been correctly calculated on the free-body diagram in Fig. 6–9b. We can now begin the analysis at joint C. Why?

Joint C. From the free-body diagram, Fig. 6–9c,

$$\xrightarrow{+}\Sigma F_x = 0; \qquad -F_{CD} \cos 30° + F_{CB} \sin 45° = 0$$

$$+\uparrow\Sigma F_y = 0; \qquad 1.5 + F_{CD} \sin 30° - F_{CB} \cos 45° = 0$$

These two equations must be solved *simultaneously* for each of the two unknowns. Note, however, that a *direct solution* for one of the unknown forces may be obtained by applying a force summation along an axis that is *perpendicular* to the direction of the other unknown force. For example, summing forces along the y' axis, which is perpendicular to the direction of $\mathbf{F}_{CD}$, Fig. 6–9d, yields a direct solution for F_{CB}.

$$+\nearrow\Sigma F_{y'} = 0;$$

$$1.5 \cos 30° - F_{CB} \sin 15° = 0 \qquad F_{CB} = 5.02 \text{ kN} \quad (C) \quad Ans.$$

In a similar fashion, summing forces along the y'' axis, Fig. 6–9e, yields a direct solution for F_{CD}.

$$+\nearrow\Sigma F_{y''} = 0;$$

$$1.5 \cos 45° - F_{CD} \sin 15° = 0 \qquad F_{CD} = 4.10 \text{ kN} \quad (T) \quad Ans.$$

(c) 1.5 kN   (d) 1.5 kN

Joint D. We can now proceed to analyze joint D. The free-body diagram is shown in Fig. 6–9f.

$$\xrightarrow{+}\Sigma F_x = 0; \qquad -F_{DA} \cos 30° + 4.10 \cos 30° = 0$$

$$F_{DA} = 4.10 \text{ kN} \quad (T) \qquad\qquad Ans.$$

$$+\uparrow\Sigma F_y = 0; \qquad F_{DB} - 2(4.10 \sin 30°) = 0$$

$$F_{DB} = 4.10 \text{ kN} \quad (T) \qquad\qquad Ans.$$

The force in the last member, BA, can be obtained from joint B or joint A. As an exercise, draw the free-body diagram of joint B, sum the forces in the horizontal direction, and show that $F_{BA} = 0.777$ kN (C).

(a)

(b)

(e)

(f)

Fig. 6–9

Example 6–3

Determine the force in each member of the truss shown in Fig. 6–10a. Indicate whether the members are in tension or compression.

(a)   (b)

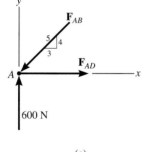

(c)

Fig. 6–10(a–f)

SOLUTION

Support Reactions.   No joint can be analyzed until the support reactions are determined. Why? A free-body diagram of the entire truss is given in Fig. 6–10b. Applying the equations of equilibrium, we have

$$\xrightarrow{+} \Sigma F_x = 0; \qquad\qquad\qquad 600 - C_x = 0 \qquad C_x = 600 \text{ N}$$
$$\zeta + \Sigma M_C = 0; \quad -A_y(6) + 400(3) + 600(4) = 0 \qquad A_y = 600 \text{ N}$$
$$+ \uparrow \Sigma F_y = 0; \qquad\qquad\qquad 600 - 400 - C_y = 0 \qquad C_y = 200 \text{ N}$$

The analysis can now start at either joint A or C. The choice is arbitrary, since there are one known and two unknown member forces acting on the pin at each of these joints.

Joint A (Fig. 6–10c).   As shown on the free-body diagram, there are three forces that act on the pin at joint A. The inclination of $\mathbf{F}_{AB}$ is determined from the geometry of the truss. By inspection, can you see why this force is assumed to be compressive and $\mathbf{F}_{AD}$ tensile? Applying the equations of equilibrium, we have

$$+ \uparrow \Sigma F_y = 0; \quad 600 - \tfrac{4}{5}F_{AB} = 0 \qquad F_{AB} = 750 \text{ N} \quad (\text{C}) \qquad\qquad Ans.$$
$$\xrightarrow{+} \Sigma F_x = 0; \quad F_{AD} - \tfrac{3}{5}(750) = 0 \qquad F_{AD} = 450 \text{ N} \quad (\text{T}) \qquad\qquad Ans.$$

Joint D (Fig. 6–10*d*).   The pin at this joint is chosen next since, by inspection of Fig. 6–10*a*, the force in *AD* is known and the unknown forces in *DB* and *DC* can be determined. Summing forces in the horizontal direction, Fig. 6–10*d*, we have

$$\xrightarrow{+} \Sigma F_x = 0; \quad -450 + \tfrac{3}{5}F_{DB} + 600 = 0 \quad F_{DB} = -250 \text{ N}$$

The negative sign indicates that $\mathbf{F}_{DB}$ acts in the *opposite sense* to that shown in Fig. 6–10*d*.* Hence,

$$F_{DB} = 250 \text{ N} \quad (\text{T}) \qquad \qquad \textit{Ans.}$$

To determine $\mathbf{F}_{DC}$, we can either correct the sense of $\mathbf{F}_{DB}$ and then apply $\Sigma F_y = 0$, or apply this equation and retain the negative sign for F_{DB}, i.e.,

$$+\uparrow \Sigma F_y = 0; \quad -F_{DC} - \tfrac{4}{5}(-250) = 0 \quad F_{DC} = 200 \text{ N} \quad (\text{C}) \qquad \textit{Ans.}$$

Joint C (Fig. 6–10*e*)

$$\xrightarrow{+} \Sigma F_x = 0; \quad F_{CB} - 600 = 0 \quad F_{CB} = 600 \text{ N} \quad (\text{C}) \qquad \textit{Ans.}$$
$$+\uparrow \Sigma F_y = 0; \quad 200 - 200 \equiv 0 \quad (\text{check})$$

The analysis is summarized in Fig. 6–10*f*, which shows the correct free-body diagram for each pin and member.

(d)

(e)

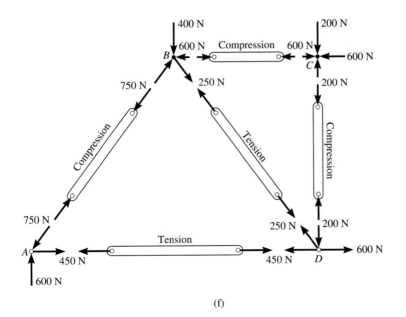

(f)

*The proper sense could have been determined by inspection, prior to applying $\Sigma F_x = 0$.

6.3 Zero-Force Members

Truss analysis using the method of joints is greatly simplified if one is able to first determine those members which support *no loading*. These *zero-force members* are used to increase the stability of the truss during construction and to provide support if the applied loading is changed.

The zero-force members of a truss can generally be determined *by inspection* of each of its joints. For example, consider the truss shown in Fig. 6–11*a*. If a free-body diagram of the pin at joint *A* is drawn, Fig. 6–11*b*, it is seen that members *AB* and *AF* are zero-force members. On the other hand, notice that we could not have come to this conclusion if we had considered the free-body diagrams of joints *F* or *B*, simply because there are five unknowns at each of these joints. In a similar manner, consider the free-body diagram of joint *D*, Fig. 6–11*c*. Here again it is seen that *DC* and *DE* are zero-force members. As a general rule, then, *if only two members form a truss joint and no external load or support reaction is applied to the joint, the members must be zero-force members.* The load on the truss in Fig. 6–11*a* is therefore supported by only five members as shown in Fig. 6–11*d*.

(a)

(d)

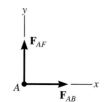

$$\xrightarrow{+} \Sigma F_x = 0; \; F_{AB} = 0$$
$$+\uparrow \Sigma F_y = 0; \; F_{AF} = 0$$

(b)

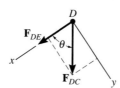

$$+ \searrow \Sigma F_y = 0; \; F_{DC} \sin \theta = 0; \; F_{DC} = 0 \text{ since } \sin \theta \neq 0$$
$$+ \swarrow \Sigma F_x = 0; \; F_{DE} + 0 = 0; \; F_{DE} = 0$$

(c)

Fig. 6–11

Now consider the truss shown in Fig. 6–12*a*. The free-body diagram of the pin at joint *D* is shown in Fig. 6–12*b*. By orienting the *y* axis along members *DC* and *DE* and the *x* axis along member *DA*, it is seen that *DA* is a zero-force member. Note that this is also the case for member *CA*, Fig. 6–12*c*. In general, then, *if three members form a truss joint for which two of the members are collinear, the third member is a zero-force member provided no external force or support reaction is applied to the joint.* The truss shown in Fig. 6–12*d* is therefore suitable for supporting the load **P.**

(a)

(d)

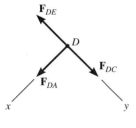

$+\swarrow \Sigma F_x = 0; \quad F_{DA} = 0$

$+\searrow \Sigma F_y = 0; \quad F_{DC} = F_{DE}$

(b)

$+\swarrow \Sigma F_x = 0; \quad F_{CA} \sin \theta = 0; \quad F_{CA} = 0$ since $\sin \theta \neq 0$;

$+\searrow \Sigma F_y = 0; \quad F_{CB} = F_{CD}$

(c)

Fig. 6–12

■ Example 6–4

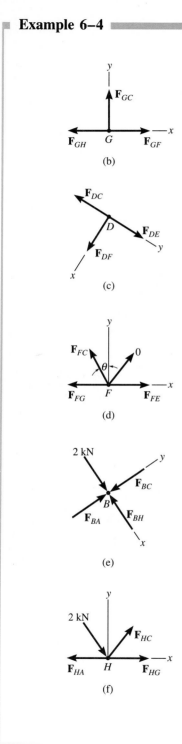

(b)

(c)

(d)

(e)

(f)

Using the method of joints, determine all the zero-force members of the *Fink truss* shown in Fig. 6–13a.

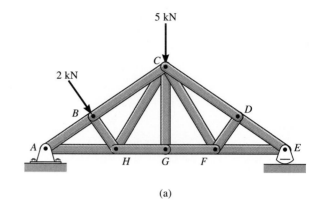

(a)

Fig. 6–13

SOLUTION

Looking for joint geometries that are similar to those outlined in Figs. 6–11 and 6–12, we have

Joint G (Fig. 6–13b)

$$+\uparrow \Sigma F_y = 0; \qquad\qquad F_{GC} = 0 \qquad\qquad\qquad \textit{Ans.}$$

Realize that we could not conclude that GC is a zero-force member by considering joint C, where there are five unknowns. The fact that GC is a zero-force member means that the 5-kN load at C must be supported by members CB, CH, CF, and CD.

Joint D (Fig. 6–13c)

$$+\swarrow \Sigma F_x = 0; \qquad\qquad F_{DF} = 0 \qquad\qquad\qquad \textit{Ans.}$$

Joint F (Fig. 6–13d)

$$+\uparrow \Sigma F_y = 0; \quad F_{FC} \cos \theta = 0, \text{ since } \theta \neq 0; \quad F_{FC} = 0 \qquad \textit{Ans.}$$

Note that if joint B is analyzed, Fig. 6–13e,

$$+\searrow \Sigma F_x = 0; \qquad 2 - F_{BH} = 0 \qquad F_{BH} = 2 \text{ kN} \quad (C)$$

Consequently, the numerical value of F_{HC} must satisfy $\Sigma F_y = 0$, Fig. 6–13f, and therefore HC is *not* a zero-force member.

PROBLEMS

6–1. Determine the force in each member of the truss and indicate whether the members are in tension or compresssion.

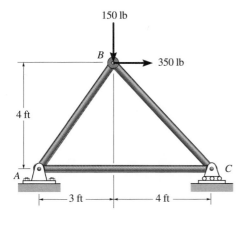

Prob. 6–1

6–2. Determine the force in each member of the truss and indicate whether the members are in tension or compression.

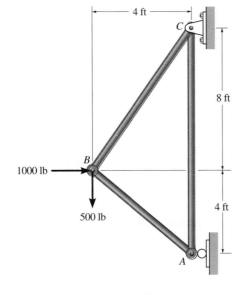

Prob. 6–2

6–3. Determine the force in each member of the truss and indicate whether the members are in tension or compression. Assume that all the members are pin-connected.

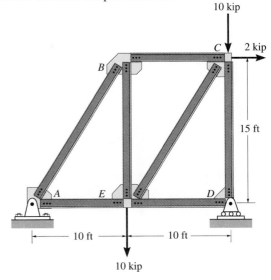

Prob. 6–3

*6–4. Determine the force in each member of the *A*-frame truss which is used to support the bridge-deck loading shown, and indicate whether the members are in tension or compression.

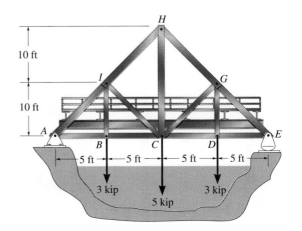

Prob. 6–4

6–5.   Determine the force in each member of the truss and indicate whether the members are in tension or compression.

6–6.   Each member of the truss is uniform and has a mass of 8 kg/m. Remove the external loads and determine the approximate force in each member due to the weight of the truss. Indicate whether the members are in tension or compression. Solve the problem by *assuming* that the weight of each member can be represented as a vertical force, half of which is applied at each end of the member.

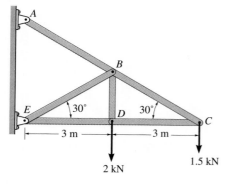

Probs. 6–5/6–6

6–7.   Determine the force in each member of the truss and indicate whether the members are in tension or compression.

*6–8.   Each member of the truss is uniform and has a mass of 8 kg/m. Remove the external loads and determine the approximate force in each member due to the weight of the truss. Indicate whether the members are in tension or compression. Solve the problem by *assuming* that the weight of each member can be represented as a vertical force, half of which is applied at each end of the member. Also, assume that all members are pin-connected.

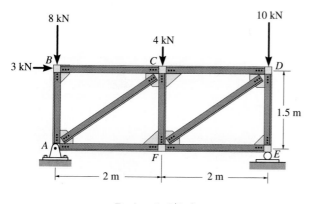

Probs. 6–7/6–8

6–9.   The sign is partially supported using the side truss shown. The wind load on the face of the sign transmits the loading of 5 kN to joints D and E of the truss. Determine the force in each member and indicate whether the members are in tension or compression.

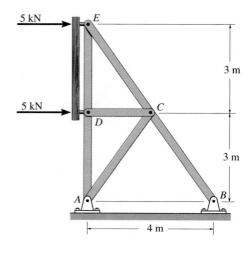

Prob. 6–9

6–10.   Determine the force in each member of the truss and indicate whether the members are in tension or compression. Assume that all members are pin-connected.

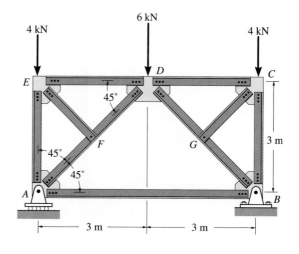

Prob. 6–10

6–11. Determine the force in each member of the truss and indicate whether the members are in tension or compression.

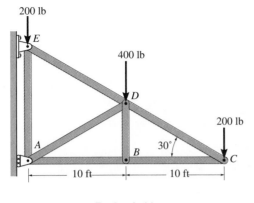

Prob. 6–11

*6–12. Determine the force in each member of the truss and indicate whether the members are in tension or compression.

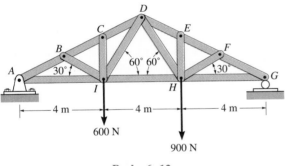

Prob. 6–12

6–13. Determine the force in each member of the truss and indicate whether the members are in tension or compression.

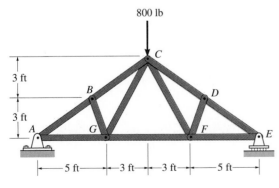

Prob. 6–13

6–14. The beam supported by the truss is subjected to a uniform distributed load of 4 kN/m. Assuming that the central 2 m of load is supported by joint C and the remaining portions by joints B and D, show that the vertical force at C due to the distributed load is 8 kN and that at B and D the force is 4 kN. Then calculate the force in members BC, GB, GC, GA, and AB, and indicate whether the members are in tension or compression. *Hint:* Start the analysis at joint G and show that the force in member GB is zero.

Prob. 6–14

6–15. Determine the force in each member of the truss in terms of the load P, and indicate whether the members are in tension or compression.

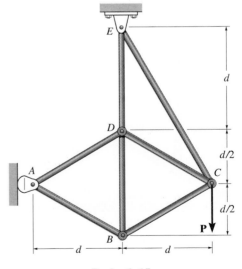

Prob. 6–15

*6–16.   Determine the force in each member and indicate whether the members are in tension or compression.

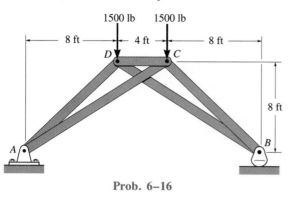

Prob. 6–16

6–17.   Determine the force in each member of the truss and indicate whether the members are in tension or compression. Assume that all members are pin-connected.

Prob. 6–17

6–18.   Determine the force in each member of the truss and indicate whether the members are in tension or compression.

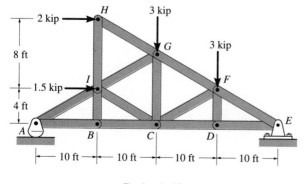

Prob. 6–18

6–19.   Determine the force in each member of the roof truss and indicate whether the members are in tension or compression.

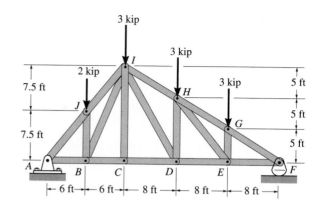

Prob. 6–19

*6–20.   Determine the force in each member of the truss and indicate whether the members are in tension or compression.

Prob. 6–20

6.4 The Method of Sections

The *method of sections* is used to determine the loadings acting within a body. It is based on the principle that if a body is in equilibrium, then any part of the body is also in equilibrium. To apply this method, one passes an *imaginary section* through the body, thus cutting it into two parts. When a free-body diagram of one of the parts is drawn, the loads acting at the section must be *included* on the free-body diagram. One then applies the equations of equilibrium to the part in order to determine the loading at the section. For example, consider the two truss members shown colored in Fig. 6–14. The internal loads at the section indicated by the dashed line can be obtained using one of the free-body diagrams shown on the right. Clearly, it can be seen that equilibrium requires that the member in tension be subjected to a "pull" **T** at the section, whereas the member in compression is subjected to a "push" **C**.

The method of sections can also be used to "cut" or section several members of an entire truss. If either of the two parts of the truss is isolated as a free-body diagram, we can then apply the equations of equilibrium to that part to determine the member forces at the "cut section." Since only *three* independent equilibrium equations ($\Sigma F_x = 0$, $\Sigma F_y = 0$, $\Sigma M_O = 0$) can be applied to the isolated part of the truss, one should try to select a section that, in general, passes through not more than *three* members in which the forces are unknown. For example, consider the truss in Fig. 6–15a. If the force in member *GC* is to be determined, section *aa* would be appropriate. The free-body diagrams of the two parts are shown in Figs. 6–15b and 6–15c. In particular, note that the line of action of each cut member force is specified from the *geometry* of the truss, since the force in a member passes along its axis. Also, the member forces acting on one part of the truss are equal but opposite to those acting on the other part—Newton's third law. As noted above, members assumed to be in *tension* (*BC* and *GC*) are subjected to a "pull," whereas the member in *compression* (*GF*) is subjected to a "push."

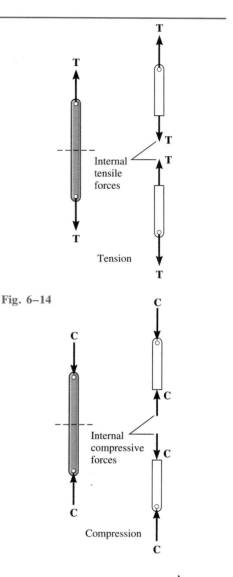

Fig. 6–14

Tension

Compression

Fig. 6–15

(a)

(b)

(c)

(a)

(b)

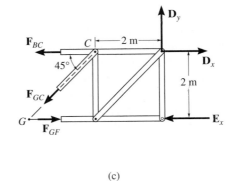

(c)

Fig. 6–15 (Repeated)

The three unknown member forces $\mathbf{F}_{BC}$, $\mathbf{F}_{GC}$, and $\mathbf{F}_{GF}$ can be obtained by applying the three equilibrium equations to the free-body diagram in Fig. 6–15b. If, however, the free-body diagram in Fig. 6–15c is considered, the three support reactions $\mathbf{D}_x$, $\mathbf{D}_y$, and $\mathbf{E}_x$ will have to be determined *first*. Why? (This, of course, is done in the usual manner by considering a free-body diagram of the *entire truss*.) When applying the equilibrium equations, one should consider ways of writing the equations so as to yield a *direct solution* for each of the unknowns, rather than having to solve simultaneous equations. For example, summing moments about C in Fig. 6–15b would yield a direct solution for $\mathbf{F}_{GF}$ since $\mathbf{F}_{BC}$ and $\mathbf{F}_{GC}$ create zero moment about C. Likewise, $\mathbf{F}_{BC}$ can be directly obtained by summing moments about G. Finally, $\mathbf{F}_{GC}$ can be found directly from a force summation in the vertical direction since $\mathbf{F}_{GF}$ and $\mathbf{F}_{BC}$ have no vertical components. This ability to *determine directly* the force in a particular truss member is one of the main advantages of using the method of sections.*

---

*By comparison, if the method of joints were used to determine, say, the force in member GC, it would be necessary to analyze joints A, B, and G in sequence.

As in the method of joints, there are two ways in which one can determine the correct sense of an unknown member force.

1. *Always assume* that the unknown member forces at the cut section are in *tension,* i.e., "pulling" on the member. By doing this, the numerical solution of the equilibrium equations will yield *positive scalars for members in tension and negative scalars for members in compression.*

2. The correct sense of an unknown member force can in many cases be determined "by inspection." For example, $\mathbf{F}_{BC}$ is a tensile force as represented in Fig. 6–15b, since moment equilibrium about G requires that $\mathbf{F}_{BC}$ create a moment opposite to that of the 1000-N force. Also, $\mathbf{F}_{GC}$ is tensile since its vertical component must balance the 1000-N force acting downward. In more complicated cases, the sense of an unknown member force may be *assumed*. If the solution yields a *negative* scalar, it indicates that the force's sense is *opposite* to that shown on the free-body diagram. This is the method we will use in the example problems which follow.

PROCEDURE FOR ANALYSIS

The following procedure provides a means for applying the method of sections to determine the forces in the members of a truss.

Free-Body Diagram. Make a decision as to how to "cut" or section the truss through the members where forces are to be determined. Before isolating the appropriate section, it may first be necessary to determine the truss's *external* reactions, so that the three equilibrium equations are used *only* to solve for member forces at the cut section. Draw the free-body diagram of that part of the sectioned truss which has the least number of forces acting on it. Use one of the two methods described above for establishing the sense of an unknown member force.

Equations of Equilibrium. Try to apply the three equations of equilibrium such that simultaneous solution of equations is avoided. In this regard, moments should be summed about a point that lies at the intersection of the lines of action of two unknown forces, so that the third unknown force is determined directly from the moment equation. If two of the unknown forces are *parallel,* forces may be summed *perpendicular* to the direction of these unknowns to determine *directly* the third unknown force.

The following examples illustrate these concepts numerically.

■ Example 6–5

(a)

(b)

Determine the force in members GE, GC, and BC of the truss shown in Fig. 6–16a. Indicate whether the members are in tension or compression.

SOLUTION

Section aa in Fig. 6–16a has been chosen since it cuts through the *three* members whose forces are to be determined. In order to use the method of sections, however, it is *first* necessary to determine the external reactions at A or D. Why? A free-body diagram of the entire truss is shown in Fig. 6–16b. Applying the equations of equilibrium, we have

$$\xrightarrow{+}\Sigma F_x = 0; \qquad 400\ N - A_x = 0 \qquad A_x = 400\ N$$
$$\curvearrowleft+\Sigma M_A = 0; \quad -1200\ N(8\ m) - 400\ N(3\ m) + D_y(12\ m) = 0$$
$$D_y = 900\ N$$
$$+\uparrow\Sigma F_y = 0; \quad A_y - 1200\ N + 900\ N = 0 \qquad A_y = 300\ N$$

Free-Body Diagrams. The free-body diagrams of the sectioned truss are shown in Figs. 6–16c and 6–16d. For the analysis the free-body diagram in Fig. 6–16c will be used since it involves the least number of forces.

(c)

Equations of Equilibrium. Summing moments about point G eliminates $\mathbf{F}_{GE}$ and $\mathbf{F}_{GC}$ and yields a direct solution for F_{BC}.

$$\curvearrowleft+\Sigma M_G = 0; \quad -300\ N(4\ m) - 400\ N(3\ m) + F_{BC}(3\ m) = 0$$
$$F_{BC} = 800\ N \quad (T) \qquad\qquad Ans.$$

In the same manner, by summing moments about point C we obtain a direct solution for F_{GE}.

$$\curvearrowleft+\Sigma M_C = 0; \qquad -300\ N(8\ m) + F_{GE}(3\ m) = 0$$
$$F_{GE} = 800\ N \quad (C) \qquad\qquad Ans.$$

(d)

Since $\mathbf{F}_{BC}$ and $\mathbf{F}_{GE}$ have no vertical components, summing forces in the y direction directly yields F_{GC}, i.e.,

$$+\uparrow\Sigma F_y = 0; \qquad 300\ N - \tfrac{3}{5}F_{GC} = 0$$
$$F_{GC} = 500\ N \quad (T) \qquad\qquad Ans.$$

Obtain these results by applying the equations of equilibrium to the free-body diagram shown in Fig. 6–16d.

Fig. 6–16

Example 6–6

Determine the force in member CF of the truss shown in Fig. 6–17a. Indicate whether the member is in tension or compression.

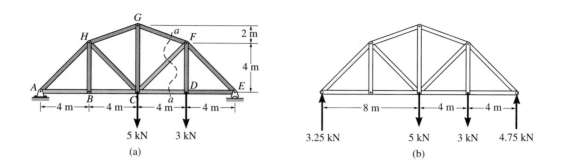

(a)

(b)

SOLUTION

Free-Body Diagram.   Section aa in Fig. 6–17a will be used since this section will "expose" the internal force in member CF as "external" on the free-body diagram of either the right or left portion of the truss. It is first necessary, however, to determine the external reactions on either the left or right side of the truss. Verify the results shown on the free-body diagram in Fig. 6–17b.

The free-body diagram of the right portion of the truss, which is the easiest to analyze, is shown in Fig. 16–17c. There are three unknowns, F_{FG}, F_{CF}, and F_{CD}.

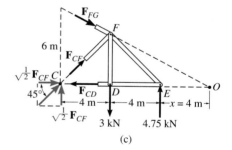

(c)

Fig. 6–17

Equations of Equilibrium.   The most direct method for solving this problem requires application of the moment equation about a point that eliminates two of the unknown forces. Hence, to obtain $\mathbf{F}_{CF}$, we will eliminate $\mathbf{F}_{FG}$ and $\mathbf{F}_{CD}$ by summing moments about point O, Fig. 6–17c. Note that the location of point O measured from E is determined from proportional triangles, i.e., $4/(4 + x) = 6/(8 + x)$, $x = 4$ m. Or, stated in another manner, the slope of member GF has a drop of 2 m to a horizontal distance of $CD = 4$ m. Since FD is 4 m, Fig. 6–17c, then from D to O the distance must be 8 m.

An easy way to determine the moment of $\mathbf{F}_{CF}$ about point O is to resolve $\mathbf{F}_{CF}$ into its two rectangular components and then use the principle of transmissibility to move $\mathbf{F}_{CF}$ to point C. We have

$$\zeta+\Sigma M_O = 0; \quad -\frac{1}{\sqrt{2}}F_{CF}(12 \text{ m}) + (3 \text{ kN})(8 \text{ m}) - (4.75 \text{ kN})(4 \text{ m}) = 0$$

$$F_{CF} = 0.589 \text{ kN} \quad \text{(C)} \qquad\qquad \textit{Ans.}$$

Example 6–7

1000 N

3000 N   1000 N

1000 N

4000 N   2000 N

(a)

1000 N

3000 N

1000 N

F_{FB} F_{EB} F_{ED}

F_{AB}

|— 2 m —|— 2 m —|—— 4 m ——|

4000 N F_{ED} sin 30°

(b)

y

1000 N

F_{EF} F_{ED} = 3000 N

F_{EB}

(c)

Fig. 6–18

Determine the force in member EB of the truss shown in Fig. 6–18a. Indicate whether the member is in tension or compression.

SOLUTION

Free-Body Diagrams.   By the method of sections, any imaginary vertical section that cuts through EB, Fig. 6–18a, will also have to cut through three other members for which the forces are unknown. For example, section aa cuts through ED, EB, FB, and AB. If the components of reaction at A are calculated first ($A_x = 0$, $A_y = 4000$ N) and a free-body diagram of the left side of this section is considered, Fig. 6–18b, it is possible to obtain F_{ED} by summing moments about B to eliminate the other three unknowns; however, F_{EB} cannot be determined from the remaining two equilibrium equations. One possible way of obtaining F_{EB} is first to determine F_{ED} from section aa, then use this result on section bb, Fig. 6–18a, which is shown in Fig. 6–18c. Here the force system is concurrent and our sectioned free-body diagram is the same as the free-body diagram for the pin at E (method of joints).

Equations of Equilibrium.   In order to determine the moment of $\mathbf{F}_{ED}$ about point B, Fig. 6–18b, we will resolve the force into its rectangular components and, by the principle of transmissibility, extend it to point C as shown. The moments of 1000 N, F_{AB}, F_{FB}, F_{EB}, and F_{ED} cos 30° are all zero about B. Therefore,

$$\zeta + \Sigma M_B = 0; \quad 1000(4) + 3000(2) - 4000(4) + F_{ED} \sin 30°(4) = 0$$
$$F_{ED} = 3000 \text{ N} \quad \text{(C)}$$

Considering now the free-body diagram of section bb, Fig. 6–18c, we have

$$\xrightarrow{+} \Sigma F_x = 0; \quad F_{EF} \cos 30° - 3000 \cos 30° = 0$$
$$F_{EF} = 3000 \text{ N} \quad \text{(C)}$$

$$+ \uparrow \Sigma F_y = 0; \quad 2(3000 \sin 30°) - 1000 - F_{EB} = 0$$
$$F_{EB} = 2000 \text{ N} \quad \text{(T)} \qquad \qquad Ans.$$

PROBLEMS

6–21. Determine the force in members *GF*, *FC*, and *CD* of the bridge truss and indicate whether the members are in tension or compression.

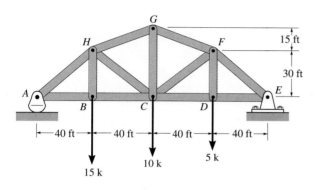

Prob. 6–21

6–22. Determine the force in members *HG*, *GC*, and *HC* of the truss and indicate whether the members are in tension or compression.

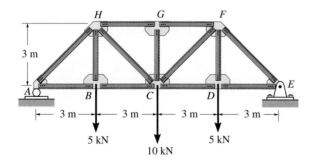

Prob. 6–22

6–23. Determine the force in members *CF* and *GC* of the truss and indicate whether the members are in tension or compression.

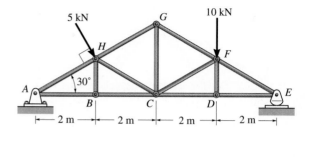

Prob. 6–23

*6–24. Determine the force in members *BE* and *EF* of the truss and indicate whether the members are in tension or compression. After sectioning the truss, solve for each force *directly*, using a single equation of equilibrium to determine each force.

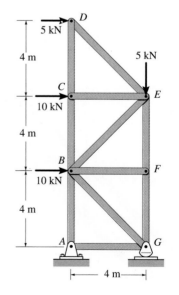

Prob. 6–24

6–25. The *Howe* truss is subjected to the loading shown. Determine the force in members *GF*, *CD*, and *GC*, and indicate whether the members are in tension or compression.

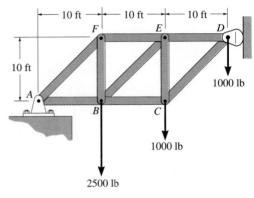

Prob. 6–25

6–26. The *Warren* truss is used to support a staircase. Determine the force in members *CE*, *ED*, and *DF*, and indicate whether the members are in tension or compression. Assume that all joints are pinned.

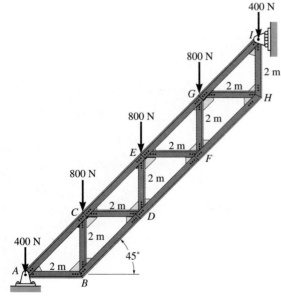

Prob. 6–26

6–27. Determine the force in members *CE*, *FE*, and *CD*, and indicate whether the members are in tension or compression.

*6–28. Determine the force in members *BC*, *EB*, and *AF*, and indicate whether the members are in tension or compression.

Probs. 6–27/6–28

6–29. Determine the force developed in members *BC* and *CH* of the roof truss and indicate whether the members are in tension or compression.

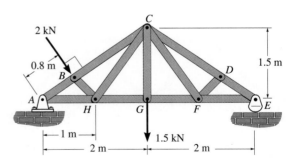

Prob. 6–29

6–30. Determine the force in members *FE*, *FB*, and *BC* of the truss and indicate whether the members are in tension or compression. After sectioning the truss, solve for each force *directly* using a single equilibrium equation to obtain each force. Assume that all joints are pin-connected.

*6–32. Determine the force in members *BG*, *HG*, and *BC* of the truss and indicate whether the members are in tension or compression.

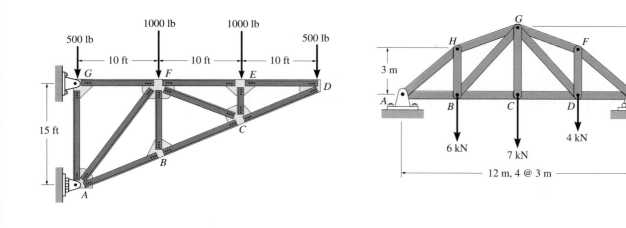

Prob. 6–30

Prob. 6–32

6–33. Indicate all zero-force members, then determine the force in members *IC* and *CG* of the truss and indicate whether the members are in tension or compression.

6–31. For the given loading, determine the force in members *CD*, *CJ*, and *KJ* of the *Howe roof truss*. Indicate whether the members are in tension or compression.

6–34. Indicate all zero-force members, then determine the force in members *JE* and *GF* of the truss and indicate whether the members are in tension or compression.

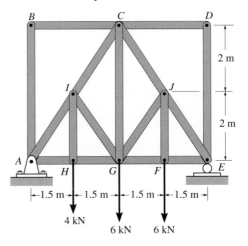

Prob. 6–31

Probs. 6–33/6–34

6–35. Determine the force in members *BC*, *HC*, and *HG*, and indicate whether the members are in tension or compression. After the truss is sectioned, use a single equation of equilibrium to obtain each force.

*6–36. Determine the force in members *CD*, *CF*, and *CG*, and indicate whether the members are in tension or compression.

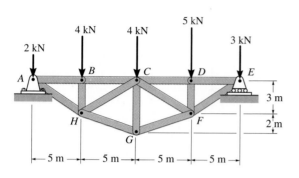

Probs. 6–35/6–36

6–37. Determine the force in members *KJ*, *JN*, and *CD*, and indicate whether the members are in tension or compression.

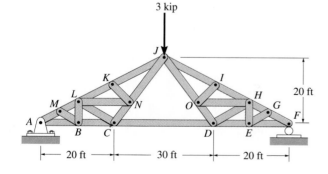

Prob. 6–37

6–38. Determine the force in members *GF* and *GD* of the truss, and indicate whether the members are in tension or compression.

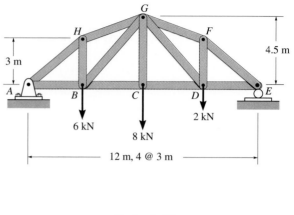

Prob. 6–38

6–39. Determine the force in members *DE*, *JI*, and *DO* of the *K truss*, and indicate whether the members are in tension or compression. *Hint:* Use sections *aa* and *bb*.

*6–40. Determine the force in members *CD* and *KJ* of the *K truss*, and indicate whether the members are in tension or compression. *Hint:* Note section *aa* can be used to find the force in members *DE* and *JI*.

Probs. 6–39/6–40

*6.5 Space Trusses

A *space truss* consists of members joined together at their ends to form a stable three-dimensional structure. The simplest element of a space truss is a *tetrahedron,* formed by connecting six members together, as shown in Fig. 6–19. Any additional members added to this basic element would be redundant in supporting the force **P.** A *simple space truss* can be built from this basic tetrahedral element by adding three additional members and a joint forming a system of multiconnected tetrahedrons.

Assumptions for Design. The members of a space truss may be treated as two-force members provided the external loading is applied at the joints and the joints consist of ball-and-socket connections. These assumptions are justified if the welded or bolted connections of the joined members intersect at a common point and the weight of the members can be neglected. In cases where the weight of a member is to be included in the analysis, it is generally satisfactory to apply it as a vertical force, half of its magnitude applied at each end of the member.

Fig. 6–19

PROCEDURE FOR ANALYSIS

Either the method of sections or the method of joints can be used to determine the forces developed in the members of a simple space truss.

Method of Sections. If only a *few* member forces are to be determined, the method of sections may be used. When an imaginary section is passed through a truss, and the truss is separated into two parts, the force system acting on one of the parts must satisfy the *six* scalar equilibrium equations: $\Sigma F_x = 0$, $\Sigma F_y = 0$, $\Sigma F_z = 0$, $\Sigma M_x = 0$, $\Sigma M_y = 0$, $\Sigma M_z = 0$ (Eqs. 5–6). By proper choice of the section and axes for summing forces and moments, many of the unknown member forces in a space truss can be computed *directly,* using a single equilibrium equation.

Method of Joints. Generally, if the forces in *all* the members of the truss must be determined, the method of joints is most suitable for the analysis. When using the method of joints, it is necessary to solve the three scalar equilibrium equations $\Sigma F_x = 0$, $\Sigma F_y = 0$, $\Sigma F_z = 0$ at each joint. The solution of many simultaneous equations can be avoided if the force analysis begins at a joint having at least one known force and at most three unknown forces. If the three-dimensional geometry of the force system at the joint is hard to visualize, it is recommended that a Cartesian vector analysis be used for the solution.

The following example illustrates this procedure numerically.

Example 6–8

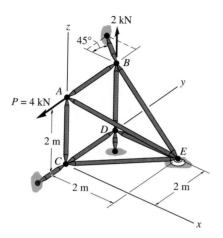

(a)

Determine the forces acting in the members of the space truss shown in Fig. 6–20a. Indicate whether the members are in tension or compression.

SOLUTION

Since there are one known force and three unknown forces acting at joint A, the force analysis of the truss will begin at this joint.

Joint A (Fig. 6–20b).   Expressing each force that acts on the free-body diagram of joint A in vector notation, we have

$$\mathbf{P} = \{-4\mathbf{j}\} \text{ kN}, \qquad \mathbf{F}_{AB} = F_{AB}\mathbf{j}, \qquad \mathbf{F}_{AC} = -F_{AC}\mathbf{k},$$

$$\mathbf{F}_{AE} = F_{AE}\left(\frac{\mathbf{r}_{AE}}{r_{AE}}\right) = F_{AE}(0.577\mathbf{i} + 0.577\mathbf{j} - 0.577\mathbf{k})$$

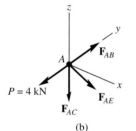

(b)

For equilibrium,

$$\Sigma\mathbf{F} = \mathbf{0}; \qquad \mathbf{P} + \mathbf{F}_{AB} + \mathbf{F}_{AC} + \mathbf{F}_{AE} = \mathbf{0}$$

$$-4\mathbf{j} + F_{AB}\mathbf{j} - F_{AC}\mathbf{k} + 0.577F_{AE}\mathbf{i} + 0.577F_{AE}\mathbf{j} - 0.577F_{AE}\mathbf{k} = \mathbf{0}$$

$$\Sigma F_x = 0; \qquad\qquad 0.577F_{AE} = 0$$

$$\Sigma F_y = 0; \qquad -4 + F_{AB} + 0.577F_{AE} = 0$$

$$\Sigma F_z = 0; \qquad\qquad -F_{AC} - 0.577F_{AE} = 0$$

$$F_{AC} = F_{AE} = 0 \qquad\qquad\qquad\qquad \textit{Ans.}$$

$$F_{AB} = 4 \text{ kN} \quad \text{(T)} \qquad\qquad \textit{Ans.}$$

Since F_{AB} is known, joint B may be analyzed next.

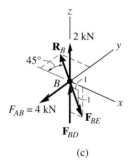

(c)

Fig. 6–20

Joint B (Fig. 6–20c)

$$\Sigma F_x = 0; \qquad -R_B \cos 45° + 0.707F_{BE} = 0$$

$$\Sigma F_y = 0; \qquad -4 + R_B \sin 45° = 0$$

$$\Sigma F_z = 0; \qquad 2 + F_{BD} - 0.707F_{BE} = 0$$

$$R_B = F_{BE} = 5.66 \text{ kN} \quad \text{(T)}, \qquad F_{BD} = 2 \text{ kN} \quad \text{(C)} \qquad \textit{Ans.}$$

The *scalar* equations of equilibrium may also be applied directly to the force systems on the free-body diagrams of joints D and C, since the force components are easily determined. Show that

$$F_{DE} = F_{DC} = F_{CE} = 0 \qquad\qquad\qquad \textit{Ans.}$$

PROBLEMS

6–41. Determine the force in each member of the space truss and indicate whether the members are in tension or compression. The truss is supported by ball-and-socket joints at D, C, and E. *Hint:* The support reaction at E acts along member EB. Why?

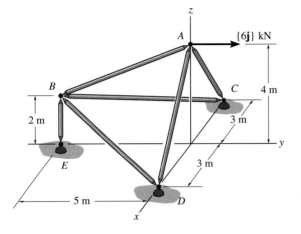

Prob. 6–41

6–42. Determine the force in each member of the space truss and indicate whether the members are in tension or compression. The support reactions at A, B, and C are to be determined.

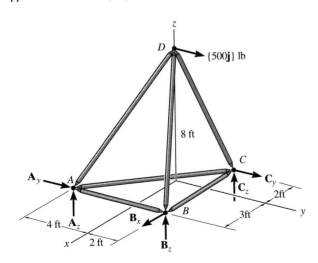

Prob. 6–42

6–43. Determine the force in each member of the space truss and indicate whether the members are in tension or compression. The crate has a mass of 50 kg.

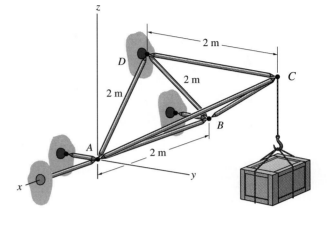

Prob. 6–43

*6–44. The space truss is supported by a ball-and-socket joint at A and short links at D, E, and F. Determine the force in each member and indicate whether the members are in tension or compression.

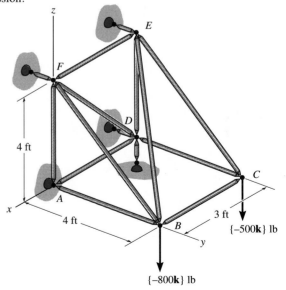

Prob. 6–44

6–45. Determine the force in each member of the space truss when joint A is subjected to the force of 600 N. Indicate whether the members are in tension or compression. The support reactions at B, C, and D are indicated.

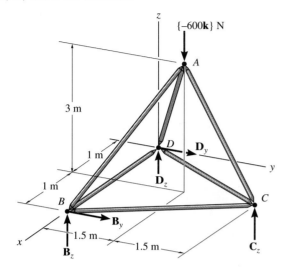

Prob. 6–45

6–46. The space truss is used to support vertical forces at joints D, E, and F. Determine the force developed in each member and indicate whether the members are in tension or compression.

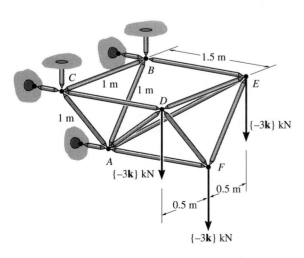

Prob. 6–46

6–47. Determine the force in each member of the space truss and indicate whether the members are in tension or compression.

Prob. 6–47

*6–48. Determine the force in members FE and ED of the space truss and indicate whether the members are in tension or compression. The support reactions at A, B, and C are indicated.

Prob. 6–48

6.6 Frames and Machines

Frames and machines are two common types of structures which are often composed of pin-connected *multiforce members,* i.e., members that are subjected to more than two forces. *Frames* are generally stationary and are used to support loads, whereas *machines* contain moving parts and are designed to transmit and alter the effect of forces. Provided a frame or machine is properly constrained and contains no more supports or members than are necessary to prevent collapse, the forces acting at the joints and supports can be determined by applying the equations of equilibrium to each member. Once the forces at the joints are obtained, it is then possible to *design* the size of the members, connections, and supports using the theory of mechanics of materials and an appropriate engineering design code.

Free-Body Diagrams. In order to determine the forces acting at the joints and supports of a frame or machine, the structure must be disassembled and the free-body diagrams of its parts must be drawn. In this regard, the following important points *must* be observed:

1. Isolate each part by drawing its *outlined shape*. Then show all the forces and/or couple moments that act on the part. Make sure to *label* or *identify* each known and unknown force and couple moment and indicate any dimensions used for taking moments. Most often the equations of equilibrium are easier to apply if the forces are represented by their rectangular components. As usual, the sense of an unknown force or couple moment can be assumed.
2. Identify all the two-force members in the structure, and represent their free-body diagrams as having two equal but opposite forces acting at their points of application. The line of action of the forces is defined by the line joining the two points where the forces act (see Sec. 5.4). By recognizing the two-force members, we can avoid solving an unnecessary number of equilibrium equations. (See Example 6–14.)
3. Forces common to any two *contacting* members act with equal magnitudes but opposite sense on the respective members. If the two members are treated as a *"system" of connected members,* then these forces are *"internal"* and are *not shown* on the *free-body diagram of the system;* however, if the free-body diagram of *each member* is drawn, the forces are *"external"* and *must* be shown on each of the free-body diagrams.

The following examples graphically illustrate application of these points in drawing the free-body diagrams of a dismembered frame or machine. In all cases, the weight of the members is neglected.

Example 6–9

For the frame shown in Fig. 6–21a, draw the free-body diagram of (a) each member, (b) the pin at B, and (c) the two members connected together.

(a)

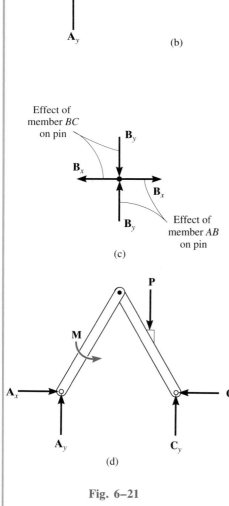

(b)

(c)

(d)

Fig. 6–21

SOLUTION

Part (a).   By inspection, the members are *not* two-force members. Instead, as shown on the free-body diagrams, Fig. 6–21b, BC is subjected to *not* five but *three forces,* namely, the *resultants* of the two components of reaction at pins B and C and the external force **P.** Likewise, AB is subjected to the *resultant* pin-reactive forces at A and B and the external couple moment **M.**

Part (b).   It can be seen in Fig. 6–21a that the pin at B is subjected to only *two forces,* i.e., the force of member BC on the pin and the force of member AB on the pin. For *equilibrium* these forces and therefore their respective components must be equal but opposite, Fig. 6–21c. Notice carefully how Newton's third law is applied between the pin and its contacting members, i.e., the effect of the pin on the two members, Fig. 6–21b, and the equal but opposite effect of the two members on the pin, Fig. 6–21c. Also note that B_x and B_y shown equal but opposite in Fig. 6–21b on members AB and BC is *not* the effect of Newton's third law; instead, this results from the *equilibrium* analysis of the pin, Fig. 6–21c.

Part (c).   The free-body diagram of both members connected together, yet removed from the supporting pins at A and C, is shown in Fig. 6–21d. The force components B_x and B_y are *not shown* on this diagram since they form equal but opposite collinear pairs of *internal* forces (Fig. 6–21b) and therefore cancel out.* Also, to be consistent when later applying the equilibrium equations, the unknown force components at A and C must act in the *same sense* as those shown in Fig. 6–21b.

*This is similar to not including internal forces exerted between adjacent particles of a rigid body when drawing the free-body diagram of the entire rigid body.

Example 6–10

For the frame shown in Fig. 6–22a, draw the free-body diagrams of (a) each of the three members, and (b) members *ABC* and *BD* together.

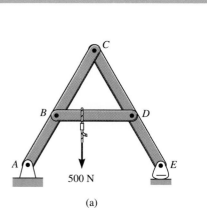

(a)

SOLUTION

Part (a).   By inspection, none of the three members of the frame are two-force members. Instead, each is subjected to *three* forces. The components of these forces are shown on the free-body diagrams in Fig. 6–22b. Notice that equal but opposite force reactions occur at *B*, *C*, and *D*. Draw a free-body diagram of one of the pins at *B*, *C*, or *D* and show why this is so.

Part (b).   The free-body diagram of *ABC* and *BD* together is shown in Fig. 6–22c. Since the entire frame is in equilibrium, the force system on these two members also satisfies the equilibrium equations. Why not show the force components $\mathbf{B}_x$ and $\mathbf{B}_y$ on this diagram?

(b)

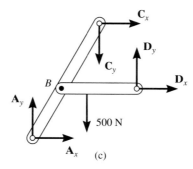

(c)

Fig. 6–22

Example 6–11

Draw the free-body diagram of each part of the smooth piston and link mechanism shown in Fig. 6–23a.

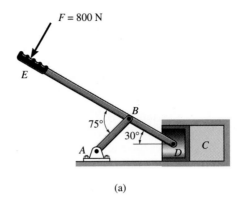

(a)

SOLUTION

By inspection, member AB is a two-force member. The free-body diagrams of the parts are shown in Fig. 6–23b. Since the pins at B and D *connect only two parts together,* the forces there are shown as equal but opposite on the separate free-body diagrams of their connected members. In particular, four components of force act on the piston: $\mathbf{D}_x$ and $\mathbf{D}_y$ represent the effect of the pin (or lever EBD), $\mathbf{N}_w$ is the *resultant force* of the cylinder's wall, and $\mathbf{P}$ is the resultant compressive force caused by the material within the cylinder C.

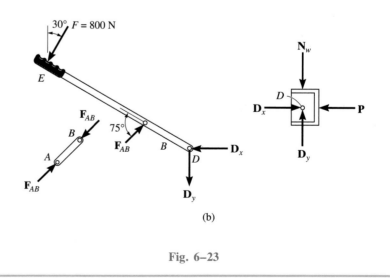

(b)

Fig. 6–23

Example 6–12

For the frame shown in Fig. 6–24a, draw the free-body diagrams of (a) the entire frame including the pulleys and cords, (b) the frame without the pulleys and cords, and (c) each of the pulleys.

(a)

SOLUTION

Part (a):   When the entire frame including the pulleys and cords is considered, the interactions at the points where the pulleys and cords are connected to the frame become pairs of *internal forces* which cancel each other and therefore are not shown on the free-body diagram, Fig. 6–24b.

Part (b):   When the cords and pulleys are removed, their effect *on the frame* must be shown, Fig. 6–24c.

Part (c):   The force components B_x, B_y, C_x, C_y of the pins on the pulleys, Fig. 6–24d, are equal but opposite to the force components exerted by the pins on the frame, Fig. 6–24c. Why?

Fig. 6–24

Example 6–13

The hydraulic truck-mounted crane shown in Fig. 6–25a is used to lift a beam that has a mass of 1 Mg. Draw the free-body diagrams of each of its parts, including the pins at *A* and *C*.

Fig. 6–25

(a)

SOLUTION

By inspection, *HF*, *EC*, and *AB* are all two-force members. The free-body diagrams are shown in Fig. 6–25b. The pin at *A* is subjected to only *two* forces, namely, the force of the link *AB* and the force of the support. For equilibrium, these forces must be equal in magnitude but opposite in direction. The pin at *C*, however, is subjected to *three* forces. The force $\mathbf{F}_{EC}$ is caused by the hydraulic cylinder, the force components $\mathbf{C}_x$ and $\mathbf{C}_y$ are caused by member *CBD*, and finally, $\mathbf{C}'_x$ and $\mathbf{C}'_y$ are caused by the support. These components can be related by the equations of force equilibrium applied to the pin.

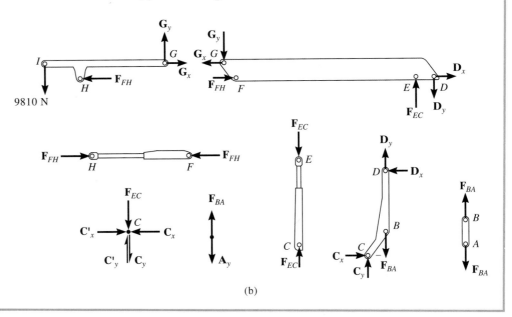

(b)

Before proceeding, it is recommended to cover the solutions to the previous examples and attempt to draw the requested free-body diagrams. When doing so, make sure the work is neat and that all the forces and couple moments are properly labeled.

Equations of Equilibrium.

Provided the structure (frame or machine) is properly supported and contains no more supports or members than are necessary to prevent its collapse, then the unknown forces at the supports and connections can be determined from the equations of equilibrium. If the structure lies in the x–y plane, then for *each* free-body diagram drawn the loading must satisfy $\Sigma F_x = 0$, $\Sigma F_y = 0$, and $\Sigma M_O = 0$. The selection of the free-body diagrams used for the analysis is *completely arbitrary*. They may represent each of the members of the structure, a portion of the structure, or its entirety. For example, consider finding the six components of the pin reactions at A, B, and C for the frame shown in Fig. 6–26a. If the frame is dismembered, Fig. 6–26b, these unknowns can be determined by applying the three equations of equilibrium to each of the two members (total of six equations). The free-body diagram of the *entire frame* can also be used for part of the analysis, Fig. 6–26c. Hence, if so desired, all six unknowns can be determined by applying the three equilibrium equations to the entire frame, Fig. 6–26c, and also to either one of its members. Furthermore, the answers can be checked in part by applying the three equations of equilibrium to the remaining ''second'' member. In general, then, this problem can be solved by writing *at most* six equilibrium equations using free-body diagrams of the members and/or the combination of connected members. Any more than six equations written would *not* be unique from the original six and would serve only to check the results.

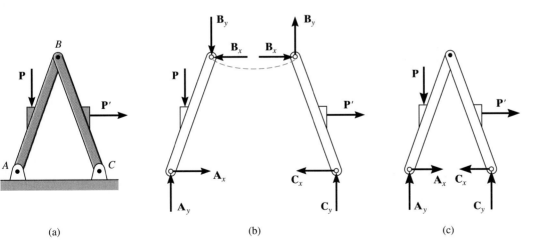

(a)   (b)   (c)

Fig. 6–26

PROCEDURE FOR ANALYSIS

The following procedure provides a method for determining the *joint reactions* of frames and machines (structures) composed of multiforce members.

Free-Body Diagrams.   Draw the free-body diagram of the entire structure, a portion of the structure, or each of its members. The choice should be made so that it leads to the most direct solution to the problem.

Forces common to two members which are in contact act with equal magnitude but opposite sense on the respective free-body diagrams of the members. Recall that all *two-force members,* regardless of their shape, have equal but opposite collinear forces acting at the ends of the member. The unknown forces acting at the joints of multiforce members should be represented by their rectangular components. In many cases it is possible to tell by inspection the proper sense of the unknown forces; however, if this seems difficult, the sense can be assumed.

Equations of Equilibrium.   Count the total number of unknowns to make sure that an equivalent number of equilibrium equations can be written for solution. Recall that in general three equilibrium equations can be written for each rigid body represented in two dimensions. Many times, the solution for the unknowns will be straightforward if moments are summed about a point that lies at the intersection of the lines of action of as many unknown forces as possible. If after obtaining the solution an unknown force magnitude is found to be negative, it means the sense of the force is the reverse of that shown on the free-body diagrams.

The examples that follow illustrate this procedure. All these examples should be *thoroughly understood* before proceeding to solve the problems.

Example 6–14

Determine the horizontal and vertical components of force which the pin at C exerts on member CB of the frame in Fig. 6–27a.

SOLUTION I

Free-Body Diagrams. By inspection it can be seen that AB is a two-force member. The free-body diagrams are shown in Fig. 6–27b.

Equations of Equilibrium. The *three unknowns, C_x, C_y,* and F_{AB}, can be determined by applying the three equations of equilibrium to member CB.

$\zeta + \Sigma M_C = 0;$ $2000(2) - (F_{AB} \sin 60°)(4) = 0$ $F_{AB} = 1154.7$ N

$\Rightarrow \Sigma F_x = 0;$ $1154.7 \cos 60° - C_x = 0$ $C_x = 577$ N   *Ans.*

$+ \uparrow \Sigma F_y = 0;$ $1154.7 \sin 60° - 2000 + C_y = 0$ $C_y = 1000$ N   *Ans.*

SOLUTION II

Free-Body Diagrams. If one does not recognize that AB is a two-force member, then more work is involved in solving this problem. The free-body diagrams are shown in Fig. 6–27c.

Equations of Equilibrium. The *six unknowns, A_x, A_y, B_x, B_y, C_x, C_y,* are determined by applying the three equations of equilibrium to each member.

   Member AB

$\zeta + \Sigma M_A = 0;$ $B_x(3 \sin 60°) - B_y(3 \cos 60°) = 0$ (1)

$\Rightarrow \Sigma F_x = 0;$ $A_x - B_x = 0$ (2)

$+ \uparrow \Sigma F_y = 0;$ $A_y - B_y = 0$ (3)

   Member BC

$\zeta + \Sigma M_C = 0;$ $2000(2) - B_y(4) = 0$ (4)

$\Rightarrow \Sigma F_x = 0;$ $B_x - C_x = 0$ (5)

$+ \uparrow \Sigma F_y = 0;$ $B_y - 2000 + C_y = 0$ (6)

The results for C_x and C_y can be determined by solving these equations in the following sequence: 4, 1, 5, then 6. The results are

$$B_y = 1000 \text{ N}$$
$$B_x = 577 \text{ N}$$
$$C_x = 577 \text{ N} \qquad Ans.$$
$$C_y = 1000 \text{ N} \qquad Ans.$$

By comparison, Solution I is simpler since the requirement that $\mathbf{F}_{AB}$ in Fig. 6–27b be equal, opposite, and collinear at the ends of member AB automatically satisfies Eqs. 1, 2, and 3 above and therefore eliminates the need to write these equations. *As a result, always identify the two-force members before starting the analysis!*

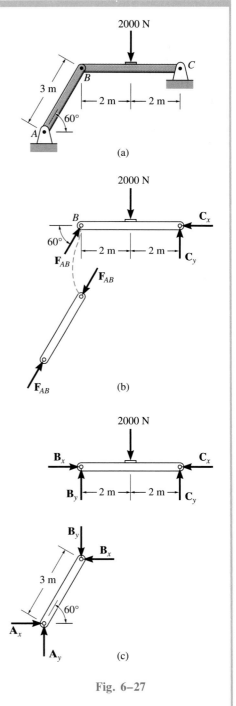

(a)

(b)

(c)

Fig. 6–27

■ Example 6–15

The beam shown in Fig. 6–28a is pin-connected at B. Determine the reactions at its supports. Neglect its weight and thickness.

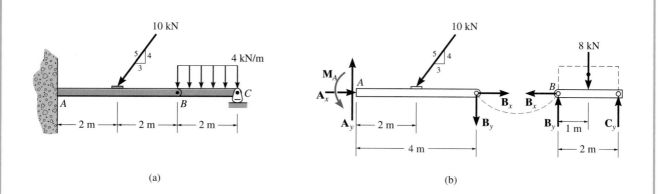

(a)

(b)

Fig. 6–28

SOLUTION

Free-Body Diagrams. By inspection, if we consider a free-body diagram of the entire beam ABC, there will be three unknown reactions at A and one at C. These four unknowns cannot all be obtained from the three equations of equilibrium, and so it will become necessary to dismember the beam into its two segments as shown in Fig. 6–28b.

Equations of Equilibrium. The six unknowns are determined as follows:

Segment BC

$$\stackrel{+}{\rightarrow}\Sigma F_x = 0; \qquad\qquad B_x = 0$$
$$\stackrel{}{\downarrow}+\Sigma M_B = 0; \qquad -8 \text{ kN}(1 \text{ m}) + C_y(2 \text{ m}) = 0$$
$$+\uparrow \Sigma F_y = 0; \qquad B_y - 8 \text{ kN} + C_y = 0$$

Segment AB

$$\stackrel{+}{\rightarrow}\Sigma F_x = 0; \qquad A_x - 10 \text{ kN}(\tfrac{3}{5}) + B_x = 0$$
$$\stackrel{}{\downarrow}+\Sigma M_A = 0; \qquad M_A - 10 \text{ kN}(\tfrac{4}{5})(2 \text{ m}) - B_y(4 \text{ m}) = 0$$
$$+\uparrow \Sigma F_y = 0; \qquad A_y - 10 \text{ kN}(\tfrac{4}{5}) - B_y = 0$$

Solving each of these equations successively, using previously calculated results, we obtain

$$A_x = 6 \text{ kN} \qquad A_y = 12 \text{ kN} \qquad M_A = 32 \text{ kN} \cdot \text{m} \qquad \textit{Ans.}$$
$$B_x = 0 \qquad\quad B_y = 4 \text{ kN}$$
$$C_y = 4 \text{ kN} \qquad\qquad\qquad\qquad\qquad\qquad\qquad \textit{Ans.}$$

Example 6–16

Determine the horizontal and vertical components of force which the pin at C exerts on member $ABCD$ of the frame shown in Fig. 6–29a.

SOLUTION

Free-Body Diagrams. By inspection, the three components of reaction that the supports exert on $ABCD$ can be determined from a free-body diagram of the entire frame, Fig. 6–29b. Also, the free-body diagram of each frame member is shown in Fig. 6–29c. Notice that member BE is a two-force member. As shown by the colored dashed lines, the forces at B, C, and E have equal magnitudes but opposite directions on the separate free-body diagrams.

Equations of Equilibrium. The six unknowns A_x, A_y, F_B, C_x, C_y, and D_x will be determined from the equations of equilibrium applied to the entire frame and then to member CEF. We have

Entire Frame

$$\curvearrowleft + \Sigma M_A = 0; \quad -981 \text{ N}(2 \text{ m}) + D_x(2.8 \text{ m}) = 0 \qquad D_x = 700.7 \text{ N}$$

$$\xrightarrow{+} \Sigma F_x = 0; \qquad A_x - 700.7 \text{ N} = 0 \qquad A_x = 700.7 \text{ N}$$

$$+ \uparrow \Sigma F_y = 0; \qquad A_y - 981 \text{ N} = 0 \qquad A_y = 981 \text{ N}$$

Member CEF

$$\curvearrowleft + \Sigma M_C = 0; \quad -981 \text{ N}(2 \text{ m}) - (F_B \sin 45°)(1.6 \text{ m}) = 0$$

$$F_B = -1734.2 \text{ N}$$

$$\xrightarrow{+} \Sigma F_x = 0; \qquad -C_x - (-1734.2 \cos 45° \text{ N}) = 0$$

$$C_x = 1230 \text{ N} \qquad\qquad\qquad Ans.$$

$$+ \uparrow \Sigma F_y = 0; \quad C_y - (-1734.2 \sin 45° \text{ N}) - 981 \text{ N} = 0$$

$$C_y = -245 \text{ N} \qquad\qquad\qquad Ans.$$

Since the magnitudes of forces $\mathbf{F}_B$ and $\mathbf{C}_y$ were calculated as negative quantities, they were assumed to be acting in the wrong sense on the free-body diagrams, Fig. 6–29c. The correct sense of these forces might have been determined "by inspection" *before* applying the equations of equilibrium to member CEF. As shown in Fig. 6–29c, moment equilibrium about point E on member CEF indicates that $\mathbf{C}_y$ must actually act *downward* to counteract the moment created by the 981-N force about point E. Similarly, summing moments about point C, it is seen that the vertical component of force $\mathbf{F}_B$ must actually act *upward*, and so $\mathbf{F}_B$ must act upward to the right.

The above calculations can be checked by applying the three equilibrium equations to member $ABCD$, Fig. 6–29c.

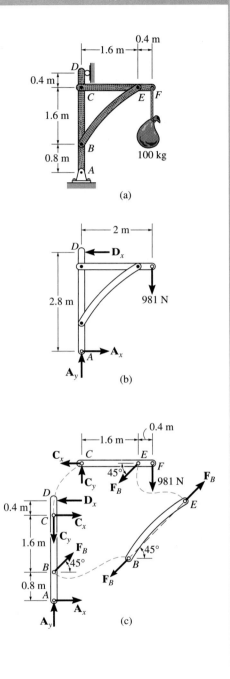

Fig. 6–29

Example 6–17

The smooth disk shown in Fig. 6–30a is pinned at D and has a weight of 20 lb. Neglecting the weights of the other members, determine the horizontal and vertical components of reaction at pins B and D.

(a)

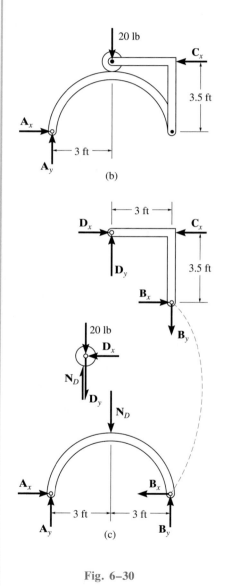

(b)

(c)

Fig. 6–30

SOLUTION

Free-Body Diagrams.   By inspection, the three components of reaction at the supports can be determined from a free-body diagram of the entire frame, Fig. 6–30b. Also, free-body diagrams of the members are shown in Fig. 6–30c.

Equations of Equilibrium.   The eight unknowns can of course be obtained by applying the eight equilibrium equations to each member—three to member AB, three to member BCD, and two to the disk. (Moment equilibrium is automatically satisfied for the disk.) If this is done, however, all the results can be obtained only from a simultaneous solution of some of the equations. (Try it and find out.) To avoid this situation, it is best to first determine the three support reactions on the *entire* frame; then, using these results, the remaining five equilibrium equations can be applied to two other parts in order to solve successively for the other unknowns.

   Entire Frame

$\zeta + \Sigma M_A = 0$; $-20\ \text{lb}(3\ \text{ft}) + C_x(3.5\ \text{ft}) = 0$ $C_x = 17.1\ \text{lb}$

$\xrightarrow{+} \Sigma F_x = 0$; $A_x - 17.1\ \text{lb} = 0$ $A_x = 17.1\ \text{lb}$

$+\uparrow \Sigma F_y = 0$; $A_y - 20\ \text{lb} = 0$ $A_y = 20\ \text{lb}$

   Member AB

$\xrightarrow{+} \Sigma F_x = 0$; $17.1\ \text{lb} - B_x = 0$ $B_x = 17.1\ \text{lb}$ *Ans.*

$\zeta + \Sigma M_B = 0$; $-20\ \text{lb}(6\ \text{ft}) + N_D(3\ \text{ft}) = 0$ $N_D = 40\ \text{lb}$

$+\uparrow \Sigma F_y = 0$; $20\ \text{lb} - 40\ \text{lb} + B_y = 0$ $B_y = 20\ \text{lb}$ *Ans.*

   Disk

$\xrightarrow{+} \Sigma F_x = 0$; $D_x = 0$ *Ans.*

$+\uparrow \Sigma F_y = 0$; $40\ \text{lb} - 20\ \text{lb} - D_y = 0$ $D_y = 20\ \text{lb}$ *Ans.*

Example 6–18

Determine the tension in the cables and also the force **P** required to support the 600-N force using the frictionless pulley system shown in Fig. 6–31*a*.

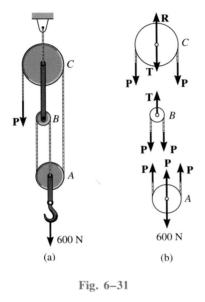

(a)   (b)

Fig. 6–31

SOLUTION

Free-Body Diagrams.   A free-body diagram of each pulley *including* its pin and a portion of the contacting cable is shown in Fig. 6–31*b*. Since the cable is *continuous* and the pulleys are frictionless, the cable has a *constant tension P* acting throughout its length (see Example 5–7). The link connection between pulleys *B* and *C* is a two-force member and therefore it has an unknown tension *T* acting on it. Notice that the *principle of action, equal but opposite reaction* must be carefully observed for forces **P** and **T** when the *separate* free-body diagrams are drawn.

Equations of Equilibrium.   The three unknowns are obtained as follows:

Pulley A

$+\uparrow \Sigma F_y = 0;$ $3P - 600 \text{ N} = 0$ $P = 200 \text{ N}$ *Ans.*

Pulley B

$+\uparrow \Sigma F_y = 0;$ $T - 2P = 0$ $T = 400 \text{ N}$ *Ans.*

Pulley C

$+\uparrow \Sigma F_y = 0;$ $R - 2P - T = 0$ $R = 800 \text{ N}$ *Ans.*

Example 6–19

A man having a weight of 150 lb supports himself by means of the cable and pulley system shown in Fig. 6–32a. If the seat has a weight of 15 lb, determine the equilibrium force that he must exert on the cable at A and the force he exerts on the seat. Neglect the weight of the cables and pulleys.

SOLUTION I

Free-Body Diagrams. The free-body diagrams of the man, seat, and pulley C are shown in Fig. 6–32b. The *two* cables are subjected to tensions T_A and T_E, respectively. The man is subjected to three forces: his weight, the tension T_A of cable AC, and the reaction N_s of the seat.

Equations of Equilibrium. The three unknowns are obtained as follows:

Man

$$+\uparrow \Sigma F_y = 0; \qquad T_A + N_s - 150 \text{ lb} = 0 \qquad (1)$$

Seat

$$+\uparrow \Sigma F_y = 0; \qquad T_E - N_s - 15 \text{ lb} = 0 \qquad (2)$$

Pulley C

$$+\uparrow \Sigma F_y = 0; \qquad 2T_E - T_A = 0 \qquad (3)$$

The magnitude of force T_E can be determined by adding Eqs. 1 and 2 to eliminate N_s and then using Eq. 3. The other unknowns are then obtained by resubstitution of T_E.

$$T_A = 110 \text{ lb} \qquad \qquad \textit{Ans.}$$
$$T_E = 55 \text{ lb}$$
$$N_s = 40 \text{ lb} \qquad \qquad \textit{Ans.}$$

SOLUTION II

Free-Body Diagrams. By using the dashed section shown in Fig. 6–32a, the man, pulley, and seat can be considered as a *single system*, Fig. 6–32c. Here N_s and T_A are *internal* forces and hence are not included on the "combined" free-body diagram.

Equations of Equilibrium. Applying $\Sigma F_y = 0$ yields a *direct* solution for T_E.

$$+\uparrow \Sigma F_y = 0; \qquad 3T_E - 15 \text{ lb} - 150 \text{ lb} = 0 \qquad T_E = 55 \text{ lb}$$

The other unknowns can be obtained from Eqs. 2 and 3.

Fig. 6–32

Example 6–20

The hand exerts a force of 8 lb on the grip of the spring compressor shown in Fig. 6–33a. Determine the force in the spring needed to maintain equilibrium of the mechanism in the position shown.

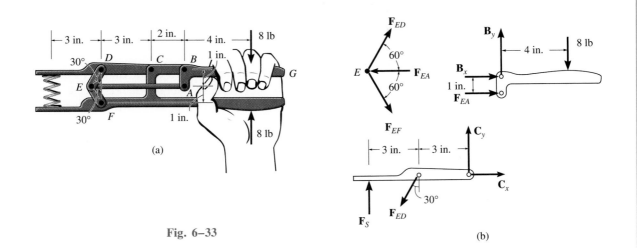

Fig. 6–33

SOLUTION

Free-Body Diagrams. By inspection, members *EA*, *ED*, and *EF* are all two-force members. As a result, the free-body diagrams for each part are shown in Fig. 6–33b. The pin at *E* has been included here since *three* force interactions occur on this pin. They represent the effects of members *ED*, *EA*, and *EF*. Note carefully how equal and opposite force reactions occur between each of the parts.

Equations of Equilibrium. By studying the free-body diagrams, the most direct way to obtain the spring force is to apply the equations of equilibrium in the following sequence:

Lever *ABG*

$$\zeta+\Sigma M_B = 0; \qquad F_{EA}(1) - 8 \text{ lb}(4 \text{ in.}) = 0 \qquad F_{EA} = 32 \text{ lb}$$

Pin *E*

$$+\uparrow \Sigma F_y = 0; \quad F_{ED} \sin 60° - F_{EF} \sin 60° = 0 \quad F_{ED} = F_{EF} = F$$

$$\xrightarrow{+}\Sigma F_x = 0; \qquad 2F \cos 60° - 32 \text{ lb} = 0 \qquad F = 32 \text{ lb}$$

Arm *EC*

$$\zeta+\Sigma M_C = 0; \qquad -F_s(6 \text{ in.}) + 32 \text{ lb} \cos 30°(3 \text{ in.}) = 0$$

$$F_s = 13.9 \text{ lb} \qquad \qquad Ans.$$

Example 6–21

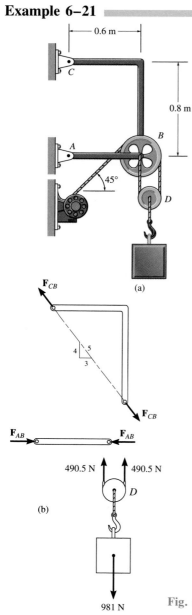

The 100-kg block is held in equilibrium by means of the pulley and continuous cable system shown in Fig. 6–34a. If the cable is attached to the pin at B, compute the forces which this pin exerts on each of its connecting members.

SOLUTION

Free-Body Diagrams. A free-body diagram of each member of the frame is shown in Fig. 6–34b. By inspection, members AB and CB are two-force members. Furthermore, the cable must be subjected to a force of 490.5 N in order to hold pulley D and the block in equilibrium. A free-body diagram of the pin at B is needed, since *four interactions* occur at this pin. These are caused by the attached cable (490.5 N), member AB ($\mathbf{F}_{AB}$), member CB ($\mathbf{F}_{CB}$), and pulley $B(\mathbf{B}_x$ and $\mathbf{B}_y)$.

Equations of Equilibrium. Applying the equations of force equilibrium to pulley B, we have

$$\xrightarrow{+}\Sigma F_x = 0; \qquad B_x - 490.5 \cos 45° \text{ N} = 0 \qquad B_x = 346.8 \text{ N} \qquad Ans.$$

$$+\uparrow\Sigma F_y = 0; \qquad B_y - 490.5 \sin 45° \text{ N} - 490.5 \text{ N} = 0$$

$$B_y = 837.3 \text{ N} \qquad Ans.$$

Using these results, equilibrium of the pin requires that

$$+\uparrow\Sigma F_y = 0; \qquad \tfrac{4}{5}F_{CB} - 837.3 \text{ N} - 490.5 \text{ N} = 0 \qquad F_{CB} = 1660 \text{ N} \quad Ans.$$

$$\xrightarrow{+}\Sigma F_x = 0; \qquad F_{AB} - \tfrac{3}{5}(1660 \text{ N}) - 346.8 \text{ N} = 0 \qquad F_{AB} = 1343 \text{ N} \quad Ans.$$

It may be noted that the two-force member CB is subjected to bending as caused by the force $\mathbf{F}_{CB}$. From the standpoint of design, it would be better to make this member *straight* (from C to B) so that the force $\mathbf{F}_{CB}$ would only create tension in the member.

Fig. 6–34

Before solving the following problems, it is suggested that a brief review be made of all the previous examples. This may be done by covering each solution and trying to locate the two-force members, drawing the free-body diagrams, and conceiving ways of applying the equations of equilibrium to obtain the solution.

PROBLEMS

6–49. Determine the reactive force at pins A and C of the two-member frame.

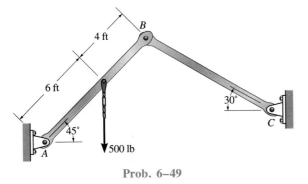

Prob. 6–49

6–50. Determine the horizontal and vertical components of force at pins A and C of the two-member frame.

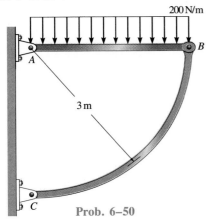

Prob. 6–50

6–51. Determine the horizontal and vertical components of force acting at A, B, and C of the frame.

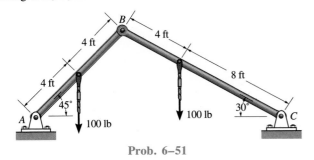

Prob. 6–51

*6–52. Determine the force in members AB and AC of the over-hang jack which is used to support the forms needed to cast the concrete slab. The forms exert a uniform load of 400 N/m on the top member of the jack assembly. There is a roller support at A and pins at A, B, C, and D.

Prob. 6–52

6–53. The two insulators AB and BC are used to support the powerline P. If the weight of the line is 450 N, determine the force in each insulator if they are pin-connected at their ends. Also, what are the components of reaction at the fixed support D?

Prob. 6–53

6–54. Determine the tension in member BC of the frame and the horizontal and vertical components of force acting on the pin at D.

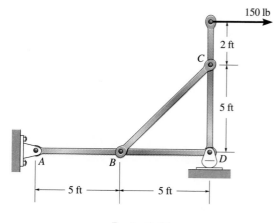

150 lb

2 ft

C

5 ft

A B D

5 ft   5 ft

Prob. 6–54

6–55. Determine the reactions at the supports A and B of the compound beam. There is a pin at C.

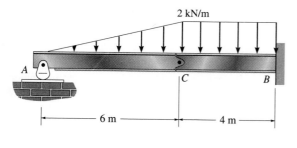

2 kN/m

A C B

6 m   4 m

Prob. 6–55

*6–56. The bridge frame consists of three segments which can be considered pinned at A, D, and E, rocker-supported at C and F, and roller-supported at B. Determine the horizontal and vertical components of reaction at all these supports due to the loading shown.

2 kip/ft

A E

15 ft B 30 ft D 15 ft

20 ft

C F

5 ft   5 ft

Prob. 6–56

6–57. The compound beam is fixed at E and supported by a roller at A and B. Determine the reactions at these supports. There are hinges (pins) at C and D.

1200 lb

800 lb   600 lb

A B C D E

5 ft   10 ft   5 ft   4 ft   6 ft   5 ft

Prob. 6–57

6–58. Each of the clamshell buckets has a weight of 350 lb and a center of gravity at G. Determine the force in link AB and the reaction at the pin C and smooth contact D. Neglect the weights of the other members.

6 in.

E A

30 in.

D

B

G

9 in.

C

12 in.

Prob. 6–58

6–59. Determine the force that must be developed in the hydraulic cylinder AB of the metal shear in order to develop a normal force of 4 MN in the grip. The jaws are pin-connected at C. Also, determine the magnitude of the force developed in the pin at C.

6–61. Determine the horizontal and vertical components of force which the pins at C and D exert on member ECD of the frame.

6–62. Determine the horizontal and vertical components of force which the pins at B and D exert on member ABD of the frame.

Probs. 6–61/6–62

Prob. 6–59

6–63. Determine the clamping force exerted on the smooth pipe at B if a force of 100 N is applied to the handles of the pliers. The pliers are pinned together at A.

Prob. 6–63

*6–60. The foot pump is used to fill a tire. If the force required to develop the necessary pressure in the cylinder is 70 lb, determine the vertical force **P** that must be applied to the pedal. The assembly is pin-connected at A, B, and C.

*6–64. Determine the compressive force exerted on the specimen by a vertical load of 50 N applied to the toggle press.

Prob. 6–64

Prob. 6–60

6–65. The bucket of the backhoe carries a load having a mass of 500 kg, and has a center of gravity at *G*. Determine the force in link *DH* and the hydraulic cylinder *AB*, and the horizontal and vertical components of force at pins *C* and *E* needed to hold the bucket in the position shown. Neglect the weights of the members.

Prob. 6–65

6–66. If block *A* weighs 50 lb and the platform *B* weighs 30 lb, determine the tension in the cables and the normal force which the block exerts on the platform.

Prob. 6–66

6–67. A 5-lb force is applied to the handles of the vice grip. Determine the compressive force developed on the smooth bolt shank *A* at the jaws.

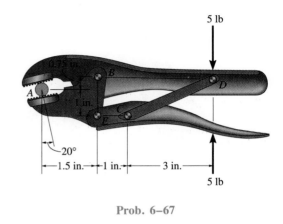

Prob. 6–67

*6–68. If a force of 10 lb is applied to the grip of the clamp, determine the compressive force *F* that the wood block exerts on the clamp.

Prob. 6–68

6–69. A force of 4 lb is applied to the handle of the toggle clamp. Determine the normal force F that the bolt exerts on the clamp.

1.5 in.

4 in.

4 lb

D

$30°$

A

B

C

E

F

6 in.

1.5 in.

Prob. 6–69

6–70. Determine the horizontal and vertical components of force acting on the pins A and C of the frame. The pulley at D is frictionless.

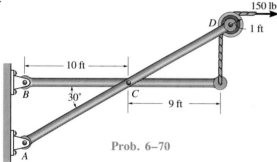

150 lb

D

1 ft

10 ft

B

$30°$

C

9 ft

A

Prob. 6–70

6–71. Determine the horizontal and vertical components of force which the pins exert on member ABD.

50 lb

1 ft

D

5 ft

B

4 ft

A

C

3 ft

Prob. 6–71

*6–72. The shovel of the tractor supports the 400-kg load which has a center of mass at G. Determine the horizontal and vertical components of reaction at pin A and the force in the hydraulic cylinder EF. All labeled points are pin connections. The mechanism is the same on both sides and supports the load equally on both sides.

0.4 m

D 20°

0.5 m

F

C

E

G

0.5 m

D

A

B

0.6 m 0.4 m

Prob. 6–72

6–73. Determine the horizontal and vertical components of force at C which member ABC exerts on member CEF.

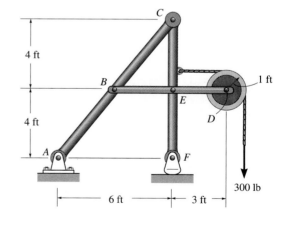

C

4 ft

B

1 ft

E

4 ft

D

A

F

300 lb

6 ft

3 ft

Prob. 6–73

6–74. Determine the horizontal and vertical components of force which the connecting pins at B, E, and D exert on member BED.

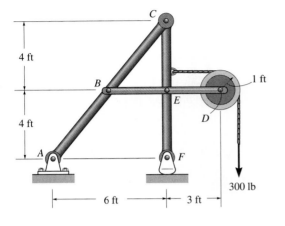

Prob. 6–74

6–75. The clamp has a rated load capacity of 1500 lb. Determine the compressive force this creates in the portion AB of the screw and the magnitude of force exerted at the pin C. The screw is pin-connected at its end B and passes through the pin-connected (swivel) block at A.

Prob. 6–75

*6–76. The lift is operated using the hydraulic cylinder CD. Determine the horizontal and vertical components of reaction at the pins, A and B, and the compressive force in the cylinder when it is used to hold the lift in the position shown. The man has a weight of 230 lb and a center of gravity at G.

Prob. 6–76

6–77. The crane is pin-connected at A and fits through the smooth collar at B. Determine the horizontal and vertical components of reaction at these supports and the resultant force that the pin at C exerts on the crane when it supports the crate, which has a mass of 175 kg.

Prob. 6–77

6–78. Determine the reactions at pins A and B of the frame needed to support the 200-lb load. The large pulley is pinned at C and has a radius of 0.5 ft. The small pulley at D has a radius of 0.25 ft.

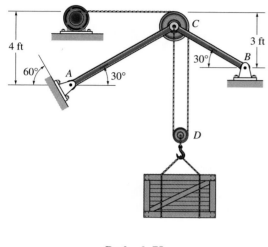

Prob. 6–78

*6–80. Determine the vertical reactions at the smooth ground and the horizontal and vertical components of reaction at the pin B when the man, who weighs 170 lb, stands upright on the ladder in the position shown. The two segments of the ladder are connected together by the cable DE.

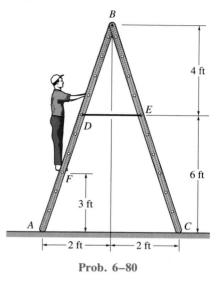

Prob. 6–80

6–79. Determine the horizontal and vertical components of force which the pins exert on member $ABCD$. Note that the pin at C is attached to member $ABCD$ and passes through the smooth slot in member ECF.

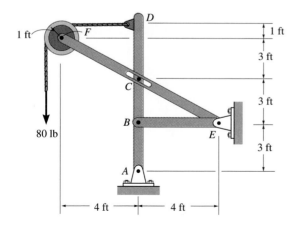

Prob. 6–79

6–81. Two smooth tubes A and B, each having the same weight W, are suspended from a common point O by means of equal-length cords. A third tube, C, is placed between A and B. Determine the greatest weight of C without upsetting equilibrium.

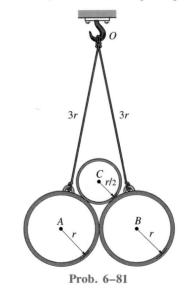

Prob. 6–81

6–82. Determine the tension T in the cord and the angle θ that the pulley-supporting link AB makes with the vertical. Neglect the mass of the pulleys and the link. The block has a mass of 30 kg, and the cord is attached to the pin at B.

Prob. 6–82

*6–84. If each of the three links of the mechanism has a weight of 25 lb, determine the angle θ for equilibrium. The spring, which always remains vertical, is unstretched when $\theta = 0°$.

Prob. 6–84

6–83. The bracket is used to support a smooth cylinder having a weight of 40 lb. If the bracket is pinned at D, determine the normal reactions at points B and C and the horizontal and vertical components of reaction at D.

Prob. 6–83

6–85. The piston C moves vertically between the two smooth walls. If the spring has a stiffness of $k = 15$ lb/in. and is unstretched when $\theta = 0°$, determine the couple moment M that must be applied to AB to hold the mechanism in equilibrium when $\theta = 30°$.

Prob. 6–85

6–86. The weight W is placed on the table such that its center of gravity G is directly over the pin at F. Determine the compressive force in the hydraulic cylinder DE in terms of W and the geometry of the platform. Each member has a length $2L$ and is pinned at its center.

*6–88. The spring mechanism is used as a shock absorber for a load applied to the drawbar AB. Determine the equilibrium length of each spring when the 50-N force is applied. Each spring has an unloaded length of 200 mm, and the drawbar slides along the smooth guide posts CG and EF. The ends of all springs are attached to their respective members.

Prob. 6–88

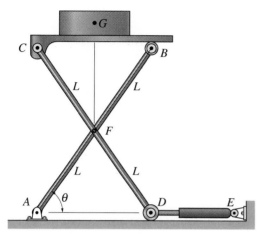

Prob. 6–86

6–87. The lever mechanism for a machine press serves as a toggle which develops a large force at E when a small force is applied at the handle H. To show that this is the case, determine the force at E if someone applies a vertical force of 80 N at H. The smooth head at D is able to slide freely downward. All members are pin-connected.

6–89. Determine the couple moment $\mathbf{M}$ that must be applied to member DC for equilibrium of the quick-return mechanism. Express the result in terms of the angles ϕ and θ, dimension L, and the applied vertical load P. The block at C is confined to slide within the slot of member AB.

6–90. Solve Prob. 6–89 if the applied load $\mathbf{P}$ is horizontal and directed to the right.

Prob. 6–87

Probs. 6–89/6–90

6–91. The tongs consist of two jaws pinned to links at A, B, C, and D. Determine the horizontal and vertical components of force exerted on the 500-lb stone at F and G in order to lift it.

Prob. 6–91

6–93. The two disks each have a weight W and are connected to each other by an inextensible cord. If they rest against smooth planes, one of which is inclined at an angle ϕ as shown and the other is vertical, determine the angle θ for equilibrium (a) in terms of the parameters shown, and (b) with $W = 10$ lb, $\phi = 60°$.

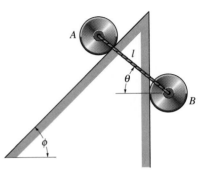

Prob. 6–93

*6–92. The man weighs 150 lb and is standing on the uniform plank, which weighs 40 lb. Determine the force he exerts on the plank if he pulls with just enough force to lift the plank off the support at B. The plank is pin-connected at A.

6–94. The structure is subjected to the loading shown. Member AD is supported by a cable AB and roller at C and fits through a smooth circular hole at D. Member ED is supported by a roller at D and a pole that fits in a smooth snug circular hole at E. Determine the x, y, z components of reaction at E and the tension in cable AB.

Prob. 6–92

Prob. 6–94

6–95. The structure is subjected to the loadings shown. Member *AB* is supported by a ball-and-socket at *A* and smooth collar at *B*. Member *CD* is supported by a pin at *C*. Determine the *x*, *y*, *z* components of reaction at *A* and *C*.

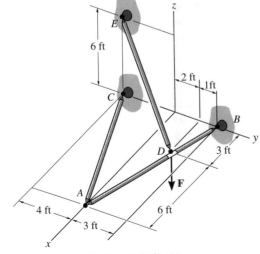

Prob. 6–95

6–98. The tripod rests on the smooth surface at *A*, *B*, and *C*. Its three legs, each having the same length, are held in position by the struts *DE*, *DF*, and *FE*. If the members are connected by ball-and-socket joints, determine the force in each strut for equilibrium.

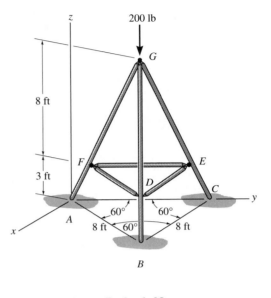

Prob. 6–98

***6–96.** The three-member frame is connected at its ends using ball-and-socket joints. Determine the *x*, *y*, *z* components of reaction at *B* and the tension in member *DE*. The force acting at *D* is $\mathbf{F} = \{250\mathbf{i} - 350\mathbf{k}\}$ lb.

6–97. The three-member frame is connected at its ends using ball-and-socket joints. Determine the *x*, *y*, *z* components of reaction at *B* and the tension in member *DE*. The force acting at *D* is $\mathbf{F} = \{135\mathbf{i} + 200\mathbf{j} - 180\mathbf{k}\}$ lb.

6–99. The two-member frame is connected at its ends using ball-and-socket joints. Determine the components of reaction at the supports *A* and *C* and the tension in the cable *DE*.

Prob. 6–99

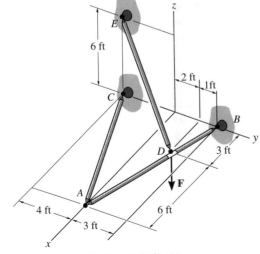

Probs. 6–96/6–97

REVIEW PROBLEMS

*6–100.   Determine the reactions at the supports of the compound beam. There is a short vertical link at C.

Prob. 6–100

6–101.   The two-bar mechanism consists of a lever arm AB and smooth link CD, which has a fixed collar at its end C and a roller at the other end D. Determine the force **P** needed to hold the lever in the position θ. The spring has a stiffness k and unstretched length $2L$. The roller contacts either the top or bottom portion of the horizontal guide.

Prob. 6–101

6–102.   Determine the force in each member of the truss and indicate whether the members are in tension or compression.

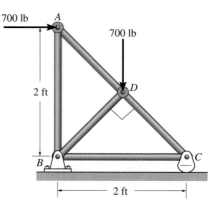

Prob. 6–102

6–103.   The *Howe bridge truss* is subjected to the loading shown. Determine the force in members HD, CD, and GD, and indicate whether the members are in tension or compression.

*6–104.   The *Howe bridge truss* is subjected to the loading shown. Determine the force in members HI, HB, and BC, and indicate whether the members are in tension or compression.

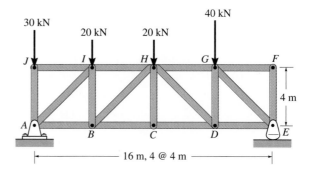

Probs. 6–103/6–104

6–105. Determine the horizontal and vertical components of force at pins A, B, and C of the two-member frame.

Prob. 6–105

6–106. Determine the force in members AB, AD, and AC of the space truss and indicate whether the members are in tension or compression.

{$-600\mathbf{k}$} lb

Prob. 6–106

6–107. The compound beam is supported by a rocker at B and fixed to the wall at A. If it is hinged (pinned) together at C, determine the reactions at the supports.

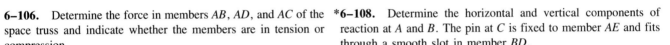

Prob. 6–107

*6–108. Determine the horizontal and vertical components of reaction at A and B. The pin at C is fixed to member AE and fits through a smooth slot in member BD.

Prob. 6–108

6–109. In each case, determine the force P required to maintain equilibrium.

6–110. Determine the force in each member of the truss and indicate whether the members are in tension or compression.

Prob. 6–109

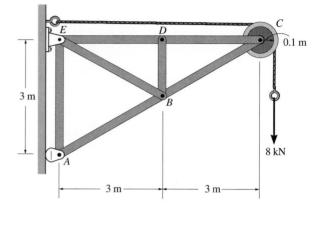

Prob. 6–110

7 Internal Forces

In this chapter we will develop a technique for finding the internal loading at specific *points* within a structural member, and then we will generalize this method to find the point-to-point variation of loading along the axis of the member. A graph showing this variation of internal load will allow us to find the critical points where the *maximum* internal loading occurs. The analysis of cables will be treated in the last part of the chapter.

7.1 Internal Forces Developed in Structural Members

The design of a structural member requires an investigation of the loadings acting *within* the member which are necessary to balance the loadings acting external to it. The *method of sections* can be used for this purpose. In order to illustrate the procedure, consider the "simply supported" beam shown in Fig. 7–1a, which is subjected to the forces $\mathbf{F}_1$ and $\mathbf{F}_2$. The *support reactions* $\mathbf{A}_x$, $\mathbf{A}_y$, and $\mathbf{B}_y$ can be determined by applying the equations of equilibrium, using the free-body diagram of the *entire beam,* Fig. 7–1b. If the *internal loadings* at point C are to be determined, it is necessary to pass an imaginary section through the beam, cutting it into two segments at that point, Fig. 7–1a. Doing this "exposes" the internal loadings as *external* on the free-body diagram of

(a)

Fig. 7–1

(b)

(a)   (b)

(c)

Fig. 7–1

each segment, Fig. 7–1c. Since both segments (*AC* and *CB*) were prevented from translating and rotating *before* the beam was sectioned, equilibrium of each segment is maintained if rectangular force components N_C and V_C and a resultant couple moment M_C are developed at the cut section. Note that these loadings must be equal in magnitude and opposite in direction on each of the segments (Newton's third law). The magnitude of each unknown can now be determined by applying the three equations of equilibrium to either segment *AC* or *CB*. A *direct solution* for N_C is obtained by applying $\Sigma F_x = 0$; V_C is obtained directly from $\Sigma F_y = 0$; and M_C is determined by summing moments about point C, $\Sigma M_C = 0$, in order to eliminate the moments of the unknowns N_C and V_C.

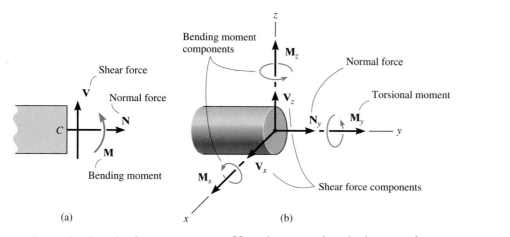

Bending moment components

Shear force

Normal force

$\mathbf{M}_z$

Normal force

$\mathbf{V}_z$

Torsional moment

$\mathbf{N}_y$ $\mathbf{M}_y$

Shear force

$\mathbf{N}$

$\mathbf{M}$

$\mathbf{M}_x$ $\mathbf{V}_x$

Bending moment

Shear force components

(a)   (b)

Fig. 7–2

In mechanics, the force components **N**, acting normal to the beam at the cut section, and **V**, acting tangent to the section, are termed the *normal or axial force* and the *shear force,* respectively. The couple moment **M** is referred to as the *bending moment,* Fig. 7–2a. In three dimensions, a general internal force and couple moment resultant will act at the section. The x, y, z components of these loadings are shown in Fig. 7–2b. Here $\mathbf{N}_y$ is the *normal force,* $\mathbf{V}_x$ and $\mathbf{V}_z$ are *shear force components,* $\mathbf{M}_y$ is a *torsional or twisting moment,* and $\mathbf{M}_x$ and $\mathbf{M}_z$ are *bending moment components*. For most applications, these *resultant loadings* will act at the geometric center or centroid (C) of the section's cross-sectional area. Although the magnitude for each loading generally will be different at various points along the axis of the member, the method of sections can always be used to determine their values.

Free-Body Diagrams. Since frames and machines are composed of *multiforce members,* each of these members will generally be subjected to internal normal, shear, and bending loadings. For example, consider the frame shown in Fig. 7–3a. If the dashed section is passed through the frame to determine the internal loadings at points H, G, and F, the resulting free-body diagram of the top portion of this section is shown in Fig. 7–3b. At each point where a member is sectioned there is an unknown normal force, shear force, and bending moment. As a result, we cannot apply the *three* equations of equilibrium to this section in order to obtain these *nine unknowns.** Instead, to solve this problem we must *first dismember* the frame and determine the reactions at the connections of the members using the techniques of Sec. 6.6. Once this is done, *each member* may then be sectioned at its appropriate point and the three equations of equilibrium can be applied to determine **N, V,** and **M.** For example, the free-body diagram of segment DG, Fig. 7–3c, can be used to determine the internal loadings at G provided the reactions of the pin, $\mathbf{D}_x$ and $\mathbf{D}_y$, are known.

*Recall that this method of analysis worked well for trusses, since truss members are *straight two-force members* which support only an axial or normal load.

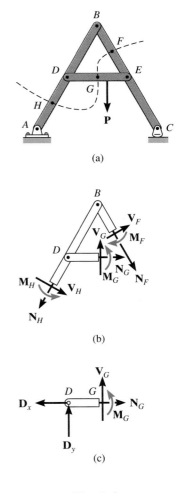

(a)

(b)

(c)

Fig. 7–3

PROCEDURE FOR ANALYSIS

The following procedure provides a means for applying the method of sections to determine the internal loadings at a specific location in a member.

Support Reactions.   Before the member is "cut" or sectioned, it may first be necessary to determine the member's support reactions, so that the equilibrium equations are used only to solve for the internal loadings when the member is sectioned. If the member is part of a frame or machine, the reactions at its connections are determined using the methods of Sec. 6.6.

Free-Body Diagram.   Keep all distributed loadings, couple moments, and forces acting on the member in their *exact location,* then pass an imaginary section through the member, perpendicular to its axis at the point where the internal loading is to be determined. Draw a free-body diagram of one of the "cut" segments on either side of the section and indicate the *x, y, z* components of the force and couple moment resultants at the section. In particular, if the member is subjected to a *coplanar* system of forces, only **N, V,** and **M** act at the section. In many cases it may be possible to tell by inspection the proper sense of the unknown loadings; however, if this seems difficult, the sense can be assumed.

Equations of Equilibrium.   Apply the equations of equilibrium to obtain the unknown internal loadings. Generally, moments should be summed at the section about axes passing through the *centroid* or geometric center of the member's cross-sectional area, in order to eliminate the unknown normal and shear forces and thereby obtain direct solutions for the moment components. If the solution of the equilibrium equations yields a negative scalar, the assumed sense of the quantity is opposite to that shown on the free-body diagram.

The following examples numerically illustrate this procedure.

Example 7–1

The bar is fixed at its end and is loaded as shown in Fig. 7–4a. Determine the internal normal force at points B and C.

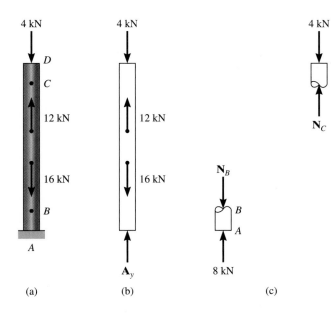

Fig. 7–4

(a)   (b)   (c)

SOLUTION

Support Reactions.   A free-body diagram of the entire bar is shown in Fig. 7–4b. By inspection, only a normal force A_y acts at the fixed support, since the loads are applied symmetrically along the bar's axis.

$$+\uparrow\ \Sigma F_z = 0; \qquad A_y - 16\ \text{kN} + 12\ \text{kN} - 4\ \text{kN} = 0 \qquad A_y = 8\ \text{kN}$$

Free-Body Diagrams.   The internal forces at B and C will be found using the free-body diagrams of the sectioned bar shown in Fig. 7–4c. In particular, segments AB and DC will be chosen here, since they contain the *least* number of forces.

Equations of Equilibrium
   Segment AB

$$+\uparrow\ \Sigma F_z = 0; \qquad 8\ \text{kN} - N_B = 0 \qquad N_B = 8\ \text{kN} \qquad\qquad Ans.$$

   Segment DC

$$+\uparrow\ \Sigma F_z = 0; \qquad N_C - 4\ \text{kN} = 0 \qquad N_C = 4\ \text{kN} \qquad\qquad Ans.$$

Try working this problem in the following manner: Determine N_B from segment BD. (Note that this approach does *not require* solution for the support reaction at A.) Using the result for N_B, isolate segment BC to determine N_C.

Example 7–2

The circular shaft is subjected to three concentrated torques as shown in Fig. 7–5a. Determine the internal torques at points B and C.

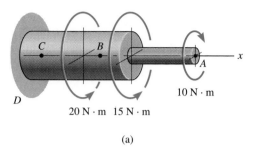

(a)

SOLUTION

Support Reactions. Since the shaft is subjected only to collinear torques, a torque reaction occurs at the support, Fig. 7–5b. Using the right-hand rule to define the positive directions of the torques, we require

$$\Sigma M_x = 0; \quad -10 \text{ N} \cdot \text{m} + 15 \text{ N} \cdot \text{m} + 20 \text{ N} \cdot \text{m} - T_D = 0$$
$$T_D = 25 \text{ N} \cdot \text{m}$$

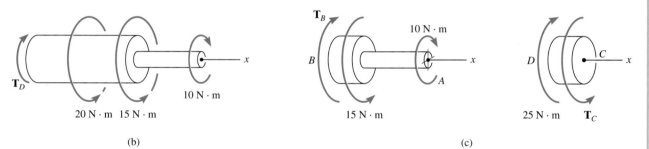

(b)

(c)

Fig. 7–5

Free-Body Diagrams. The internal torques at B and C will be found using the free-body diagrams of the shaft segments AB and CD shown in Fig. 7–5c.

Equations of Equilibrium. Applying the equation of moment equilibrium along the shaft's axis, we have

Segment AB

$$\Sigma M_x = 0; \quad -10 \text{ N} \cdot \text{m} + 15 \text{ N} \cdot \text{m} - T_B = 0 \quad T_B = 5 \text{ N} \cdot \text{m} \quad \textit{Ans.}$$

Segment CD

$$\Sigma M_x = 0; \quad T_C - 25 \text{ N} \cdot \text{m} = 0 \quad T_C = 25 \text{ N} \cdot \text{m} \quad \textit{Ans.}$$

Try to solve for T_C by using segment CA. Note that this approach does not require a solution for the support reaction at D.

Example 7–3

The beam supports the loading shown in Fig. 7–6a. Determine the internal normal force, shear force, and bending moment acting just to the left, point B, and just to the right, point C, of the 6-kN force.

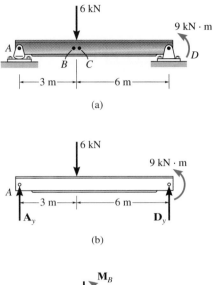

(a)

SOLUTION

Support Reactions.   The free-body diagram of the beam is shown in Fig. 7–6b. When determining the external reactions, note that the 9-kN · m couple moment is a free vector and therefore it can be placed anywhere on the free-body diagram. Here we will only determine $\mathbf{A}_y$, since segments AB and AC will be used for the analysis.

$$\zeta+\Sigma M_D = 0;\quad 9 \text{ kN} \cdot \text{m} + (6 \text{ kN})(6 \text{ m}) - A_y(9 \text{ m}) = 0$$

$$A_y = 5 \text{ kN}$$

(b)

Free-Body Diagrams.   The free-body diagrams of the left segments AB and AC of the beam are shown in Figs. 7–6c and 7–6d. In this case the 9-kN · m couple moment is *not included* on these diagrams, since it must be kept in its *original position* until *after* the section is made and the appropriate body isolated. In other words, the internal loadings, when determined from the left segments of the beam, are not influenced by the effect of the couple moment since this moment does not actually act on these segments.

(c)

Equations of Equilibrium
Segment AB

$\xrightarrow{+}\Sigma F_x = 0;$	$N_B = 0$	*Ans.*
$+\uparrow \Sigma F_y = 0;$	$5 \text{ kN} - V_B = 0$	$V_B = 5 \text{ kN}$ *Ans.*
$\zeta+\Sigma M_B = 0;$	$-(5 \text{ kN})(3 \text{ m}) + M_B = 0$	$M_B = 15 \text{ kN} \cdot \text{m}$ *Ans.*

Segment AC

$\xrightarrow{+}\Sigma F_x = 0;$	$N_C = 0$	*Ans.*
$+\uparrow \Sigma F_y = 0;$	$5 \text{ kN} - 6 \text{ kN} - V_C = 0$	$V_C = -1 \text{ kN}$ *Ans.*
$\zeta+\Sigma M_C = 0;$	$-(5 \text{ kN})(3 \text{ m}) + M_C = 0$	$M_C = 15 \text{ kN} \cdot \text{m}$ *Ans.*

(d)

Fig. 7–6

Here the moment arm for the 5-kN force in both cases is approximately 3 m, since B and C are ''almost'' coincident.

■ Example 7–4

(a)

(b)

Fig. 7–7

Determine the internal normal force, shear force, and bending moment acting at point B of the two-member frame shown in Fig. 7–7a.

SOLUTION

Support Reactions. A free-body diagram of each member is shown in Fig. 7–7b. Since CD is a two-force member, the equations of equilibrium need to be applied only to member AC.

$$\zeta+\Sigma M_A = 0; \qquad -400 \text{ lb}(4 \text{ ft}) + (\tfrac{3}{5})F_{DC}(8 \text{ ft}) = 0 \qquad F_{DC} = 333.3 \text{ lb}$$

$$\xrightarrow{+}\Sigma F_x = 0; \qquad -A_x + (\tfrac{4}{5})(333.3 \text{ lb}) = 0 \qquad A_x = 266.7 \text{ lb}$$

$$+\uparrow \Sigma F_y = 0; \qquad A_y - 400 \text{ lb} + \tfrac{3}{5}(333.3 \text{ lb}) = 0 \qquad A_y = 200 \text{ lb}$$

Free-Body Diagrams. Passing an imaginary section perpendicular to the axis of member AC through point B yields the free-body diagrams of segments AB and BC shown in Fig. 7–7c. When constructing these diagrams it is important to keep the distributed loading exactly as it is until *after* the section is made. Only then can it be replaced by a single resultant force. Why? Also, notice that $\mathbf{N}_B$, $\mathbf{V}_B$, and $\mathbf{M}_B$ act with equal magnitude but opposite direction on each segment—Newton's third law.

Equations of Equilibrium. Applying the equations of equilibrium to segment AB, we have

$$\xrightarrow{+}\Sigma F_x = 0; \qquad N_B - 266.7 \text{ lb} = 0 \qquad N_B = 267 \text{ lb} \quad \textit{Ans.}$$

$$+\uparrow \Sigma F_y = 0; \qquad 200 \text{ lb} - 200 \text{ lb} - V_B = 0 \qquad V_B = 0 \quad \textit{Ans.}$$

$$\zeta+\Sigma M_B = 0; \qquad M_B - 200 \text{ lb}(4 \text{ ft}) + 200 \text{ lb}(2 \text{ ft}) = 0$$

$$M_B = 400 \text{ lb} \cdot \text{ft} \qquad\qquad \textit{Ans.}$$

As an exercise, try to obtain these same results using segment BC.

(c)

Example 7–5

Determine the normal force, shear force, and bending moment acting at point E of the frame loaded as shown in Fig. 7–8a.

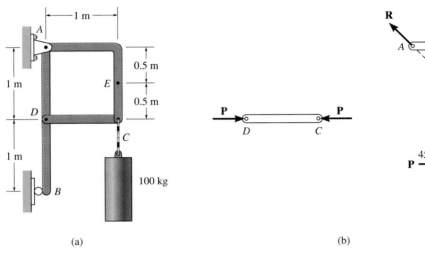

(a)

(b)

Fig. 7–8

(c)

SOLUTION

Support Reactions. By inspection, members AC and CD are two-force members, Fig. 7–8b. In order to determine the internal loadings at E, we must first determine the force **R** at the end of member AC. To do this we must analyze the equilibrium of the pin at C. Why?

Summing forces in the vertical direction on the pin, Fig. 7–8b, we have

$$+\uparrow \Sigma F_y = 0; \qquad R \sin 45° - 981 \text{ N} = 0 \qquad R = 1387.6 \text{ N}$$

Free-Body Diagram. The free-body diagram of segment CE is shown in Fig. 7–8c.

Equations of Equilibrium

$$\xrightarrow{+} \Sigma F_x = 0; \qquad 1387.6 \cos 45° \text{ N} - V_E = 0 \qquad V_E = 981 \text{ N} \quad \textit{Ans.}$$

$$+\uparrow \Sigma F_y = 0; \qquad -1387.6 \sin 45° \text{ N} + N_E = 0 \qquad N_E = 981 \text{ N} \quad \textit{Ans.}$$

$$\downarrow+\Sigma M_E = 0; \quad 1387.6 \cos 45° \text{ N}(0.5 \text{ m}) - M_E = 0 \quad M_E = 490 \text{ N} \cdot \text{m} \quad \textit{Ans.}$$

Example 7–6

A force of $\mathbf{F} = \{-3\mathbf{i} + 7\mathbf{j} - 4\mathbf{k}\}$ kN acts at the corner of a beam extended from a fixed wall as shown in Fig. 7–9a. Determine the internal loadings at A.

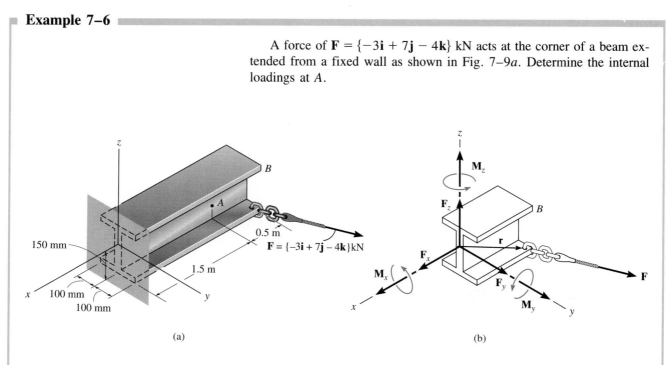

(a)   (b)

Fig. 7–9

SOLUTION

This problem will be solved by considering segment AB of the beam, since it does *not* involve the support reactions.

Free-Body Diagram.   The free-body diagram of segment AB is shown in Fig. 7–9b. The components of the resultant force $\mathbf{F}_A$ and moment $\mathbf{M}_A$ pass through the *centroid* or geometric center of the cross-sectional area at A.

Equations of Equilibrium

$$\Sigma \mathbf{F} = \mathbf{0}; \qquad \mathbf{F}_A - 3\mathbf{i} + 7\mathbf{j} - 4\mathbf{k} = \mathbf{0}$$
$$\mathbf{F}_A = \{3\mathbf{i} - 7\mathbf{j} + 4\mathbf{k}\} \text{ kN} \qquad Ans.$$

$$\Sigma \mathbf{M}_A = \mathbf{0}; \quad \mathbf{M}_A + \mathbf{r} \times \mathbf{F} = \mathbf{M}_A + \begin{vmatrix} \mathbf{i} & \mathbf{j} & \mathbf{k} \\ -0.5 & 0.1 & -0.15 \\ -3 & 7 & -4 \end{vmatrix} = \mathbf{0}$$

$$\mathbf{M}_A = \{-0.650\mathbf{i} + 1.55\mathbf{j} + 3.20\mathbf{k}\} \text{ kN} \cdot \text{m} \qquad Ans.$$

Here $\mathbf{F}_x = \{3\mathbf{i}\}$ kN represents the normal force N, whereas $\mathbf{F}_y = \{-7\mathbf{j}\}$ kN and $\mathbf{F}_z = \{4\mathbf{k}\}$ kN are components of the shear force $V = \sqrt{F_y^2 + F_z^2}$. Also, the torsional moment is $\mathbf{M}_x = \{-0.65\mathbf{i}\}$ kN $\cdot$ m, and the bending moment is determined from its components $\mathbf{M}_y = \{1.55\mathbf{j}\}$ kN $\cdot$ m and $\mathbf{M}_z = \{3.20\mathbf{k}\}$ kN $\cdot$ m; i.e., $M_b = \sqrt{M_y^2 + M_z^2}$.

PROBLEMS

7–1. Determine the normal force, shear force, and moment in the beam at points C and D. Point C is located just to the right of the 8-kip load.

Prob. 7–1

7–2. Determine the normal force, shear force, and moment in the shaft at points B and D. These points lie just to the right of the 150-lb force and the bearing at C, respectively. There is a thrust bearing at A and a journal bearing at C.

Prob. 7–2

7–3. Three torques act on the shaft as shown. Determine the torque at points A, B, C, and D.

Prob. 7–3

*7–4. The shaft is supported by smooth bearings at A and B and subjected to the torques shown. Determine the torque at points C, D, and E.

Prob. 7–4

7–5. Determine the normal force, shear force, and moment in the beam at point B.

Prob. 7–5

7–6. Determine the normal force, shear force, and moment acting at point C.

Prob. 7–6

7–7. Determine the normal force, shear force, and moment at point D of the beam.

500 N/m

A D B C

45°

0.5 m 0.5 m 1.5 m

0.5 m

Prob. 7–7

7–10. The bolt shank is subjected to a tension of 80 lb. Determine the normal force, shear force, and moment acting at point C.

C

90° 6 in

A B

Prob. 7–10

*7–8. Determine the normal force, shear force, and moment at points C and D of the beam.

300 lb/ft

A C B D

6000 lb·ft

12 ft 3 ft

15 ft 6 ft

Prob. 7–8

7–11. Determine the normal force, shear force, and moment acting at point C. The cooling unit has a total weight of 52 kip and a center of gravity at G.

F

D 30° 30° E

0.2 m

A C B

3 m 3 m

G

Prob. 7–11

7–9. Determine the normal force, shear force, and moment at point C.

4 kN/m

A C B

3 m 6 m

Prob. 7–9

*7–12.   Determine the normal force, shear force, and moment acting at point C. Neglect the thickness of the beam.

Prob. 7–12

7–13.   Determine the normal force, shear force, and moment at (*a*) point C, which is just to the right of the roller at A, and (*b*) point D, which is just to the left of the 3000-lb concentrated force.

Prob. 7–13

7–14.   Determine the normal force, shear force, and moment at point C.

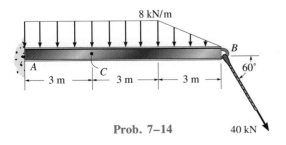

Prob. 7–14

7–15.   Determine the normal force, shear force, and moment in the beam at points C and D.

Prob. 7–15

*7–16.   The carriage of the crane exerts forces of 2000 lb on the beam. Determine the moment developed in the beam at point C when the carriage is in the position shown. Neglect the weight of the beam and carriage.

Prob. 7–16

7–17. The beam-column is fixed to the floor and supports the load shown. Determine the normal force, shear force, and moment at points A and B due to this loading.

Prob. 7–17

7–18. Determine the normal force, shear force, and moment at points C and D of the beam.

Prob. 7–18

7–19. Determine the normal force, shear force, and moment at points D and E of the frame.

Prob. 7–19

*7–20. Determine the normal force, shear force, and moment in the beam at points D and E. Point E is just to the right of the 4-kip load. The splice at B is a pin.

Prob. 7–20

7–21. Determine the normal force, shear force, and moment at point C.

Prob. 7–21

7–22. Determine the normal force, shear force, and moment at point D.

Prob. 7–22

7–23. Determine the normal force, shear force, and moment at points F and E.

Prob. 7–23

*7–24. Determine the normal force, shear force, and moment in the beam at points D and E. Point E is just to the right of the 3-kip load.

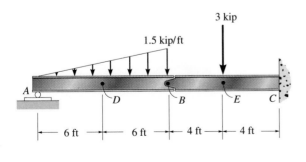

Prob. 7–24

7–25. Determine the normal force, shear force, and moment acting at points B and C on the curved member.

Prob. 7–25

7–26. Determine the normal force, shear force, and moment in the curved member as a function of θ.

Prob. 7–26

■7–27. Determine the shear force and moment acting at point C of the beam. Use Simpson's rule to evaluate the integrals used for the calculation.

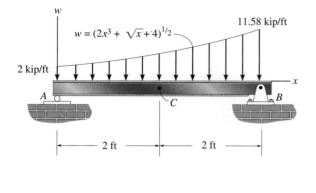

$w = (2x^3 + \sqrt{x} + 4)^{1/2}$

11.58 kip/ft

2 kip/ft

A C B

2 ft   2 ft

Prob. 7–27

*7–28. Determine the x, y, z components of force and moment at point C in the pipe assembly. Neglect the weight of the pipe. Take $\mathbf{F}_1 = \{350\mathbf{i} - 400\mathbf{j}\}$ lb and $\mathbf{F}_2 = \{-300\mathbf{j} + 150\mathbf{k}\}$ lb.

7–29. Determine the x, y, z components of force and moment at point C in the pipe assembly. Neglect the weight of the pipe. Take $\mathbf{F}_1 = \{200\mathbf{i} - 300\mathbf{j} - 80\mathbf{k}\}$ lb and $\mathbf{F}_2 = \{-150\mathbf{i} - 200\mathbf{j} + 250\mathbf{k}\}$ lb.

7–30. Determine the x, y, z components of force and moment at point C in the pipe assembly. Neglect the weight of the pipe. The load acting at (0, 3.5 ft, 3 ft) is $\mathbf{F}_1 = \{-24; -10\mathbf{k}\}$ lb and $\mathrm{M} = \{-30\mathbf{k}\}$ lb · ft and at point (0, 3.5 ft, 0) $\mathbf{F}_2 = \{-80\mathbf{i}\}$ lb.

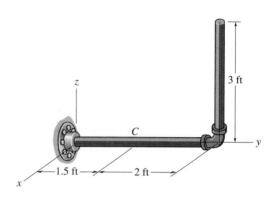

z

C

3 ft

y

1.5 ft   2 ft

x

Prob. 7–30

7–31. Determine the x, y, z components of force and moment at point D of the rod. Set $\mathbf{F} = \{7\mathbf{i} - 12\mathbf{j} - 5\mathbf{k}\}$ kN. The supports at A, B, and C are journal bearings.

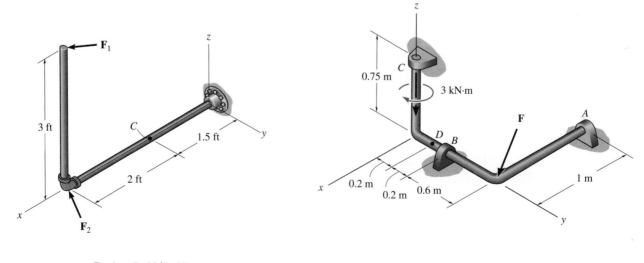

$\mathbf{F}_1$

z

3 ft

C

1.5 ft

y

2 ft

x

$\mathbf{F}_2$

Probs. 7–28/7–29

z

0.75 m

C

3 kN·m

$\mathbf{F}$

A

D B

x 0.2 m

0.2 m   0.6 m

1 m

y

Prob. 7–31

*7.2 Shear and Moment Equations and Diagrams

Beams are structural members which are designed to support loadings applied perpendicular to their axes. In general, beams are long, straight bars having a constant cross-sectional area. Often they are classified as to how they are supported. For example, a *simply supported beam* is pinned at one end and roller-supported at the other, Fig. 7–10, whereas a *cantilevered beam* is fixed at one end and free at the other. The actual design of a beam requires a detailed knowledge of the *variation* of the internal shear force V and bending moment M acting at *each point* along the axis of the beam. After this force and bending-moment analysis is complete, one can then use the theory of mechanics of materials and an appropriate engineering design code to determine the beam's required cross-sectional area.

The *variations* of V and M as a function of the position x along the beam's axis can be obtained by using the method of sections discussed in Sec. 7.1. Here, however, it is necessary to section the beam at an arbitrary distance x from one end rather than at a specified point. If the results are plotted, the graphical variations of V and M as a function of x are termed the *shear diagram* and *bending-moment diagram,* respectively.

In general, the internal shear and bending-moment functions will be discontinuous, or their slope will be discontinuous at points where a distributed load changes or where concentrated forces or couple moments are applied. Because of this, these functions must be determined for *each segment* of the beam located between any two discontinuities of loading. For example, sections located at x_1, x_2, and x_3 will have to be used to describe the variation of V and M throughout the length of the beam in Fig. 7–10. These functions will be valid *only* within regions from O to a for x_1, from a to b for x_2, and from b to L for x_3.

The internal normal force will not be considered in the following discussion for two reasons. In most cases, the loads applied to a beam act perpendicular to the beam's axis and hence produce only an internal shear force and bending moment. For design purposes, the beam's resistance to shear, and particularly to bending, is more important than its ability to resist a normal force.

Fig. 7–10

Positive shear

Positive moment

Beam sign convention

Fig. 7–11

Sign Convention. Before presenting a method for determining the shear and bending moment as functions of x and later plotting these functions (shear and bending-moment diagrams), it is first necessary to establish a *sign convention* so as to define a "positive" and "negative" shear force and bending moment acting in the beam. [This is analogous to assigning coordinate directions x positive to the right and y positive upward when plotting a function $y = f(x)$.] Although the choice of a sign convention is arbitrary, here we will choose the one used in the majority of books on mechanics. It is illustrated in Fig. 7–11. Here the positive directions are denoted by an internal *shear force* that causes *clockwise rotation* of the member on which it acts, and an internal *moment* that causes *compression in the upper part* of the member. Loadings that are opposite to these are considered negative.

PROCEDURE FOR ANALYSIS

The following procedure provides a method for constructing the shear and bending-moment diagrams for a beam.

Support Reactions.   Determine all the reactive forces and couple moments acting on the beam, and resolve all the forces into components acting perpendicular and parallel to the beam's axis.

Shear and Moment Functions.   Specify separate coordinates x having an origin at the beam's *left end* and extending to regions of the beam *between* concentrated forces and/or couple moments, or where there is no discontinuity of distributed loading. Section the beam perpendicular to its axis at each distance x. On the free-body diagram, be sure **V** and **M** are shown acting in their *positive sense,* in accordance with the sign convention given in Fig. 7–11. V is obtained by summing forces perpendicular to the beam's axis, and M is obtained by summing moments about the sectioned end of the segment.

Shear and Moment Diagrams.   Plot the shear diagram (V versus x) and the moment diagram (M versus x). If computed values of the functions describing V and M are *positive,* the values are plotted above the x axis, whereas *negative* values are plotted below the x axis. Generally, it is convenient to plot the shear and bending-moment diagrams directly below the free-body diagram of the beam.

The following examples illustrate this procedure numerically.

Example 7–7

Draw the shear and bending-moment diagrams for the beam shown in Fig. 7–12a.

SOLUTION

Support Reactions. The support reactions have been computed, as shown on the beam's free-body diagram, Fig. 7–12d.

Shear and Moment Functions. The beam is sectioned at an arbitrary distance x from point A, extending within the region AB, and the free-body diagram of the left segment is shown in Fig. 7–12b. The unknowns **V** and **M** are assumed to act in the *positive sense* on the right-hand face of the segment according to the established sign convention. Why? Applying the equilibrium equations yields

$$+\uparrow \Sigma F_y = 0; \qquad\qquad V = 25 \text{ kN} \qquad\qquad (1)$$

$$\downarrow +\Sigma M = 0; \qquad\qquad M = 25x \text{ kN} \cdot \text{m} \qquad\qquad (2)$$

A free-body diagram for a left segment of the beam extending a distance x within the region BC is shown in Fig. 7–12c. As always, **V** and **M** are shown acting in the positive sense. Hence,

$$+\uparrow \Sigma F_y = 0; \qquad\qquad 25 - 50 - V = 0$$
$$V = -25 \text{ kN} \qquad\qquad (3)$$

$$\downarrow +\Sigma M = 0; \qquad M + 50(x - 2) - 25(x) = 0$$
$$M = (100 - 25x) \text{ kN} \cdot \text{m} \qquad\qquad (4)$$

Shear and Moment Diagrams. When Eqs. 1 through 4 are plotted within the regions in which they are valid, the shear and bending-moment diagrams shown in Fig. 7–12d are obtained. The shear diagram indicates that the internal shear force is always 25 kN (positive) within beam segment AB. Just to the right of point B, the shear force changes sign and remains at a constant value of -25 kN for segment BC. The moment diagram starts at zero, increases linearly to point C at $x = 2$ m, where $M_{\max} = 25$ kN(2 m) $= 50$ kN $\cdot$ m, and thereafter decreases back to zero.

It is seen in Fig. 7–12d that the graph of the shear and moment diagrams is discontinuous at points of concentrated force, i.e., points A, B, and C. For this reason, as stated earlier, it is necessary to express both the shear and bending-moment functions separately for regions between concentrated loads. It should be realized, however, that all loading discontinuities are mathematical, arising from the *idealization of a concentrated force and couple moment*. Physically, loads are always applied over a finite area, and if this load variation could be accounted for, the shear and bending-moment diagrams would actually be continuous over the beam's entire length.

(a)

(b)

(c)

(d)

Fig. 7–12

Example 7–8

Draw the shear and bending-moment diagrams for the beam shown in Fig. 7–13a.

6 kN/m

9 m

(a)

$\frac{1}{3}x^2$ kN

$\frac{2}{3}x$ kN/m

V

M

$\frac{x}{3}$

x

9 kN

(b)

6 kN/m

9 kN

V (kN)

9

$V = 9 - \frac{x^2}{3}$

18 kN

5.20 m

x

M (kN·m)

−18

$M = 9x - \frac{x^3}{9}$

$M_{max} = 31.2$

x

(c)

Fig. 7–13

SOLUTION

Support Reactions.   The support reactions have been computed as shown on the beam's free-body diagram, Fig. 7–13c.

Shear and Moment Functions.   A free-body diagram for a left segment of the beam having a length x is shown in Fig. 7–13b. The distributed loading acting on this segment has an intensity of $\frac{2}{3}x$ at its end and is replaced by a resultant force *after* the segment is isolated as a free-body diagram. The *magnitude* of the resultant force is equal to $\frac{1}{2}(x)(\frac{2}{3}x) = \frac{1}{3}x^2$. This force *acts through the centroid* of the distributed loading area, a distance $\frac{1}{3}x$ from the right end. Applying the two equations of equilibrium yields

$$+ \uparrow \Sigma F_y = 0; \qquad 9 - \frac{1}{3}x^2 - V = 0$$

$$V = \left(9 - \frac{x^2}{3}\right) \text{ kN} \qquad (1)$$

$$\zeta + \Sigma M = 0; \qquad M + \frac{1}{3}x^2\left(\frac{x}{3}\right) - 9x = 0$$

$$M = \left(9x - \frac{x^3}{9}\right) \text{ kN} \cdot \text{m} \qquad (2)$$

Shear and Moment Diagrams.   The shear and bending-moment diagrams shown in Fig. 7–13c are obtained by plotting Eqs. 1 and 2.

The point of *zero shear* can be found using Eq. 1:

$$V = 9 - \frac{x^2}{3} = 0$$

$$x = 5.20 \text{ m}$$

This value of x happens to represent the point on the beam where the *maximum moment* occurs (see Sec. 7.3). Using Eq. (2), we have

$$M_{max} = \left(9(5.20) - \frac{(5.20)^3}{9}\right) \text{ kN} \cdot \text{m}$$

$$= 31.2 \text{ kN} \cdot \text{m}$$

PROBLEMS

For each of the following problems, establish the x axis with the origin at the left side of the beam, and obtain the internal shear and moment as a function of x. Use these results to plot the shear and moment diagrams.

*7–32. Draw the shear and moment diagrams for the shaft.

Prob. 7–32

7–33. The suspender bar supports the 600-lb engine. Draw the shear and moment diagrams for the bar.

Prob. 7–33

7–34. Draw the shear and moment diagrams for the beam.

Prob. 7–34

7–35. Draw the shear and moment diagrams for the beam in terms of the parameters shown.

Prob. 7–35

*7–36. Draw the shear and moment diagrams for the beam.

Prob. 7–36

7–37. Draw the shear and moment diagrams for the shaft. The supports at A and C are journal bearings.

Prob. 7–37

7–38. Draw the shear and moment diagrams for the beam.

Prob. 7–38

7–39. Draw the shear and moment diagrams for the shaft. The supports at *A* and *B* are journal bearings.

Prob. 7–39

*7–40. The jib crane supports a load of 750 lb. If the boom *AB* has a uniform weight of 60 lb/ft, draw the shear and moment diagrams for the boom.

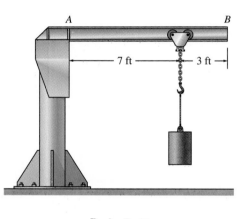

Prob. 7–40

7–41. Draw the shear and moment diagrams for beam *ABC*. Note that *A* is a fixed support and there is a pin at *B*. Solve the problem using the parameters shown.

Prob. 7–41

7–42. Draw the shear and moment diagrams for the beam.

Prob. 7–42

7–43. Draw the shear and moment diagrams for the beam.

Prob. 7–43

*7–44. Draw the shear and moment diagrams for the beam.

Prob. 7–44

7–45. Draw the shear and moment diagrams for the beam.

Prob. 7–45

7–46. Draw the shear and moment diagrams for the beam.

Prob. 7–46

7–47. The cantilevered beam is made of material having a specific weight γ. Determine the shear and moment in the beam as a function of x.

Prob. 7–47

*7–48. The beam will fail when the maximum moment developed in it is M_{max}. Determine the position x of the concentrated force **P** and its smallest magnitude that will cause the beam to fail.

Prob. 7–48

7–49. Determine the distance a between the supports in terms of the shaft's length L so that the bending moment in the *symmetric* shaft is zero at the shaft's center. The intensity of the distributed load at the center of the shaft is w_o. The supports are journal bearings.

Prob. 7–49

7–50. Determine the distance a between the supports in terms of the beam's length L so that the moment in the *symmetric* beam is zero at the beam's center.

Prob. 7–50

7–51. Determine the placement a of roller B from C so that the maximum negative moment and maximum positive moment are equal in magnitude.

Prob. 7–51

*7–52. Express the shear and moment components acting in the pipe as a function of y, where $0 \le y \le 4$ ft.

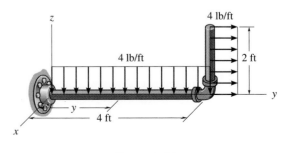

Prob. 7–52

*7.3 Relations Between Distributed Load, Shear, and Moment

In cases where a beam is subjected to several concentrated forces, couple moments, and distributed loads, the method of constructing the shear and bending-moment diagrams discussed in Sec. 7.2 may become quite tedious. In this section a simpler method for constructing these diagrams is discussed—a method based on differential relations that exist between the load, shear, and bending moment.

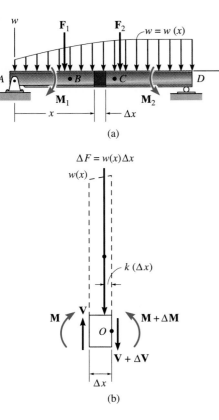

(a)

(b)

Fig. 7–14

Distributed Load. Consider the beam AD shown in Fig. 7–14a, which is subjected to an arbitrary distributed load $w = w(x)$ and a series of concentrated forces and couple moments. In the following discussion, the *distributed load* will be considered *positive* when the *loading acts downward* as shown. The free-body diagram for a small segment of the beam having a length Δx is shown in Fig. 7–14b. Since this segment has been chosen at a point x along the beam which is *not* subjected to a concentrated force or couple moment, any results obtained will not apply at points of concentrated loading. The internal shear force and bending moment shown on the free-body diagram are assumed to act in the *positive sense* according to the established sign convention, Fig. 7–14b. Note that both the shear force and moment acting on the right-hand face must be increased by a small, finite amount in order to keep the segment in equilibrium. The distributed loading has been replaced by a resultant force $\Delta F = w(x)\,\Delta x$ that acts at a fractional distance $k\,(\Delta x)$ from the right end, where $0 < k < 1$ [for example, if $w(x)$ is *uniform*, $k = \frac{1}{2}$]. Applying the equations of equilibrium, we have

$$+\uparrow \Sigma F_y = 0; \qquad V - w(x)\,\Delta x - (V + \Delta V) = 0$$
$$\Delta V = -w(x)\,\Delta x$$

$$\zeta + \Sigma M_O = 0; \quad -V\,\Delta x - M + w(x)\,\Delta x[k(\Delta x)] + (M + \Delta M) = 0$$
$$\Delta M = V\,\Delta x - w(x)k(\Delta x)^2$$

Dividing by Δx and taking the limit as $\Delta x \to 0$, these two equations become

$$\frac{dV}{dx} = -w(x) \qquad (7\text{–}1)$$

$$\begin{array}{c} \text{Slope of} \\ \text{shear diagram} \end{array} = \begin{array}{c} \text{Negative of distributed} \\ \text{load intensity} \end{array}$$

$$\frac{dM}{dx} = V \qquad (7-2)$$

$$\text{Slope of} \atop \text{moment diagram} = \text{Shear}$$

These two equations provide a convenient means for plotting the shear and moment diagrams for a beam. At a specific point in a beam, Eq. 7–1 states that the *slope of the shear diagram is equal to the negative of the intensity of the distributed load,* while Eq. 7–2 states that the *slope of the moment diagram is equal to the shear.* In particular, if the shear is equal to zero, $dM/dx = 0$, and therefore *a point of zero shear corresponds to a point of maximum (or possibly minimum) moment.*

Equations 7–1 and 7–2 may also be rewritten in the form $dV = -w(x)\ dx$ and $dM = V\ dx$. Noting that $w(x)\ dx$ and $V\ dx$ represent differential areas under the distributed-loading and shear diagrams, respectively, we can integrate these areas between two points B and C along the beam, Fig. 7–14a, and write

$$\Delta V_{BC} = -\int w(x)\ dx \qquad (7-3)$$

$$\text{Change} \atop \text{in shear} = \text{Negative of area under} \atop \text{loading curve}$$

and

$$\Delta M_{BC} = \int V\ dx \qquad (7-4)$$

$$\text{Change} \atop \text{in moment} = \text{Area under} \atop \text{shear diagram}$$

Equation 7–3 states that the *change in shear between points B and C is equal to the negative of the area under the distributed-loading curve between these points.* Similarly, from Eq. 7–4, the *change in moment between B and C is equal to the area under the shear diagram within region BC.* Because two integrations are involved, first to determine the change in shear, Eq. 7–3, then to determine the change in moment, Eq. 7–4, we can state that if the loading curve $w = w(x)$ is a polynomial of degree n, then $V = V(x)$ will be a curve of degree $n + 1$, and $M = M(x)$ will be a curve of degree $n + 2$.

As stated previously, the above equations do not apply at points where a *concentrated* force or couple moment acts. These two special cases create *discontinuities* in the shear and moment diagrams, and as a result, each deserves separate treatment.

(a)

(b)

Fig. 7–15

Force. A free-body diagram of a small segment of the beam in Fig. 7–14*a*, taken from under one of the forces, is shown in Fig. 7–15*a*. Here it can be seen that force equilibrium requires

$$+ \uparrow \Sigma F_y = 0; \qquad\qquad \Delta V = -F \qquad\qquad (7\text{–}5)$$

Thus, the *change in shear is negative,* so that on the shear diagram the shear "jumps" *downward when* **F** *acts downward* on the beam. Likewise, the jump in shear (ΔV) is upward when **F** acts upward.

Couple Moment. If we remove a segment of the beam in Fig. 7–14*a* that is located at the couple moment, the free-body diagram shown in Fig. 7–15*b* results. In this case letting $\Delta x \rightarrow 0$, moment equilibrium requires

$$\zeta + \Sigma M = 0; \qquad\qquad \Delta M = M \qquad\qquad (7\text{–}6)$$

Thus, the *change in moment is positive,* or the moment diagram "jumps" *upward if* **M**$_O$ *is clockwise.* Likewise, the jump ΔM is downward when **M**$_O$ is counterclockwise.

The following examples illustrate application of the above equations for the construction of the shear and moment diagrams. After working through these examples, it is recommended that Examples 7–7 and 7–8 be solved using this method.

Example 7–9

Draw the shear and bending-moment diagrams for the beam shown in Fig. 7–16a.

(a)

Fig. 7–16

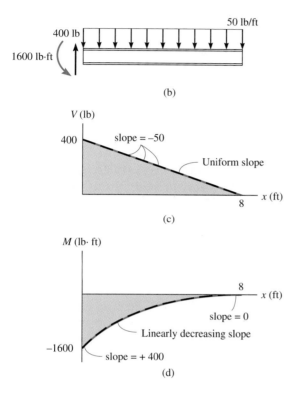

(b)

(c)

(d)

SOLUTION

Support Reactions.   The reactions at the fixed support have been calculated and are shown on the free-body diagram of the beam, Fig. 7–16b.

Shear Diagram.   The shear at the end points is plotted first, Fig. 7–16c. From the sign convention, Fig. 7–11, $V = +400$ at $x = 0$ and $V = 0$ at $x = 8$. Since $dV/dx = -w = -50$, a straight, *negative* sloping line connects the end points.

Moment Diagram.   From our sign convention, Fig. 7–11, the moments at the beam's end points, $M = -1600$ at $x = 0$ and $M = 0$ at $x = 8$, are plotted first, Fig. 7–16d. Successive values of shear taken from the shear diagram, Fig. 7–16c, indicate that the *slope* $dM/dx = V$ of the moment diagram, Fig. 7–16d, is always positive yet *linearly decreasing* from $dM/dx = 400$ at $x = 0$ to $dM/dx = 0$ at $x = 8$. Thus, due to the integrations, w a constant becomes V a sloping line (first-degree curve) and M a parabola (second-degree curve).

Example 7–10

Draw the shear and bending-moment diagrams for the beam shown in Fig. 7–17a.

SOLUTION

Support Reactions.   The reactions at the fixed support have been calculated and are shown on the free-body diagram of the beam, Fig. 7–17b.

Shear Diagram.   Using the established sign convention, Fig. 7–11, the shear at the ends of the beam is plotted first; i.e., $x = 0$, $V = +1080$; $x = 20$, $V = +600$, Fig. 7–17c.

Since the uniform distributed load is downward and *constant,* the slope of the shear diagram is $dV/dx = -w = -40$ for $0 \le x \le 12$ as indicated.

The magnitude of shear at $x = 12$ is $V = +600$. This can be determined by first finding the area under the load diagram between $x = 0$ and $x = 12$. This represents the change in shear. That is, $\Delta V = -\int w(x)\, dx = -40(12) = -480$.   Thus $V|_{x=12} = V|_{x=0} + (-480) = 1080 - 480 = 600$. Also, we can obtain this value by using the method of sections, Fig. 7–17e, where for equilibrium $V = +600$.

Since the load between $12 < x \le 20$ is $w = 0$, the slope $dV/dx = 0$ as indicated. This brings the shear to the required value of $V = +600$ at $x = 20$.

Moment Diagram.   Again, using the established sign convention, Fig. 7–11, the moments at the ends of the beam are plotted first; i.e., $x = 0$, $M = -15{,}880$; $x = 20$, $M = -1000$, Fig. 7–17d.

Each value of shear gives the slope of the moment diagram since $dM/dx = V$. As indicated, at $x = 0$, $dM/dx = +1080$; and at $x = 12$, $dM/dx = +600$. For $0 < x < 12$, specific values of the shear diagram are positive but linearly decreasing. Hence, the moment diagram is parabolic with a linear, decreasing, positive slope.

The magnitude of moment at $x = 12$ ft is -5800. This can be found by first determining the trapezoidal area under the shear diagram, which represents the change in moment, $\Delta M = \int V dx = 600(12) + \frac{1}{2}(1080 - 600)(12) = +10{,}080$.   Thus, $M|_{x=12} = M|_{x=0} + 10{,}080 = -15{,}880 + 10{,}080 = -5800$. The more "basic" method of sections can also be used, where equilibrium at $x = 12$ requires $M = -5800$, Fig. 7–17e.

The moment diagram has a constant slope for $12 < x \le 20$ since, from the shear diagram, $dM/dx = V = +600$. This brings the value of $M = -1000$ at $x = 20$, as required.

Fig. 7–17

Example 7–11

Draw the shear and moment diagrams for the beam in Fig. 7–18a.

(a)

SOLUTION

Support Reactions.   The reactions at the supports are shown on the free-body diagram in Fig. 7–18b.

Shear Diagram.   The end points $x = 0$, $V = +35$ and $x = 8$, $V = -35$ are plotted first, as shown in Fig. 7–18c.

Since there is no distributed load on the beam, the slope of the shear diagram throughout the beam's length is zero; i.e., $dV/dx = -w = 0$. There is a discontinuity or "jump" of the shear diagram, however, at each concentrated force. From Eq. 7–5, $\Delta V = -F$, the change in shear is negative when the force acts downward and positive when the force acts upward. Stated another way, the "jump" follows the force, i.e., a downward force causes a downward jump, and vice versa. Thus, the 20-kN force at $x = 2$ m changes the shear from 35 kN to 15 kN; the 30-kN force at $x = 4$ m changes the shear from 15 kN to -15 kN, etc. We can *also* obtain numerical values for the shear at a specified point in the beam by using the method of sections, as for example, $x = 2^+$ m, $V = 15$ kN in Fig. 7–18e.

Moment Diagram.   The end points $x = 0$, $M = 0$ and $x = 8$, $M = 0$ are plotted first, as shown in Fig. 7–18d.

Since the shear is constant in each region of the beam, the moment diagram has a corresponding constant positive or negative slope as indicated on the diagram. Numerical values for the change in moment at any point can be computed from the *area* under the shear diagram. For example, at $x = 2$ m, $\Delta M = \int V\,dx = 35(2) = 70$. Thus, $M|_{x=2} = M|_{x=0} + 70 = 0 + 70 = 70$. Also, by the method of sections, we can determine the moment at a specified point, as for example, $x = 2^+$ m, $M = 70$ kN · m, Fig. 7–18e.

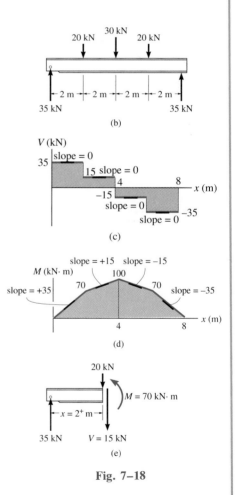

Fig. 7–18

Example 7–12

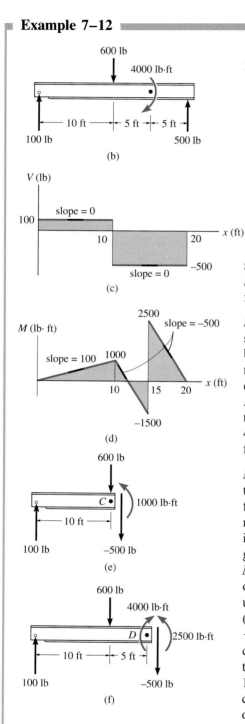

(b)

(c)

(d)

(e)

(f)

Fig. 7–19

(a)

Sketch the shear and bending-moment diagrams for the beam shown in Fig. 7–19*a*.

SOLUTION

Support Reactions. The reactions are calculated and indicated on the free-body diagram, Fig. 7–19*b*.

Shear Diagram. As in Example 7–11, the shear diagram can be constructed by "following the load" on the free-body diagram. In this regard, beginning at A, $V_A = +100$ lb. No load acts between A and C, so the shear remains constant; i.e., $dV/dx = -w(x) = 0$. At C the 600-lb force acts downward, so the shear jumps down 600 lb, from 100 lb to -500 lb. Again the shear is constant (no load) and ends at -500 lb, point B. Notice that no jump or discontinuity in shear occurs at D, the point where the 4000-lb · ft couple moment is applied, Fig. 7–19*a*. This is because, for force equilibrium, $\Delta V = 0$ in Fig. 7–15*b*.

Moment Diagram. The moment at each end of the beam is zero. These two points are plotted first, Fig. 7–19*d*. The slope of the moment diagram from A to C is constant since $dM/dx = V = +100$. The value of the moment at C can be determined by the method of sections or by first computing the rectangular area under the shear diagram between A and C. This gives the change in moment $\Delta M_{AC} = (100$ lb$)(10$ ft$) = 1000$ lb · ft. Since $M_A = 0$, then $M_C = 0 + 1000$ lb · ft $= 1000$ lb · ft. From C to D the slope of the moment diagram is $dM/dx = V = -500$, Fig. 7–19*c*. The area under the shear diagram between points C and D is $\Delta M_{CD} = M_D - M_C = (-500$ lb$)(5$ ft$) = -2500$ lb · ft, so that $M_D = 1000 - 2500 = -1500$ lb · ft. A jump in the moment diagram occurs at point D, which is caused by the concentrated couple moment of 4000 lb · ft. From Eq. 7–6, the jump is *positive* since the couple moment is *clockwise*. Thus, at $x = 15^+$ ft, the moment is $M_D = -1500 + 4000 = 2500$ lb · ft. This value can *also* be determined by the method of sections. From point D the slope of $dM/dx = -500$ is maintained until the diagram closes to zero at B, Fig. 7–19*d*.

PROBLEMS

7–53. Draw the shear and moment diagrams for the beam in Prob. 7–34.

7–54. Draw the shear and moment diagrams for the beam in Prob. 7–37.

7–55. Draw the shear and moment diagrams for the beam in Prob. 7–38.

*7–56. Draw the shear and moment diagrams for the beam in Prob. 7–44.

7–57. Draw the shear and moment diagrams for the beam in Prob. 7–43.

7–58. Draw the shear and moment diagrams for the beam in Prob. 7–45.

7–59. The girder *AB* has a uniform weight of 500 lb/ft and is subjected to the concentrated forces caused by the floor beams. Draw the shear and moment diagrams for the girder.

Prob. 7–59

*7–60. The member *ABC* supports a sign that transmits a wind loading of 2 kN/m along the sign. Draw the shear and moment diagrams for the member. The sign is assumed to be pin-connected to the member at *A* and roller-supported at *B*.

Prob. 7–60

7–61. Each of the street lights has a weight of 20 lb, and the supporting arm weighs 5 lb/ft. Draw the shear and moment diagrams for the arm.

Prob. 7–61

7–62. Draw the shear and moment diagrams for the cantilever beam.

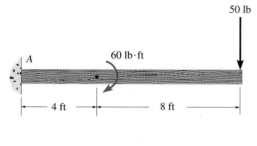

Prob. 7–62

7–63. Draw the shear and moment diagrams for the beam. There is a pin at C. And A is a fix support.

Prob. 7–63

*7–64. Draw the shear and moment diagrams for the shaft. The supports at A and B are journal bearings.

Prob. 7–64

7–65. The beam has a weight of 150 lb/ft. Draw the shear and moment diagrams for the beam.

Prob. 7–65

7–66. Draw the shear and moment diagrams for the beam.

Prob. 7–66

7–67. Draw the shear and moment diagrams for the shaft. The supports at A and B are journal bearings.

Prob. 7–67

*7–68. Draw the shear and moment diagrams for the beam.

Prob. 7–68

7–69. Draw the shear and moment diagrams for the beam.

Prob. 7–69

7–70. Draw the shear and moment diagrams for the beam.

Prob. 7–70

7–71. Draw the shear and moment diagrams for the beam in terms of the parameters shown.

Prob. 7–71

*7–72. Draw the shear and moment diagrams for the beam.

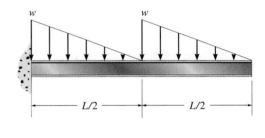

Prob. 7–72

7–73. The beam consists of two segments pin-connected at B. Draw the shear and moment diagrams for the beam.

Prob. 7–73

7–74. Draw the shear and moment diagrams for the beam.

7–75. Draw the shear and moment diagrams for the beam.

Prob. 7–74

Prob. 7–75

*7.4 Cables

Flexible cables and chains are often used in engineering structures for support and to transmit loads from one member to another. When used to support suspension bridges and trolley wheels, cables form the main load-carrying element of the structure. In the force analysis of such systems, the weight of the cable itself may be neglected; however, when cables are used as transmission lines and guys for radio antennas and derricks, the cable weight may become important and must be included in the structural analysis. Three cases will be considered in the analysis that follows: (1) a cable subjected to concentrated loads; (2) a cable subjected to a distributed load; and (3) a cable subjected to its own weight. Regardless of which loading conditions are present, provided the loading is coplanar with the cable, the requirements for equilibrium are formulated in an identical manner.

When deriving the necessary relations between the force in the cable and its slope, we will make the assumption that the cable is *perfectly flexible* and *inextensible*. Due to its flexibility, the cable offers no resistance to bending, and therefore, the tensile force acting in the cable is always tangent to the cable at points along its length. Being inextensible, the cable has a constant length both before and after the load is applied. As a result, once the load is applied, the geometry of the cable remains fixed, and the cable or a segment of it can be treated as a rigid body.

Cable Subjected to Concentrated Loads. When a cable of negligible weight supports several concentrated loads, the cable takes the form of several straight-line segments, each of which is subjected to a constant tensile force. Consider, for example, the cable shown in Fig. 7–20, where the distances h, L_1, L_2, and L_3 and the loads P_1 and P_2 are known. The problem here is to determine the *nine unknowns* consisting of the tension in each of the *three* segments, the *four* components of reaction at A and B, and the sags y_C and y_D at the *two* points C and D. For the solution we can write *two* equations of force equilibrium at each of points A, B, C, and D. This results in a total of *eight equations.** To complete the solution, it will be necessary to know something about the geometry of the cable in order to obtain the necessary ninth equation. For example, if the cable's total *length L* is specified, then the Pythagorean theorem can be used to relate each of the three segmental lengths, written in terms of h, y_C, y_D, L_1, L_2, and L_3, to the total length L. Unfortunately, this type of problem cannot be solved easily by hand. Another possibility, however, is to specify one of the sags, either y_C or y_D, instead of the cable length. By doing this, the equilibrium equations are then sufficient for obtaining the unknown forces and the remaining sag. Once the sag is obtained, the length of the cable can be determined by trigonometry. The following example illustrates a procedure for performing the equilibrium analysis for a problem of this type.

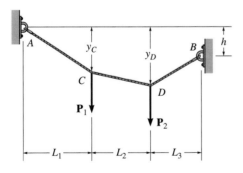

Fig. 7–20

*As will be shown in the following example, the eight equilibrium equations can *also* be written for the entire cable, or any part thereof. But *no more* than *eight* equations are available.

■ **Example 7–13**

Determine the tension in each segment of the cable shown in Fig. 7–21a.

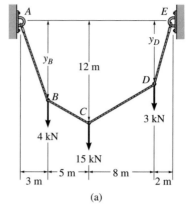

(a)

SOLUTION

By inspection, there are four unknown external reactions (A_x, A_y, E_x, and E_y) and four unknown cable tensions, one in each cable segment. These eight unknowns along with the two unknown sags y_B and y_D can be determined from *ten* available equilibrium equations. One method is to apply these equations as force equilibrium ($\Sigma F_x = 0$, $\Sigma F_y = 0$) to each of the five points A through E. Here, however, we will take a more direct approach.

Consider the free-body diagram for the entire cable, Fig. 7–21b. Thus,

$\xrightarrow{+}\Sigma F_x = 0;$ $\qquad\qquad -A_x + E_x = 0$

$\downarrow+\Sigma M_E = 0;$ $\quad -A_y(18) + 4(15) + 15(10) + 3(2) = 0 \quad A_y = 12$ kN

$+\uparrow\Sigma F_y = 0;$ $\qquad\qquad 12 - 4 - 15 - 3 + E_y = 0 \quad E_y = 10$ kN

Since the sag $y_C = 12$ m is known, we will now consider the leftmost section, which cuts cable BC, Fig. 7–21c.

$\downarrow+\Sigma M_C = 0;$ $\quad A_x(12) - 12(8) + 4(5) = 0 \qquad A_x = E_x = 6.33$ kN

$\xrightarrow{+}\Sigma F_x = 0;$ $\qquad\qquad T_{BC} \cos \theta_{BC} - 6.33 = 0$

$+\uparrow\Sigma F_y = 0;$ $\qquad\qquad 12 - 4 - T_{BC} \sin \theta_{BC} = 0$

Thus,

$$\theta_{BC} = 51.6°$$
$$T_{BC} = 10.2 \text{ kN} \qquad\qquad Ans.$$

(b)

(c)

Fig. 7–21

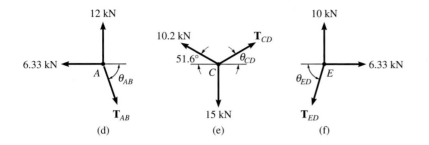

(d)   (e)   (f)

Proceeding now to analyze the equilibrium of points A, C, and E in sequence, we have

Point A   (Fig. 7–21*d*)

$\xrightarrow{+} \Sigma F_x = 0;$ $T_{AB} \cos \theta_{AB} - 6.33 = 0$

$+ \uparrow \Sigma F_y = 0;$ $-T_{AB} \sin \theta_{AB} + 12 = 0$

$\theta_{AB} = 62.2°$

$T_{AB} = 13.6 \text{ kN}$ *Ans.*

Point C   (Fig. 7–21*e*)

$\xrightarrow{+} \Sigma F_x = 0;$ $T_{CD} \cos \theta_{CD} - 10.2 \cos 51.6° = 0$

$+ \uparrow \Sigma F_y = 0;$ $T_{CD} \sin \theta_{CD} + 10.2 \sin 51.6° - 15 = 0$

$\theta_{CD} = 47.9°$

$T_{CD} = 9.44 \text{ kN}$ *Ans.*

Point E   (Fig. 7–21*f*)

$\xrightarrow{+} \Sigma F_x = 0;$ $6.33 - T_{ED} \cos \theta_{ED} = 0$

$+ \uparrow \Sigma F_y = 0;$ $10 - T_{ED} \sin \theta_{ED} = 0$

$\theta_{ED} = 57.7°$

$T_{ED} = 11.8 \text{ kN}$ *Ans.*

By comparison, the maximum cable tension is in segment AB, since this segment has the greatest slope (θ) and it is required that for any left-hand segment the horizontal component $T \cos \theta = A_x$ (a constant). Also, since the slope angles that the cable segments make with the horizontal have now been determined, it is possible to determine the sags y_B and y_D, Fig. 7–21*a*, using trigonometry.

Cable Subjected to a Distributed Load. Consider the weightless cable shown in Fig. 7–22a, which is subjected to a loading function $w = w(x)$ *as measured in the x direction*. The free-body diagram of a small segment of the cable having a length Δs is shown in Fig. 7–22b. Since the tensile force in the cable changes continuously in both magnitude and direction along the cable's length, this change is denoted on the free-body diagram by $\Delta \mathbf{T}$. The distributed load is represented by its resultant force $w(x)(\Delta x)$, which acts at a fractional distance $k(\Delta x)$ from point O, where $0 < k < 1$. Applying the equations of equilibrium yields

$$\xrightarrow{+}\Sigma F_x = 0; \qquad -T \cos \theta + (T + \Delta T) \cos (\theta + \Delta \theta) = 0$$
$$+ \uparrow \Sigma F_y = 0; \quad -T \sin \theta - w(x)(\Delta x) + (T + \Delta T) \sin (\theta + \Delta \theta) = 0$$
$$\zeta + \Sigma M_O = 0; \quad w(x)(\Delta x)k(\Delta x) - T \cos \theta \, \Delta y + T \sin \theta \, \Delta x = 0$$

Dividing each of these equations by Δx and taking the limit as $\Delta x \rightarrow 0$, and hence $\Delta y \rightarrow 0$, $\Delta \theta \rightarrow 0$, and $\Delta T \rightarrow 0$, we obtain

$$\frac{d(T \cos \theta)}{dx} = 0 \qquad\qquad (7\text{–}7)$$

$$\frac{d(T \sin \theta)}{dx} - w(x) = 0 \qquad\qquad (7\text{–}8)$$

$$\frac{dy}{dx} = \tan \theta \qquad\qquad (7\text{–}9)$$

Integrating Eq. 7–7, we have

$$T \cos \theta = \text{constant} = F_H \qquad\qquad (7\text{–}10)$$

Here F_H represents the horizontal component of tensile force at *any point* along the cable.

Integrating Eq. 7–8 gives

$$T \sin \theta = \int w(x) \, dx \qquad\qquad (7\text{–}11)$$

Dividing Eq. 7–11 by Eq. 7–10 eliminates T. Then, using Eq. 7–9, we can obtain the slope

$$\tan \theta = \frac{dy}{dx} = \frac{1}{F_H} \int w(x) \, dx$$

Performing a second integration yields

$$y = \frac{1}{F_H} \int \left(\int w(x) \, dx \right) dx \qquad (7\text{–}12)$$

This equation is used to determine the curve for the cable, $y = f(x)$. The horizontal force component F_H and the two constants, say C_1 and C_2, resulting from the integration are determined by applying the boundary conditions for the cable.

(a)

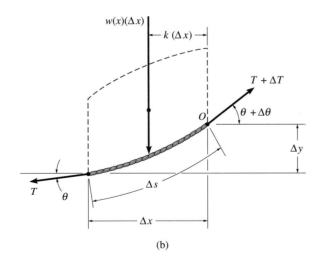

(b)

Fig. 7–22

Example 7–14

The cable of a suspension bridge supports half of the uniform road surface between the two columns at A and B, as shown in Fig. 7–23a. If this loading is w_o N/m, determine the maximum force developed in the cable and the cable's required length. The span length L and sag h are known.

(a)

Fig. 7–23

SOLUTION

We can determine the unknowns in the problem by first finding the curve that defines the shape of the cable by using Eq. 7–12. For reasons of symmetry, the origin of coordinates has been placed at the cable's center. Noting that $w(x) = w_o$, we have

$$y = \frac{1}{F_H} \int \left(\int w_o \, dx \right) dx$$

Integrating this equation twice gives

$$y = \frac{1}{F_H} \left(\frac{w_o x^2}{2} + C_1 x + C_2 \right) \tag{1}$$

The constants of integration may be determined by using the boundary conditions $y = 0$ at $x = 0$ and $dy/dx = 0$ at $x = 0$. Substituting into Eq. 1 yields $C_1 = C_2 = 0$. The curve then becomes

$$y = \frac{w_o}{2F_H} x^2 \tag{2}$$

This is the equation of a *parabola*. The constant F_H may be obtained by using the boundary condition $y = h$ at $x = L/2$. Thus,

$$F_H = \frac{w_o L^2}{8h} \tag{3}$$

Therefore, Eq. 2 becomes

$$y = \frac{4h}{L^2} x^2 \tag{4}$$

Since F_H is known, the tension in the cable may be determined using Eq. 7–10, written as $T = F_H/\cos \theta$. For $0 \le \theta < \pi/2$, the maximum tension will occur when θ is *maximum*, i.e., at point B, Fig. 7–23a. From Eq. 2, the slope at this point is

$$\left. \frac{dy}{dx} \right|_{x=L/2} = \tan \theta_{max} = \left. \frac{w_o}{F_H} x \right|_{x=L/2}$$

or

$$\theta_{max} = \tan^{-1} \left(\frac{w_o L}{2F_H} \right) \tag{5}$$

Therefore,

$$T_{max} = \frac{F_H}{\cos (\theta_{max})} \tag{6}$$

Using the triangular relationship shown in Fig. 7–23b, which is based on Eq. 5, Eq. 6 may be written as

$$T_{max} = \frac{\sqrt{4F_H^2 + w_o^2 L^2}}{2}$$

(b)

Substituting Eq. 3 into the above equation yields

$$T_{max} = \frac{w_o L}{2} \sqrt{1 + \left(\frac{L}{4h} \right)^2} \qquad \textit{Ans.}$$

For a differential segment of cable length ds, we can write

$$ds = \sqrt{(dx)^2 + (dy)^2} = \sqrt{1 + \left(\frac{dy}{dx} \right)^2} \, dx$$

Hence, the total length of the cable, $\mathscr{L}$, can be determined by integration. Using Eq. 4, we have

$$\mathscr{L} = \int ds = 2 \int_0^{L/2} \sqrt{1 + \left(\frac{8h}{L^2} x \right)^2} \, dx \tag{7}$$

Integrating and substituting the limits yields

$$\mathscr{L} = \frac{L}{2} \left[\sqrt{1 + \left(\frac{4h}{L} \right)^2} + \frac{L}{4h} \sinh^{-1} \left(\frac{4h}{L} \right) \right] \qquad \textit{Ans.}$$

Cable Subjected to Its Own Weight. When the weight of the cable becomes important in the force analysis, the loading function along the cable becomes a function of the arc length s rather than the projected length x. A generalized loading function $w = w(s)$ acting along the cable is shown in Fig. 7–24a. The free-body diagram for a segment of the cable is shown in Fig. 7–24b. Applying the equilibrium equations to the force system on this diagram, one obtains relationships identical to those given by Eqs. 7–7 through 7–9, but with ds replacing dx. Therefore, it may be shown that

$$T \cos \theta = F_H \tag{7-13}$$

$$T \sin \theta = \int w(s) \, ds$$

$$\frac{dy}{dx} = \frac{1}{F_H} \int w(s) \, ds \tag{7-14}$$

To perform a direct integration of Eq. 7–14, it is necessary to replace dy/dx by ds/dx. Since

$$ds = \sqrt{dx^2 + dy^2}$$

then

$$\frac{dy}{dx} = \sqrt{\left(\frac{ds}{dx}\right)^2 - 1}$$

Therefore,

$$\frac{ds}{dx} = \left\{1 + \frac{1}{F_H^2}\left(\int w(s) \, ds\right)^2\right\}^{1/2}$$

Separating the variables and integrating yields

$$x = \int \frac{ds}{\left\{1 + \frac{1}{F_H^2}\left(\int w(s) \, ds\right)^2\right\}^{1/2}} \tag{7-15}$$

The two constants of integration, say C_1 and C_2, are found using the boundary conditions for the cable.

(a)

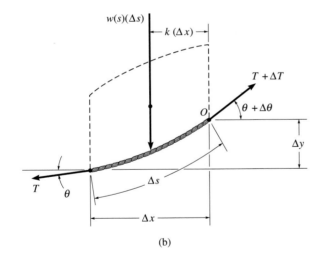

(b)

Fig. 7–24

Example 7–15

Determine the deflection curve, the length, and the maximum tension in the uniform cable shown in Fig. 7–25. The cable weighs $w_o = 5$ N/m.

SOLUTION

By reasons of symmetry, the origin of coordinates is located at the center of the cable. The deflection curve is expressed as $y = f(x)$. We can determine it by first applying Eq. 7–15, where $w(s) = w_o$.

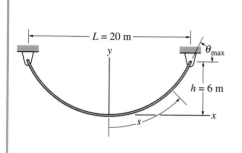

$$x = \int \frac{ds}{[1 + (1/F_H^2)(\int w_o \, ds)^2]^{1/2}}$$

Integrating the term under the integral sign in the denominator, we have

$$x = \int \frac{ds}{[1 + (1/F_H^2)(w_o s + C_1)^2]^{1/2}}$$

Substituting $u = (1/F_H)(w_o s + C_1)$ so that $du = (w_o/F_H) \, ds$, a second integration yields

$$x = \frac{F_H}{w_o}(\sinh^{-1} u + C_2)$$

or

$$x = \frac{F_H}{w_o}\left\{ \sinh^{-1}\left[\frac{1}{F_H}(w_o s + C_1) \right] + C_2 \right\} \tag{1}$$

To evaluate the constants note that, from Eq. 7–14,

$$\frac{dy}{dx} = \frac{1}{F_H} \int w_o \, ds \qquad \text{or} \qquad \frac{dy}{dx} = \frac{1}{F_H}(w_o s + C_1)$$

Since $dy/dx = 0$ at $s = 0$, then $C_1 = 0$. Thus,

$$\frac{dy}{dx} = \frac{w_o s}{F_H} \tag{2}$$

The constant C_2 may be evaluated by using the condition $s = 0$ at $x = 0$ in Eq. 1, in which case $C_2 = 0$. To obtain the deflection curve, solve for s in Eq. 1, which yields

$$s = \frac{F_H}{w_o} \sinh\left(\frac{w_o}{F_H} x \right) \tag{3}$$

Now substitute into Eq. 2, in which case

$$\frac{dy}{dx} = \sinh\left(\frac{w_o}{F_H} x \right)$$

Hence

Fig. 7–25

$$y = \frac{F_H}{w_o} \cosh \left(\frac{w_o}{F_H} x \right) + C_3 \tag{4}$$

If the boundary condition $y = 0$ at $x = 0$ is applied, the constant $C_3 = -F_H/w_o$, and therefore the deflection curve becomes

$$y = \frac{F_H}{w_o} \left[\cosh \left(\frac{w_o}{F_H} x \right) - 1 \right]$$

This equation defines the shape of a *catenary curve*. The constant F_H is obtained by using the boundary condition that $y = h$ at $x = L/2$, in which case

$$h = \frac{F_H}{w_o} \left[\cosh \left(\frac{w_o L}{2 F_H} \right) - 1 \right] \tag{5}$$

Since $w_o = 5$ N/m, $h = 6$ m, and $L = 20$ m, Eqs. 4 and 5 become

$$y = \frac{F_H}{5 \text{ N/m}} \left[\cosh \left(\frac{5 \text{ N/m}}{F_H} x \right) - 1 \right] \tag{6}$$

$$6 \text{ m} = \frac{F_H}{5 \text{ N/m}} \left[\cosh \left(\frac{50 \text{ N}}{F_H} \right) - 1 \right] \tag{7}$$

Equation 7 can be solved for F_H by using a trial-and-error procedure. The result is

$$F_H = 45.8 \text{ N}$$

and therefore the deflection curve, Eq. 6, becomes

$$y = 9.16[\cosh (0.109x) - 1] \text{ m} \qquad \textit{Ans.}$$

Using Eq. 3, with $x = 10$ m, the half-length of the cable is

$$\frac{\mathcal{L}}{2} = \frac{45.8 \text{ N}}{5 \text{ N/m}} \sinh \left[\frac{5 \text{ N/m}}{45.8 \text{ N}} (10 \text{ m}) \right] = 12.1 \text{ m}$$

Hence,

$$\mathcal{L} = 24.2 \text{ m} \qquad \textit{Ans.}$$

Since $T = F_H/\cos \theta$, Eq. 7–13, the maximum tension occurs when θ is maximum, i.e., at $s = \mathcal{L}/2 = 12.1$ m. Using Eq. 2 yields

$$\left. \frac{dy}{dx} \right|_{s=12.1 \text{ m}} = \tan \theta_{max} = \frac{5 \text{ N/m}(12.1 \text{ m})}{45.8 \text{ N}} = 1.32$$

$$\theta_{max} = 52.9°$$

Thus, $$T_{max} = \frac{F_H}{\cos \theta_{max}} = \frac{45.8 \text{ N}}{\cos 52.9°} = 75.9 \text{ N} \qquad \textit{Ans.}$$

PROBLEMS

Neglect the weight of the cable in the following problems, *unless* specified.

*7–76. Determine the tension in each segment of the cable and the cable's total length.

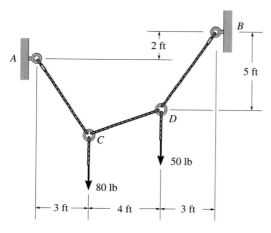

Prob. 7–76

7–77. The cable supports the loading shown. Determine the distance x_B from the wall to point B.

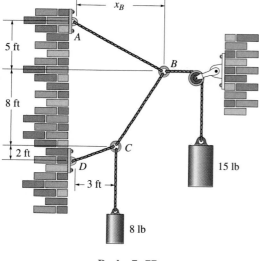

Prob. 7–77

7–78. The cable supports the three loads shown. Determine the sags y_B and y_D of points B and D. Take $P_1 = 400$ lb, $P_2 = 250$ lb.

7–79. The cable supports the three loads shown. Determine the magnitude of P_1 if $P_2 = 300$ lb and $y_B = 8$ ft. Also find the sag y_D.

Probs. 7–78/7–79

*7–80. The cable supports the loading shown. Determine the distance x_B from the force at B to point A. Set $P = 40$ lb.

7–81. The cable supports the loading shown. Determine the magnitude of the horizontal force P so that $x_B = 6$ ft.

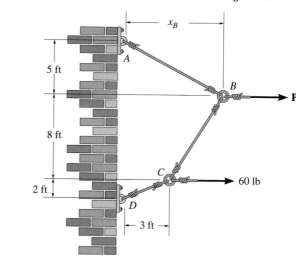

Probs. 7–80/7–81

7–82. Cable $ABCD$ supports the 10-kg lamp E and the 15-kg lamp F. Determine the maximum tension in the cable and the sag of point B.

*7–84. Determine the maximum uniform loading w N/m that the cable can support if it is capable of sustaining a maximum tension of 60 kN before it will break.

Prob. 7–82

Prob. 7–84

7–83. Determine the tension in each segment of the cable and the cable's total length.

7–85. If the side loading on the suspension bridge in Example 7–14 is $w_o = 0.75$ kN/m and $L = 190$ m, $h = 20$ m, determine the maximum tension in the cable.

7–86. The cable will break when the maximum tension reaches $T_{max} = 12$ kN. Determine the uniform distributed load w required to develop this maximum tension. Set $h = 6$ m.

Prob. 7–83

Prob. 7–86

7–87. The cable *AB* is subjected to a uniform loading of 200 N/m. If the weight of the cable is neglected and the slope angles at points *A* and *B* are 30° and 60°, respectively, determine the deflection curve of the cable and the maximum tension developed in the cable.

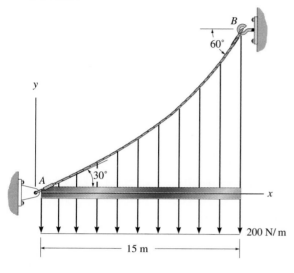

Prob. 7–87

*7–88. The weightless cable is subjected to the triangular loading. If the slope of the cable at point *A* is zero, determine the deflection curve of the cable and the maximum tension developed in the cable. *L*, *h*, and w_o are known.

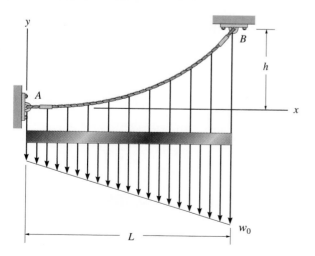

Prob. 7–88

7–89. The cable supports a girder which weighs 850 lb/ft. Determine the tension in the cable at points *A*, *B*, and *C*.

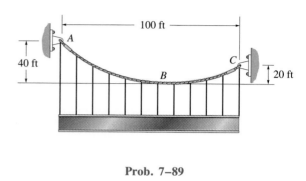

Prob. 7–89

7–90. A uniform cord is suspended between two points having the same elevation. Determine the sag-to-span ratio so that the maximum tension in the cord equals the cord's total weight.

7–91. If the chain has a mass of 3 kg/m, determine the maximum tension in the cable. What is the cable's length?

Prob. 7–91

*7–92. A 50-m-long chain has a mass of 100 kg and is suspended between two points 10 m apart at the same elevation. Determine the maximum tension and the sag in the chain.

7–93. A 100-lb cable is attached between two points having equal elevations at a distance of 50 ft apart. If the maximum tension developed in the cable is 75 lb, determine the length of the cable and the sag.

7–94. Show that the deflection curve of the cable discussed in Example 7–15 reduces to Eq. (4) in Example 7–14 when the *hyperbolic cosine function* is expanded in terms of a series and only the first two terms are retained. (The answer indicates that the *catenary* may be replaced by a *parabola* in the analysis of problems in which the sag is small. In this case, the cable weight is assumed to be uniformly distributed along the horizontal.)

7–95. A 100-ft chain has a weight of 150 lb and is suspended between two points 25 ft apart, at the same elevation. Determine the maximum tension and the sag in the chain.

*7–96. The chain has a weight of 3 lb/ft. Determine the tension at points A, B, and C necessary for equilibrium.

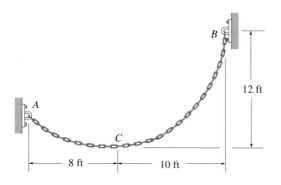

Prob. 7–96

7–97. The power line is supported at A by the tower. If the cable weighs 0.75 lb/ft, and the sag $s = 3$ ft, determine the resultant horizontal force the cable exerts at A.

7–98. The power line is supported at A by the tower. If the cable weighs 0.75 lb/ft, determine the total length of the cable, BAC. Set $s = 3$ ft.

7–99. The power line is supported at A by the tower. If the cable weighs 0.75 lb/ft, determine the required sag s so that the resultant horizontal force the cable exerts at A is zero.

Probs. 7–97/7–98/7–99

*7–100. The 80-ft-long chain is fixed at its ends and hoisted at its midpoint B using a crane. If the chain has a weight of 0.5 lb/ft, determine the minimum height h of the hook in order to lift the chain *completely* off the ground. What is the horizontal force at pin A or C when the chain is in this position? *Hint:* When h is a minimum, the slope at A and C is zero.

Prob. 7–100

7–101. The chain is suspended between points A and B. If it has a weight of 0.5 lb/ft and the sag is 3 ft, determine the maximum tension in the chain.

Prob. 7–101

7–102. The uniform beam weighs 80 lb/ft and is held in the horizontal position by means of the cable AB, which has a weight of 10 lb/ft. If the slope of the cable at A is 15°, determine the length of the cable.

Prob. 7–102

7–103. The cable has a weight of 8 N/m and length of 10 m. If the tension in the cable at C is 300 N, determine the distance from A to C.

7–105. The balloon is held in place using a 400-ft cord that weighs 0.8 lb/ft and makes a 60° angle with the horizontal. If the tension in the cord at point A is 150 lb, determine the length of the cord, l, that is lying on the ground and the height h. *Hint:* Establish the coordinate system at B as shown.

Prob. 7–103

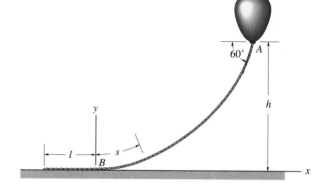

Prob. 7–105

*7–104. The cable has a mass of 0.5 kg/m and is 25 m long. Determine the vertical and horizontal components of force it exerts on the top of the tower.

Prob. 7–104

REVIEW PROBLEMS

7–106. Draw the shear and moment diagrams for the beam.

Prob. 7–106

7–107. Determine the normal force, shear force, and moment at point C. Assume the support at A is a roller and the support at B is a pin.

Prob. 7–107

*7–108. A 50-ft cable is suspended between two points a distance of 15 ft apart and at the same elevation. If the minimum tension in the cable is 200 lb, determine the total weight of the cable and the maximum tension developed in the cable.

7–109. Draw the shear and moment diagrams for the beam.

Prob. 7–109

7–110. Draw the shear and moment diagrams for the beam.

Prob. 7–110

7–111. The beam is supported by a pin at C and a rod AB. Determine the normal force, shear force, and moment at point D.

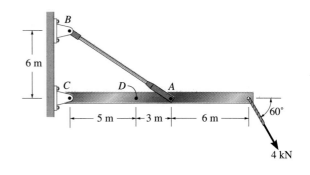

Prob. 7–111

*7–112.   Draw the shear and moment diagrams for the shaft. The supports at *A* and *B* are journal bearings.

7–114.   Draw the shear and moment diagrams for the beam.

Prob. 7–112

Prob. 7–114

7–113.   Determine the normal force, shear force, and moment at points *D* and *E* of the frame.

Prob. 7–113

8 Friction

In the previous chapters the surfaces of contact between two bodies were considered to be perfectly *smooth*. Because of this, the force of interaction between the bodies always acts *normal* to the surface at points of contact. In reality, however, all surfaces are *rough,* and depending on the nature of the problem, the ability of a body to support a *tangential* as well as a *normal* force at its contacting surface must be considered. The tangential force is caused by friction, and in this chapter we will show how to analyze problems involving frictional forces. Specific application will include frictional forces on screws, bearings, disks, and belts. The analysis of rolling resistance is given in the last part of the chapter.

8.1 Characteristics of Dry Friction

Friction may be defined as a force of resistance acting on a body which prevents or retards slipping of the body relative to a second body or surface with which it is in contact. This force always acts *tangent* to the surface at points of contact with other bodies and is directed so as to oppose the possible or existing motion of the body relative to these points.

Rough surface

(a)

(b)

(c)

Resultant Normal
and Frictional Forces

(d)

Fig. 8–1

In general, two types of friction can occur between surfaces. *Fluid friction* exists when the contacting surfaces are separated by a film of fluid (gas or liquid). The nature of fluid friction is studied in fluid mechanics, since it depends on knowledge of the velocity of the fluid and the fluid's ability to resist shear force. In this book only the effects of *dry friction* will be presented. This type of friction is often called *Coulomb friction*, since its characteristics were studied extensively by C. A. Coulomb in 1781. Specifically, dry friction occurs between the contacting surfaces of bodies in the absence of a lubricating fluid.

Theory of Dry Friction. The theory of dry friction can best be explained by considering what effects are caused by pulling horizontally on a block of uniform weight W which is resting on a rough horizontal surface, Fig. 8–1a. To properly develop a full understanding of the nature of friction, it is necessary to consider the surfaces of contact to be *nonrigid or deformable*. The other portion of the block, however, will be considered rigid. As shown on the free-body diagram of the block, Fig. 8–1b, the floor exerts a *distribution* of both *normal force* ΔN_n and *frictional force* ΔF_n along the contacting surface. For equilibrium, the normal forces must act *upward* to balance the block's weight W, and the frictional forces act to the left to prevent the applied force P from moving the block to the right. Close examination of the contacting surfaces between the floor and block reveals how these frictional and normal forces develop, Fig. 8–1c. It can be seen that many microscopic irregularities exist between the two surfaces and, as a result, reactive forces ΔR_n are developed at each of the protuberances.* These forces act at all points of contact and, as shown, each reactive force contributes both a frictional component ΔF_n and a normal component ΔN_n.

Equilibrium. For simplicity in the following analysis, the effect of the distributed normal and frictional loadings will be indicated by their *resultants* N and F, which are represented on the free-body diagram of the block as shown in Fig. 8–1d. Clearly, the distribution of ΔF_n in Fig. 8–1b indicates that F always acts *tangent to the contacting surface, opposite* to the direction of P. The normal force N is determined from the distribution of ΔN_n in Fig. 8–1b and is directed upward to balance the block's weight W. Notice that it acts a distance x to the right of the line of action of W, Fig. 8–1d. This location, which coincides with the centroid of the loading diagram in Fig. 8–1b, is necessary in order to balance the "tipping effect" caused by P. For example, if P is applied at a height h from the surface, Fig. 8–1d, then moment equilibrium about point O is satisfied if $Wx = Ph$ or $x = Ph/W$. In particular, note that the block will be on the verge of *tipping* if $x = a/2$.

---

*Besides mechanical interactions as explained here, a detailed treatment of the nature of frictional forces must also include the effects of temperature, density, cleanliness, and atomic or molecular attraction between the contacting surfaces. See D. Tabor, *Journal of Lubrication Technology*, 103, 169, 1981.

Impending Motion. In cases where h is small or the surfaces of contact are rather "slippery," the frictional force **F** may *not* be great enough to balance the magnitude of **P**, and consequently the block will tend to slip *before* it can tip. In other words, as the magnitude of **P** is slowly increased, the magnitude of **F** correspondingly increases until it attains a certain *maximum value* F_s, called the *limiting static frictional force*, Fig. 8–1e. When this value is reached, the block is in *unstable equilibrium*, since any further increase in P will cause deformations and fractures at the points of surface contact and consequently the block will begin to move. Experimentally, it has been determined that the magnitude of the limiting static frictional force F_s is *directly proportional* to the magnitude of the resultant normal force **N**. This may be expressed mathematically as

$$F_s = \mu_s N \qquad (8\text{–}1)$$

where the constant of proportionality, μ_s (mu "sub" s), is called the *coefficient of static friction*.

Typical values for μ_s, found in many engineering handbooks, are given in Table 8–1. Although this coefficient is generally less than 1, be aware that in some cases it is possible, as in the case of aluminum on aluminum, for μ_s to be greater than 1. Physically this means, of course, that in this case the frictional force is greater than the corresponding normal force. Furthermore, it should be noted that μ_s is dimensionless and depends only on the characteristics of the two surfaces in contact. A wide range of values is given for each value of μ_s, since experimental testing was done under variable conditions of roughness and cleanliness of the contacting surfaces. For applications, therefore, it is important that both caution and judgment be exercised when selecting a coefficient of friction for a given set of conditions. When an exact calculation of F_s is required, the coefficient of friction should be determined directly by an experiment that involves the two materials to be used.

Motion. If the magnitude of **P** acting on the block is increased so that it becomes greater than F_s, the frictional force at the contacting surfaces drops slightly to a smaller value F_k, called the *kinetic frictional force*. The block will *not* be held in equilibrium ($P > F_k$); instead, it will begin to slide with increasing speed, Fig. 8–1f. The drop made in the frictional force magnitude, from F_s (static) to F_k (kinetic), can be explained by again examining the surfaces of contact, Fig. 8–1g. Here it is seen that when $P > F_s$, then P has the capacity to shear off the peaks at the contact surfaces and cause the block to "lift" somewhat out of its settled position and "ride" on top of the peaks. Once the block begins to slide, high local temperatures at the points of contact cause momentary adhesion (welding) of these points. The continued shearing of these welds is the dominant mechanism creating friction. Since the resultant contact forces $\Delta \mathbf{R}_n$ are aligned slightly more in the vertical direction than before, Fig. 8–1c, they thereby contribute *smaller* frictional components, $\Delta \mathbf{F}_n$, as when the irregularities are meshed.

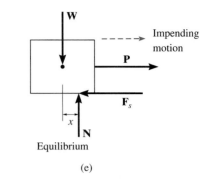

Equilibrium

(e)

Table 8–1 Typical Values for μ_s

Contact Materials	Coefficient of Static Friction (μ_s)
Metal on ice	0.03–0.05
Wood on wood	0.30–0.70
Leather on wood	0.20–0.50
Leather on metal	0.30–0.60
Aluminum on aluminum	1.10–1.70

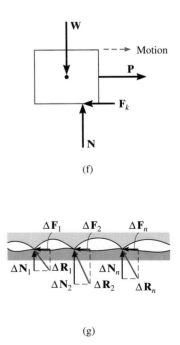

(f)

(g)

Experiments with sliding blocks indicate that the magnitude of the resultant frictional force $\mathbf{F}_k$ is directly proportional to the magnitude of the resultant normal force $\mathbf{N}$. This may be expressed mathematically as

$$F_k = \mu_k N \qquad (8-2)$$

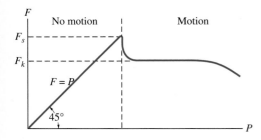

Fig. 8–2

where the constant of proportionality, μ_k, is called the *coefficient of kinetic friction*. Typical values for μ_k are approximately 25 percent *smaller* than those listed in Table 8–1 for μ_s.

The above effects regarding friction can be summarized by reference to the graph in Fig. 8–2, which shows the variation of the frictional force $\mathbf{F}$ versus the applied load $\mathbf{P}$. Here the frictional force is categorized in three different ways: namely, $\mathbf{F}$ is a *static-frictional force* if equilibrium is maintained; $\mathbf{F}$ is a *limiting static-frictional force* $\mathbf{F}_s$ when its magnitude reaches a maximum value needed to maintain equilibrium; and finally, $\mathbf{F}$ is termed a *kinetic-frictional force* $\mathbf{F}_k$ when sliding occurs at the contacting surface. Notice also from the graph that for very large values of $\mathbf{P}$ or for high speeds, because of aerodynamic effects, $\mathbf{F}_k$ and likewise μ_k begin to decrease.

Characteristics of Dry Friction. As a result of *experiments* that pertain to the foregoing discussion, the following rules which apply to bodies subjected to dry friction may be stated.

1. The frictional force acts *tangent* to the contacting surfaces in a direction *opposed* to the *relative motion* or tendency for motion of one surface against another.
2. The magnitude of the maximum static frictional force $\mathbf{F}_s$ that can be developed is independent of the area of contact, provided the normal pressure is not very low nor great enough to severely deform or crush the contacting surfaces of the bodies.
3. The magnitude of the maximum static frictional force is generally greater than the magnitude of the kinetic frictional force for any two surfaces of contact. However, if one of the bodies is moving with a *very low velocity* over the surface of another, F_k becomes approximately equal to F_s, i.e., $\mu_s \approx \mu_k$.
4. When *slipping* at the surface of contact is *about to occur,* the magnitude of the maximum static frictional force is proportional to the magnitude of the normal force, such that $F_s = \mu_s N$, Eq. 8–1.
5. When *slipping* at the surface of contact is *occurring,* the magnitude of the kinetic frictional force is proportional to the magnitude of the normal force, such that $F_k = \mu_k N$, Eq. 8–2.

Angle of Friction. It should be observed that Eqs. 8–1 and 8–2 have a specific yet *limited* use in the solution of friction problems. In particular, the frictional force acting at a contacting surface is determined from $F_k = \mu_k N$ *only* if *relative motion* is occurring between the two surfaces. Furthermore, if two bodies are *stationary,* the magnitude of the frictional force, F, *does not necessarily* equal $\mu_s N$; instead, F must satisfy the inequality $F \leq \mu_s N$. Only when *impending motion* occurs does F reach its upper limit, $F = F_s = \mu_s N$. This situation may be better understood by considering the block shown in Fig. 8–3a, which is acted upon by a force **P**. In this case consider $P = F_s$, so that the block is on the *verge of sliding.* For equilibrium, the normal force **N** and frictional force $\mathbf{F}_s$ combine to create a resultant $\mathbf{R}_s$. The angle ϕ_s that $\mathbf{R}_s$ makes with **N** is called the *angle of static friction.* From the figure,

$$\phi_s = \tan^{-1}\left(\frac{F_s}{N}\right) = \tan^{-1}\left(\frac{\mu_s N}{N}\right) = \tan^{-1}\mu_s$$

Provided the block is *not in motion,* any horizontal force $P < F_s$ causes a resultant **R** which has a line of action directed at an angle ϕ from the vertical such that $\phi \leq \phi_s$. If **P** creates uniform *motion* of the block, then $P = F_k$. In this case, the resultant $\mathbf{R}_k$ has a line of action defined by ϕ_k, Fig. 8–3b. This angle is referred to as the *angle of kinetic friction,* where

$$\phi_k = \tan^{-1}\left(\frac{F_k}{N}\right) = \tan^{-1}\left(\frac{\mu_k N}{N}\right) = \tan^{-1}\mu_k$$

By comparison, $\phi_s \geq \phi_k$.

Angle of Repose. An experimental method which can be used to measure the coefficient of friction between two contacting surfaces consists of placing a block of one material having a weight W on a plane made of another material, Fig. 8–4a. The plane is inclined to the angle θ_s, at which point the block is on the *verge of sliding* and therefore $F_s = \mu_s N$. The free-body diagram of the block at this instant is shown in Fig. 8–4b. Applying the force equations of equilibrium the normal force $N = W\cos\theta_s$, and the frictional force $F_s = W\sin\theta_s$. Since $F_s = \mu_s N$, then $W\sin\theta_s = \mu_s(W\cos\theta_s)$ or

$$\theta_s = \tan^{-1}\mu_s$$

The angle θ_s is referred to as the *angle of repose,* and by comparison it is equal to the angle of static friction ϕ_s. Once it is measured, the coefficient of static friction is obtained from $\mu_s = \tan\theta_s$. Note that this calculation is independent of the weight of the block, and so for the experiment W does not have to be known.

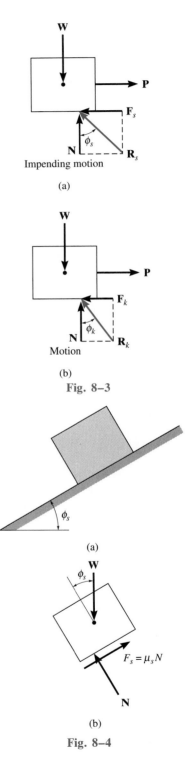

Fig. 8–3

Fig. 8–4

8.2 Problems Involving Dry Friction

(a)

(b)

Fig. 8–5

(a)

(b)

Fig. 8–6

If a rigid body is in equilibrium when it is subjected to a system of forces that includes the effect of friction, the force system must satisfy not only the equations of equilibrium but *also* the laws that govern the frictional forces.

Types of Friction Problems. In general, there are three types of mechanics problems involving dry friction. They can easily be classified once the free-body diagrams are drawn and the total number of unknowns are identified and compared with the total number of available equilibrium equations. Each type of problem will now be explained and illustrated graphically by examples. In all these cases the geometry and dimensions for the problem are assumed to be known.

Equilibrium. Problems in this category are strictly equilibrium problems which require *the total number of unknowns to be equal to the total number of available equilibrium equations*. Once the frictional forces are determined from the solution, however, their numerical values must be checked to be sure they satisfy the inequality $F \leq \mu_s N$; otherwise, slipping will occur and the body will not remain in equilibrium. A problem of this type is shown in Fig. 8–5a. Here we must determine the frictional forces at A and C to check if the equilibrium position of the bars can be maintained. If the bars are uniform and have known weights of 100 N each, then the free-body diagrams are as shown in Fig. 8–5b. There are six unknown force components which can be determined *strictly* from the six equilibrium equations (three for each member). Once $\mathbf{F}_A$, $\mathbf{N}_A$, $\mathbf{F}_C$, and $\mathbf{N}_C$ are determined, then the bars will remain in equilibrium provided $F_A \leq 0.3N_A$ and $F_C \leq 0.5N_C$ are satisfied.

Impending Motion at All Points. In this case *the total number of unknowns will equal the total number of available equilibrium equations plus the total number of available frictional equations, $F = \mu N$*. In particular, if *motion is impending* at the points of contact, then $F_s = \mu_s N$; whereas if the body is *slipping*, then $F_k = \mu_k N$. For example, consider the problem of finding the smallest angle θ at which the 100-N bar in Fig. 8–6a can be placed against the wall without slipping. The free-body diagram is shown in Fig. 8–6b. Here there are *five* unknowns: F_A, N_A, F_B, N_B, θ. For the solution there are *three* equilibrium equations and *two* static frictional equations which apply at *both* points of contact, so that $F_A = 0.3N_A$ and $F_B = 0.4N_B$. (It should also be noted that the bar will not be in a state where motion impends *unless* the bar slips at *both* points A and B simultaneously.)

Fig. 8–7

(a)　　　　　(b)

Tipping or Impending Motion at Some Points. Here *the total number of unknowns will be less than the number of available equilibrium equations plus the total number of frictional equations or conditional equations for tipping.* As a result, several possibilities for motion or impending motion will exist and the problem will involve a determination of the kind of motion which actually occurs. For example, consider the two-member frame shown in Fig. 8–7a. In this problem we wish to determine the horizontal force **P** needed to cause movement of the frame. If each member has a weight of 100 N, then the free-body diagrams are as shown in Fig. 8–7b. There are *seven* unknowns: N_A, F_A, N_C, F_C, B_x, B_y, P. For a unique solution we must satisfy the *six* equilibrium equations (three for each member) and only *one* of two possible static frictional equations. This means that as **P** increases its magnitude it will either cause slipping at A and no slipping at C, so that $F_A = 0.3N_A$ and $F_C \le 0.5N_C$; or slipping occurs at C and no slipping at A, in which case $F_C = 0.5N_C$ and $F_A \le 0.3N_A$. The actual situation can be determined by calculating P for each case and then choosing the case for which P is *smallest*. If in both cases the *same value* for P is calculated, which in practice would be highly improbable, then slipping at both points occurs simultaneously; i.e., the *seven unknowns* will satisfy *eight equations*. As a second example, consider a block having a width b, height h, and weight W which is resting on a rough surface, Fig. 8–8a. The force **P** needed to cause motion is to be determined. Inspection of the free-body diagram, Fig. 8–8b, indicates that there are *four unknowns*, namely, P, F, N, and x. For a unique solution, however, we must satisfy the *three* equilibrium equations and either *one* static friction equation or *one* conditional equation which requires the block not to tip. Hence two possibilities of motion exist. Either the block will *slip*, Fig. 8–8b, in which case $F = \mu_s N$ and the value obtained for x must satisfy $0 \le x \le b/2$; or the block will *tip*, Fig. 8–8c, in which case $x = b/2$ and the frictional force will satisfy the inequality $F \le \mu_s N$. The solution yielding the *smallest* value of P will define the type of motion the block undergoes. If it happens that the same value of P is calculated for both cases, although this would be very improbable, then slipping and tipping will occur simultaneously; i.e., the *four unknowns* will satisfy *five equations*.

Fig. 8–8

Equilibrium Versus Frictional Equations. It was stated earlier that the frictional force *always* acts so as to either oppose the relative motion or impede the motion of a body over its contacting surface. Realize, however, that we can *assume* the sense of the frictional force in problems which require F to be an "equilibrium force" and satisfy the inequality $F < \mu_s N$. The correct sense is made known *after* solving the equations of equilibrium for F. For example, if F is a negative scalar, the sense of **F** is the reverse of that which was assumed. This convenience of *assuming* the sense of **F** is possible because the equilibrium equations equate to zero the *components of vectors* acting in the *same direction*. In cases where the frictional equation $F = \mu N$ is used in the solution of a problem, however, the convenience of *assuming* the sense of **F** is *lost,* since the frictional equation relates only the *magnitudes* of two perpendicular vectors. Consequently, **F** *must always* be shown acting with its *correct sense* on the free-body diagram whenever the frictional equation is used for the solution of a problem.

PROCEDURE FOR ANALYSIS

The following procedure provides a method for solving equilibrium problems involving dry friction.

Free-Body Diagrams. Draw the necessary free-body diagrams and determine the number of unknowns or equations required for a complete solution. Unless stated in the problem, *always* show the frictional forces as *unknowns;* i.e., *do not assume that $F = \mu N$.* Recall that only three equations of coplanar equilibrium can be written for each body. Consequently, if there are more unknowns than equations of equilibrium, it will be necessary to apply the frictional equation at some, if not all, points of contact to obtain the extra equations needed for a complete solution.

Equations of Friction and Equilibrium. Apply the equations of equilibrium and the necessary frictional equations (or conditional equations if tipping is involved) and solve for the unknowns. If the problem involves a three-dimensional force system such that it becomes difficult to obtain the force components or the necessary moment arms, apply the equations of equilibrium using Cartesian vectors.

The following example problems illustrate this procedure numerically.

Example 8–1

The uniform crate shown in Fig. 8–9a has a mass of 20 kg. If a force $P = 80$ N is applied to the crate, determine if it remains in equilibrium. The coefficient of static friction is $\mu_s = 0.3$.

$P = 80$ N

$30°$

0.8 m

0.4 m

0.2 m

(a)

Fig. 8–9

SOLUTION

Free-Body Diagram.   As shown in Fig. 8–9b, the *resultant* normal force N_C must act a distance x from the crate's center line in order to counteract the tipping effect caused by **P**. There are *three unknowns: F, N_C,* and *x*, which can be determined strictly from the *three* equations of equilibrium.

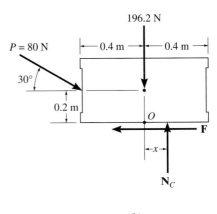

196.2 N

$P = 80$ N

0.4 m   0.4 m

$30°$

0.2 m

O

F

x

N_C

(b)

Equations of Equilibrium

$$\xrightarrow{+}\Sigma F_x = 0; \qquad 80 \cos 30° - F = 0$$
$$+\uparrow\Sigma F_y = 0; \qquad -80 \sin 30° + N_C - 196.2 = 0$$
$$\downarrow+\Sigma M_O = 0; \qquad 80 \sin 30°(0.4) - 80 \cos 30°(0.2) + N_C(x) = 0$$

Solving,

$$F = 69.3 \text{ N}$$
$$N_C = 236 \text{ N}$$
$$x = -0.00908 \text{ m} = -9.08 \text{ mm}$$

Since x is negative it indicates the *resultant* normal force acts (slightly) to the *left* of the crate's center line. No tipping will occur since $x \le 0.4$ m. Also, the *maximum* frictional force which can be developed at the surface of contact is $F_{max} = \mu_s N_C = 0.3(236 \text{ N}) = 70.8$ N. Since $F = 69.3$ N < 70.8 N, the crate will *not slip*, although it is very close to doing so.

Example 8–2

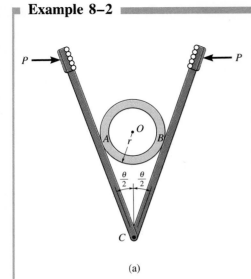

(a)

The pipe shown in Fig. 8–10a is gripped between two levers that are pinned together at C. If the coefficient of static friction between the levers and the pipe is $\mu = 0.3$, determine the maximum angle θ at which the pipe can be gripped without slipping. Neglect the weight of the pipe.

SOLUTION

Free-Body Diagram.   As shown in Fig. 8–10b, there are five unknowns: N_A, F_A, N_B, F_B, and θ. The *three* equations of equilibrium and *two* frictional equations at A and B apply. The frictional forces act toward C to prevent upward motion of the pipe.

Equations of Friction and Equilibrium.   The frictional equations are

$$F_s = \mu_s N; \qquad\qquad\qquad F_A = \mu N_A$$
$$F_B = \mu N_B$$

Using these results, and applying the equations of equilibrium, yields

$\xrightarrow{+}\Sigma F_x = 0;$

$$N_A \cos\left(\frac{\theta}{2}\right) + \mu N_A \sin\left(\frac{\theta}{2}\right) - N_B \cos\left(\frac{\theta}{2}\right) - \mu N_B \sin\left(\frac{\theta}{2}\right) = 0 \quad (1)$$

$\curvearrowright +\Sigma M_O = 0;$

$$-\mu N_B(r) + \mu N_A(r) = 0 \qquad\qquad (2)$$

$+\uparrow\Sigma F_y = 0;$

$$N_A \sin\left(\frac{\theta}{2}\right) - \mu N_A \cos\left(\frac{\theta}{2}\right) + N_B \sin\left(\frac{\theta}{2}\right) - \mu N_B \cos\left(\frac{\theta}{2}\right) = 0 \quad (3)$$

From either Eq. 1 or 2 it is seen that $N_A = N_B$. This could also have been determined directly from the symmetry of *both* geometry and loading. Substituting the result into Eq. 3, we obtain

$$\sin\left(\frac{\theta}{2}\right) - \mu \cos\left(\frac{\theta}{2}\right) = 0$$

so that

$$\tan\left(\frac{\theta}{2}\right) = \frac{\sin(\theta/2)}{\cos(\theta/2)} = \mu = 0.3$$
$$\theta = 2 \tan^{-1} 0.3 = 33.4° \qquad\qquad\qquad\qquad Ans.$$

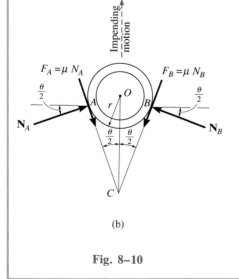

(b)

Fig. 8–10

Example 8–3

The uniform rod having a weight W and length l is supported at its ends A and B, where the coefficient of static friction is μ, Fig. 8–11a. Determine the greatest angle θ so the rod does not slip. Neglect the thickness of the rod for the calculation.

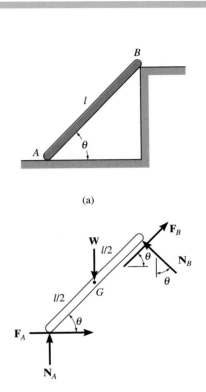

(a)

SOLUTION

Free-Body Diagram. As shown in Fig. 8–11b, there are *five* unknowns: F_A, N_A, F_B, N_B, and θ. These can be determined from the *three* equilibrium equations and *two* frictional equations applied at points A and B. The frictional forces must be drawn with their correct sense so that they oppose the tendency for motion of the rod. Why?

Equations of Friction and Equilibrium. Writing the frictional equations,

$$F = \mu_s N; \qquad\qquad F_A = \mu N_A$$
$$F_B = \mu N_B$$

Using these results and applying the equations of equilibrium yields

$$\xrightarrow{+}\Sigma F_x = 0; \qquad \mu N_A + \mu N_B \cos\theta - N_B \sin\theta = 0 \qquad (1)$$
$$+\uparrow\Sigma F_y = 0; \qquad N_A - W + N_B \cos\theta + \mu N_B \sin\theta = 0 \qquad (2)$$
$$\zeta+\Sigma M_G = 0; \qquad -N_A\left(\frac{l}{2}\cos\theta\right) + \mu N_A\left(\frac{l}{2}\sin\theta\right) + N_B\left(\frac{l}{2}\right) = 0 \qquad (3)$$

Moments were summed about the center of the rod G in order to eliminate W. We can solve Eqs. 1 and 3, which reduce to

$$\mu N_A = N_B(\sin\theta - \mu\cos\theta)$$
$$N_B = N_A(\cos\theta - \mu\sin\theta)$$

Thus

Fig. 8–11

$$\mu N_A = N_A(\cos\theta - \mu\sin\theta)(\sin\theta - \mu\cos\theta)$$
$$\mu = \sin\theta\cos\theta - \mu\cos^2\theta - \mu\sin^2\theta + \mu^2\sin\theta\cos\theta$$
$$\mu = (1 + \mu^2)\sin\theta\cos\theta - \mu(\sin^2\theta + \cos^2\theta)$$

Since $\sin^2\theta + \cos^2\theta = 1$ and $\sin 2\theta = 2\sin\theta\cos\theta$, then

$$2\mu = \left(\frac{1 + \mu^2}{2}\right)\sin 2\theta$$

Solving for θ, we have

$$\theta = \frac{1}{2}\sin^{-1}\left(\frac{4\mu}{1 + \mu^2}\right) \qquad\qquad \textit{Ans.}$$

(b)

Example 8–4

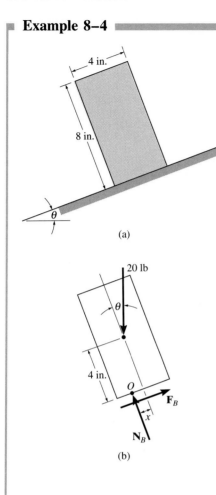

(a)

(b)

Fig. 8–12

The homogeneous block shown in Fig. 8–12a has a weight of 20 lb and rests on the incline for which $\mu_s = 0.55$. Determine the largest angle of tilt, θ, of the plane before the block moves.

SOLUTION

Free-Body Diagram.   As shown in Fig. 8–12b, the dimension x is used to locate the position of the resultant normal force $\mathbf{N}_B$ under the block. There are *four* unknowns, θ, N_B, F_B, and x. *Three* equations of equilibrium are available. The *fourth* equation is obtained by investigating the conditions for tipping or sliding of the block.

Equations of Equilibrium.   Applying the equations of equilibrium yields

$$+\swarrow \Sigma F_x = 0; \qquad\qquad 20 \sin\theta - F_B = 0 \qquad\qquad (1)$$

$$+\nwarrow \Sigma F_y = 0; \qquad\qquad N_B - 20 \cos\theta = 0 \qquad\qquad (2)$$

$$\downarrow + \Sigma M_O = 0; \qquad 20 \sin\theta(4) - 20 \cos\theta(x) = 0 \qquad (3)$$

(Impending Motion of Block.)   This requires use of the frictional equation

$$F_s = \mu_s N; \qquad\qquad F_B = 0.55 N_B \qquad\qquad (4)$$

Solving Eqs. 1 through 4 yields

$$N_B = 17.5 \text{ lb} \qquad F_B = 9.64 \text{ lb} \qquad \theta = 28.8° \qquad x = 2.2 \text{ in.}$$

Since $x = 2.2$ in. > 2 in., the block will tip *before* sliding.

(Tipping of Block.)   This requires

$$x = 2 \text{ in.} \qquad\qquad (5)$$

Solving Eqs. 1 through 3 using Eq. 5 yields

$$N_B = 17.9 \text{ lb} \qquad F_B = 8.94 \text{ lb}$$

$$\theta = 26.6° \qquad\qquad\qquad Ans.$$

Note: If we *first* assumed that the block tips, then the results for F_B would have to be checked with the maximum *possible* static frictional force; i.e.,

$$F_B = 8.94 \overset{?}{<} (0.55)(17.9 \text{ lb}) = 9.84 \text{ lb}$$

Since the inequality holds, indeed the block will tip before it slips.

Example 8–5

Beam AB is subjected to a uniform load of 200 N/m and is supported at B by a post BC, Fig. 8–13a. If the coefficients of static friction at B and C are $\mu_B = 0.2$ and $\mu_C = 0.5$, determine the force **P** needed to pull the post out from under the beam. Neglect the weight of the members and the thickness of the post.

(a)

(b)

(c)

Fig. 8–13

SOLUTION

Free-Body Diagrams. The free-body diagram of beam AB is shown in Fig. 8–13b. Applying $\Sigma M_A = 0$, we obtain $N_B = 400$ N. This result is shown on the free-body diagram of the post, Fig. 8–13c. Referring to this member, the *four* unknowns F_B, P, F_C, and N_C are determined from the *three* equations of equilibrium and *one* frictional equation applied either at B or C.

Equations of Equilibrium and Friction.

$$\xrightarrow{+} \Sigma F_x = 0; \qquad\qquad P - F_B - F_C = 0 \qquad (1)$$
$$+\uparrow \Sigma F_y = 0; \qquad\qquad N_C - 400 = 0 \qquad (2)$$
$$\downarrow + \Sigma M_C = 0; \qquad\qquad -P(0.25) + F_B(1) = 0 \qquad (3)$$

(Post Slips Only at B.)   This requires $F_C \le \mu N_C$ and

$$F_B = \mu_B N_B; \qquad\qquad F_B = 0.2(400) = 80 \text{ N}$$

Using this result and solving Eqs. 1 through 3, we obtain

$$P = 320 \text{ N}$$
$$F_C = 240 \text{ N}$$
$$N_C = 400 \text{ N}$$

Since $F_C = 240 \text{ N} > \mu_C N_C = 0.5(400 \text{ N}) = 200 \text{ N}$, the other case of movement must be investigated.

(Post Slips Only at C.)   Here $F_B \le \mu_B N_B$ and

$$F_C = \mu_C N_C; \qquad\qquad F_C = 0.5 N_C \qquad (4)$$

Solving Eqs. 1 through 4 yields

$$P = 267 \text{ N} \qquad\qquad Ans.$$
$$N_C = 400 \text{ N}$$
$$F_C = 200 \text{ N}$$
$$F_B = 66.7 \text{ N}$$

Obviously, this case occurs first, since it requires a *smaller* value for P.

Example 8–6

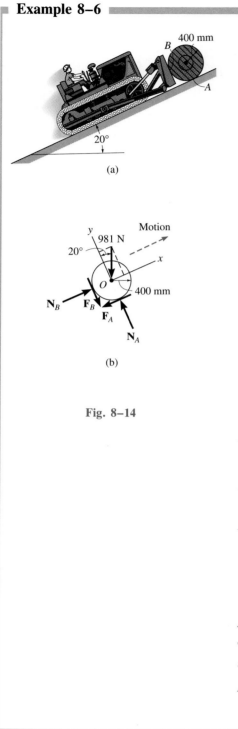

(a)

(b)

Fig. 8–14

Determine the normal force that must be exerted on the 100-kg spool shown in Fig. 8–14a to push it up the 20° incline at constant velocity. The coefficients of static and kinetic friction at the points of contact are $(\mu_s)_A = 0.18$, $(\mu_k)_A = 0.15$ and $(\mu_s)_B = 0.45$, $(\mu_k)_B = 0.4$.

SOLUTION

Free-Body Diagram. As shown in Fig. 8–14b, there are four unknowns N_A, F_A, N_B, and F_B acting on the spool. These can be determined from the *three* equations of equilibrium and *one* frictional equation, which applies either at A or B. If slipping only occurs at B, the spool *rolls* up the incline; whereas if slipping only occurs at A, the spool will *slide* up the incline. Here we must calculate N_B.

Equations of Equilibrium and Friction

$$+\nearrow \Sigma F_x = 0; \qquad -F_A + N_B - 981 \sin 20° = 0 \qquad (1)$$

$$+\nwarrow \Sigma F_y = 0; \qquad N_A - F_B - 981 \cos 20° = 0 \qquad (2)$$

$$\zeta + \Sigma M_O = 0; \qquad F_B(400 \text{ mm}) - F_A(400 \text{ mm}) = 0 \qquad (3)$$

(Spool Rolls up Incline.) In this case $F_A \le 0.18 N_A$ and

$$(F_k)_B = (\mu_k)_B N_B; \qquad F_B = 0.40 N_B \qquad (4)$$

The direction of the frictional force at B must be specified correctly. Why? Since the spool is being forced up the plane, $\mathbf{F}_B$ acts downward to prevent the clockwise rolling motion of the spool, Fig. 8–14b. Solving Eqs. 1 through 4, we have

$$N_A = 1146 \text{ N} \qquad F_A = 224 \text{ N} \qquad N_B = 559 \text{ N} \qquad F_B = 224 \text{ N}$$

The assumption regarding no slipping at A should be checked.

$$F_A \le (\mu_s)_A N_A; \qquad 224 \text{ N} \overset{?}{\le} 0.18(1146) = 206 \text{ N}$$

The inequality does *not apply*, and therefore slipping occurs at A and not at B. Hence, the other case of motion must be investigated.

(Spool Slides up Incline.) In this case, $F_B \le 0.45 N_B$ and

$$(F_k)_A = (\mu_k)_A N_A; \qquad F_A = 0.15 N_A \qquad (5)$$

Solving Eqs. 1 through 3 and 5 yields

$$N_A = 1084 \text{ N} \qquad F_A = 163 \text{ N} \qquad N_B = 498 \text{ N} \qquad F_B = 163 \text{ N}$$

The validity of the solution ($N_B = 498$ N) can be checked by testing the assumption that indeed no slipping occurs at B.

$$F_B \le (\mu_s)_B N_B; \qquad 163 \text{ N} < 0.45(498 \text{ N}) = 224 \text{ N} \qquad \text{(check)}$$

PROBLEMS

8–1. The footing of a ladder rests on a concrete surface for which $\mu_s = 0.8$ for the two materials. If the magnitude of a force directed along the axis of the ladder is 600 lb, determine the frictional force acting at the bottom of the ladder. What is the minimum coefficient of static friction that will prevent the ladder from slipping?

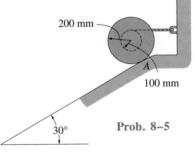

600 lb

55°

Prob. 8–1

8–2. Determine the horizontal force P needed to just start moving the 300-lb crate up the plane. Take $\mu_s = 0.3$.

8–3. Determine the range of values for which the horizontal force P will prevent the crate from slipping down or up the inclined plane. Take $\mu_s = 0.1$.

P

20°

Probs. 8–2/8–3

*8–4. The refrigerator has a weight of 200 lb and a center of gravity at G. Determine the force **P** required to move it. Will the refrigerator tip or slip? Take $\mu_s = 0.4$.

3 ft

P

G

1.5 ft

4 ft

3 ft

Prob. 8–4

8–5. The spool has a mass of 20 kg. If a cord is wrapped around its inner core and is attached to the wall, and the coefficient of static friction at A is $\mu_s = 0.15$, determine if the spool remains in equilibrium when it is released.

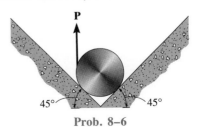

200 mm

100 mm

A

30°

Prob. 8–5

8–6. The 100-lb cylinder rests between the two inclined planes. When $P = 15$ lb, the cylinder is on the verge of impending motion. Determine the coefficient of static friction μ_s between the surfaces of contact and the cylinder.

P

45°   45°

Prob. 8–6

8–7. The bracket can move freely along the pole until a load P is placed on it. When this occurs it contacts the pole at A and B. If the coefficient of static friction is $\mu_s = 0.6$, determine the smallest distance d at which the load can be applied so that the bracket does not slip. Neglect the weight of the bracket.

Prob. 8–7

8–9. Two blocks A and B rest on the inclined planes. If the weight of block A is 60 lb, determine the weight of block B so that (a) block B is on the verge of sliding up the plane, and (b) once static friction is overcome, block B moves at constant velocity down the plane. Set $\mu_s = 0.5$ and $\mu_k = 0.25$ for both surfaces.

Prob. 8–9

*8–8. The car has a weight of 4000 lb and a center of gravity at G. If it pulls off the side of a road, determine the greatest angle of tilt, θ, it can have without slipping or tipping over. The coefficient of static friction between its wheels and the ground is $\mu_s = 0.4$.

8–10. The square block has a weight W and rests on the board. If the board is tipped at an angle θ, the block will tip over. Determine θ and the smallest coefficient of static friction that will allow this to happen.

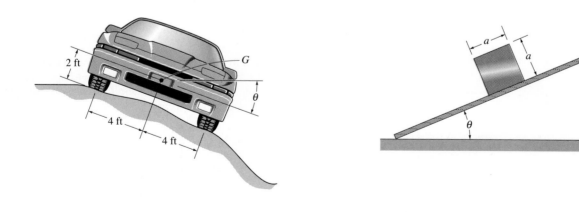

Prob. 8–8

Prob. 8–10

8–11. The uniform ladder has a weight W and rests against the wall at B and the floor at A. If the coefficient of static friction is μ, determine the smallest angle θ at which the ladder can be placed against the wall and not slip.

8–13. Determine the greatest angle of placement for the stick that will lock the door in place when any horizontal force P is applied to the handle. Neglect the weight of the stick and assume no slipping occurs at B. The coefficient of static friction at A is μ.

Prob. 8–11

Prob. 8–13

*8–12. If the coefficient of static friction between the incline and the 200-lb crate is $\mu_s = 0.3$, determine the force P that must be applied to the rope to begin moving the crate up the incline.

8–14. The machine has a weight of 500 lb. Determine the force P that must be applied to the handle of the crowbar to lift it. The coefficient of static friction at A and B is $\mu_s = 0.3$.

8–15. The machine has a weight of 500 lb. Determine the force P that must be applied to the handle of the crowbar to lift it. The coefficient of static friction at A is $\mu_A = 0.3$ and at B, $\mu_B = 0.2$.

Prob. 8–12

Probs. 8–14/8–15

*8–16. The man has a mass of 40 kg. He plans to scale the vertical crevice using the method shown. If the coefficient of static friction between his shoes and the rock is $\mu_s = 0.4$ and between his backside and the rock, $\mu_s' = 0.3$, determine the smallest horizontal force his body must exert on the rock in order to do this.

0.2 m

1.25 m

Prob. 8–16

8–17. The truck has a mass of 1.25 Mg and a center of mass at G. Determine the greatest load it can pull if (a) the truck has rear-wheel drive while the front wheels are free to roll, and (b) the truck has four-wheel drive. The coefficient of static friction between the wheels and the ground is $\mu_s = 0.5$, and between the crate and the ground, it is $\mu_s' = 0.4$.

8–18. Solve Prob. 8–17 if the truck and crate are traveling up a 10° incline.

8–19. The uniform dresser has a weight of 80 lb and rests on a tile floor for which $\mu_s = 0.25$. If the man pushes on it in the direction shown, determine the smallest magnitude of force **F** needed to move the dresser. Also, if the man has a weight of 150 lb, determine the smallest coefficient of static friction between his shoes and the floor so he doesn't slip.

30° F

Prob. 8–19

*8–20. The tractor pulls on the fixed tree stump. Determine the torque that must be applied by the engine to the rear wheels to cause them to slip. The front wheels are free to roll. The tractor weighs 3500 lb and has a center of gravity at G. The coefficient of static friction between the rear wheels and the ground is $\mu_s = 0.5$.

8–21. The tractor pulls on the fixed tree stump. If the coefficient of static friction between the rear wheels and the ground is $\mu_s = 0.6$, determine if the rear wheels slip or the front wheels lift off the ground as the engine provides torque to the rear wheels. What is the torque needed to cause the motion? The front wheels are free to roll. The tractor weighs 2500 lb and has a center of gravity at G.

800 mm

600 mm

A

G

B

1.5 m   1 m

Probs. 8–17/8–18

G

2 ft O

2 ft

A   5 ft   3 ft   B

Probs. 8–20/8–21

8–22. The cabinet has a weight of 300 lb and due to the items inside the drawers, a center of gravity at G. If the coefficient of static friction at A and B is $\mu_s = 0.2$, determine if the 200-lb man can push the cabinet to the left. The coefficient of static friction between his shoes and the floor is $\mu_s' = 0.4$. Assume the man exerts only a horizontal force on the cabinet.

8–23. The cabinet has a weight of 300 lb and due to the items inside the drawers, a center of gravity at G. If the coefficient of static friction at A and B is $\mu_s = 0.2$, determine if the 200-lb man can pull the cabinet to the right. The coefficient of static friction between his shoes and the floor is $\mu_s' = 0.35$. Assume the man exerts only a horizontal force on the cabinet.

8–25. Determine the greatest load W the boy can pull on the sled S, if the sled has a weight of 8 lb and $(\mu_S)_s = 0.1$ between the runners and the ice. The boy weighs 70 lb and $(\mu_B)_s = 0.4$ between his boots and the ice. Set $\theta = 30°$.

8–26. Determine the angle θ at which the cord must be directed so that the boy can pull the sled S, which supports a load of 400 lb. The sled weighs 8 lb and $(\mu_S)_s = 0.1$ between the runners and the ice. The boy weighs 70 lb, and $(\mu_B)_s = 0.4$ between his boots and the ice.

Probs. 8–25/8–26

Probs. 8–22/8–23

8–27. Two blocks A and B, each weighing 50 lb, are joined by a light 3-ft rod having a pin at each end. Determine the maximum height h at which block B may be placed up the inclined plane without causing the assembly to slip down the plane. Neglect the size of the blocks in the calculations. Set $(\mu_A)_s = 0.3$, $(\mu_B)_s = 0.15$.

*8–24. The 2-m-long uniform rod has a weight W and is placed along the ledge. If the smallest angle at which it can lean without slipping is $\theta = 60°$, determine the coefficient of static friction μ between the contact surfaces A and B and the rod.

Prob. 8–24

Prob. 8–27

*8–28. Two boys, each weighing 60 lb, sit at the ends of a uniform board, which has a weight of 30 lb. If the board rests at its center on a post having a coefficient of static friction of $\mu_s = 0.6$ with the board, determine the greatest angle of tilt θ before slipping occurs. Neglect the size of the post and the thickness of the board in the calculations.

8–30. A man attempts to lift the uniform 40-lb ladder to an upright position by applying a force **F** perpendicular to the ladder at rung R. Determine the coefficient of static friction between the ladder and the ground at A if the ladder begins to slip on the ground when his hands reach a height of 6 ft.

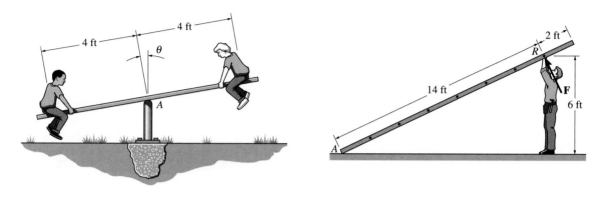

Prob. 8–28

Prob. 8–30

8–29. Determine the minimum force **F** needed to push the tube E up the incline. The tube has a mass of 75 kg and the roller D has a mass of 100 kg. The force acts parallel to the plane and the coefficients of static friction at the contacting surfaces are $\mu_A = 0.3$, $\mu_B = 0.25$, and $\mu_C = 0.4$. Each cylinder has a radius of 150 mm.

8–31. The man has a mass of 60 kg and the crate has a mass of 100 kg. If the coefficient of static friction between his shoes and the ground is $\mu_S = 0.4$ and between the crate and the ground $\mu_C = 0.3$, determine if the man is able to move the crate using the rope-and-pulley system shown.

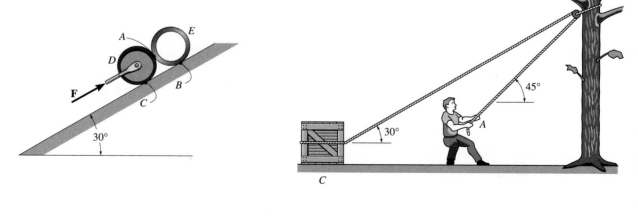

Prob. 8–29

Prob. 8–31

*8–32.   A 50-lb beam rests on two surfaces at points A and B. Determine the maximum distance x to which a boy can walk up the beam before the beam begins to slip. The boy weighs 120 lb and assume he walks up the beam with a constant velocity. Neglect the thickness of the beam.

8–34.   The rider has a weight of 170 lb and a center of gravity at G. If $\mu_s = 0.5$, is it possible to move the bicycle forward? Neglect friction at the front wheel and the weight of the bicycle.

Prob. 8–32

Prob. 8–34

8–33.   The man attempts to move the block, which has a weight of 800 lb and a center of gravity at G, over the edge of the step. If the coefficient of static friction at A is $\mu_A = 0.3$, determine the angle θ at which the block begins to slip. Also, with what horizontal force $\mathbf{P}$ must the man pull for this to occur?

8–35.   The device is used to hold brooms and mops. It consists of two small rollers having a negligible weight that can slide freely along the smooth inclines. If the coefficient of static friction between the broomstick and each roller is $\mu_s = 0.5$, determine the angle of inclination θ at which the inclines should be designed so that any weight of broom can be supported.

Prob. 8–33

Prob. 8–35

*8–36. The refrigerator has a mass of 75 kg and a center of mass at *G*. Determine the force **P** required to move it. Does it tip or slide? The coefficient of static friction between the refrigerator and the incline is $\mu_s = 0.3$.

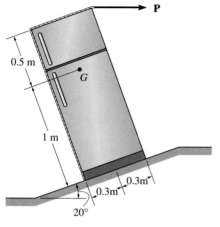

Prob. 8–36

8–37. Determine the maximum weight *W* the man can lift using the pulley system, without and then with the "leading block" or pulley at *A*. The man has a weight of 200 lb and the coefficient of static friction between his feet and the ground is $\mu_s = 0.6$.

(a)

(b)

Prob. 8–37

8–38. The beam is supported by a pin at *A* and a roller at *B* which has negligible weight and a radius of 15 mm. If the coefficient of static friction at *B* and *C* is $\mu_B = \mu_C = 0.3$, respectively, determine the largest angle θ of the incline so that the roller does not slip for any force **P** applied to the beam.

Prob. 8–38

■8–39. The homogeneous semicylinder of weight *W* rests on the surface, for which the coefficient of static friction is $\mu_s = 0.2$. If a force **P** is applied at its edge, determine the maximum angle θ the top of the semicylinder tilts before it begins to slip. The center of gravity is located at *G* as shown.

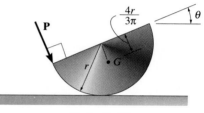

Prob. 8–39

*8–40. The man leans against the wall in the position shown. If he has a mass of 70 kg and center of mass at *G*, determine if he will slip at *A*. The coefficients of static friction at *A* and *B* are $\mu_A = 0.4$ and $\mu_B = 0.5$, respectively.

8–41. The man stands against the wall in the position shown. If he has a mass of 70 kg and center of mass at *G*, determine if he slips at *A* or *B* if he begins to straighten out his legs. The coefficients of static friction at *A* and *B* are $\mu_A = 0.4$ and $\mu_B = 0.5$, respectively.

Probs. 8–40/8–41

8–42.   Two blocks A and B, each having a mass of 6 kg, are connected by the linkage shown. If the coefficients of static friction at the contacting surfaces are $\mu_A = 0.2$ and $\mu_B = 0.3$, determine the largest vertical force **P** that may be applied to pin C without causing the blocks to slip. Neglect the weight of the links.

Prob. 8–42

8–43.   The uniform rod has a weight W. Determine the largest angle θ at which the rod can be inclined such that it will not slip off the two fixed pegs at A and B. The coefficient of static friction between the rod and the pegs is μ.

Prob. 8–43

*8–44.   The semicircular disk has a weight W and a center of gravity at G. Determine the smallest angle of placement, θ, so that the disk does not slip at the wall or floor. The coefficient of static friction at the contacting surfaces is μ.

Prob. 8–44

8–45.   The uniform rod has a weight W and rests on a smooth peg at A and against a wall at B for which $\mu_s = 0.25$. Determine the greatest angle θ for placement of the rod so that it does not slip.

Prob. 8–45

8–46.   The beam has a mass of 20 kg and is subjected to a force of 250 N. It is supported at one end by a pin and at the other end by a spool having a mass of 35 kg. Determine the minimum cable force **P** needed to move the spool from under the beam. The coefficients of static friction at B and D are $\mu_B = 0.4$ and $\mu_D = 0.2$, respectively.

Prob. 8–46

8–47. The wheel of a shopping cart comes in contact with a circular obstruction having a negligible weight. If the coefficient of static friction at the points of contact A and B is $\mu_s = 0.4$, determine the largest radius r of the obstruction which can lock the wheel and keep it from rotating when any horizontal force is applied to the shopping cart.

Prob. 8–47

8–50. The uniform 5-lb rod rests on the ground at B and against the wall at A. If the rod does not slip at B, determine the smallest coefficient of static friction at A which will prevent slipping.

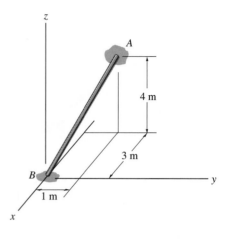

Prob. 8–50

***8–48.** The log of weight W is to be pulled up the inclined plane of slope α using a force **P**. If **P** acts at the angle ϕ as shown, show that for slipping to occur $P = W \sin(\alpha + \theta)/\cos(\phi - \theta)$, where θ is the angle of static friction; $\theta = \tan^{-1}\mu_s$.

8–49. Determine the angle ϕ at which the applied force **P** should act on the log so that the magnitude of **P** is as small as possible for pulling the log up the incline. What is the corresponding value of P? The log weighs W and the slope α is known. Express the answer in terms of the angle of kinetic friction, $\theta = \tan^{-1}\mu$.

Probs. 8–48/8–49

8–51. The 50-lb block is on the verge of slipping when placed on the inclined plane. Determine the coefficient of static friction between the block and the plane.

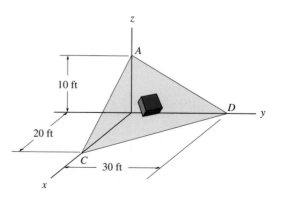

Prob. 8–51

8.3 Wedges

A *wedge* is a simple machine which is often used to transform an applied force into much larger forces, directed at approximately right angles to the applied force. Also, wedges can be used to give small displacements or adjustments to heavy loads.

Consider, for example, the wedge shown in Fig. 8–15a, which is used to *lift* a block of weight **W** by applying a force **P** to the wedge. Free-body diagrams of the wedge and block are shown in Fig. 8–15b. Here we have excluded the weight of the wedge since it is usually *small* compared to the weight of the block. Also, note that the frictional forces F_1 and F_2 must oppose the motion of the wedge. Likewise, the frictional force F_3 of the wall on the block must act downward so as to oppose the block's upward motion. The locations of the resultant normal forces are not important in the force analysis, since neither the block nor wedge will "tip." Hence the moment equilibrium equations will not be considered. There are seven unknowns consisting of the applied force **P,** needed to cause motion of the wedge, and six normal and frictional forces. The seven available equations consist of two force equilibrium equations ($\Sigma F_x = 0$, $\Sigma F_y = 0$) applied to the wedge and block (four equations total) and the frictional equation $F = \mu N$ applied at each surface of contact (three equations total).

If the block is to be *lowered,* the frictional forces will all act in a sense opposite to that shown in Fig. 8–15b. The applied force **P** will act to the right as shown if the coefficient of friction is small or the wedge angle θ is large. Otherwise, **P** may have to *pull* on the wedge in order to remove it. If **P** is *removed,* or $\mathbf{P} = \mathbf{0},$ and friction forces hold the block in place, then the wedge is referred to as *self-locking*.

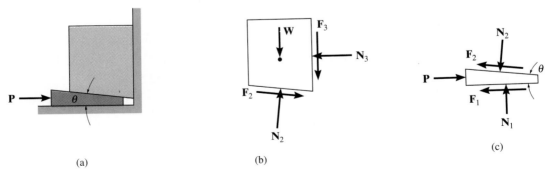

(a) (b) (c)

Fig. 8–15

Example 8–7

The uniform stone has a mass of 500 kg and is held in the horizontal position using a wedge at B as shown in Fig. 8–16a. If the coefficient of static friction is $\mu_s = 0.3$ at the surface in contact with the wedge, determine the force $\mathbf{P}$ needed to remove the wedge. Is the wedge self-locking? Assume that the stone does not slip at A.

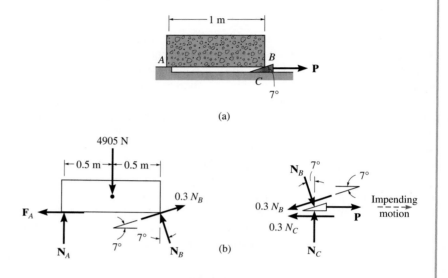

Fig. 8–16

SOLUTION

Since the wedge is to be removed, slipping is about to occur at the surfaces of contact. Thus, $F = \mu_s N$, and the free-body diagrams are shown in Fig. 8–16b. Note that on the stone at A, $F_A \le \mu_s N_A$, since slipping does not occur there. From the free-body diagram of the stone,

$$\zeta + \Sigma M_A = 0; \quad -4905(0.5) + (N_B \cos 7°)(1) + (0.3N_B \sin 7°)(1) = 0$$
$$N_B = 2383.1 \text{ N}$$

Using this result for the wedge, we have

$$\xrightarrow{+} \Sigma F_x = 0; \quad 2383.1 \sin 7° - 0.3(2383.1) \cos 7° + P - 0.3N_C = 0$$
$$+\uparrow \Sigma F_y = 0; \quad N_C - 2383.1 \cos 7° - 0.3(2383.1) \sin 7° = 0$$
$$N_C = 2452.5 \text{ N}$$
$$P = 1154.9 \text{ N} = 1.15 \text{ kN} \qquad \textit{Ans.}$$

Since P is positive, indeed the wedge must be pulled out. Obviously, if P is zero, the wedge will remain in place (self-locking) and the frictional forces $\mathbf{F}_C$ and $\mathbf{F}_B$ developed at the points of contact will satisfy $F_B < \mu_s N_B$ and $F_C < \mu_s N_C$.

*8.4 Frictional Forces on Screws

In most cases screws are used as fasteners; however, in many types of machines they are incorporated to transmit power or motion from one part of the machine to another. A *square-threaded screw* is commonly used for the latter purpose, especially when large forces are applied along its axis. In this section we will analyze the forces acting on square-threaded screws. The analysis of other types of screws, such as the V-thread, is based on the same principles.

A *screw* may be thought of simply as an inclined plane or wedge wrapped around a cylinder. A nut initially at position A on the screw shown in Fig. 8–17a will move up to B when rotated 360° around the screw. This rotation is equivalent to translating the nut up an inclined plane of height l and length $2\pi r$, where r is the mean radius of the thread, Fig. 8–17b. The rise l for a single revolution is referred to as the *lead* of the screw, where the *lead angle* is given by $\theta = \tan^{-1}(l/2\pi r)$.

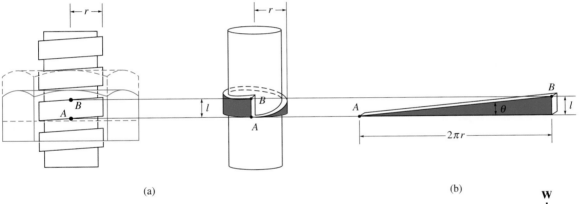

(a)   (b)

Fig. 8–17

Frictional Analysis. When a screw is subjected to large axial loads, the frictional forces developed on the thread become important if we are to determine the moment **M*** needed to turn the screw. Consider, for example, the square-threaded jack screw shown in Fig. 8–18, which supports the vertical load **W**. The reactive forces of the jack to this load are actually distributed over the circumference of the screw thread in contact with the screw hole in the jack, that is, within region h shown in Fig. 8–18. For simplicity, this

Fig. 8–18

---

*For applications, **M** is developed by applying a horizontal force **P** at a right angle to the end of a lever that would be fixed to the screw.

portion of thread can be imagined as being unwound from the screw and represented as a simple block resting on an inclined plane having the screw's lead angle θ, Fig. 8–19a. Here the inclined plane represents the inside *supporting thread* of the jack base. Three forces act on the block or screw. The force **W** is the total axial load applied to the screw. The horizontal force **S** is caused by the applied moment **M,** such that the magnitudes of these loads can be related by summing moments about the axis of the screw. We require $M = Sr$, where r is the screw's mean radius. As a result of **W** and **S,** the inclined plane exerts a resultant force **R** on the block, which is shown to have components acting normal, **N,** and tangent, **F,** to the contacting surfaces.

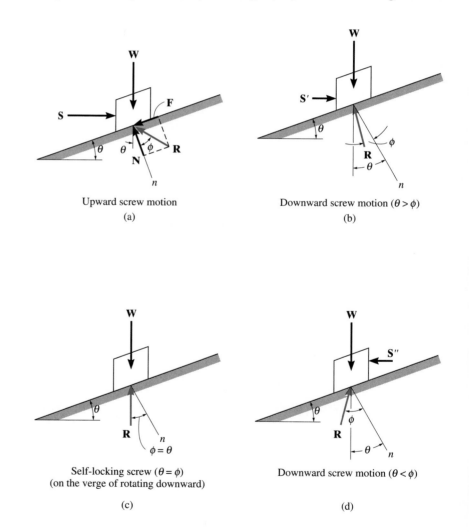

Upward screw motion
(a)

Downward screw motion ($\theta > \phi$)
(b)

Self-locking screw ($\theta = \phi$)
(on the verge of rotating downward)

(c)

Downward screw motion ($\theta < \phi$)

(d)

Fig. 8–19

Upward Screw Motion. Provided M is great enough, the screw (and hence the block) can either be brought to the verge of upward impending motion or motion can be occurring. Under these conditions, $\mathbf{R}$ acts at an angle $(\theta + \phi)$ from the vertical as shown in Fig. 8–19a, where $\phi = \tan^{-1}(F/N) = \tan^{-1}(\mu N/N) = \tan^{-1}\mu$. Applying the two force equations of equilibrium to the block, we obtain

$$\xrightarrow{+}\Sigma F_x = 0; \qquad S - R \sin(\theta + \phi) = 0$$
$$+\uparrow\Sigma F_y = 0; \qquad R \cos(\theta + \phi) - W = 0$$

Eliminating R and solving for S, then substituting this value into the equation $M = Sr$, yields

$$\boxed{M = Wr \tan(\theta + \phi)} \qquad (8\text{–}3)$$

As indicated, M is the moment necessary to cause upward impending motion of the screw, provided $\phi = \phi_s = \tan^{-1}\mu_s$ (the angle of static friction). If ϕ is replaced by $\phi_k = \tan^{-1}\mu_k$ (the angle of kinetic friction), Eq. 8–3 will give a smaller value M necessary to maintain uniform upward motion of the screw.

Downward Screw Motion ($\theta > \phi$). If the surface of the screw is very *slippery*, it may be possible for the screw to rotate downward if the magnitude, *not the direction,* of the moment is reduced to, say, $M' < M$. As shown in Fig. 8–19b, this causes the effect of $\mathbf{M}'$ to become $\mathbf{S}'$, and it requires the angle ϕ (ϕ_s or ϕ_k) to lie on the opposite side of the normal n to the plane supporting the block, such that $\theta > \phi$. For this case, Eq. 8–3 becomes

$$\boxed{M' = Wr \tan(\theta - \phi)} \qquad (8\text{–}4)$$

Self-Locking Screw. If the moment $\mathbf{M}$ (or its effect $\mathbf{S}$) is *removed,* the screw will remain *self-locking;* i.e., it will support the load $\mathbf{W}$ by *friction forces alone* provided $\phi \geq \theta$. To show this, consider the necessary limiting case when $\phi = \theta$, Fig. 8–19c. Here vertical equilibrium is maintained since $\mathbf{R}$ is vertical and thus balances $\mathbf{W}$.

Downward Screw Motion ($\theta < \phi$). When the surface of the screw is *very rough*, the screw will not rotate downward as stated above. Instead, the direction of the applied moment must be *reversed* in order to cause the motion. The free-body diagram shown in Fig. 8–19d is representative of this case. Here $\mathbf{S}''$ is caused by the applied (reverse) moment $\mathbf{M}''$. Hence Eq. 8–3 becomes

$$\boxed{M'' = Wr \tan(\phi - \theta)} \qquad (8\text{–}5)$$

Each of the above cases should be thoroughly understood before proceeding to solve problems.

Example 8–8

The turnbuckle shown in Fig. 8–20 has a square thread with a mean radius of 5 mm and a lead of 2 mm. If the coefficient of static friction between the screw and the turnbuckle is $\mu_s = 0.25$, determine the moment **M** that must be applied to draw the end screws closer together. Is the turnbuckle self-locking?

2 kN

M

2 kN

Fig. 8–20

SOLUTION

The moment may be obtained by using Eq. 8–3. Why? Since friction at *two screws* must be overcome, this requires

$$M = 2[Wr \tan{(\theta + \phi)}] \qquad (1)$$

Here $W = 2000$ N, $r = 5$ mm, $\phi_s = \tan^{-1} \mu_s = \tan^{-1}(0.25) = 14.04°$, and $\theta = \tan^{-1}(l/2\pi r) = \tan^{-1}(2 \text{ mm}/[2\pi(5 \text{ mm})]) = 3.64°$. Substituting these values into Eq. 1 and solving gives

$$M = 2[(2000 \text{ N})(5 \text{ mm}) \tan(14.04° + 3.64°)]$$
$$= 6375.1 \text{ N} \cdot \text{mm} = 6.38 \text{ N} \cdot \text{m} \qquad \textit{Ans.}$$

When the moment is *removed*, the turnbuckle will be self-locking; i.e., it will not unscrew, since $\phi_s > \theta$.

PROBLEMS

*8–52. The two stone blocks have weights of $W_A = 600$ lb and $W_B = 500$ lb. Determine the smallest horizontal force **P** that must be applied to block A in order to move it. The coefficient of static friction between the blocks is $\mu_s = 0.3$ and between the floor and each block $\mu_s' = 0.5$.

Prob. 8–52

8–53. The tool is used to lift very heavy machinery into position. If the machine to be lifted requires a vertical force of 60,000 lb at A, determine the force **P** the hydraulic cylinder must exert on the wedge in order to push it forward. The coefficient of static friction between the wedge and members AB and CD is $\mu_s = 0.2$.

Prob. 8–53

8–54. The uniform roll of paper has a mass of 120 kg and rests on a 30° wedge of negligible mass. If the coefficient of static friction at all contacting surfaces is $\mu_s = 0.3$, determine the horizontal force **P** that must be applied to the wedge in order to lift the paper. Neglect the weight of the wedge.

Prob. 8–54

8–55. Determine the smallest force P needed to lift the 3000-lb load. The coefficient of static friction between A and C and B and D is $\mu_s = 0.3$, and between A and B, $\mu_s' = 0.4$. Neglect the weight of each wedge.

*8–56. Determine the reversed horizontal force $-\mathbf{P}$ needed to pull out wedge A. The coefficient of static friction between A and C and B and D is $\mu_s = 0.2$, and between A and B, $\mu_s' = 0.1$. Neglect the weight of each wedge.

Probs. 8–55/8–56

8–57. Column D is subjected to a vertical load of 8000 lb. It is supported on two identical wedges A and B, for which the coefficient of static friction at the contacting surfaces AB and BC is $\mu_s = 0.4$. Determine the force **P** needed to move wedge B to the right and the smallest equilibrium force **P′** needed to hold wedge A stationary. The contacting surfaces AD are smooth. Neglect the weight of the wedges.

8–58. Column D is subjected to a vertical load of 8000 lb. It is supported on two identical wedges A and B, for which the coefficient of static friction at the contacting surfaces AB and BC is $\mu_s = 0.4$. If the forces **P** and **P′** are removed, are the wedges self-locking? The contacting surfaces AD are smooth. Neglect the weight of the wedges.

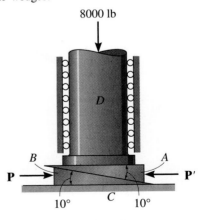

Probs. 8–57/8–58

8–59. The wedge is used to level the member. For the loading shown, determine the horizontal force **P** that must be applied to move the wedge to the right. The coefficient of static friction between the wedge and the two surfaces of contact is $\mu_s = 0.25$. Neglect the size and weight of the wedge.

*8–60. The wedge is used to level the member. For the loading shown, determine the reversed horizontal force $-\mathbf{P}$ that must be applied to pull the wedge out to the left. The coefficient of static friction between the wedge and the two surfaces of contact is $\mu_s = 0.15$. Neglect the weight of the wedge.

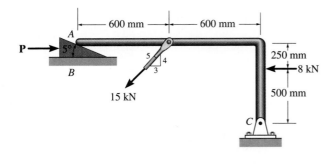

Probs. 8–59/8–60

8–61. An ax is driven into the tree stump. If the coefficient of static friction between the ax and the wood is $\mu_s = 0.2$, determine the smallest angle α of the blade which will cause the ax to be "self-locking." Neglect the weight of the ax.

Prob. 8–61

8–62. The wedge blocks are used to hold the specimen in a tension-testing machine. Determine the design angle θ of the wedges so that the specimen will not slip regardless of the applied load. The coefficients of static friction are $\mu_A = 0.1$ at A and $\mu_B = 0.6$ at B. Neglect the weight of the blocks.

Prob. 8–62

8–63. The wedge is used to level the floor of a building. For the floor loading shown, determine the horizontal force **P** that must be applied to move the wedge forward. The coefficient of static friction between the wedge and the two surfaces of contact is $\mu_s = 0.25$. Neglect the weight of the wedge.

Prob. 8–63

*8–64.   A 50-lb wedge is placed in the grooved slot of an inclined plane. Determine the maximum angle θ for the inclined plane without causing the wedge to slip. The coefficient of static friction between the wedge and surface is $\mu_s = 0.2$.

60°

θ

Prob. 8–64

8–65.   The load bar of negligible weight is screwed in place between the walls of the truck bed in order to keep loads from shifting. Determine the torsional moment M that must be applied to the bar so that it produces a compression in the bar of 80 lb. The square-threaded screw at A has a lead of 0.2 in. and a radius of 0.5 in., and the coefficient of static friction between the bar and the screw is $\mu_s = 0.3$. Also, what amount of centrally applied force F must be applied to pull the bar out? The coefficient of static friction between the walls of the truck and the end pads is $\mu_s = 0.7$.

A B M

Prob. 8–65

8–66.   Determine the clamping force on the board at A if the screw of the "C" clamp is tightened with a twist of $M = 8$ N · m. The single square-threaded screw has a mean radius of 10 mm, a lead of 3 mm, and the coefficient of static friction is $\mu_s = 0.35$.

8–67.   If the required clamping force at the board A is to be 500 N, determine the torque M that must be applied to the handle of the "C" clamp. The single square-threaded screw has a mean radius of 10 mm, a lead of 3 mm, and the coefficient of static friction is $\mu_s = 0.35$.

M

A

Probs. 8–66/8–67

*8–68.   The lolly column is used to support the upper floor. If the force applied perpendicular to the handle is $F = 80$ N to tighten the screw, determine the compressive force in the column. The square-threaded screw on the jack has a coefficient of static friction of $\mu_s = 0.4$, mean diameter of 25 mm, and a lead of 3 mm.

0.5 m

F

Prob. 8–68

8–69. If the required clamping force at the board A is to be 50 N, determine the force **F** that must be applied to the handle of the hold-down clamp to tighten it down. The single square-threaded screw has a mean radius of 10 mm, a lead of 3 mm, and the coefficient of static friction is $\mu_s = 0.35$.

8–70. Determine the clamping force on the board A if the screw of the hold-down clamp is tightened with a force of $F = 400$ N. The single square-threaded screw has a mean radius of 8 mm, a lead of 2 mm, and the coefficient of static friction is $\mu_s = 0.3$.

150 mm

F

A

Probs. 8–69/8–70

8–71. The square-threaded bolt is used to join two plates together. If the bolt has a mean diameter of $\frac{3}{4}$ in. and a lead of $\frac{3}{16}$ in., determine the torque required to loosen the bolt if the tension in the bolt is 35,000 lb. The coefficient of static friction between the threads and the bolt is $\mu_s = 0.15$.

***8–72.** The square-threaded screw of the clamp has a mean diameter of 14 mm and a lead of 6 mm. If $\mu_s = 0.2$ for the threads, and the torque applied to the handle is 1.5 N · m, determine the compressive force F on the block.

1.5 N·m

–F

F

Prob. 8–72

8–73. The shaft has a square-threaded screw with a lead of 9 mm and a mean radius of 15 mm. If it is in contact with a plate gear having a mean radius of 20 mm, determine the resisting torque **M** on the plate gear needed to prevent the screw from turning if a torque of 7 N · m is applied to the shaft. The coefficient of static friction at the screw is $\mu_s = 0.2$. Neglect friction at the screw bearings located at A and B.

0.75 in.

Prob. 8–71

7 N · m

A   B

15 mm

M

20 mm

Prob. 8–73

8–74. The device is used to pull the battery cable terminal C from the post of a battery. If the required pulling force is 85 lb, determine the torque **M** that must be applied to the handle on the screw to tighten it. The screw has square threads, a mean diameter of 0.2 in., a lead of 0.08 in., and the coefficient of static friction is $\mu_s = 0.5$.

Prob. 8–74

8–75. The braking mechanism consists of two pinned arms and a square-threaded screw with left- and right-hand threads. Thus, when turned, the screw draws the two arms together. If the lead of the screw is 4 mm, the mean diameter 12 mm, and the coefficient of static friction is $\mu_s = 0.35$, determine the tension in the screw when a torque of 5 N · m is applied to the screw. If the coefficient of static friction between the brake pads A and B and the circular shaft is $\mu_s' = 0.5$, determine the maximum torque **M** the brake can resist.

Prob. 8–75

*8–76. A turnbuckle, similar to that shown in Fig. 8–20, is used to tension member AB of the truss. The coefficient of static friction between the square-threaded screws and the turnbuckle is $\mu_s = 0.5$. The screws have a mean radius of 6 mm and a lead of 3 mm. If a torque of $M = 10$ N · m is applied to the turnbuckle, to draw the screws closer together, determine the force in each member of the truss. No external forces act on the truss.

8–77. A turnbuckle, similar to that shown in Fig. 8–20, is used to tension member AB of the truss. The coefficient of static friction between the square-threaded screws and the turnbuckle is $\mu_s = 0.5$. The screws have a mean radius of 6 mm and a lead of 3 mm. Determine the torque M which must be applied to the turnbuckle to draw the screws closer together, so that a compressive force of 500 N is developed in member BC.

Probs. 8–76/8–77

8–78. The fixture clamp consists of a square-threaded screw having a coefficient of static friction of $\mu_s = 0.3$, mean diameter of 3 mm, and a lead of 1 mm. The five points indicated are pin connections. Determine the clamping force at the smooth blocks D and E when a torque of $M = 0.08$ N · m is applied to the handle of the screw.

Prob. 8–78

*8.5 Frictional Forces on Flat Belts

Whenever belt drives or band brakes are designed, it is necessary to determine the frictional forces developed between the belt and its contacting surface. In this section we will analyze the frictional forces acting on a flat belt, although the analysis of other types of belts, such as the V-belt, is based on similar principles.

Here we will consider the flat belt shown in Fig. 8–21a, which passes over a fixed curved surface, such that the total angle of belt to surface contact in radians is β and the coefficient of friction between the two surfaces is μ. We will determine the tension T_2 in the belt which is needed to pull the belt counterclockwise over the surface and thereby overcome both the frictional forces at the surface of contact and the known tension T_1. Obviously, $T_2 > T_1$.

Frictional Analysis. A free-body diagram of the belt segment in contact with the surface is shown in Fig. 8–21b. Here the normal force $\mathbf{N}$ and the frictional force $\mathbf{F}$, acting at different points along the belt, will vary both in magnitude and direction. Due to this *unknown* force distribution, the analysis of the problem will proceed on the basis of initially studying the forces acting on a differential element of the belt.

A free-body diagram of an element having a length ds is shown in Fig. 8–21c. Assuming either impending motion or motion of the belt, the magni-

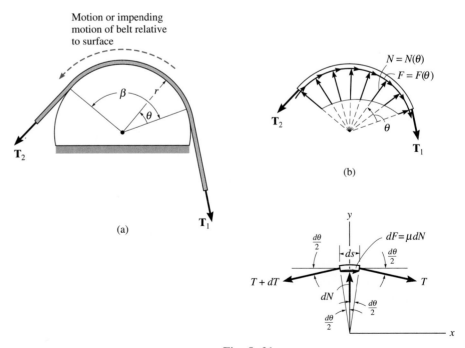

Fig. 8–21

tude of the frictional force $dF = \mu\ dN$. This force opposes the sliding motion of the belt and thereby increases the magnitude of the tensile force acting in the belt by dT. Applying the two force equations of equilibrium, we have

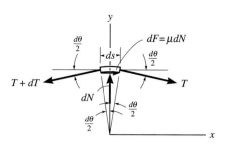

(c)

$$\xrightarrow{+} \Sigma F_x = 0; \quad T \cos\left(\frac{d\theta}{2}\right) + \mu\ dN - (T + dT) \cos\left(\frac{d\theta}{2}\right) = 0$$

$$+\uparrow \Sigma F_y = 0; \quad dN - (T + dT) \sin\left(\frac{d\theta}{2}\right) - T \sin\left(\frac{d\theta}{2}\right) = 0$$

Since $d\theta$ is of *infinitesimal size,* $\sin(d\theta/2)$ and $\cos(d\theta/2)$ can be replaced by $d\theta/2$ and 1, respectively. Also, the *product* of the two infinitesimals dT and $d\theta/2$ may be neglected when compared to infinitesimals of the first order. The above two equations therefore reduce to

$$\mu\ dN = dT$$

and

$$dN = T\ d\theta$$

Eliminating dN yields

$$\frac{dT}{T} = \mu\ d\theta$$

Integrating this equation between all the points of contact that the belt makes with the drum, and noting that $T = T_1$ at $\theta = 0$ and $T = T_2$ at $\theta = \beta$, yields

$$\int_{T_1}^{T_2} \frac{dT}{T} = \mu \int_0^\beta d\theta \quad \text{or} \quad \ln\frac{T_2}{T_1} = \mu\beta$$

Solving for T_2, we obtain

$$\boxed{T_2 = T_1 e^{\mu\beta}} \qquad (8\text{–}6)$$

where T_2, T_1 = belt tensions; $\mathbf{T}_1$ opposes the direction of motion (or impending motion) of the belt measured relative to the surface, while $\mathbf{T}_2$ acts in the direction of the relative belt motion (or impending motion); because of friction, $T_2 > T_1$

  μ = coefficient of static or kinetic friction between the belt and the surface of contact

  β = angle of belt to surface contact, measured in radians

  $e = 2.718. \ . \ .$, base of the natural logarithm

Note that Eq. 8–6 is *independent* of the *radius* of the drum and instead depends on the angle of belt to surface contact, β. Furthermore, as indicated by the integration, this equation is valid for flat belts placed on *any shape* of contacting surface. For application, however, keep in mind that Eq. 8–6 is valid only when *impending motion* occurs.

Example 8–9

The maximum tension that can be developed in the belt shown in Fig. 8–22a is 500 N. If the pulley at A is free to rotate and the coefficient of static friction at the fixed drums B and C is $\mu_s = 0.25$, determine the largest mass of the cylinder that can be lifted by the belt. Assume that the force **T** applied at the end of the belt is directed vertically downward, as shown.

(a)

SOLUTION

Lifting the cylinder, which has a weight $W = mg$, causes the belt to move counterclockwise over the drums at B and C; hence, the maximum tension T_2 in the belt occurs at D. Thus, $T_2 = 500$ N. A section of the belt passing over the drum at B is shown in Fig. 8–22b. Since $180° = \pi$ rad, the angle of contact between the drum and the belt is $\beta = (135°/180°)\pi = 3\pi/4$ rad. Using Eq. 8–6, we have

$$T_2 = T_1 e^{\mu_s \beta}; \qquad 500 \text{ N} = T_1 e^{0.25[(3/4)\pi]}$$

Hence,

$$T_1 = \frac{500 \text{ N}}{e^{0.25[(3/4)\pi]}} = \frac{500}{1.80} = 277.4 \text{ N}$$

Since the pulley at A is free to rotate, equilibrium requires that the tension in the belt remains the *same* on both sides of the pulley.

The section of the belt passing over the drum at C is shown in Fig. 8–22c. The weight $W < 277.4$ N. Why? Applying Eq. 8–6, we obtain

$$T_2 = T_1 e^{\mu_s \beta}; \qquad 277.4 = W e^{0.25[(3/4)\pi]}$$

$$W = 153.9 \text{ N}$$

so that

$$m = \frac{W}{g} = \frac{153.9}{9.81}$$

$$= 15.7 \text{ kg} \qquad \qquad Ans.$$

(b)

(c)

Fig. 8–22

PROBLEMS

8–79.   Determine the variation of the normal force in Fig. 8–21*b* as a function of *θ*.

*8–80.   A "hawser" is wrapped around a fixed "capstan" to secure a ship for docking. If the tension in the rope, caused by the ship, is 1500 lb, determine the least number of complete turns the rope must be wrapped around the capstan in order to prevent slipping of the rope. The greatest horizontal force that a longshoreman can exert on the rope is 50 lb. The coefficient of static friction is $\mu_s = 0.3$.

8–82.   A cylinder having a mass of 250 kg is to be supported by the cord which wraps over the pipe. Determine the smallest vertical force **F** needed to support the load if the cord passes *(a)* once over the pipe, $\beta = 180°$, and *(b)* two times over the pipe, $\beta = 540°$. Take $\mu_s = 0.2$.

8–83.   A cylinder having a mass of 250 kg is to be supported by the cord which wraps over the pipe. Determine the largest vertical force **F** that can be applied to the cord without moving the cylinder. The cord passes *(a)* once over the pipe, $\beta = 180°$, and *(b)* two times over the pipe, $\beta = 540°$. Take $\mu_s = 0.2$.

Probs. 8–82/8–83

Prob. 8–80

8–81.   Determine the *minimum* tension in the rope at points *A* and *B* which is necessary to maintain equilibrium of the rope. The coefficient of static friction between the rope and the fixed post *D* is $\mu_s = 0.3$. The rope is wrapped only once around the post.

*8–84.   Two boys have a tug-of-war using a rope. In the process the rope rubs against a post such that the rope is deflected 3° on each side of the post. If boy *A* weighs 60 lb and boy *B* weighs 50 lb, determine if it is possible for one of them to pull the rope toward him. The coefficient of static friction between the rope and the post is $\mu_s = 0.3$, and the coefficient of static friction between the boys' shoes and the ground is $\mu'_s = 0.4$.

8–85.   Two boys have a tug-of-war using a rope. In the process the rope rubs against a post such that the rope is deflected 3° on each side of the post. If boy *A* weighs 60 lb and boy *B* weighs 50 lb, determine if it is possible for one of them to pull the rope toward him. The coefficient of static friction between the rope and the post is $\mu_s = 0.3$, and the coefficient of static friction between the shoes of the boy at *A* and the ground is $\mu_A = 0.4$, whereas for boy *B* it is $\mu_B = 0.5$.

Prob. 8–81

Probs. 8–84/8–85

8–86. The cord supporting the 6-kg block passes around three pegs, A, B, C, where $\mu_s = 0.2$. Determine the range of values for the magnitude of the horizontal force P for which the block will not move up or down.

Prob. 8–86

8–87. A cord having a weight of 0.5 lb/ft and a total length of 10 ft is suspended over the peg P. If the coefficient of static friction between the peg and the cord is $\mu_s = 0.5$, determine the longest length h which one side of the suspended cord can have without causing motion. Neglect the size of the peg and the length of cord draped over it.

Prob. 8–87

*8–88. Show that the frictional relationship between the belt tensions, the coefficient of friction μ, and the angular contacts α and β for the V-belt is $T_2 = T_1 e^{\mu\beta/\sin(\alpha/2)}$.

Prob. 8–88

8–89. Blocks A and B have masses of 100 kg and 150 kg, respectively. If the coefficient of static friction between A and B and between B and C is $\mu_s = 0.25$, and between the ropes and the pegs D and E, $\mu_s' = 0.5$, determine the smallest force F needed to cause motion of block B if $P = 30$ N.

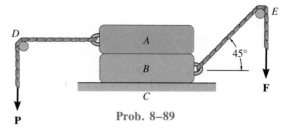

Prob. 8–89

8–90. Blocks A and B weigh 50 lb and 30 lb, respectively. Using the coefficients of static friction indicated, determine the greatest weight of block D without causing motion.

8–91. Blocks A, B, and D weigh 50 lb, 30 lb, and 12 lb, respectively. Using the coefficients of static friction indicated, determine the frictional force between blocks A and B and between block A and the floor C.

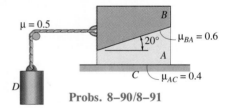

Probs. 8–90/8–91

*8–92. The load can be supported by two boys when the cord is suspended over the pipe (a half-turn). If each boy can pull with a force of 125 lb, determine the weight of the load. Can one boy support the load when the cord is wrapped one-and-a-half times around the pipe? The coefficient of friction is $\mu_s = 0.3$.

Prob. 8–92

8–93. A boy weighing 100 lb attempts to pull himself up a tree using a rope which is draped over the limb at *B*. If $\mu_s = 0.4$, determine the smallest force at which he must pull on the rope to lift himself.

Prob. 8–93

8–94. If a torque of $M = 150$ N · m is applied to the disk, determine the minimum magnitude of the vertical force **P** that must be applied to the band brake to prevent the disk from rotating. The coefficient of static friction between the band and the disk is $\mu_s = 0.45$.

Prob. 8–94

■**8–95.** The 50-lb block is attached to a cord which passes over the drum. If the coefficient of static friction at the drum is $\mu_s = 0.3$, determine the angle θ of the cord if a vertical load of 100 lb is applied to the roller. When $\theta = 0°$, the tension in the cord is zero since the drum rests on the floor.

Prob. 8–95

*****8–96.** Determine the force the man must exert on the rope to hoist the 20-kg sign upward at constant velocity. Also, if the man has a mass of 70 kg, determine the frictional force he exerts on the ground. The coefficient of kinetic friction between the rope and the fixed peg at *B* is $\mu_B = 0.2$, and the coefficient of static friction between the man's shoes and the ground is $\mu = 0.5$. Take $\theta = 30°$.

■**8–97.** Determine the greatest angle θ at which the man can position the rope without slipping in order to support the sign. The sign has a mass of 50 kg, and the man has a mass of 70 kg. The coefficient of static friction between the rope and the fixed peg at *B* is $\mu_B = 0.2$, and between the man's shoes and the ground $\mu = 0.5$.

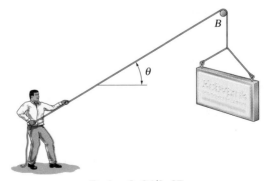

Probs. 8–96/8–97

8–98. The cylinder weighs 10 lb and is held in equilibrium by the belt and wall. If slipping does not occur at the wall, determine the minimum vertical force P which must be applied to the belt for equilibrium. The coefficient of static friction between the belt and cylinder is $\mu_B = 0.25$.

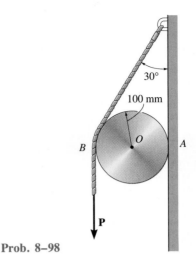

Prob. 8–98

*8.6 Frictional Forces on Collar Bearings, Pivot Bearings, and Disks

Pivot and *collar bearings* are commonly used in machines to support an *axial load* on a rotating shaft. These two types of support are shown in Fig. 8–23. Provided the bearings are not lubricated, or are only partially lubricated, the laws of dry friction may be applied to determine the moment **M** needed to turn the shaft when it supports an axial force **P**.

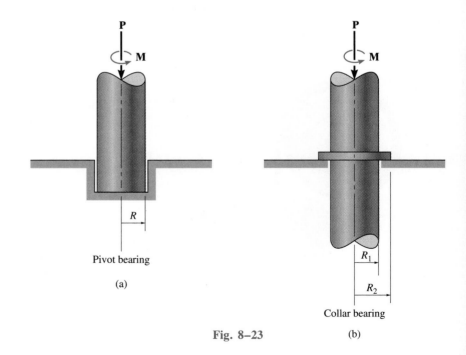

Pivot bearing

(a)

Collar bearing

(b)

Fig. 8–23

Frictional Analysis. The collar bearing on the shaft shown in Fig. 8–24 is subjected to an axial force **P** and has a total bearing or contact area $\pi(R_2^2 - R_1^2)$. In the following analysis, the normal pressure p is considered to be *uniformly distributed* over this area—a reasonable assumption provided the bearing is new and evenly supported. Since $\Sigma F_z = 0$, p, measured as a force per unit area, is $p = P/\pi(R_2^2 - R_1^2)$.

The moment needed to cause impending rotation of the shaft can be determined from moment equilibrium of the frictional forces $d\mathbf{F}$ developed at the bearing surface by applying $\Sigma M_z = 0$. A small area element $dA = (r\, d\theta)(dr)$, shown in Fig. 8–24, is subjected to both a normal force $dN = p\, dA$ and an associated frictional force,

$$dF = \mu_s\, dN = \mu_s\, p\, dA = \frac{\mu_s P}{\pi(R_2^2 - R_1^2)}\, dA$$

The normal force does not create a moment about the z axis of the shaft; however, the frictional force does: namely, $dM = r\, dF$. Integration is needed to compute the total moment created by all the frictional forces acting on differential areas dA. Therefore, for impending rotational motion,

Fig. 8–24

$$\Sigma M_z = 0; \qquad\qquad M - \int_A r\, dF = 0$$

Substituting for dF and dA and integrating over the entire bearing area yields

$$M = \int_{R_1}^{R_2} \int_0^{2\pi} r\left[\frac{\mu_s P}{\pi(R_2^2 - R_1^2)}\right](r\, d\theta\, dr) = \frac{\mu_s P}{\pi(R_2^2 - R_1^2)} \int_{R_1}^{R_2} r^2\, dr \int_0^{2\pi} d\theta$$

or

$$M = \tfrac{2}{3}\mu_s P\left(\frac{R_2^3 - R_1^3}{R_2^2 - R_1^2}\right) \qquad\qquad (8\text{–}7)$$

This equation gives the magnitude of moment required for impending rotation of the shaft. The frictional moment developed at the end of the shaft, when it is *rotating* at constant speed, can be found by substituting μ_k for μ_s in Eq. 8–7.

When $R_2 = R$ and $R_1 = 0$, as in the case of a pivot bearing, Fig. 8–23a, Eq. 8–7 reduces to

$$M = \tfrac{2}{3}\mu_s PR \qquad\qquad (8\text{–}8)$$

Recall from the initial assumption that both Eqs. 8–7 and 8–8 apply only for bearing surfaces subjected to *constant pressure*. If the pressure is not uniform, a variation of the pressure as a function of the bearing area must be determined before integrating to obtain the moment. The following example illustrates this concept.

Example 8–10

(a)

(b)

Fig. 8–25

The uniform bar shown in Fig. 8–25a has a total mass m. If it is assumed that the normal pressure acting at the contacting surface varies linearly along the length of the bar as shown, determine the couple moment M required to rotate the bar. Assume that the bar's width a is negligible in comparison to its length l. The coefficient of static friction is equal to μ.

SOLUTION

A free-body diagram of the bar is shown in Fig. 8–25b. Since the bar has a total weight of $W = mg$, the intensity w_o of the distributed load at the center ($x = 0$) is determined from vertical force equilibrium, Fig. 8–25a.

$$+ \uparrow \Sigma F_z = 0; \quad -mg + 2\left[\frac{1}{2}\left(\frac{l}{2}\right)w_o\right] = 0 \qquad w_o = \frac{2mg}{l}$$

Since $w = 0$ at $x = l/2$, the distributed load expressed as a function of x is

$$w = w_o\left(1 - \frac{2x}{l}\right) = \frac{2mg}{l}\left(1 - \frac{2x}{l}\right)$$

The magnitude of the normal force acting on a segment of area having a length dx is therefore

$$dN = w\,dx = \frac{2mg}{l}\left(1 - \frac{2x}{l}\right)dx$$

The magnitude of the frictional force acting on the same element of area is

$$dF = \mu\,dN = \frac{2\mu mg}{l}\left(1 - \frac{2x}{l}\right)dx$$

Hence, the moment created by this force about the z axis is

$$dM = x\,dF = \frac{2\mu mg}{l}x\left(1 - \frac{2x}{l}\right)dx$$

The summation of moments about the z axis of the bar is determined by integration, which yields

$$\Sigma M_z = 0; \qquad M - 2\int_0^{l/2}\frac{2\mu mg}{l}x\left(1 - \frac{2x}{l}\right)dx = 0$$

$$M = \frac{4\mu mg}{l}\left(\frac{x^2}{2} - \frac{2x^3}{3l}\right)\Bigg|_0^{l/2}$$

$$M = \frac{\mu mgl}{6} \qquad\qquad \textit{Ans.}$$

*8.7 Frictional Forces on Journal Bearings

When a shaft or axle is subjected to lateral loads, a *journal bearing* is commonly used for support. Well-lubricated journal bearings are subjected to the laws of fluid mechanics, in which the viscosity of the lubricant, the speed of rotation, and the amount of clearance between the shaft and bearing are needed to determine the frictional resistance of the bearing. When the bearing is not lubricated or is only partially lubricated, however, a reasonable analysis of the frictional resistance can be based on the laws of dry friction.

Frictional Analysis. A typical journal-bearing support is shown in Fig. 8–26a. As the shaft rotates in the direction shown in the figure, it rolls up against the wall of the bearing to some point A where slipping occurs. If the lateral load acting at the end of the shaft is **W**, it is necessary that the bearing reactive force **R** acting at A be equal and opposite to **W**, Fig. 8–26b. The moment needed to maintain constant rotation of the shaft can be found by summing moments about the z axis of the shaft; i.e.,

$$\Sigma M_z = 0; \qquad -M + (R \sin \phi_k)r = 0$$

or

$$M = Rr \sin \phi_k \qquad (8–9)$$

where ϕ_k is the angle of kinetic friction defined by $\tan \phi_k = F/N = \mu_k N/N = \mu_k$. In Fig. 8–26c, it is seen that $r \sin \phi_k = r_f$. The dashed circle with radius r_f is called the *friction circle*, and as the shaft rotates, the reaction **R** will always be tangent to it. If the bearing is partially lubricated, μ_k is small, and therefore $\mu_k = \tan \phi_k \approx \sin \phi_k \approx \phi_k$. Under these conditions, a reasonable *approximation* to the moment needed to overcome the frictional resistance becomes

$$M \approx Rr\mu_k \qquad (8–10)$$

The following example illustrates a common application of this analysis.

Rotation z

(a)

Fig. 8–26

(b)

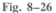

(c)

■ **Example 8–11** ■

The 100-mm-diameter pulley shown in Fig. 8–27a fits loosely on a 10-mm-diameter shaft for which the coefficient of static friction is $\mu_s = 0.4$. Determine the minimum tension T in the belt needed to (a) raise the 100-kg block and (b) lower the block. Assume that no slipping occurs between the belt and pulley and neglect the weight of the pulley.

50 mm

$r = 5$ mm

100 kg   T

(a)

SOLUTION

Part (a). A free-body diagram of the pulley is shown in Fig. 8–27b. When the pulley is subjected to belt tensions of 981 N each, it makes contact with the shaft at point P_1. As the tension T is *increased*, the pulley will roll around the shaft to point P_2 before motion impends. From the figure, the friction circle has a radius $r_f = r \sin \phi$. Using the simplification $\sin \phi \approx \phi$, $r_f \approx r\mu_s = (5 \text{ mm})(0.4) = 2$ mm, so that summing moments about P_2 gives

(b)

$\zeta+\Sigma M_{P_2} = 0;$ 981 N(52 mm) $-$ T(48 mm) $= 0$

$$T = 1063 \text{ N} = 1.06 \text{ kN} \qquad \textbf{\textit{Ans.}}$$

If a more exact analysis is used, then $\phi = \tan^{-1} 0.4 = 21.8°$. Thus, the radius of the friction circle would be $r_f = r \sin \phi = 5 \sin 21.8° = 1.86$ mm. Therefore,

$\zeta+\Sigma M_{P_2} = 0;$

981 N(50 mm $+$ 1.86 mm) $-$ T(50 mm $-$ 1.86 mm) $= 0$

$$T = 1057 \text{ N} = 1.06 \text{ kN} \qquad \textbf{\textit{Ans.}}$$

Part (b). When the block is lowered, the resultant force **R** acting on the shaft passes through point P_3, as shown in Fig. 8–27c. Summing moments about this point yields

(c)

Fig. 8–27

$\zeta+\Sigma M_{P_3} = 0;$ 981 N(48 mm) $-$ T(52 mm) $= 0$

$$T = 906 \text{ N} \qquad \textbf{\textit{Ans.}}$$

*8.8 Rolling Resistance

If a *rigid* cylinder of weight **W** rolls at constant velocity along a *rigid* surface, the normal force exerted by the surface on the cylinder acts at the tangent point of contact, as shown in Fig. 8–28a. Under these conditions, provided the cylinder does not encounter frictional resistance from the air, motion will continue indefinitely. Actually, however, no materials are perfectly rigid, and therefore the reaction of the surface on the cylinder consists of a distribution of normal pressure. For example, consider the cylinder to be made of a very hard material, and the surface on which it rolls to be soft. Due to its weight, the cylinder compresses the surface underneath it, Fig. 8–28b. As the cylinder rolls, the surface material in front of the cylinder *retards* the motion since it is being *deformed*, whereas the material in the rear is *restored* from the deformed state and therefore tends to *push* the cylinder forward. The normal pressures acting on the cylinder in this manner are represented in Fig. 8–28b by their resultant forces $\mathbf{N}_d$ and $\mathbf{N}_r$. Unfortunately, the magnitude of the force of *deformation* and its horizontal component is *always greater* than that of *restoration,* and consequently a horizontal driving force **P** must be applied to the cylinder to maintain the motion, Fig. 8–28b.*

Rolling resistance is caused primarily by this effect, although it is also, to a smaller degree, the result of surface adhesion and relative micro-sliding between the surfaces of contact. Because the actual force **P** needed to overcome these effects is difficult to determine, a simplified method will be developed here to explain one way engineers have analyzed this phenomenon. To do this, we will consider the resultant of the *entire* normal pressure, $\mathbf{N} = \mathbf{N}_d + \mathbf{N}_r$, acting on the cylinder, Fig. 8–28c. As shown in Fig. 8–28d, this force acts at an angle θ with the vertical. To keep the cylinder in equilibrium, i.e., rolling at a constant rate, it is necessary that **N** be *concurrent* with the driving force **P** and the weight **W**. Summing moments about point A gives $Wa = P(r \cos \theta)$. Since the deformations are generally very small in relation to the cylinder's radius, $\cos \theta \approx 1$; hence,

$$Wa \approx Pr$$

or

$$\boxed{P \approx \frac{Wa}{r}} \qquad (8\text{–}11)$$

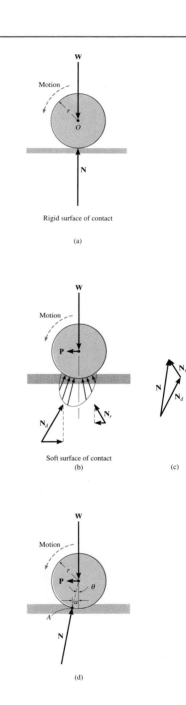

Rigid surface of contact

(a)

Soft surface of contact

(b)   (c)

(d)

Fig. 8–28

---

*Actually, the deformation force $\mathbf{N}_d$ causes *energy* to be stored in the material as its magnitude is increased, whereas the restoration force $\mathbf{N}_r$, as its magnitude is decreased, allows some of this energy to be released. The remaining energy is *lost* since it is used to heat up the surface, and if the cylinder's weight is very large, it accounts for permanent deformation of the surface. Work must be done by the horizontal force **P** to make up for this loss.

The distance a is termed the *coefficient of rolling resistance*, which has the dimension of length. For instance, $a \approx 0.5$ mm for a wheel rolling on a rail, both of which are made of mild steel. For hardened steel ball bearings on steel, $a \approx 0.1$ mm. Experimentally, though, this factor is difficult to measure, since it depends on such parameters as the rate of rotation of the cylinder, the elastic properties of the contacting surfaces, and the surface finish. For this reason, little reliance is placed on the data for determining a. The analysis presented here does, however, indicate why a heavy load (W) offers greater resistance to motion (P) than a light load under the same conditions. Furthermore, since the force needed to *roll* the cylinder over the surface will be much less than that needed to *slide* the cylinder across the surface, the analysis indicates why roller or ball bearings are often used to minimize the frictional resistance between moving parts.

Example 8–12

A 10-kg steel wheel shown in Fig. 8–29a has a radius of 100 mm and rests on an inclined plane made of wood. If θ is increased so that the wheel begins to roll down the incline with constant velocity when $\theta = 1.2°$, determine the coefficient of rolling resistance.

(a)

SOLUTION

As shown on the free-body diagram, Fig. 8–29b, when the wheel has impending motion, the normal reaction **N** acts at point A defined by the dimension a. Resolving the weight into components parallel and perpendicular to the incline, and summing moments about point A, yields (approximately)

$$\curvearrowright + \Sigma M_A = 0; \quad 98.1 \cos 1.2°(a) - 98.1 \sin 1.2°(100) = 0$$

Solving, we obtain

$$a = 2.1 \text{ mm} \qquad \textit{Ans.}$$

98.1 N

98.1 cos 1.2° N

1.2°

98.1 sin 1.2° N

O 100 mm

1.2°

A ‹–a–›

N

(b)

Fig. 8–29

PROBLEMS

8–99. The collar bearing uniformly supports an axial force of $P = 500$ lb. If the coefficient of static friction is $\mu_s = 0.3$, determine the torque **M** required to overcome friction.

*8–100. The collar bearing uniformly supports an axial force of $P = 500$ lb. If a torque of $M = 3$ lb · ft is applied to the shaft and causes it to rotate at constant velocity, determine the coefficient of kinetic friction at the surface of contact.

8–102. The *disk clutch* is used in standard transmissions of automobiles. If four springs are used to force the two plates A and B together, determine the force in each spring required to transmit a moment of $M = 600$ lb · ft across the plates. The coefficient of static friction between A and B is $\mu_s = 0.3$.

Prob. 8–102

Probs. 8–99/8–100

8–101. The annular ring bearing is subjected to a thrust of 800 lb. If $\mu_s = 0.35$, determine the torque **M** that must be applied to the shaft to turn it.

8–103. The conical bearing is subjected to a constant pressure distribution at its surface of contact. If the coefficient of static friction is μ, determine the torque **M** required to turn the shaft if it supports an axial force **P**.

Prob. 8–101

Prob. 8–103

*8–104. The pivot bearing is subjected to a pressure distribution at its surface of contact which varies as shown. If the coefficient of static friction is μ, determine the torque **M** required to overcome friction if the shaft supports an axial force **P**.

Prob. 8–104

8–105. Because of wearing at the edges, the pivot bearing is subjected to a conical pressure distribution at its surface of contact. Determine the torque **M** required to overcome friction and turn the shaft which supports an axial force **P**. The coefficient of static friction is μ. For the solution, it is necessary to determine the peak pressure p_o in terms of P and the bearing radius R.

Prob. 8–105

8–106. The pivot bearing is subjected to a parabolic pressure distribution at its surface of contact. If the coefficient of static friction is μ, determine the torque **M** required to overcome friction and turn the shaft if it supports an axial force **P**.

Prob. 8–106

8–107. The 4-in.-diameter shaft is held in the hole such that the normal pressure acting around the shaft varies linearly with its depth as shown. Determine the frictional torque that must be overcome to rotate the shaft. Take $\mu_s = 0.2$.

Prob. 8–107

*8–108.   The axle of the 200-mm-diameter pulley fits loosely in a 50-mm-diameter pin hole. If the coefficient of kinetic friction is $\mu_k = 0.30$, determine the minimum tension **T** required to raise the 40-kg load at constant velocity. Neglect the weight of the pulley and assume that the cord does not slip on the pulley.

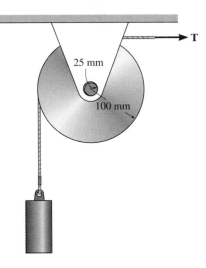

25 mm

100 mm

T

Prob. 8–108

8–109.   A pulley having a diameter of 80 mm and a mass of 1.25 kg is supported loosely on a shaft having a diameter of 20 mm. Determine the torque **M** that must be applied to the pulley to cause it to rotate with constant motion. The coefficient of kinetic friction between the shaft and pulley is $\mu_k = 0.4$. Also calculate the angle θ which the normal force at the point of contact makes with the horizontal. The shaft itself cannot rotate.

M

Prob. 8–109

8–110.   The 5-kg pulley has a diameter of 240 mm and the axle has a diameter of 40 mm. If the coefficient of kinetic friction between the axle and the pulley is $\mu_k = 0.15$, determine the tension **T** in the rope required to lift the 80-kg block.

8–111.   Solve Prob. 8–110 if the tension **T** is applied to the rope horizontally to the right.

120 mm

T

Probs. 8–110/8–111

*8–112.   Determine the tension **T** in the belt needed to overcome the tension of 200 lb created on the other side. Also, what are the normal and frictional components of force developed on the collar bushing? The coefficient of static friction is $\mu_s = 0.21$.

8–113.   If a tension force $T = 215$ lb is required to pull the 200-lb force around the collar bushing, determine the coefficient of static friction at the contacting surface.

2 in.

1.125 in.

200 lb   **T**

Probs. 8–112/8–113

8–114. The collar fits *loosely* around a fixed shaft that has a radius of 2 in. If the coefficient of kinetic friction between the shaft and the collar is $\mu_k = 0.3$, determine the tension **T** in the horizontal segment of the belt so that the collar rotates counterclockwise with a constant angular speed. Assume that the belt does not slip on the collar; rather, the collar slips on the shaft. Neglect the weight and thickness of the belt and collar. The radius, measured from the center of the collar to the mean thickness of the belt, is 2.25 in.

8–115. The collar fits *loosely* around a fixed shaft that has a radius of 2 in. If the coefficient of kinetic friction between the shaft and the collar is $\mu_k = 0.3$, determine the tension **T** in the horizontal segment of the belt so that the collar rotates clockwise with a constant angular speed. Assume that the belt does not slip on the collar; rather, the collar slips on the shaft. Neglect the weight and thickness of the belt and collar. The radius, measured from the center of the collar to the mean thickness of the belt, is 2.25 in.

20 lb

Probs. 8–114/8–115

*8–116. The collar bushing *B* fits loosely over the fixed shaft *S*, which has a radius of 20 mm. Determine the smallest force $\mathbf{T}_A$ needed to start pulling the belt downward at *A*. Also determine the normal and frictional components of force developed on the collar bushing if $\mu_s = 0.3$ between the collar and the shaft. The belt does not slip on the collar; rather, the collar slips on the shaft.

8–117. If the smallest tension force $T_A = 500$ N is required to start pulling the belt downward at *A* over the shaft *S*, determine the coefficient of static friction between the loosely fitting collar bushing *B* and the shaft. The belt does not slip on the collar; rather, the collar slips on the shaft.

Probs. 8–116/8–117

8–118. A disk having an outer diameter of 120 mm fits loosely over a fixed shaft having a diameter of 30 mm. If the coefficient of static friction between the disk and the shaft is $\mu_s = 0.15$ and the disk has a mass of 50 kg, determine the smallest vertical force **F** acting on the rim which must be applied to the disk to cause it to slip over the shaft.

Prob. 8–118

8–119. Experimentally, it is found that a cylinder having a diameter of 150 mm rolls with a constant speed down an inclined plane having a slope of 18 mm/m. Determine the coefficient of rolling resistance for the cylinder.

*8–120. A tank car has a mass of 30 Mg and is supported by eight wheels, each of which has a diameter of 500 mm. If the coefficient of rolling resistance is 0.4 mm between the tracks and each wheel, determine the magnitude of horizontal force **P** required to overcome the rolling resistance of the wheels.

8–121. A carriage having eight wheels is used to support a crane load of $F = 1.25(10^6)$ lb. Determine the horizontal force **P** that must be applied to the carriage in order to overcome the resistance to rolling. The coefficient of rolling resistance is 0.007 in. and each wheel has a diameter of 36 in.

8–122. The cylinder is subjected to a load that has a weight W. If the coefficients of rolling resistance for the cylinder's top and bottom surfaces are a_A and a_B, respectively, show that a force having a magnitude of $P = [W(a_A + a_B)]/2r$ is required to move the load and thereby roll the cylinder forward. Neglect the weight of the cylinder.

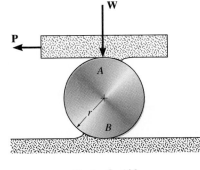

Prob. 8–122

8–123. A large crate having a mass of 200 kg is moved along the floor using a series of 150-mm-diameter rollers for which the coefficient of rolling resistance is 3 mm at the ground and 7 mm at the bottom surface of the crate. Determine the horizontal force **P** needed to push the crate forward at a constant speed. *Hint:* Use the result of Prob. 8–122.

Prob. 8–123

*8–124. A large stone having a mass of 500 kg is moved along the incline using a series of 150-mm-diameter rollers for which the coefficient of rolling resistance is 3 mm at the ground and 4 mm at the bottom surface of the stone. Determine the magnitude of force **T** needed to tow the stone up the plane at a constant speed. *Hint:* Use the result of Prob. 8–122.

8–125. A large stone having a mass of 500 kg is moved along the incline using a series of 150-mm-diameter rollers for which the coefficient of rolling resistance is 3 mm at the ground and 4 mm at the bottom surface of the stone. Determine the magnitude of force **T** needed to allow the stone to descend the plane at a constant speed. *Hint:* Use the result of Prob. 8–122.

Probs. 8–124/8–125

8–126. The car has a weight of 2600 lb and center of gravity at G. Determine the horizontal force **P** that must be applied to overcome the rolling resistance of the wheels. The coefficient of rolling resistance is 0.5 in. The tires have a diameter of 2.75 ft.

Prob. 8–126

8–127. The lawn roller weighs 300 lb. If the rod BA is held at an angle of 30° from the horizontal and the coefficient of rolling resistance for the roller is 2 in., determine the force **F** needed to push the roller at constant speed. Neglect friction developed at the axle and assume that the resultant force acting on the handle is applied along BA.

Prob. 8–127

REVIEW PROBLEMS

*8–128. The cone has a weight W and center of gravity at G. If a horizontal force **P** is gradually applied to the string attached to its vertex, determine the maximum coefficient of static friction for slipping to occur.

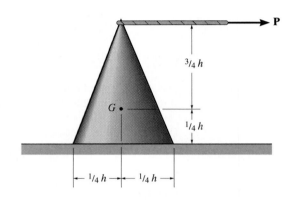

Prob. 8–128

8–130. Blocks A and B have a mass of 7 kg and 10 kg, respectively. Using the coefficients of static friction indicated, determine the largest force **P** which can be applied to the cord without causing motion.

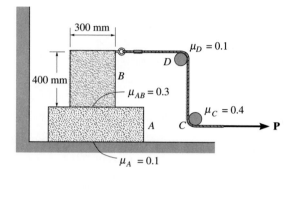

Prob. 8–130

8–129. If the smallest tension force $T_A = 300$ lb is required to pull the belt at A downward with constant velocity over the shaft S, determine the coefficient of kinetic friction between the loosely fitting collar bushing and the shaft. The belt does not slip on the collar; rather, the collar slips on the shaft.

8–131. The wheel weighs 20 lb and rests on a surface for which $\mu_B = 0.2$. A cord wrapped around it is attached to the 30-lb homogeneous block for which $\mu_D = 0.3$. Determine the smallest couple moment which can be applied to the wheel that will cause motion to impend.

Prob. 8–129

Prob. 8–131

*8–132. A 100-lb disk rests on an inclined plane for which $\mu_s = 0.2$. Determine the maximum vertical force **P** that can be applied to the link AB without causing the disk to slip.

8–134. A roofer, having a mass of 70 kg, walks slowly in an upright position down along the surface of a dome that has a radius of curvature of $r = 20$ m. If the coefficient of static friction between his shoes and the dome is $\mu = 0.7$, determine the angle θ at which he first begins to slip.

Prob. 8–132

Prob. 8–134

8–133. The trailer has a total weight of 850 lb and center of gravity at G which is directly over its axle. If the axle has a diameter of 1 in., the radius of the wheel is $r = 1.5$ ft, and the coefficient of kinetic friction at the bearing is $\mu_k = 0.08$, determine the horizontal force **P** needed to pull the trailer.

8–135. The symmetrical crab hook is used to lift packages by means of friction developed between the shoes A and B and a package. Determine the smallest coefficient of static friction at the shoes so that the package of weight W can be lifted.

Prob. 8–133

Prob. 8–135

*8–136. The tapered shank of angle θ is used on a lathe. If the contact pressure between the socket B and the shank is uniform, and the coefficient of static friction is μ, determine the torque **M** needed to overcome friction if the shank is subjected to an axial force **P.**

■8–137. The empty drum has a weight W and center of gravity at G. If the coefficient of static friction between it and the inclined plane is $\mu_s = 0.4$, determine the maximum angle θ of the plane without causing the drum to slip.

Prob. 8–136

Prob. 8–137

9 Center of Gravity and Centroid

In this chapter we will discuss the method used to determine the location of the center of gravity and center of mass for a system of discrete particles, and then we will expand its application to include a body of arbitrary shape. The same method of analysis will also be used to determine the geometric center, or centroid, of lines, areas, and volumes. Once the centroid has been located, we will then show how to obtain the area and volume of a surface of revolution and determine the resultants of various types of distributed loadings.

9.1 Center of Gravity and Center of Mass for a System of Particles

Center of Gravity. Consider the system of n particles fixed within a region of space as shown in Fig. 9–1a. The weights of the particles comprise a system of parallel forces* which can be replaced by a single (equivalent) resultant weight and a defined point of application. This point is called the *center of gravity G*. To find its $\bar{x}$, $\bar{y}$, $\bar{z}$ coordinates, we must use the principles outlined in Sec. 4.9. This requires that the resultant weight be equal to the

---

*This is not true in the exact sense, since the weights are not parallel to each other; rather they are all *concurrent* at the earth's center. Furthermore, the acceleration of gravity g is actually different for each particle, since it depends on the distance from the earth's center to the particle. For all practical purposes, however, both of these effects can be neglected.

(a)

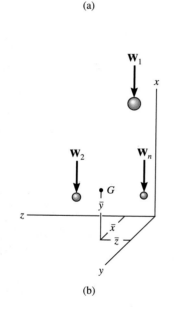

(b)

Fig. 9–1

total weight of all n particles; that is,

$$W_R = \Sigma W$$

The sum of the moments of the weights of all the particles about the x, y, and z axes is then equal to the moment of the resultant weight about these axes. Thus, to determine the $\bar{x}$ coordinate of G, we can sum moments about the y axis. This yields

$$\bar{x} W_R = \widetilde{x}_1 W_1 + \widetilde{x}_2 W_2 + \cdots + \widetilde{x}_n W_n$$

Likewise, summing moments about the x axis, we can obtain the $\bar{y}$ coordinate; i.e.,

$$\bar{y} W_R = \widetilde{y}_1 W_1 + \widetilde{y}_2 W_2 + \cdots + \widetilde{y}_n W_n$$

Although the weights do not produce a moment about the z axis, we can obtain the z coordinate of G by imagining the coordinate system, with the particles fixed in it, as being rotated 90° about either the x (or y) axis, Fig. 9–1b. Summing moments about the x axis, we have

$$\bar{z} W_R = \widetilde{z}_1 W_1 + \widetilde{z}_2 W_2 + \cdots + \widetilde{z}_n W_n$$

We can generalize these formulas, and write them symbolically in the form

$$\bar{x} = \frac{\Sigma \widetilde{x} W}{\Sigma W} \qquad \bar{y} = \frac{\Sigma \widetilde{y} W}{\Sigma W} \qquad \bar{z} = \frac{\Sigma \widetilde{z} W}{\Sigma W} \qquad (9\text{–}1)$$

Here

$\bar{x}, \bar{y}, \bar{z}$ represent the coordinates of the center of gravity G of the system of particles, regardless of the orientation of the x, y, z axes.

$\widetilde{x}, \widetilde{y}, \widetilde{z}$ represent the coordinates of each particle in the system.

$W_R = \Sigma W$ is the total sum of the weights of all the particles in the system.

Formulas having this same form will be presented throughout this chapter to represent other "quantities" for a system. In all cases, however, keep in mind that they simply represent a balance between the sum of the moments of the "quantity" for *each part* of the system and the moment of the *resultant* "quantity" for the system.

Center of Mass. To study problems concerning the motion of *matter* under the influence of force, i.e., dynamics, it is necessary to locate a point called the *center of mass*. Provided the acceleration of gravity g for every particle is constant, then $W = mg$. Substituting into Eqs. 9–1 and canceling g from both the numerator and denominator yields

$$\bar{x} = \frac{\Sigma \widetilde{x} m}{\Sigma m} \qquad \bar{y} = \frac{\Sigma \widetilde{y} m}{\Sigma m} \qquad \bar{z} = \frac{\Sigma \widetilde{z} m}{\Sigma m} \qquad (9\text{–}2)$$

By comparison, then, the location of the center of gravity *coincides* with that of the center of mass. Recall, however, that particles have "weight" only when under the influence of a gravitational attraction, whereas the center of mass is independent of gravity. For example, it would be meaningless to define the center of gravity of a system of particles representing the planets of our solar system, while the center of mass of this system is important.

9.2 Center of Gravity, Center of Mass, and Centroid for a Body

Center of Gravity. When the principles used to determine Eqs. 9–1 are applied to a system of particles composing a rigid body, one obtains the same form as these equations except that each particle located at $(\tilde{x}, \tilde{y}, \tilde{z})$ is thought to have a *differential weight dW*. As a result, *integration* is required rather than a discrete summation of the terms. The resulting equations are

$$\bar{x} = \frac{\int \tilde{x} \, dW}{\int dW} \qquad \bar{y} = \frac{\int \tilde{y} \, dW}{\int dW} \qquad \bar{z} = \frac{\int \tilde{z} \, dW}{\int dW} \qquad (9\text{–}3)$$

In order to use these equations properly, the differential weight dW must be expressed in terms of its associated volume dV. If γ represents the *specific weight* of the body, measured as a weight per unit volume, then $dW = \gamma \, dV$ and therefore

$$\bar{x} = \frac{\int_V \tilde{x}\gamma \, dV}{\int_V \gamma \, dV} \qquad \bar{y} = \frac{\int_V \tilde{y}\gamma \, dV}{\int_V \gamma \, dV} \qquad \bar{z} = \frac{\int_V \tilde{z}\gamma \, dV}{\int_V \gamma \, dV} \qquad (9\text{–}4)$$

Here integration must be performed throughout the entire volume of the body.

Center of Mass. The *density* ρ, or mass per unit volume, is related to γ by the equation $\gamma = \rho g$, where g is the acceleration of gravity. Substituting this relationship into Eqs. 9–4 and canceling g from both the numerators and denominators yields similar equations (with ρ replacing γ) that can be used to determine the body's *center of mass*.

Centroid. The *centroid* is a point which defines the *geometric center* of an object. Its location can be determined from formulas similar to those used to determine the body's center of gravity or center of mass. In particular, if the material composing a body is uniform or *homogeneous*, the *density or specific weight* will be *constant* throughout the body, and therefore this term will factor out of the integrals and *cancel* from both the numerators and denominators of Eqs. 9–4. The resulting formulas define the centroid of the body since they are independent of the body's weight and instead depend only on the body's geometry. Three specific cases will be considered.

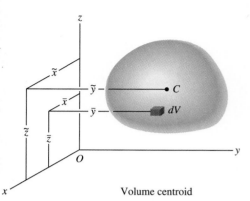

Volume centroid

Fig. 9–2

Volume. If an object is subdivided into volume elements dV, Fig. 9–2, the location of the centroid C $(\bar{x}, \bar{y}, \bar{z})$ for the volume of the object can be determined by computing the "moments" of the elements about the coordinate axes. The resulting formulas are

$$\bar{x} = \frac{\int_V \tilde{x}\, dV}{\int_V dV} \qquad \bar{y} = \frac{\int_V \tilde{y}\, dV}{\int_V dV} \qquad \bar{z} = \frac{\int_V \tilde{z}\, dV}{\int_V dV} \qquad (9\text{–}5)$$

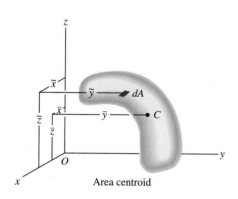

Area centroid

Fig. 9–3

Area. In a similar manner, the centroid for the surface area of an object, such as a plate or shell, Fig. 9–3, can be found by subdividing the area into differential elements dA and computing the "moments" of these area elements about the coordinate axes, namely,

$$\bar{x} = \frac{\int_A \tilde{x}\, dA}{\int_A dA} \qquad \bar{y} = \frac{\int_A \tilde{y}\, dA}{\int_A dA} \qquad \bar{z} = \frac{\int_A \tilde{z}\, dA}{\int_A dA} \qquad (9\text{–}6)$$

Line centroid

Fig. 9–4

Line. If the geometry of the object, such as a thin rod or wire, takes the form of a line, Fig. 9–4, the manner of finding its centroid is identical to the procedure outlined above. The results are

$$\bar{x} = \frac{\int_L \tilde{x}\, dL}{\int_L dL} \qquad \bar{y} = \frac{\int_L \tilde{y}\, dL}{\int_L dL} \qquad \bar{z} = \frac{\int_L \tilde{z}\, dL}{\int_L dL} \qquad (9\text{–}7)$$

Notice that in all of the above cases the location of C does not necessarily have to be within the object; rather, it can be located off the object in space. Also, the *centroids* of some shapes may be partially or completely specified by using conditions of *symmetry*. In cases where the shape has an axis of symmetry, the centroid of the shape will lie along that axis. For example, the centroid C for the line shown in Fig. 9–5 must lie along the y axis, since for every elemental length dL at a distance $+\widetilde{x}$ to the right of the y axis, there is an identical element at a distance $-\widetilde{x}$ to the left. The total moment for all the elements about the axis of symmetry will therefore cancel; i.e., $\int \widetilde{x}\, dL = 0$ (Eq. 9–7), so that $\bar{x} = 0$. In cases where a shape has two or three axes of symmetry, it follows that the centroid lies at the intersection of these axes, Fig. 9–6.

It is important to remember that the terms $\widetilde{x}, \widetilde{y}, \widetilde{z}$ in Eqs. 9–4 through 9–7 refer to the "moment arms" or coordinates of the *center of gravity or centroid for the differential element* used in the equations. If possible, this differential element should be chosen such that it has a differential size or thickness in only *one direction*. When this is done, only a single integration is required to cover the entire region.

Fig. 9–5

PROCEDURE FOR ANALYSIS

The following procedure provides a method for determining the center of gravity or centroid of an object or shape using a single integration.

Differential Element.   Specify the coordinate axes and choose an appropriate differential element for integration. For lines this element dL is represented as a differential line segment; for areas the element dA is generally a rectangle having a finite length and differential width; and for volumes the element dV is either a circular disk having a finite radius and differential thickness, or a shell having a finite length and radius and differential thickness. Locate the element so that it intersects the *boundary* of the shape at an *arbitrary point* (x, y, z).

Size and Moment Arms.   Express the length dL, area dA, or volume dV of the element in terms of the coordinates used to define the boundary of the shape. Determine the coordinates or moment arms $\widetilde{x}, \widetilde{y}, \widetilde{z}$ for the centroid or center of gravity of the element.

Integrations.   Substitute the data computed above into the appropriate equations (Eqs. 9–4 through 9–7) and perform the integrations.* Note that integration can be accomplished only when the function in the integrand is expressed in terms of the *same variable as the differential thickness of the element*. The limits of the integral are then defined from the two extreme locations of the element's differential thickness, so that when the elements are "summed" or the integration performed, the entire region is covered.

Fig. 9–6

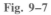

Fig. 9–7

The following examples illustrate this procedure numerically.

*Formulas for integration are given in Appendix A.

Example 9–1

Locate the centroid of the rod bent into the shape of a parabolic arc, shown in Fig. 9–7.

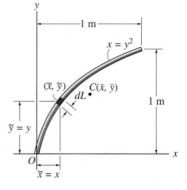

Fig. 9–7

SOLUTION

Differential Element.   The differential element is shown in Fig. 9–7. It is located on the curve at the *arbitrary point* (x, y).

Length and Moment Arms.   The differential length of the element dL can be expressed in terms of the differentials dx and dy by using the Pythagorean theorem.

$$dL = \sqrt{(dx)^2 + (dy)^2} = \sqrt{\left(\frac{dx}{dy}\right)^2 + 1}\ dy$$

Since $x = y^2$, then $dx/dy = 2y$. Therefore, expressing dL in terms of y and dy, we have

$$dL = \sqrt{(2y)^2 + 1}\ dy$$

The centroid is located at $\tilde{x} = x$, $\tilde{y} = y$.

Integrations.   Applying Eqs. 9–7 and integrating with respect to y using the formulas in Appendix A, we have

$$\bar{x} = \frac{\int_L \tilde{x}\ dL}{\int_L dL} = \frac{\int_0^1 x\sqrt{4y^2 + 1}\ dy}{\int_0^1 \sqrt{4y^2 + 1}\ dy} = \frac{\int_0^1 y^2\sqrt{4y^2 + 1}\ dy}{\int_0^1 \sqrt{4y^2 + 1}\ dy}$$

$$= \frac{0.739}{1.479} = 0.500\ \text{m} \qquad\qquad \textit{Ans.}$$

$$\bar{y} = \frac{\int_L \tilde{y}\ dL}{\int_L dL} = \frac{\int_0^1 y\sqrt{4y^2 + 1}\ dy}{\int_0^1 \sqrt{4y^2 + 1}\ dy} = \frac{0.848}{1.479} = 0.573\ \text{m} \qquad \textit{Ans.}$$

Example 9–2

Locate the centroid of the triangular area shown in Fig. 9–8a.

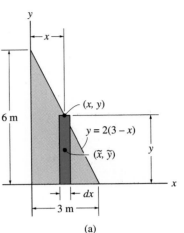

SOLUTION I

Differential Element. A rectangular differential element of thickness dx is shown in Fig. 9–8a. Since the element intersects the curve at the *arbitrary point* (x, y), it has a height y.

Area and Moment Arms. The area of the element is $dA = y\,dx$, and its centroid is located at $\widetilde{x} = x$, $\widetilde{y} = \frac{1}{2}y$.

Integrations. Applying Eqs. 9–6 and integrating with respect to x yields

$$\bar{x} = \frac{\displaystyle\int_A \widetilde{x}\,dA}{\displaystyle\int_A dA} = \frac{\displaystyle\int_0^3 xy\,dx}{\displaystyle\int_0^3 y\,dx} = \frac{\displaystyle\int_0^3 x\,2(3-x)\,dx}{\displaystyle\int_0^3 2(3-x)\,dx} = \frac{\frac{6}{2}x^2 - \frac{2}{3}x^3 \Big|_0^3}{6x - \frac{2}{2}x^2 \Big|_0^3} = 1\ \text{m} \quad Ans.$$

$$\bar{y} = \frac{\displaystyle\int_A \widetilde{y}\,dA}{\displaystyle\int_A dA} = \frac{\displaystyle\int_0^3 (\frac{1}{2}y)y\,dx}{\displaystyle\int_0^3 y\,dx} = \frac{\displaystyle\int_0^3 \frac{4}{2}(3-x)^2\,dx}{\displaystyle\int_0^3 2(3-x)\,dx} = \frac{2(9x - \frac{6}{2}x^2 + \frac{1}{3}x^3)\Big|_0^3}{6x - \frac{2}{2}x^2 \Big|_0^3} = 2\ \text{m} \quad Ans.$$

SOLUTION II

Differential Element. A rectangular differential element of thickness dy is shown in Fig. 9–8b. The element intersects the curve at the *arbitrary point* (x, y), and so it has a length x.

Area and Moment Arms. The area of the element is $dA = x\,dy$, and its centroid is located at $\widetilde{x} = \frac{1}{2}x$, $\widetilde{y} = y$.

Integrations. Applying Eqs. 9–6 and integrating with respect to y yields

$$\bar{x} = \frac{\displaystyle\int_A \widetilde{x}\,dA}{\displaystyle\int_A dA} = \frac{\displaystyle\int_0^6 (\frac{1}{2}x)x\,dy}{\displaystyle\int_0^6 x\,dy} = \frac{\displaystyle\int_0^6 \frac{1}{2}(3-\frac{1}{2}y)^2\,dy}{\displaystyle\int_0^6 (3-\frac{1}{2}y)\,dy} = \frac{\frac{1}{2}(9y - \frac{3}{2}y^2 + \frac{1}{12}y^3)\Big|_0^6}{3y - \frac{1}{4}y^2 \Big|_0^6} = 1\ \text{m} \quad Ans.$$

$$\bar{y} = \frac{\displaystyle\int_A \widetilde{y}\,dA}{\displaystyle\int_A dA} = \frac{\displaystyle\int_0^6 yx\,dy}{\displaystyle\int_0^6 x\,dy} = \frac{\displaystyle\int_0^6 y(3-\frac{1}{2}y)\,dy}{\displaystyle\int_0^6 (3-\frac{1}{2}y)\,dy} = \frac{\frac{3}{2}y^2 - \frac{1}{6}y^3 \Big|_0^6}{3y - \frac{1}{4}y^2 \Big|_0^6} = 2\ \text{m} \quad Ans.$$

Fig. 9–8

Example 9–3

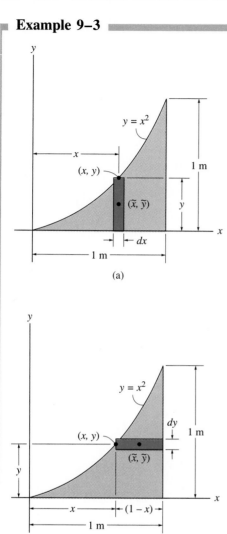

(a)

(b)

Fig. 9–9

Locate the centroid of the area shown in Fig. 9–9a.

SOLUTION I

Differential Element.   A differential element of thickness dx is shown in Fig. 9–9a. The element intersects the curve at the *arbitrary point* (x, y), and so it has a height y.

Area and Moment Arms.   The area of the element is $dA = y\,dx$, and its centroid is located at $\widetilde{x} = x$, $\widetilde{y} = y/2$.

Integrations.   Applying Eqs. 9–6 and integrating with respect to x yields

$$\bar{x} = \frac{\displaystyle\int_A \widetilde{x}\,dA}{\displaystyle\int_A dA} = \frac{\displaystyle\int_0^1 xy\,dx}{\displaystyle\int_0^1 y\,dx} = \frac{\displaystyle\int_0^1 x^3\,dx}{\displaystyle\int_0^1 x^2\,dx} = \frac{0.250}{0.333} = 0.75 \text{ m} \qquad Ans.$$

$$\bar{y} = \frac{\displaystyle\int_A \widetilde{y}\,dA}{\displaystyle\int_A dA} = \frac{\displaystyle\int_0^1 (y/2)y\,dx}{\displaystyle\int_0^1 y\,dx} = \frac{\displaystyle\int_0^1 (x^2/2)x^2\,dx}{\displaystyle\int_0^1 x^2\,dx} = \frac{0.100}{0.333} = 0.3 \text{ m} \;\; Ans.$$

SOLUTION II

Differential Element.   The differential element of thickness dy is shown in Fig. 9–9b. The element intersects the curve at the *arbitrary point* (x, y), and so it has a length $(1 - x)$.

Area and Moment Arms.   The area of the element is $dA = (1 - x)\,dy$, and its centroid is located at

$$\widetilde{x} = x + \left(\frac{1 - x}{2}\right) = \frac{1 + x}{2}, \; \widetilde{y} = y$$

Integrations.   Applying Eqs. 9–6 and integrating with respect to y, we obtain

$$\bar{x} = \frac{\displaystyle\int_A \widetilde{x}\,dA}{\displaystyle\int_A dA} = \frac{\displaystyle\int_0^1 [(1 + x)/2](1 - x)\,dy}{\displaystyle\int_0^1 (1 - x)\,dy} = \frac{\dfrac{1}{2}\displaystyle\int_0^1 (1 - y)\,dy}{\displaystyle\int_0^1 (1 - \sqrt{y})\,dy} = \frac{0.250}{0.333} = 0.75 \text{ m} \;\; Ans.$$

$$\bar{y} = \frac{\displaystyle\int_A \widetilde{y}\,dA}{\displaystyle\int_A dA} = \frac{\displaystyle\int_0^1 y(1 - x)\,dy}{\displaystyle\int_0^1 (1 - x)\,dy} = \frac{\displaystyle\int_0^1 (y - y^{3/2})\,dy}{\displaystyle\int_0^1 (1 - \sqrt{y})\,dy} = \frac{0.100}{0.333} = 0.3 \text{ m} \;\; Ans.$$

Example 9–4

Locate the $\bar{x}$ centroid of the shaded area bounded by the two curves $y = x$ and $y = x^2$, Fig. 9–10.

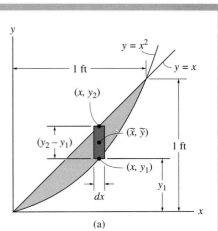

SOLUTION I

Differential Element.   A differential element of thickness dx is shown in Fig. 9–10a. The element intersects the curves at *arbitrary points* (x, y_1) and (x, y_2), and so it has a height $(y_2 - y_1)$.

Area and Moment Arm.   The area of the element is $dA = (y_2 - y_1)\, dx$, and its centroid is located at $\widetilde{x} = x$.

Integrations.   Applying Eq. 9–6, we have

$$\bar{x} = \frac{\displaystyle\int_A \widetilde{x}\, dA}{\displaystyle\int_A dA} = \frac{\displaystyle\int_0^1 x(y_2 - y_1)\, dx}{\displaystyle\int_0^1 (y_2 - y_1)\, dx} = \frac{\displaystyle\int_0^1 x(x - x^2)\, dx}{\displaystyle\int_0^1 (x - x^2)\, dx} = \frac{\frac{1}{12}}{\frac{1}{6}} = 0.5 \text{ ft} \quad \textit{Ans.}$$

(a)

SOLUTION II

Differential Element.   A differential element having a thickness dy is shown in Fig. 9–10b. The element intersects the curves at the *arbitrary points* (x_2, y) and (x_1, y), and so it has a length $(x_1 - x_2)$.

Area and Moment Arm.   The area of the element is $dA = (x_1 - x_2)\, dy$, and its centroid is located at

$$\widetilde{x} = x_2 + \frac{x_1 - x_2}{2} = \frac{x_1 + x_2}{2}$$

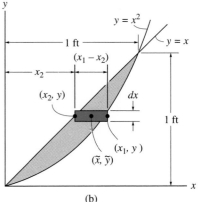

Integrations.   Applying Eq. 9–6, we have

$$\bar{x} = \frac{\displaystyle\int_A \widetilde{x}\, dA}{\displaystyle\int_A dA} = \frac{\displaystyle\int_0^1 [(x_1 + x_2)/2](x_1 - x_2)\, dy}{\displaystyle\int_0^1 (x_1 - x_2)\, dy} = \frac{\displaystyle\int_0^1 [(\sqrt{y} + y)/2](\sqrt{y} - y)\, dy}{\displaystyle\int_0^1 (\sqrt{y} - y)\, dy}$$

Fig. 9–10

$$= \frac{\frac{1}{2}\displaystyle\int_0^1 (y - y^2)\, dy}{\displaystyle\int_0^1 (\sqrt{y} - y)\, dy} = \frac{\frac{1}{12}}{\frac{1}{6}} = 0.5 \text{ ft} \qquad \textit{Ans.}$$

Example 9–5

Locate the $\bar{y}$ centroid for the paraboloid of revolution, which is generated by revolving the shaded area shown in Fig. 9–11a about the y axis.

SOLUTION I

Differential Element.   An element having the shape of a *thin disk* is chosen, Fig. 9–11a. This element has a thickness dy. In this "disk" method of analysis, the element of planar area, dA, is always taken *perpendicular* to the axis of revolution. Here the element intersects the generating curve at the *arbitrary point* (0, y, z), and so its radius is $r = z$.

Volume and Moment Arm.   The volume of the element is $dV = (\pi z^2)\,dy$, and its centroid is located at $\tilde{y} = y$.

Integrations.   Applying the second of Eqs. 9–5 and integrating with respect to y yields

$$\bar{y} = \frac{\displaystyle\int_V \tilde{y}\,dV}{\displaystyle\int_V dV} = \frac{\displaystyle\int_0^{100} y(\pi z^2)\,dy}{\displaystyle\int_0^{100} (\pi z^2)\,dy} = \frac{100\pi \displaystyle\int_0^{100} y^2\,dy}{100\pi \displaystyle\int_0^{100} y\,dy} = 66.7 \text{ mm} \quad \textit{Ans.}$$

(a)

Fig. 9–11

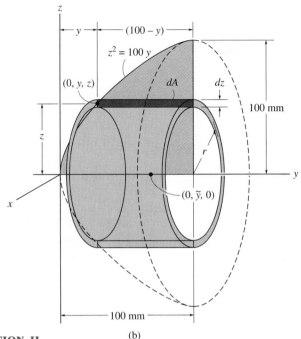

(b)

SOLUTION II

Differential Element.   As shown in Fig. 9–11b, the volume element can be chosen in the form of a *thin cylindrical shell,* where the shell's thickness is dz. In this "shell" method of analysis, the element of planar area, dA, is always taken *parallel* to the axis of revolution. Here the element intersects the generating curve at point $(0, y, z)$, and so the radius of the shell is $r = z$.

Volume and Moment Arm.   The volume of the element is $dV = 2\pi r \, dA = 2\pi z(100 - y) \, dz$, and its centroid is located at $\tilde{y} = y + (100 - y)/2 = (100 + y)/2$.

Integrations.   Applying the second of Eqs. 9–5 and integrating with respect to z yields

$$\bar{y} = \frac{\displaystyle\int_V \tilde{y} \, dV}{\displaystyle\int_V dV} = \frac{\displaystyle\int_0^{100} [(100 + y)/2] 2\pi z (100 - y) \, dz}{\displaystyle\int_0^{100} 2\pi z (100 - y) \, dz}$$

$$= \frac{\pi \displaystyle\int_0^{100} z(10^4 - 10^{-4} z^4) \, dz}{2\pi \displaystyle\int_0^{100} z(100 - 10^{-2} z^2) \, dz} = 66.7 \text{ mm} \qquad \textit{Ans.}$$

Example 9–6

(a)

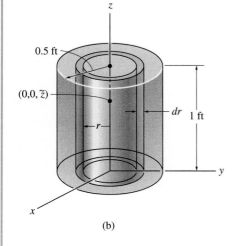

(b)

Fig. 9–12

Determine the location of the center of mass of the cylinder shown in Fig. 9–12a if its density varies directly with its distance from the base, such that $\rho = 200z$ kg/m^3.

SOLUTION

For reasons of material symmetry,

$$\bar{x} = \bar{y} = 0 \qquad\qquad Ans.$$

Differential Element. A disk element of radius 0.5 ft and thickness dz is chosen for integration, Fig. 9–12a, since the *density of the entire element is constant* for a given value of z. The element is located along the z axis at the *arbitrary point* $(0, 0, z)$.

Volume and Moment Arm. The volume of the element is $dV = \pi(0.5)^2\,dz$, and its centroid is located at $\tilde{z} = z$.

Integrations. Using an equation similar to the third of Eqs. 9–4 and integrating with respect to z, noting that $\rho = 200z$, we have

$$\bar{z} = \frac{\displaystyle\int_V \tilde{z}\rho\,dV}{\displaystyle\int_V \rho\,dV} = \frac{\displaystyle\int_0^1 z(200z)\pi(0.5)^2\,dz}{\displaystyle\int_0^1 (200z)\pi(0.5)^2\,dz}$$

$$= \frac{\displaystyle\int_0^1 z^2\,dz}{\displaystyle\int_0^1 z\,dz} = 0.667 \text{ m} \qquad\qquad Ans.$$

Note: It is not possible to use a shell element for integration such as shown in Fig. 9–12b, since the density of the material composing the shell would *vary* along the shell's height, and hence the location of $\tilde{z}$ for the element cannot be specified.

PROBLEMS

9–1. The density of the rod having a uniform cross-sectional area can be expressed as $\rho = 2x^3$ kg/m³. Determine the distance $\bar{x}$ to the rod's center of mass.

Prob. 9–1

9–2. Determine the distance $\bar{x}$ to the centroid of the rod, bent into a circular arc. From this, derive the results for a quartercircular and a semicircular arc.

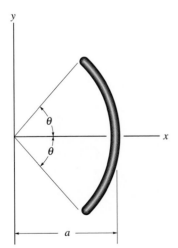

Prob. 9–2

9–3. Determine the location $(\bar{x}, \bar{y})$ of the centroid of the wire.

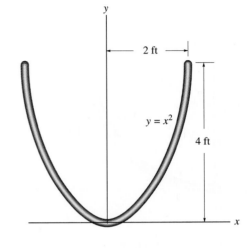

Prob. 9–3

*9–4. Determine the distance $\bar{x}$ to the center of mass of the rod. If the rod has a mass per unit length of 0.5 kg/m, determine the components of reaction at the fixed support O.

9–5. Determine the distance $\bar{y}$ to the center of mass of the homogeneous rod.

Probs. 9–4/9–5

9–6.   Determine the distance $\bar{x}$ to the center of mass of the rod. If the rod has a mass per unit length of 0.5 kg/m, determine the components of reaction at the fixed support O.

■*9–8.   Determine the location $(\bar{x}, \bar{y})$ of the center of gravity of the rod. What is its weight if it has a mass of 3 kg/m? Solve the problem by evaluating the integrals using Simpson's rule.

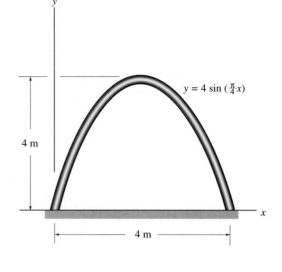

$y = 4 \sin \left(\frac{\pi}{4}x\right)$

4 m

4 m

Prob. 9–8

Prob. 9–6

$y^2 = x^3$

1 m

1 m

O

9–7.   Determine the distance $\bar{x}$ to the center of gravity of the rod bent in the form of a parabola. If the rod has a weight per unit length of 0.5 lb/ft, determine the components of reaction at the fixed support A.

9–9.   Determine the location $(\bar{x}, \bar{y})$ of the centroid of the triangular area.

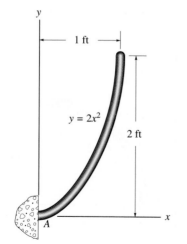

1 ft

$y = 2x^2$

2 ft

A

Prob. 9–7

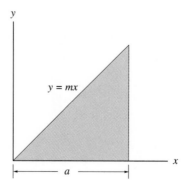

$y = mx$

a

Prob. 9–9

9–10. Determine the location $(\bar{x}, \bar{y})$ of the centroid of the shaded area.

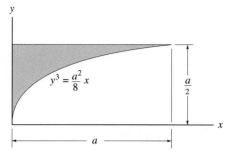

$$y^3 = \frac{a^2}{8} x$$

Prob. 9–10

9–11. Determine the location $(\bar{x}, \bar{y})$ of the centroid of the shaded area.

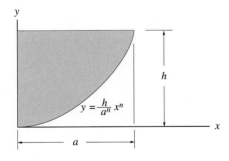

$$y = \frac{h}{a^n} x^n$$

Prob. 9–11

*9–12. Determine the location $(\bar{x}, \bar{y})$ of the centroid of the shaded area.

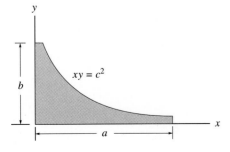

$$xy = c^2$$

Prob. 9–12

9–13. Determine the location $(\bar{x}, \bar{y})$ of the centroid of the shaded area.

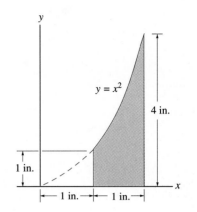

$$y = x^2$$

Prob. 9–13

9–14. Determine the location $(\bar{x}, \bar{y})$ of the centroid of the shaded area.

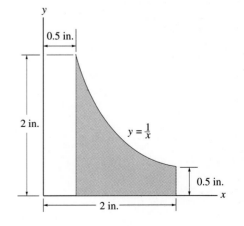

$$y = \frac{1}{x}$$

Prob. 9–14

9–15. Determine the location $(\bar{x}, \bar{y})$ of the centroid of the shaded area.

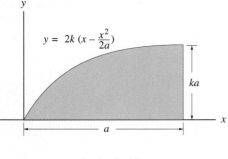

$$y = 2k\left(x - \frac{x^2}{2a}\right)$$

Prob. 9–15

*9–16. Determine the location $(\bar{x}, \bar{y})$ of the centroid of the quartercircular plate. Also compute the force in each of the supporting wires. The plate has a weight of 5 lb/ft².

$x^2 + y^2 = 16$

4 ft

Prob. 9–16

9–17. Determine the location $(\bar{x}, \bar{y})$ of the centroid of the circular area.

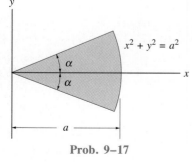

$x^2 + y^2 = a^2$

Prob. 9–17

9–18. Determine the location $(\bar{x}, \bar{y})$ of the centroid of the quarterelliptical plate.

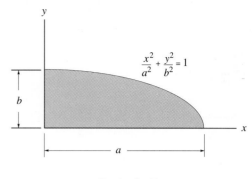

$$\frac{x^2}{a^2} + \frac{y^2}{b^2} = 1$$

Prob. 9–18

9–19. Determine the location $(\bar{x}, \bar{y})$ of the centroid of the shaded area.

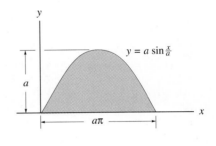

$y = a \sin\frac{x}{a}$

$a\pi$

Prob. 9–19

*9–20. Determine the location $(\bar{x}, \bar{y})$ of the centroid of the shaded area.

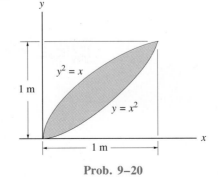

$y^2 = x$

1 m

$y = x^2$

1 m

Prob. 9–20

9–21. Determine the location $(\bar{x}, \bar{y})$ of the centroid of the shaded area.

$y^2 = 4x$

2 ft

$y = 2x$

1 ft

Prob. 9–21

9–23. Determine the location $(\bar{x}, \bar{y})$ of the centroid of the shaded area.

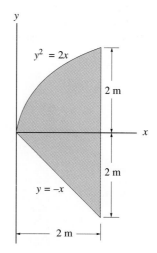

$y^2 = 2x$

2 m

2 m

$y = -x$

2 m

Prob. 9–23

9–22. Determine the location $(\bar{x}, \bar{y})$ of the centroid of the shaded area. *Hint:* Choose elements of thickness dy and length $[(2 - y) - y^2]$.

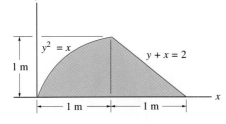

$y^2 = x$

1 m

$y + x = 2$

1 m   1 m

Prob. 9–22

*9–24. Determine the distance $\bar{y}$ to the centroid of the cone.

9–25. Determine the distance $\bar{y}$ to the center of mass of the cone. The density of the material varies linearly from zero at the origin O to ρ_0 at $x = h$.

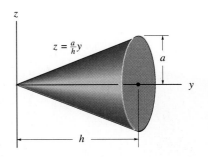

$z = \frac{a}{h}y$

a

h

Probs. 9–24/9–25

9–26. Determine the distance $\bar{y}$ to the centroid of the "bell-shaped" volume.

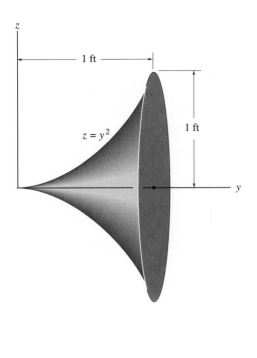

$z = y^2$

1 ft

1 ft

y

Prob. 9–26

9–27. Determine the distance $\bar{z}$ to the centroid of the volume.

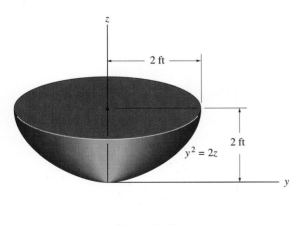

2 ft

2 ft

$y^2 = 2z$

y

Prob. 9–27

*9–28. Determine the distance $\bar{y}$ to the centroid of the paraboloid.

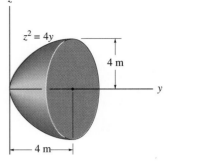

$z^2 = 4y$

4 m

y

4 m

Prob. 9–28

9–29. Determine the distance $\bar{y}$ to the center of gravity of the volume. The material is homogeneous.

$y^2 = 100z$

100 mm

25 mm

y

−50 mm−−50 mm−

Prob. 9–29

9–30. Determine the distance $\bar{z}$ to the centroid of the right-elliptical cone.

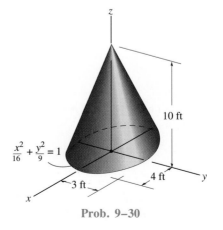

$\dfrac{x^2}{16} + \dfrac{y^2}{9} = 1$

10 ft

3 ft

4 ft

y

x

Prob. 9–30

9–31. Determine the distance $\bar{z}$ to the centroid of the hemisphere.

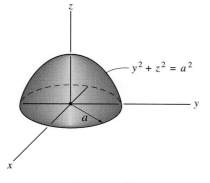

Prob. 9–31

9–33. Determine the distance $\bar{z}$ to the centroid of the solid.

Prob. 9–33

*9–32. Determine the distance $\bar{y}$ to the centroid of the semiellipsoid.

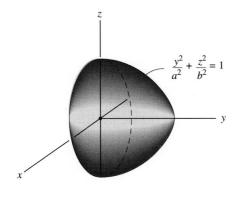

Prob. 9–32

9–34. Determine the distance $\bar{z}$ to the centroid of the spherical segment.

Prob. 9–34

9.3 Composite Bodies

A *composite body* consists of a series of connected "simpler" shaped bodies, which may be rectangular, triangular, semicircular, etc. Such a body can often be sectioned or divided into its composite parts, and provided the *weight* and location of the center of gravity of each of these parts are known, we can eliminate the need for integration to determine the center of gravity for the entire body. The method for doing this requires treating each composite part as a particle and following the procedure outlined in Sec. 9.1. Formulas analogous to Eqs. 9–1 result, since we must account for a finite number of weights. Rewriting these formulas, we have

$$\bar{x} = \frac{\Sigma \tilde{x} W}{\Sigma W} \qquad \bar{y} = \frac{\Sigma \tilde{y} W}{\Sigma W} \qquad \bar{z} = \frac{\Sigma \tilde{z} W}{\Sigma W} \qquad (9\text{–}8)$$

Here

$\bar{x}, \bar{y}, \bar{z}$ represent the coordinates of the center of gravity G of the composite body.

$\tilde{x}, \tilde{y}, \tilde{z}$ represent the coordinates of the center of gravity of each composite part of the body.

ΣW is the sum of the weights of all the composite parts of the body, or simply the total weight of the body.

When the body has a *constant density or specific weight,* the center of gravity *coincides* with the centroid of the body. The centroid for composite lines, areas, and volumes can be found using relations analogous to Eqs. 9–8; however, the W's are replaced by L's, A's, and V's, respectively. Centroids for common shapes of lines, areas, shells, and volumes are given in the table on the inside back cover.

PROCEDURE FOR ANALYSIS

The following procedure provides a method for determining the center of gravity of a body or the centroid of a composite geometrical object represented by a line, area, or volume.

Composite Parts.   Using a sketch, divide the body or object into a finite number of composite parts. If a composite part has a *hole,* or geometric region having no material, then consider the composite part without the hole, and the hole as an *additional* composite part having *negative* weight or size.

Moment Arms.   Establish the coordinate axes on the sketch and determine the coordinates $\tilde{x}, \tilde{y}, \tilde{z}$ of the center of gravity or centroid of each part.

Summations.   Determine $\bar{x}, \bar{y}, \bar{z}$ by applying the center of gravity equations, Eqs. 9–8, or the analogous centroid equations. If an object is *symmetrical* about an axis, recall that the centroid of the object lies on this axis.

If desired, the calculations can be arranged in tabular form, as indicated in the following three examples.

Example 9–7

Locate the centroid of the wire shown in Fig. 9–13a.

SOLUTION

Composite Parts. The wire is divided into three segments as shown in Fig. 9–13b.

Moment Arms. The location of the centroid for each piece is determined and indicated in the figure. In particular, the centroid of segment ① is determined either by integration or using the table on the inside back cover.

Summations. The calculations are tabulated as follows:

Segment	L (mm)	$\widetilde{x}$ (mm)	$\widetilde{y}$ (mm)	$\widetilde{z}$ (mm)	$\widetilde{x}L$ (mm²)	$\widetilde{y}L$ (mm²)	$\widetilde{z}L$ (mm²)
1	$\pi(60) = 188.5$	60	−38.2	0	11 310	−7200	0
2	40	0	20	0	0	800	0
3	20	0	40	−10	0	800	−200
	$\Sigma L = 248.5$				$\Sigma\widetilde{x}L = 11\ 310$	$\Sigma\widetilde{y}L = -5600$	$\Sigma\widetilde{z}L = -200$

Thus,

$$\bar{x} = \frac{\Sigma\widetilde{x}L}{\Sigma L} = \frac{11\ 310}{248.5} = 45.5 \text{ mm} \qquad Ans.$$

$$\bar{y} = \frac{\Sigma\widetilde{y}L}{\Sigma L} = \frac{-5600}{248.5} = -22.5 \text{ mm} \qquad Ans.$$

$$\bar{z} = \frac{\Sigma\widetilde{z}L}{\Sigma L} = \frac{-200}{248.5} = -0.805 \text{ mm} \qquad Ans.$$

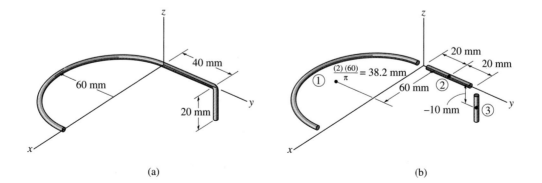

(a)   (b)

Fig. 9–13

Example 9–8

Locate the centroid of the plate area shown in Fig. 9–14a.

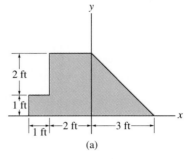

(a)

SOLUTION

Composite Parts. The plate is divided into three segments as shown in Fig. 9–14b. Here the area of the small rectangle ③ is considered "negative" since it must be subtracted from the larger one ②.

Moment Arms. The centroid of each segment is located as indicated in the figure. Note that the $\tilde{x}$ coordinates of ② and ③ are *negative*.

Summations. Taking the data from Fig. 9–14b, the calculations are tabulated as follows:

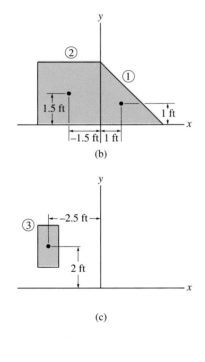

(b)

(c)

Fig. 9–14

Segment	A (ft²)	$\tilde{x}$ (ft)	$\tilde{y}$ (ft)	$\tilde{x}A$ (ft³)	$\tilde{y}A$ (ft³)
1	$\frac{1}{2}(3)(3) = 4.5$	1	1	4.5	4.5
2	$(3)(3) = 9$	-1.5	1.5	-13.5	13.5
3	$-(2)(1) = -2$	-2.5	2	5	-4
	$\Sigma A = 11.5$			$\Sigma\tilde{x}A = -4$	$\Sigma\tilde{y}A = 14$

Thus,

$$\bar{x} = \frac{\Sigma\tilde{x}A}{\Sigma A} = \frac{-4}{11.5} = -0.348 \text{ ft} \qquad Ans.$$

$$\bar{y} = \frac{\Sigma\tilde{y}A}{\Sigma A} = \frac{14}{11.5} = 1.22 \text{ ft} \qquad Ans.$$

Example 9–9

Locate the center of mass of the composite assembly shown in Fig. 9–15a. The conical frustum has a density of $\rho_c = 8$ Mg/m^3, and the hemisphere has a density of $\rho_h = 4$ Mg/m^3.

SOLUTION

Composite Parts.   The assembly can be thought of as consisting of four segments as shown in Fig. 9–15b. For the calculations, ③ and ④ must be considered as "negative" volumes in order that the four segments, when added together, yield the total composite shape shown in Fig. 9–15a.

Moment Arm.   Using the table on the inside back cover, the computations for the centroid $\widetilde{z}$ of each piece are shown in the figure.

Summations.   Because of *symmetry*, note that

$$\bar{x} = \bar{y} = 0 \qquad\qquad Ans.$$

Since $W = mg$ and g is constant, the third of Eqs. 9–8 becomes $\bar{z} = \Sigma\widetilde{z}m/\Sigma m$. The mass of each piece can be computed from $m = \rho V$ and used for the calculations. Also, 1 Mg/m^3 = 10^{-6} kg/mm^3, so that

25 mm

100 mm

50 mm

50 mm

(a)

Fig. 9–15

Segment	m (kg)	$\widetilde{z}$ (mm)	$\widetilde{z}m$ (kg · mm)
1	$8(10^{-6})(\frac{1}{3})\pi(50)^2(200) = 4.189$	50	209.440
2	$4(10^{-6})(\frac{2}{3})\pi(50)^3 = 1.047$	-18.75	-19.635
3	$-8(10^{-6})(\frac{1}{3})\pi(25)^2(100) = -0.524$	$100 + 25 = 125$	-65.450
4	$-8(10^{-6})\pi(25)^2(100) = -1.571$	50	-78.540
	$\Sigma m = 3.141$		$\Sigma\widetilde{z}m = 45.815$

Thus,

$$\bar{z} = \frac{\Sigma\widetilde{z}m}{\Sigma m} = \frac{45.815}{3.141} = 14.6 \text{ mm} \qquad\qquad Ans.$$

200 mm

$\dfrac{200 \text{ mm}}{4} = 50$ mm

50 mm

①

50 mm

②

$-\frac{3}{8}(50) = -18.75$ mm

③

$\dfrac{100 \text{ mm}}{4} = 25$ mm

100 mm

25 mm

100 mm

④

25 mm

100 mm

50 mm

(b)

PROBLEMS

9–35. If the four particles can be replaced by a single 10-kg particle acting at a distance of 2 m to the left of the origin, determine the position $\widetilde{x}_p$ and the mass m_p of particle P.

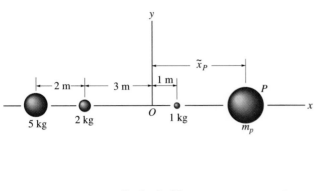

Prob. 9–35

9–37. Determine the location of the center of gravity $(\bar{x}, \bar{y}, \bar{z})$ of the four particles.

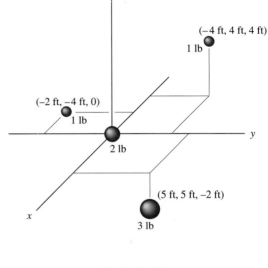

Prob. 9–37

*9–36. Determine the location of the center of mass $(\bar{x}, \bar{y}, \bar{z})$ of the three particles.

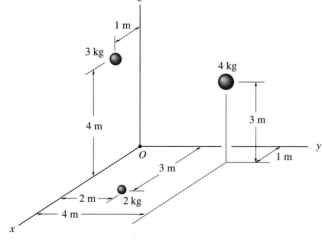

Prob. 9–36

9–38. Determine the location $(\bar{x}, \bar{y})$ of the centroid for the structural shape used in the construction of an airplane wing. Neglect the thickness of the member.

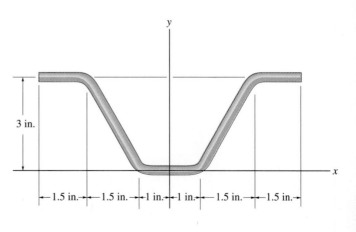

Prob. 9–38

9–39. Determine the location $(\bar{x}, \bar{y})$ of the centroid of the rod. Neglect the rod's thickness.

150 mm

100 mm

100 mm

150 mm

Prob. 9–39

9–41. Determine the location $(\bar{x}, \bar{y})$ of the centroid of the sheet-metal cross section. Neglect the thickness of the material and slight bends at the corners.

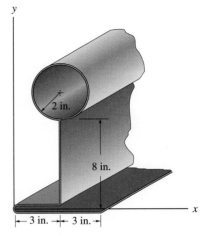

2 in.

8 in.

3 in. 3 in.

Prob. 9–41

*9–40. The "roll-formed" strut has a cross section as shown. Determine the distance $\bar{y}$ to its centroid C. Neglect the thickness of the material.

C

80 mm

$\bar{y}$

20 mm

Prob. 9–40

9–42. The truss is made from seven members, each having a mass of 6 kg/m. Determine the location $(\bar{x}, \bar{y})$ of the center of mass. Neglect the mass of the gusset plates at the joints.

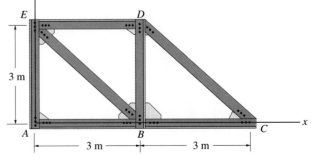

E D

3 m

A B C

3 m   3 m

Prob. 9–42

9–43. The steel and aluminum plate assembly is bolted together and fastened to the wall. Each plate has a width of 200 mm and thickness of 20 mm. If the mass density of A and B is $\rho_s = 7.85$ Mg/m³, and for C, $\rho_{al} = 2.71$ Mg/m³, determine the distance $\bar{x}$ to the center of mass. Neglect the size of the bolts.

Prob. 9–43

*9–44.** The three members of the frame each have a weight per unit length of 4 lb/ft. Determine the location $(\bar{x}, \bar{y})$ of the center of gravity. Neglect the size of the pins at the joints and the thickness of the members. Also, calculate the reactions at the fixed support A.

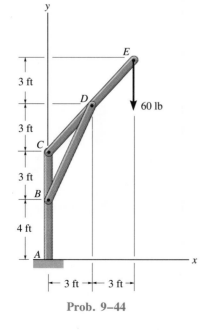

Prob. 9–44

9–45. Determine the length of the wire segment AB such that the center of gravity for the entire assembly is located at G. Each segment is made of the same material and has the same cross-sectional area.

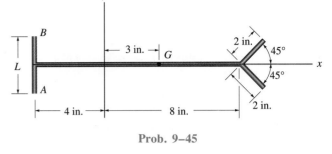

Prob. 9–45

9–46. Determine the distance $\bar{y}$ to the centroid of the area.

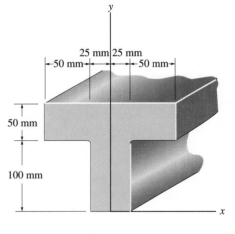

Prob. 9–46

9–47. Determine the distance $\bar{y}$ to the centroid of the beam's cross-sectional area.

Prob. 9–47

*9–48. Determine the distance $\bar{y}$ to the centroid of the beam's cross-sectional area. Neglect the size of the corner welds for the calculation.

Prob. 9–48

9–50. Determine the location $(\bar{x}, \bar{y})$ of the centroid of the area.

Prob. 9–50

9–49. Determine the distance $\bar{y}$ to the centroid of the area.

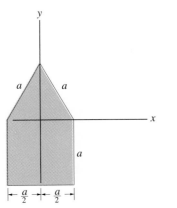

Prob. 9–49

9–51. Determine the location $(\bar{x}, \bar{y})$ of the centroid of the area.

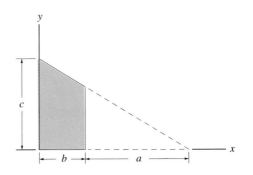

Prob. 9–51

*9–52.   Determine the location $(\bar{x}, \bar{y})$ of the centroid of the cross-sectional area.

Prob. 9–52

9–53.   Determine the distance $\bar{y}$ to the centroid of the trapezoidal area in terms of the dimensions shown.

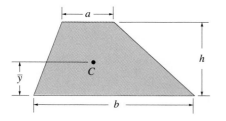

Prob. 9–53

9–54.   The gravity wall is made of concrete. Determine the location $(\bar{x}, \bar{y})$ of the center of gravity G for the wall.

Prob. 9–54

9–55.   Determine the distance $\bar{y}$ to the centroid of the shaded area.

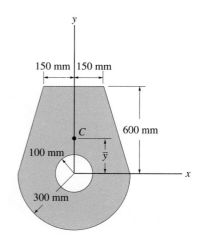

Prob. 9–55

*9–56.   Determine the distance $\bar{x}$ to the centroid of the shaded area.

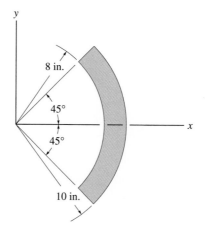

Prob. 9–56

9–57. Determine the location $(\bar{x}, \bar{y})$ of the centroid of the shaded area.

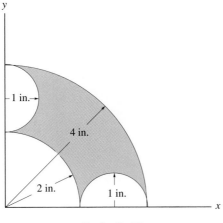

Prob. 9–57

9–58. Determine the location $(\bar{x}, \bar{y})$ of the centroid of the shaded area.

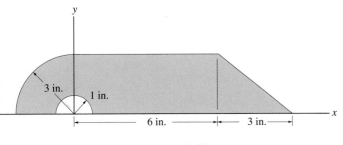

Prob. 9–58

9–59. Determine the distance $\bar{y}$ to the centroid C for the circular sector having a radius $r = 0.25$ m.

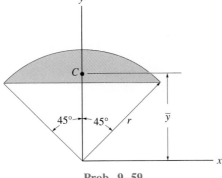

Prob. 9–59

***9–60.** The girder is to be lifted by a single cable attached to a point directly over the girder's center of gravity G. To locate this point, the girder is supported at end A and end B using two scales. The scale at A reads 5685 lb, and the one at B reads 4864 lb. Determine the distance $\bar{x}$ to G, measured from end A.

Prob. 9–60

9–61. Determine the distance $\bar{x}$ to the center of gravity of the gas dehydrator assembly. The weight and the center of gravity of each of the various components are indicated below. What are the vertical reactions at A and B needed to support the assembly?

Prob. 9–61

9–62. Determine the distance $\bar{x}$ to the center of gravity of the generator assembly. The weight and the center of gravity of each of the various components are indicated below. What are the vertical reactions at blocks A and B needed to support the assembly?

G_1(200 lb) G_2(1000 lb) G_3(250 lb)

A B

1ft 3ft 4ft 2ft

$\bar{x}$

Prob. 9–62

9–63. The wooden table is made from a square board having a weight of 15 lb. Each of the legs weighs 2 lb and is 3 ft long. Determine how high its center of gravity is from the floor. Also, what is the greatest angle, measured from the horizontal, through which its top surface can be tilted on two of its legs before it begins to overturn? Neglect the thickness of each leg.

4ft 4ft

3ft

Prob. 9–63

*9–64. Determine the distance $\bar{x}$ to the centroid of the volume.

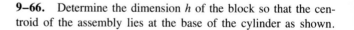

8 in. 4 in.

2 in. C 1.5 in. x

$\bar{x}$

Prob. 9–64

9–65. Determine the distance $\bar{z}$ to the center of mass of the casting that is formed from a hollow cylinder having a density of 8 Mg/m³ and a hemisphere having a density of 3 Mg/m³.

z

20 mm 40 mm

120 mm

x

40 mm

y

Prob. 9–65

9–66. Determine the dimension h of the block so that the centroid of the assembly lies at the base of the cylinder as shown.

6 in. h

6 in. C

5 in.

2 in. x

Prob. 9–66

9–67.   Determine the distance $\bar{y}$ to the center of mass of the assembly, which has a hole bored through the center. The material has a density of $\rho = 3$ Mg/m³.

Prob. 9–67

*9–68.   The solid is formed by boring a conical hole into the cylinder. Determine the distance $\bar{z}$ to the center of gravity.

Prob. 9–68

9–69.   The solid is formed by boring a conical hole into the hemisphere. Determine the distance $\bar{z}$ to the center of gravity.

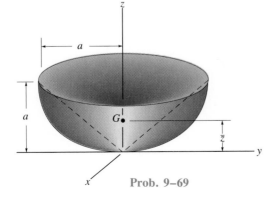

Prob. 9–69

9–70.   Determine the distance h to which a 100-mm-diameter hole must be bored into the cone so that the center of mass G of the resulting shape is located at $\bar{z} = 115$ mm. The material has a density of 8 Mg/m³.

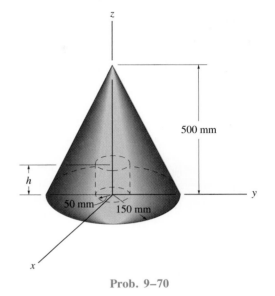

Prob. 9–70

9–71.   A hole having a radius r is to be drilled in the center of the homogeneous block. Determine the depth h of the hole so that the center of gravity G is as low as possible.

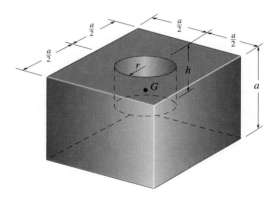

Prob. 9–71

*9.4 Theorems of Pappus and Guldinus

Fig. 9–16

Fig. 9–17

The two *theorems of Pappus and Guldinus,* which were first developed by Pappus of Alexandria during the third century A.D. and then restated at a later time by the Swiss mathematician Paul Guldin or Guldinus (1577–1643), are used to find the surface area and volume of any object of revolution.

A *surface area of revolution* is generated by revolving a *plane curve* about a nonintersecting fixed axis in the plane of the curve; whereas a *volume of revolution* is generated by revolving a *plane area* about a nonintersecting fixed axis in the plane of the area. For example, if the *line AB* shown in Fig. 9–16 is rotated about a fixed axis, it generates the *surface area* of a cone; if the triangular *area ABC* shown in Fig. 9–17 is rotated about the axis, it generates the *volume* of a cone.

The statements and proofs of the theorems of Pappus and Guldinus follow. The proofs require that the generating curves and areas do *not* cross the axis about which they are rotated; otherwise, two sections on either side of the axis would generate areas or volumes having opposite signs and hence cancel each other.

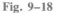

Fig. 9–18

Surface Area

The area of a surface of revolution equals the product of the length of the generating curve and the distance traveled by the centroid of the curve in generating the surface area.

Proof. When a differential length dL of the curve shown in Fig. 9–18 is revolved about an axis through a distance $2\pi r$, it generates a ring having a surface area $dA = 2\pi r\, dL$. The entire surface area, generated by revolving the entire curve about the axis, is therefore $A = 2\pi \int_L r\, dL$. This equation may be simplified, however, by noting that the location r of the centroid for the line of total length L can be determined from an equation having the form of Eqs. 9–7, namely, $\bar{r} = \int r\, dL / L$. Thus the total surface area becomes $A = 2\pi\bar{r}L$. In general, though, if the line does not undergo a complete revolution, then

$$\boxed{A = \theta\bar{r}L} \qquad (9\text{–}9)$$

where A = surface area of revolution

θ = angle of revolution measured in radians, $\theta \leqslant 2\pi$

$\bar{r}$ = perpendicular distance from the axis of revolution to the centroid of the generating curve

L = length of the generating curve

Fig. 9–19

Volume

The volume of a body of revolution equals the product of the generating area and the distance traveled by the centroid of the area in generating the volume.

Proof. When the differential area dA shown in Fig. 9–19 is revolved about an axis through a distance $2\pi r$, it generates a ring having a volume $dV = 2\pi r\, dA$. The entire volume, generated by revolving A about the axis, is therefore $V = 2\pi\int_A r\, dA$. Here the integral can be eliminated by using an equation analogous to Eqs. 9–6, $\int_A r\, dA = \bar{r}A$, where $\bar{r}$ locates the centroid C of the generating area A, and the volume becomes $V = 2\pi\bar{r}A$. In general, though,

$$V = \theta\bar{r}A \qquad\qquad (9\text{–}10)$$

where V = volume of revolution
   θ = angle of revolution measured in radians, $\theta \leqslant 2\pi$
   $\bar{r}$ = perpendicular distance from the axis of revolution to the centroid of the generating area
   A = generating area

Composite Shapes.

We may also apply the above two theorems to lines or areas that may be composed of a series of composite parts. In this case the total surface area or volume generated is the addition of the surface areas or volumes generated by each of the composite parts. Since each part undergoes the *same* angle of revolution, θ, and the distance from the axis of revolution to the centroid of each composite part is r, then

$$A = \theta\Sigma rL \qquad\qquad (9\text{–}11)$$

and

$$V = \theta\Sigma rA \qquad\qquad (9\text{–}12)$$

Application of the above theorems is illustrated numerically in the following examples.

Example 9–10

Show that the surface area of a sphere is $A = 4\pi R^2$ and its volume is $V = \frac{4}{3}\pi R^3$.

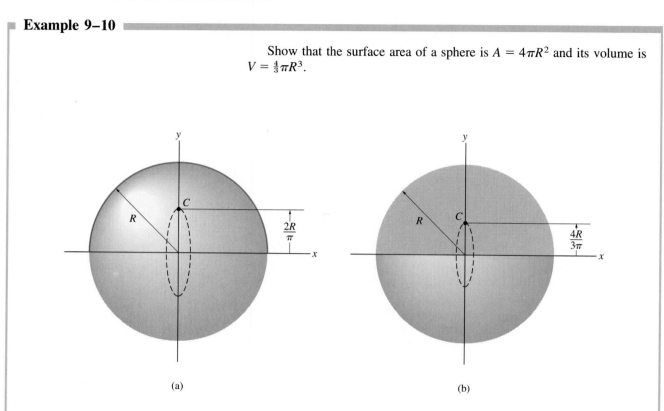

(a)

(b)

Fig. 9–20

SOLUTION

Surface Area.   The surface area of the sphere in Fig. 9–20a is generated by rotating a semicircular *arc* about the x axis. Using the table on the inside back cover, it is seen that the centroid of this arc is located at a distance $\bar{r} = 2R/\pi$ from the x axis of rotation. Since the centroid moves through an angle of $\theta = 2\pi$ rad in generating the sphere, by applying Eq. 9–9 we have

$$A = \theta\bar{r}L; \qquad\qquad A = 2\pi\left(\frac{2R}{\pi}\right)\pi R = 4\pi R^2 \qquad\qquad Ans.$$

Volume.   The volume of the sphere is generated by rotating the semicircular *area* in Fig. 9–20b about the x axis. Using the table on the inside back cover to locate the centroid and applying Eq. 9–10, we have

$$V = \theta\bar{r}A; \qquad\qquad V = 2\pi\left(\frac{4R}{3\pi}\right)\left(\frac{1}{2}\pi R^2\right) = \frac{4}{3}\pi R^3 \qquad\qquad Ans.$$

Example 9–11

Determine the mass of concrete needed to construct the arched beam shown in Fig. 9–21. The density of concrete is $\rho_c = 2.1$ Mg/m³. Also, determine the surface area of the beam.

Side view
(a)

Area Segments
(b)

Line Segments
(c)

Fig. 9–21

SOLUTION

Mass.   The mass of the arch may be determined by using the density ρ_c, provided the volume of concrete is known. The cross-sectional area of the beam is composed of two rectangles having centroids at points C_1 and C_2 as shown in Fig. 9–21b. The volume generated by each of these areas is equal to the product of the cross-sectional area and the distance traveled by its centroid. In generating the arch, the centroids move through an angle of $\theta = \pi(90°/180°) = 1.57$ rad. Point C_1 is at a distance of $\bar{r}_1 = 4.5 - 0.1 = 4.4$ m from the center of rotation, point O; whereas C_2 is at $\bar{r}_2 = 4 + 0.15 = 4.15$ m from O. Applying Eq. 9–12, we obtain the total volume:

$V = \theta \Sigma \bar{r} A$

$\quad = 1.57 \text{ rad}[(4.4 \text{ m})(1 \text{ m})(0.2 \text{ m}) + (4.15 \text{ m})(0.15 \text{ m})(0.3 \text{ m})]$

$\quad = 1.676 \text{ m}^3$

The required mass of concrete is then

$$m = \rho_c V = (2.1 \text{ Mg/m}^3)(1.676 \text{ m}^3) = 3.52 \text{ Mg} \qquad Ans.$$

Surface Area.   The surface area of the arch is determined by dividing the perimeter of its cross-sectional area into eight line segments, Fig. 9–21c, and then applying Eq. 9–11. Thus,

$A = \theta \Sigma \bar{r} L$

$\quad = 1.57 \text{ rad}[4 \text{ m}(0.15 \text{ m}) + 2(4.15 \text{ m})(0.3 \text{ m})$

$\quad\quad + (4.3 \text{ m})(1 \text{ m} - 0.15 \text{ m}) + 2(4.4 \text{ m})(0.2 \text{ m}) + (4.5 \text{ m})(1 \text{ m})]$

$\quad = 20.4 \text{ m}^2 \qquad\qquad Ans.$

PROBLEMS

*9–72. Using the second theorem of Pappus and Guldinus, determine the location of the centroid of the quartercircular area.

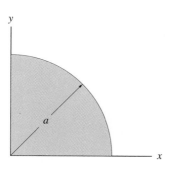

Prob. 9–72

9–74. A steel wheel has a cross-sectional view as shown. Determine the weight of the wheel if the specific gravity of the material is $\gamma = 0.25$ lb/in^3.

Prob. 9–74

9–73. Sand is piled between two walls as shown. Assume the pile to be a quarter section of a cone and use the second theorem of Pappus and Guldinus to determine its volume.

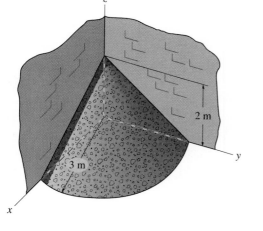

Prob. 9–73

9–75. Determine the surface area and the volume of the ring.

Prob. 9–75

*9–76. Determine the surface area and the volume of the conical solid.

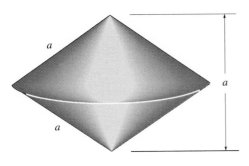

Prob. 9–76

9–77. Determine the outside surface area and the volume of the storage tank.

Prob. 9–77

9–78. The tank is fabricated from a hemisphere and cylindrical shell. Determine the vertical reactions that each of the four symmetrically placed legs exerts on the floor if the tank contains water which is 12 ft deep in the tank. The specific gravity of water is 62.4 lb/ft^3. Neglect the weight of the tank.

9–79. Determine the approximate amount of paint needed to cover the outside surface of the tank. Assume that a gallon of paint covers 400 ft^2.

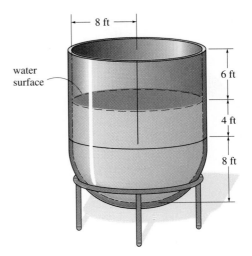

Probs. 9–78/9–79

*9–80. The *rim* of a flywheel has the cross section A–A shown. Determine the volume of material needed for its construction.

Prob. 9–80

9–81. Determine the weight of the wedge which is generated by rotating a right triangle of base 6 in. and height 3 in. through an angle of 30°. The specific weight of the material is $\gamma = 0.22$ lb/in³.

Prob. 9–81

9–82. The full circular aluminum housing is used in an automotive brake system. The cross section is shown in the figure. Determine its weight if aluminum has a specific weight of 169 lb/ft³.

Prob. 9–82

9–83. A circular V-belt has an inner radius of 600 mm and a cross-sectional area as shown. Determine the volume of material required to make the belt.

Prob. 9–83

***9–84.** Determine the approximate amount of aluminum necessary to make the funnel. It consists of a full circular part having a thickness of 2 mm. Its cross section is shown in the figure.

9–85. Determine the approximate outer surface area of the funnel. It consists of a full circular part of negligible thickness.

Probs. 9–84/9–85

9–86. Determine the surface area of the roof of the structure if it is formed by rotating the parabola about the y axis.

*9–88. Determine both the area and the distance $\bar{r}$ to the centroid of the shaded region. Using the second theorem of Pappus and Guldinus, compute the volume of the solid generated by revolving the shaded area about the a–a axis.

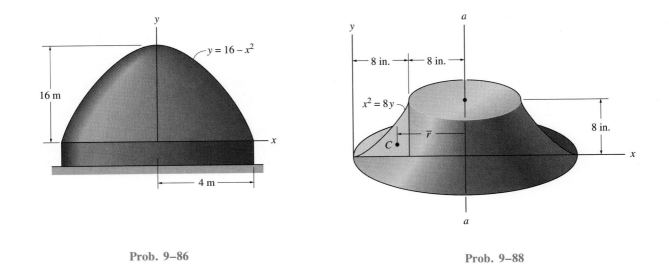

$y = 16 - x^2$

16 m

4 m

Prob. 9–86

8 in. — 8 in.

$x^2 = 8y$

$\bar{r}$

C

8 in.

Prob. 9–88

9–87. Using integration, determine the area and the distance $\bar{y}$ to the centroid of the shaded area. Then using the second theorem of Pappus and Guldinus, compute the volume of the solid generated by revolving the shaded area about the x axis.

9–89. Using Simpson's rule, determine the area and the distance $\bar{y}$ to the centroid of the shaded area. Then using the second theorem of Pappus and Guldinus, determine the volume formed by revolving the area about the x axis.

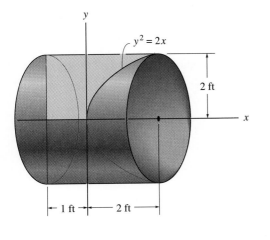

$y^2 = 2x$

2 ft

1 ft | 2 ft

Prob. 9–87

$y = x\sqrt{\ln x}$

C

$\bar{y}$

1 m — 1 m

Prob. 9–89

*9.5 Resultant of a General Distributed Force System

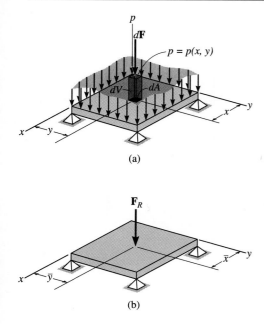

(a)

(b)

Fig. 9–22

In Sec. 4.10 we discussed the method used to simplify a distributed loading that is uniform along an axis of a rectangular surface. In this section we will generalize this method to include surfaces that have an arbitrary shape and are subjected to a variable load distribution. As a specific application, in Sec. 9.6 we will find the resultant loading acting on the surface of a body that is submerged in a fluid.

Pressure Distribution over a Surface. Consider the flat plate shown in Fig. 9–22a which is subjected to the loading function $p = p(x, y)$ Pa, where 1 pascal, Pa = 1 N/m². Knowing this function, we can determine the *magnitude* of the infinitesimal force $d\mathbf{F}$ acting on the differential area dA m² of the plate, located at the arbitrary point (x, y). This force magnitude is simply $dF = [p(x, y)\ \text{N/m}^2](dA\ \text{m}^2) = [p(x, y)\ dA]$ N. The entire loading on the plate is therefore represented as a system of *parallel forces* infinite in number and each acting on separate differential areas dA. This system of parallel forces will now be simplified to a single resultant force $\mathbf{F}_R$ acting through a unique point $(\bar{x}, \bar{y})$ on the plate, Fig. 9–22b.

Magnitude of Resultant Force. To determine the *magnitude* of $\mathbf{F}_R$, it is necessary to sum each of the differential forces $d\mathbf{F}$ acting over the plate's *entire surface area A*. This sum may be expressed mathematically as an integral:

$$F_R = \Sigma F; \qquad \boxed{F_R = \int_A p(x, y)\ dA = \int_V dV} \qquad (9\text{–}13)$$

Note that $p(x, y)\ dA = dV$, the colored differential *volume element* shown in Fig. 9–22a. Therefore, the result indicates that the *magnitude of the resultant force is equal to the total volume under the distributed-loading diagram*.

Location of Resultant Force. The location $(\bar{x}, \bar{y})$ of $\mathbf{F}_R$ is determined by setting the moments of $\mathbf{F}_R$ equal to the moments of all the forces $d\mathbf{F}$ about the respective y and x axes. From Fig. 9–22a and 9–22b, using Eq. 9–13, we have

$$\boxed{\bar{x} = \frac{\int_A x\,p(x, y)\ dA}{\int_A p(x, y)\ dA} = \frac{\int_V x\ dV}{\int_V dV} \qquad \bar{y} = \frac{\int_A y\,p(x, y)\ dA}{\int_A p(x, y)\ dA} = \frac{\int_V y\ dV}{\int_V dV}} \qquad (9\text{–}14)$$

Hence, it can be seen that the *line of action of the resultant force passes through the geometric center or centroid of the volume under the distributed loading diagram*.

*9.6 Fluid Pressure

According to Pascal's law, a fluid at rest creates a pressure p at a point that is the *same* in *all* directions. The magnitude of p, measured as a force per unit area, depends on the specific weight γ or mass density ρ of the fluid and the depth z of the point from the fluid surface.* The relationship can be expressed mathematically as

$$p = \gamma z = \rho g z \qquad (9\text{–}15)$$

where g is the acceleration of gravity. Equation 9–15 is valid only for fluids that are assumed *incompressible,* as in the case of most liquids. Gases are compressible fluids, and since their density changes significantly with both pressure and temperature, Eq. 9–15 cannot be used.

To illustrate how Eq. 9–15 is applied, consider the submerged plate shown in Fig. 9–23. Three points on the plate have been specified. Since points A and B are both at depth z_2 from the liquid surface, the *pressure* at these points has a magnitude $p_2 = \gamma z_2$. Likewise, point C is at depth z_1; hence, $p_1 = \gamma z_1$. In all cases, the pressure acts *normal* to the surface area dA located at the specified point, Fig. 9–23. Using Eq. 9–15 and the results of Sec. 9.5, it is possible to determine the resultant force of a liquid pressure distribution, and specify its location on the surface of a submerged plate. Three different shapes of plates will now be considered.

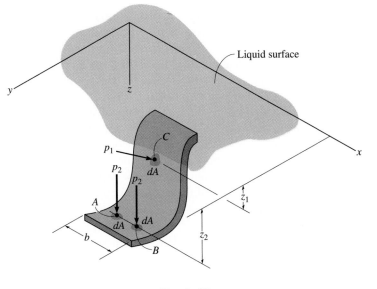

Fig. 9–23

*In particular, for water $\gamma = 62.4$ lb/ft^3, or $\gamma = 9810$ N/m^3, since $\rho = 1000$ kg/m^3 and $g = 9.81$ m/s^2.

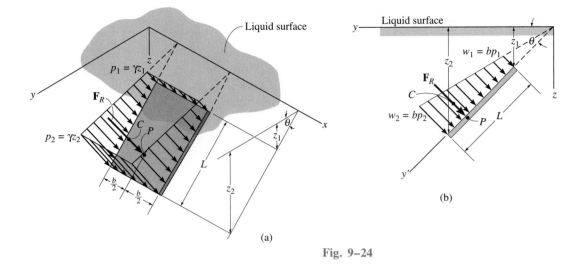

Fig. 9–24

Flat Plate of Constant Width.

A flat rectangular plate of constant width, which is submerged in a liquid having a specific weight γ, is shown in Fig. 9–24a. The plane of the plate makes an angle θ with the horizontal, such that its top edge is located at a depth z_1 from the liquid surface and its bottom edge is located at a depth z_2. Since pressure varies linearly with depth, Eq. 9–15, the distribution of pressure over the plate's surface is represented by a trapezoidal volume having an intensity of $p_1 = \gamma z_1$ at depth z_1 and $p_2 = \gamma z_2$ at depth z_2. As noted in Sec. 9.5, the magnitude of the *resultant force* $\mathbf{F}_R$ is equal to the *volume* of this loading diagram and $\mathbf{F}_R$ has a *line of action* that passes through the volume's centroid C. Hence $\mathbf{F}_R$ does *not* act at the centroid of the plate; rather, it acts at point P, called the *center of pressure*.

Since the plate has a *constant width,* the loading distribution may also be viewed in two dimensions, Fig. 9–24b. Here the loading intensity is measured as force/length and varies linearly from $w_1 = bp_1 = b\gamma z_1$ to $w_2 = bp_2 = b\gamma z_2$. The magnitude of $\mathbf{F}_R$ in this case equals the trapezoidal *area*, and $\mathbf{F}_R$ has a *line of action* that passes through the area's *centroid C*. For numerical applications, the area and location of the centroid for a trapezoid are tabulated on the inside back cover.

Curved Plate of Constant Width.

When the submerged plate is curved, the pressure acting normal to the plate continually changes direction, and therefore calculation of the magnitude of $\mathbf{F}_R$ and its location P is more difficult than for a flat plate. Three- and two-dimensional views of the loading distribution are shown in Figs. 9–25a and 9–25b, respectively. Here integration can be used to determine both F_R and the location of the centroid C or center of pressure P.

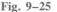

(a)

(b)

Fig. 9–25

A simpler method exists, however, for calculating the magnitude of $\mathbf{F}_R$ and its location along a curved (or flat) plate having a *constant width*. This method requires separate calculations for the horizontal and vertical *components* of $\mathbf{F}_R$. For example, the distributed loading acting on the curved plate *DB* in Fig. 9–25*b* can be represented by the *equivalent loading* shown in Fig. 9–26. Here the plate supports the weight of liquid W_f contained within the block *BDA*. This force has a magnitude $W_f = (\gamma b)(\text{area}_{BDA})$ and acts through the centroid of *BDA*. In addition, there are the pressure distributions caused by the liquid acting along the vertical and horizontal sides of the block. Along the vertical side *AD*, the force $\mathbf{F}_{AD}$ has a magnitude that equals the area under the trapezoid and acts through the centroid C_{AD} of this area. The distributed loading along the horizontal side *AB* is constant, since all points lying in this plane are at the same depth from the surface of the liquid. The magnitude of $\mathbf{F}_{AB}$ is simply the area of the rectangle. This force acts through the area's centroid C_{AB} or the midpoint of *AB*. Summing the three forces in Fig. 9–26 yields $\mathbf{F}_R = \Sigma\mathbf{F} = \mathbf{F}_{AD} + \mathbf{F}_{AB} + \mathbf{W}_f$, which is shown in Fig. 9–25. Finally, the location of the center of pressure *P* on the plate is determined by applying the equation $M_{R_O} = \Sigma M_O$, which states that the moment of the resultant force about a convenient reference point, Fig. 9–25, is equal to the sum of the moments of the three forces in Fig. 9–26 about the same point.

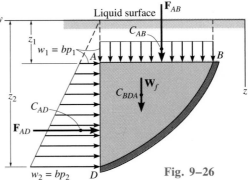

Fig. 9–26

Flat Plate of Variable Width. The pressure distribution acting on the surface of a submerged plate having a variable width is shown in Fig. 9–27. The resultant force of this loading equals the volume described by the plate area as its base and linear varying pressure distribution as its altitude. The shaded element shown in Fig. 9–27 may be used if integration is chosen to determine this volume. The element consists of a rectangular strip of area $dA = x\,dy'$ located at a depth z below the liquid surface. Since a uniform pressure $p = \gamma z$ (force/area) acts on dA, the magnitude of the differential force $d\mathbf{F}$ is equal to $dF = dV = p\,dA = \gamma z(x\,dy')$. Integrating over the entire volume yields Eq. 9–13; i.e.,

$$F_R = \int_V dV = V$$

From Eq. 9–14, the centroid of V defines the point through which $\mathbf{F}_R$ acts. The center of pressure, which lies on the surface of the plate just below C, has coordinates $P(\bar{x}, \bar{y}')$ defined by the equations

$$\bar{x} = \frac{\displaystyle\int_V \widetilde{x}\,dV}{\displaystyle\int_V dV} \qquad \bar{y}' = \frac{\displaystyle\int_V \widetilde{y}'\,dV}{\displaystyle\int_V dV}$$

This point should *not* be mistaken for the centroid of the plate's *area*.

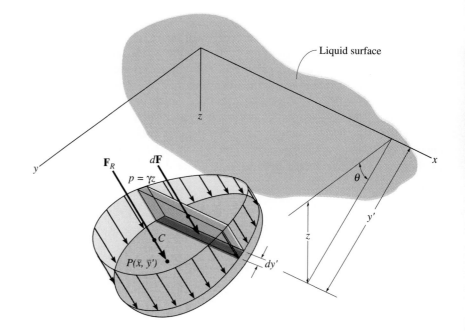

Fig. 9–27

Example 9–12

Determine the magnitude and location of the resultant hydrostatic force acting on the submerged rectangular plate AB shown in Fig. 9–28a. The plate has a width of 1.5 m; $\rho_w = 1000$ kg/m³.

SOLUTION

Since the plate has a constant width, the distributed loading can be viewed in two dimensions as shown in Fig. 9–28b. The intensity of the load at A and B is

$$w_A = b\rho_w g z_A = (1.5 \text{ m})(1000 \text{ kg/m}^3)(9.81 \text{ m/s}^2)(2 \text{ m})$$
$$= 29.4 \text{ kN/m}$$
$$w_B = b\rho_w g z_B = (1.5 \text{ m})(1000 \text{ kg/m}^3)(9.81 \text{ m/s}^2)(5 \text{ m})$$
$$= 73.6 \text{ kN/m}$$

From the table on the inside back cover, the magnitude of the resultant force $\mathbf{F}_R$ created by the distributed load is

$$F_R = \text{area of trapezoid}$$
$$= \tfrac{1}{2}(3)(29.4 + 73.6) = 154.5 \text{ kN} \qquad \textit{Ans.}$$

This force acts through the centroid of the area,

$$h = \frac{1}{3}\left(\frac{2(29.4) + 73.6}{29.4 + 73.6}\right)(3) = 1.29 \text{ m} \qquad \textit{Ans.}$$

measured upward from B, Fig. 9–28b.

The same results can be obtained by considering two components of $\mathbf{F}_R$ defined by the triangle and rectangle shown in Fig. 9–28c. Each force acts through its associated centroid and has a magnitude of

$$F_{Re} = (29.4 \text{ kN/m})(3 \text{ m}) = 88.2 \text{ kN}$$
$$F_t = \tfrac{1}{2}(44.2 \text{ kN/m})(3 \text{ m}) = 66.3 \text{ kN}$$

Hence,

$$F_R = F_{Re} + F_t = 88.2 + 66.3 = 154.5 \text{ kN} \qquad \textit{Ans.}$$

The location of $\mathbf{F}_R$ is determined by summing moments about B, Fig. 9–28b and c, i.e.,

$$\zeta + (M_R)_B = \Sigma M_B; \quad (154.5)h = 88.2(1.5) + 66.3(1)$$
$$h = 1.29 \text{ m} \qquad \textit{Ans.}$$

Fig. 9–28

Example 9–13

Determine the magnitude of the resultant hydrostatic force acting on the surface of a seawall shaped in the form of a parabola as shown in Fig. 9–29a. The wall is 5 m long; $\rho_w = 1020$ kg/m³.

(a)

(b)

Fig. 9–29

SOLUTION

The horizontal and vertical components of the resultant force will be calculated, Fig. 9–29b. Since

$$w_B = b\rho_w g z_B = 5 \text{ m}(1020 \text{ kg/m}^3)(9.81 \text{ m/s}^2)(3 \text{ m}) = 150.0 \text{ kN/m}$$

then

$$F_x = \tfrac{1}{2}(3 \text{ m})(150.0 \text{ kN/m}) = 225.0 \text{ kN}$$

The area of the parabolic sector ABC can be determined using the table on the inside back cover. Hence, the weight of water within this region is

$$F_y = (\rho_w g b)(\text{area}_{ABC})$$
$$= (1020 \text{ kg/m}^3)(9.81 \text{ m/s}^2)(5 \text{ m})[\tfrac{1}{3}(1 \text{ m})(3 \text{ m})] = 50.0 \text{ kN}$$

The resultant force is therefore

$$F_R = \sqrt{F_x^2 + F_y^2} = \sqrt{(225.0)^2 + (50.0)^2}$$
$$= 230 \text{ kN} \qquad\qquad Ans.$$

Example 9–14

Determine the magnitude and location of the resultant force acting on the triangular end plates of the water trough shown in Fig. 9–30a; $\rho_w = 1000 \text{ kg/m}^3$.

(a)

SOLUTION

The pressure distribution acting on the end plate E is shown in Fig. 9–30b. The magnitude of the resultant force $\mathbf{F}$ is equal to the volume of this loading distribution. We will solve the problem by integration. Choosing the differential volume element shown in the figure, we have

$$dF = dV = p \, dA = \rho_w gz(2x \, dz) = 19\,620zx \, dz$$

The equation of line AB is

$$x = 0.5(1 - z)$$

Hence, substituting and integrating with respect to z from $z = 0$ to $z = 1$ m yields

$$F = V = \int_V dV = \int_0^1 (19\,620)z[0.5(1 - z)] \, dz$$

$$= 9810 \int_0^1 (z - z^2) \, dz = 1635 \text{ N} = 1.64 \text{ kN} \qquad Ans.$$

This resultant passes through the *centroid of the volume*. Because of symmetry,

$$\bar{x} = 0 \qquad Ans.$$

Since $\tilde{z} = z$ for the volume element in Fig. 9–30b, then

$$\bar{z} = \frac{\displaystyle\int_V \tilde{z} \, dV}{\displaystyle\int_V dV} = \frac{\displaystyle\int_0^1 z(19\,620)z[0.5(1 - z)] \, dz}{1635} = \frac{9810 \displaystyle\int_0^1 (z^2 - z^3) \, dz}{1635}$$

$$= 0.5 \text{ m} \qquad Ans.$$

(b)

Fig. 9–30

PROBLEMS

9–90. Determine the magnitude of the resultant hydrostatic force acting on the dam and its location measured from the top surface of the water. The width of the dam is 8 m; $\rho_w = 1.0$ Mg/m^3.

*9–92. The symmetric concrete "gravity" dam is held in place by its own weight. If the density of concrete is $\rho_c = 2.5$ Mg/m^3, and water has a density of $\rho_w = 1.0$ Mg/m^3, determine the smallest width d at its base that will prevent the dam from overturning about its end A. The dam has a width of 8 m.

Prob. 9–90

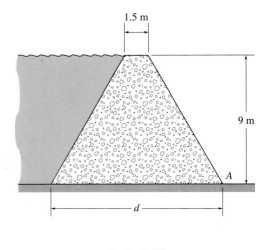

Prob. 9–92

9–91. The tank is filled with water to a depth of $d = 4$ m. Determine the resultant force the water exerts on side A and side B of the tank. If oil instead of water is placed in the tank, to what depth d should it reach so that it creates the same resultant forces? $\rho_o = 900$ kg/m^3 and $\rho_w = 1000$ kg/m^3.

9–93. The gate AB is 8 m wide. Determine the horizontal and vertical components of force acting on the pin at B and the vertical reaction at the smooth support A. $\rho_w = 1.0$ Mg/m^3.

Prob. 9–91

Prob. 9–93

9–94. Determine the resultant horizontal and vertical force components that the water exerts on the side of the dam. The dam is 25 ft long and $\gamma_w = 62.4$ lb/ft^3.

Prob. 9–94

9–95. Determine the resultant hydrostatic force acting per foot of length on the sea wall; $\gamma_w = 62.4$ lb/ft^3.

Prob. 9–95

***9–96.** The tank is used to store a liquid having a specific weight of $\gamma = 80$ lb/ft^3. If it is filled to the top, determine the magnitude of force the liquid exerts on side *ABCD*.

9–97. The tank is used to store a liquid having a specific weight of $\gamma = 80$ lb/ft^3. If it is filled to the top, determine the magnitude of force the liquid exerts on side *ABEF*.

Probs. 9–96/9–97

9–98. The tank is used to store a liquid having a density of 80 lb/ft^3. If it is filled to the top, determine the magnitude of force the liquid exerts on each of its two sides *ABDC* and *BDFE*.

Prob. 9–98

9–99. The rectangular bin is filled with coal, which creates a pressure distribution along wall A that varies as shown, i.e., $p = 4z^3$ lb/ft^2, where z is measured in feet. Determine the resultant force created by the coal, and specify its location measured from the top surface of the coal.

Prob. 9–99

*9–100. The semicircular drainage pipe is filled with water. Determine the resultant horizontal and vertical force components that the water exerts on the side AB of the pipe per foot of pipe length; $\gamma_w = 62.4$ lb/ft^3.

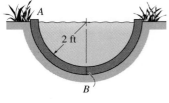

Prob. 9–100

9–101. Determine the magnitude of the resultant hydrostatic force acting per foot of length on the sea wall; $\gamma_w = 62.4$ lb/ft^3.

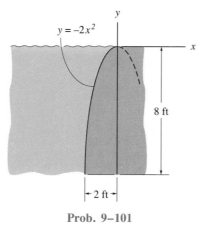

Prob. 9–101

9–102. Determine the magnitude and location of the resultant force of the parabolic loading acting on the plate.

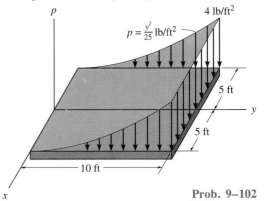

Prob. 9–102

9–103. The loading acting on a square plate is represented by a parabolic pressure distribution. Determine the magnitude and location of the resultant force. Also, what are the reactions at the rollers B and C and the ball-and-socket joint A? Neglect the weight of the plate.

Prob. 9–103

*9–104. The pressure loading on the plate is described by the function $p = [-240/(x + 1) + 340]$ Pa. Determine the magnitude of the resultant force and coordinates of the point where the line of action of the force intersects the plate.

Prob. 9–104

9–105. The support exerts the distributed loading on the pipe. If the pipe weighs 50 lb and has a length of 3 ft, determine the magnitude of p_o for equilibrium. *Hint:* The force acting on the differential segment $r\,d\theta$ of the pipe is $dF = p(3\text{ ft})r\,d\theta$. This force has horizontal and vertical components of $dF\sin\theta$ and $dF\cos\theta$, respectively.

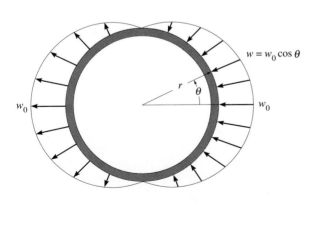

Prob. 9–105

9–106. Determine the resultant force of the distributed loading acting per unit length of the pipe. *Hint:* The force acting on the differential segment $r\,d\theta$ of the pipe is $dF = wr d\theta$. This force has horizontal and vertical components of $dF\cos\theta$ and $dF\sin\theta$, respectively.

Prob. 9–106

9–107. The curved rod is subjected to the uniform distributed load. Determine the components of reaction at the fixed support for equilibrium. *Hint:* See Prob. 9–106.

Prob. 9–107

***9–108.** The distributed load varies along the curved rod in a linear fashion such that θ is measured in radians. Determine the components of reaction at the fixed support necessary for equilibrium. *Hint:* See Prob. 9–106.

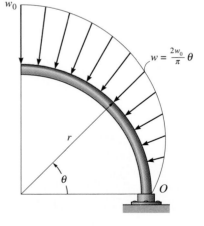

Prob. 9–108

9–109. Replace the distributed loading acting on the curved member by the horizontal and vertical components of an equivalent resultant force and a couple moment acting at point O. *Hint:* See Prob. 9–106.

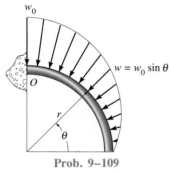

Prob. 9–109

9–110. Replace the distributed loading acting on the curved member by an equivalent resultant force and couple moment acting at point O. *Hint:* See Prob. 9–106.

Prob. 9–110

9–111. The rectangular plate is subjected to a distributed load over its *entire surface*. If the load is defined by the expression $p = p_o \sin (\pi x/a) \sin (\pi y/b)$, where p_o represents the pressure acting at the center of the plate, determine the magnitude and location of the resultant force acting on the plate.

Prob. 9–111

*9–112. The load over the plate varies linearly along the sides of the plate such that $p = \frac{2}{3}[x(4 - y)]$ kPa. Determine the resultant force and its position $(\bar{x}, \bar{y})$ on the plate.

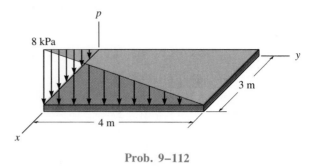

Prob. 9–112

9–113. The water trough is made from a piece of metal such that its end plates are parabolic. If the trough is filled with water having a specific weight of 62.4 lb/ft³, determine the resultant force on the ends if the level of the water is (*a*) 0.5 ft and (*b*) 1 ft.

Prob. 9–113

9–114. The tank is filled with a liquid which has a density of 900 kg/m³. Determine the resultant force that it exerts on the elliptical end plate of the tank, and the location of the center of pressure, measured from the x axis.

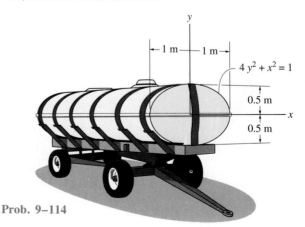

Prob. 9–114

REVIEW PROBLEMS

9–115. Determine the distance $\bar{y}$ to the centroidal axis $\bar{x}-\bar{x}$ of the beam's cross-sectional area. Neglect the size of the corner welds at A and B for the calculation.

Prob. 9–115

9–118. Determine the distance $\bar{z}$ to the center of gravity for the frustum of the paraboloid. The material is homogeneous.

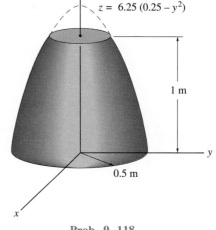

$z = 6.25 (0.25 - y^2)$

Prob. 9–118

*9–116. Determine the distance $\bar{y}$ to the centroid of the plate area.

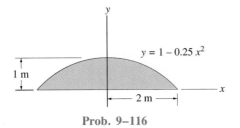

$y = 1 - 0.25\, x^2$

Prob. 9–116

9–119. Determine the location $(\bar{x}, \bar{y})$ of the center of mass for the system of four particles lying in the x–y plane.

9–117. Determine the location $(\bar{x}, \bar{y})$ of the centroid of the parabolic area.

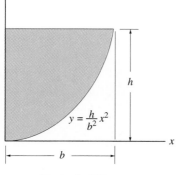

$y = \dfrac{h}{b^2} x^2$

Prob. 9–117

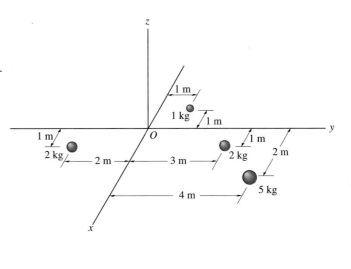

Prob. 9–119

*9–120. The form is used to cast concrete columns. Determine the resultant force that wet concrete exerts along the plate A, $0.5 \text{ m} \le z \le 3 \text{ m}$, if the pressure due to the concrete varies as shown. Specify the location of the resultant force, measured from the top of the column.

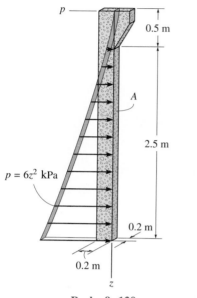

$p = 6z^2 \text{ kPa}$

0.5 m

A

2.5 m

0.2 m

0.2 m

Prob. 9–120

9–121. Determine the location $(\bar{x}, \bar{y}, \bar{z})$ of the center of gravity of the homogeneous wire.

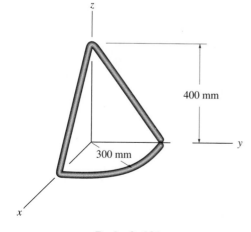

400 mm

300 mm

Prob. 9–121

9–122. Determine the location $(\bar{x}, \bar{y})$ of the centroid of the area.

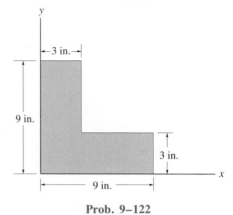

3 in.

9 in.

3 in.

9 in.

Prob. 9–122

9–123. Determine the location $(\bar{x}, \bar{y})$ of the centroid of the homogeneous plate.

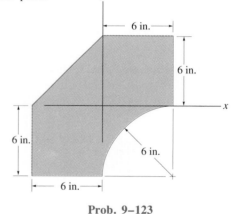

6 in.

6 in.

6 in.

6 in.

6 in.

Prob. 9–123

*9–124. Determine the volume of steel needed to produce the tapered part. The cross section is shown, although the part is 360° around. Also, compute the outside surface area of the part, excluding the ends.

100 mm

100 mm

10 mm

30 mm

25 mm

Prob. 9–124

10 Moments of Inertia

In this chapter we will develop a method for determining the moment of inertia for both an area and a body having a specified mass. The moment of inertia for an area is an important property in engineering, since it must be determined or specified if one is to analyze or design a structural member or a mechanical part. On the other hand, a body's mass moment of inertia must be known if one is studying the motion of the body.

10.1 Definition of Moments of Inertia for Areas

In the last chapter we determined the centroid for an area by considering the first moment of the area about an axis; that is, for the computation we had to evaluate an integral of the form $\int x \, dA$. Integrals of the second moment of an area, such as $\int x^2 \, dA$, are referred to as the *moment of inertia* for the area. The terminology "moment of inertia" as used here is actually a misnomer; however, it has been adopted because of the similarity with integrals of the same form related to mass.

The moment of inertia of an area originates whenever one has to compute the moment of a distributed load that varies linearly from the moment axis. A typical example of this kind of loading occurs due to the pressure of a liquid acting on the surface of a submerged plate. It was pointed out in Sec. 9.6 that the pressure, or force per unit area, exerted at a point located a distance z below the surface of a liquid is $p = \gamma z$, Eq. 9–15, where γ is the specific weight of the liquid. Thus, the magnitude of force exerted by a liquid on the

Fig. 10–1

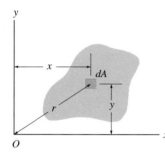

Fig. 10–2

area dA of the submerged plate shown in Fig. 10–1 is $dF = p\, dA = \gamma z\, dA$. The moment of this force about the x axis of the plate is $dM = dFz = \gamma z^2\, dA$, and therefore the moment created by the *entire* pressure distribution is $M = \gamma \int z^2\, dA$. Here the integral represents the moment of inertia of the area of the plate about the x axis. Since integrals of this form often arise in formulas used in fluid mechanics, mechanics of materials, structural mechanics, and machine design, the engineer should become familiar with the methods used for their computation.

Moment of Inertia. Consider the area A, shown in Fig. 10–2, which lies in the x–y plane. By definition, the moments of inertia of the differential planar area dA about the x and y axes are $dI_x = y^2\, dA$ and $dI_y = x^2\, dA$, respectively. For the entire area the *moments of inertia* are determined by integration; i.e.,

$$I_x = \int_A y^2\, dA$$
$$I_y = \int_A x^2\, dA$$

(10–1)

We can also formulate the second moment of the differential area dA about the pole O or z axis, Fig. 10–2. This is referred to as the polar moment of inertia, $dJ_O = r^2\, dA$. Here r is the perpendicular distance from the pole (z axis) to the element dA. For the entire area the *polar moment of inertia* is

$$J_O = \int_A r^2\, dA = I_x + I_y$$

(10–2)

The relationship between J_O and I_x, I_y is possible since $r^2 = x^2 + y^2$, Fig. 10–2.

From the above formulations it is seen that I_x, I_y, and J_O will *always* be *positive*, since they involve the product of distance squared and area. Furthermore, the units for moment of inertia involve length raised to the fourth power, e.g., m^4, mm^4, or ft^4, in^4.

10.2 Parallel-Axis Theorem for an Area

If the moment of inertia for an area is known about an axis passing through its centroid, it is convenient to determine the moment of inertia of the area about a corresponding parallel axis using the *parallel-axis theorem*. To derive this theorem, consider finding the moment of inertia of the shaded area shown in

Fig. 10–3 about the x axis. In this case, a differential element dA is located at an arbitrary distance y' from the *centroidal x'* axis, whereas the *fixed distance* between the parallel x and x' axes is defined as d_y. Since the moment of inertia of dA about the x axis is $dI_x = (y' + d_y)^2\, dA$, then for the entire area,

$$I_x = \int_A (y' + d_y)^2 \, dA$$

$$= \int_A y'^2 \, dA + 2d_y \int_A y' \, dA + d_y^2 \int_A dA$$

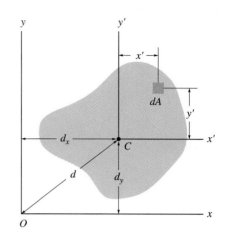

Fig. 10–3

The first integral represents the moment of inertia of the area about the centroidal axis, $\bar{I}_{x'}$. The second integral is zero since the x' axis passes through the area's centroid C; i.e., $\int y'\, dA = \bar{y} \int dA = 0$ since $\bar{y} = 0$. Realizing that the third integral represents the total area A, the final result is therefore

$$\boxed{I_x = \bar{I}_{x'} + Ad_y^2} \qquad (10\text{--}3)$$

A similar expression can be written for I_y; i.e.,

$$\boxed{I_y = \bar{I}_{y'} + Ad_x^2} \qquad (10\text{--}4)$$

And finally, for the polar moment of inertia about an axis perpendicular to the x–y plane and passing through the pole O (z axis), Fig. 10–3, we have

$$\boxed{J_O = \bar{J}_C + Ad^2} \qquad (10\text{--}5)$$

The form of each of these equations states that *the moment of inertia of an area about an axis is equal to the moment of inertia of the area about a parallel axis passing through the area's centroid plus the product of the area and the square of the perpendicular distance between the axes.*

10.3 Radius of Gyration of an Area

The *radius of gyration* of a planar area has units of length and is a quantity that is often used for the design of columns in structural mechanics. Provided the areas and moments of inertia are *known,* the radii of gyration are determined from the formulas

$$k_x = \sqrt{\frac{I_x}{A}} \qquad k_y = \sqrt{\frac{I_y}{A}} \qquad k_O = \sqrt{\frac{J_O}{A}} \qquad (10\text{--}6)$$

The form of these equations is easily remembered, since it is similar to that for finding the moment of inertia of a differential area about an axis. For example, $I_x = k_x^2 A$; whereas for a differential area, $dI_x = y^2\, dA$.

10.4 Moments of Inertia for an Area by Integration

When the boundaries for a planar area are expressed by mathematical functions, Eqs. 10–1 may be integrated to determine the moments of inertia for the area. If the element of area chosen for integration has a differential size in two directions as shown in Fig. 10–2, a double integration must be performed to evaluate the moment of inertia. Most often, however, it is easier to perform only a single integration by choosing an element having a differential size or thickness in only one direction.

PROCEDURE FOR ANALYSIS

If a single integration is performed to determine the moment of inertia of an area about an axis, it will first be necessary to specify the differential element dA. Most often this element will be rectangular, such that it will have a finite length and differential width. The element should be located so that it intersects the boundary of the area at the *arbitrary point (x, y)*. There are two possible ways to orient the element with respect to the axis about which the moment of inertia is to be determined.

Case 1.   The *length* of the element can be oriented *parallel* to the axis. This situation occurs when the rectangular element shown in Fig. 10–4 is used to determine I_y for the area. Direct application of Eq. 10–1, i.e., $I_y = \int x^2 \, dA$, can be made in this case, since the element has an infinitesimal thickness dx and therefore *all parts* of the element lie at the *same* moment-arm distance x from the y axis.*

Case 2.   The *length* of the element can be oriented *perpendicular* to the axis. Here Eq. 10–1 *does not apply*, since all parts of the element will *not* lie at the same moment-arm distance from the axis. For example, if the rectangular element in Fig. 10–4 is used for determining I_x for the area, it will first be necessary to calculate the moment of inertia of the *element* about a horizontal axis passing through the element's centroid and then determine the moment of inertia of the *element* about the x axis by using the parallel-axis theorem. Integration of this result will yield I_x.

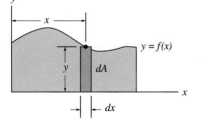

Fig. 10–4

Application of these cases is illustrated in the following examples.

*In the case of the element $dA = dx\,dy$, Fig. 10–2, the moment arms y and x are appropriate for the formulation of I_x and I_y (Eq. 10–1) since the *entire* element, because of its infinitesimal size, lies at the specified y and x perpendicular distances from the x and y axes.

Example 10–1

Determine the moment of inertia for the rectangular area shown in Fig. 10–5 with respect to (a) the centroidal x' axis, (b) the axis x_b passing through the base of the rectangle, and (c) the pole or z' axis perpendicular to the x'–y' plane and passing through the centroid C.

SOLUTION (*CASE 1*)

Part (a). The differential element shown in Fig. 10–5 is chosen for integration. Because of its location and orientation, the *entire element* is at a distance y' from the x' axis. Here it is necessary to integrate from $y' = -h/2$ to $y' = h/2$. Since $dA = b\, dy'$, then

$$\bar{I}_{x'} = \int_A y'^2\, dA = \int_{-h/2}^{h/2} y'^2(b\, dy') = b\int_{-h/2}^{h/2} y'^2\, dy'$$

$$= \frac{1}{12}\, bh^3 \qquad\qquad \textit{Ans.}$$

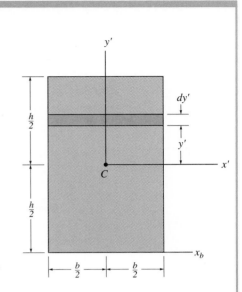

Fig. 10–5

Part (b). The moment of inertia about an axis passing through the base of the rectangle can be obtained by using the result of part (a) and applying the parallel-axis theorem, Eq. 10–3.

$$I_{x_b} = \bar{I}_{x'} + Ad_y^2$$

$$= \frac{1}{12}bh^3 + bh\left(\frac{h}{2}\right)^2 = \frac{1}{3}bh^3 \qquad\qquad \textit{Ans.}$$

Part (c). To obtain the polar moment of inertia about point C, we must first obtain $\bar{I}_{y'}$, which may be found by interchanging the dimensions b and h in the result of part (a), i.e.,

$$\bar{I}_{y'} = \frac{1}{12}hb^3$$

Using Eq. 10–2, the polar moment of inertia about C is therefore

$$\bar{J}_C = \bar{I}_{x'} + \bar{I}_{y'} = \frac{1}{12}bh(h^2 + b^2) \qquad\qquad \textit{Ans.}$$

Example 10–2

Determine the moment of inertia of the shaded area shown in Fig. 10–6a about the x axis.

(a)

(b)

Fig. 10–6

SOLUTION I (CASE 1)

A differential element of area that is *parallel* to the x axis, as shown in Fig. 10–6a, is chosen for integration. Since the element has a thickness dy and intersects the curve at the *arbitrary point* (x, y), the area is $dA = (100 - x)\, dy$. Furthermore, all parts of the element lie at the same distance y from the x axis. Hence, integrating with respect to y, from $y = 0$ to $y = 200$ mm, yields

$$I_x = \int_A y^2\, dA = \int_A y^2(100 - x)\, dy$$

$$= \int_0^{200} y^2\left(100 - \frac{y^2}{400}\right) dy = 100 \int_0^{200} y^2\, dy - \frac{1}{400}\int_0^{200} y^4\, dy$$

$$= 107(10^6)\ \text{mm}^4$$

SOLUTION II (CASE 2)

A differential element *parallel* to the y axis, as shown in Fig. 10–6b, is chosen for integration. It intersects the curve at the *arbitrary point* (x, y). In this case, all parts of the element do *not* lie at the same distance from the x axis, and therefore the parallel-axis theorem must be used to determine the *moment of inertia of the element* with respect to this axis. For a rectangle having a base b and height h, the moment of inertia about its centroidal axis has been determined in part (a) of Example 10–1. There it was found that $\bar{I}_{x'} = \frac{1}{12}bh^3$. For the differential element shown in Fig. 10–6b, $b = dx$ and $h = y$, and thus $d\bar{I}_{x'} = \frac{1}{12}dx\, y^3$. Since the centroid of the element is at $\widetilde{y} = y/2$ from the x axis, the moment of inertia of the element about this axis is

$$dI_x = d\bar{I}_{x'} + dA\,\widetilde{y}^2 = \frac{1}{12}dx\, y^3 + y\, dx\left(\frac{y}{2}\right)^2 = \frac{1}{3}y^3\, dx$$

[This result can also be concluded from part (b) of Example 10–1]. Integrating with respect to x, from $x = 0$ to $x = 100$ mm, yields

$$I_x = \int dI_x = \int_A \frac{1}{3}y^3\, dx = \int_0^{100} \frac{1}{3}(400x)^{3/2}\, dx$$

$$= 107(10^6)\ \text{mm}^4 \qquad\qquad Ans.$$

Example 10-3

Determine the moment of inertia with respect to the x axis of the circular area shown in Fig. 10–7a.

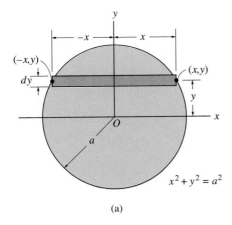

(a)

SOLUTION I *(CASE 1)*

Using the differential element shown in Fig. 10–7a, since $dA = 2x\,dy$, we have

$$I_x = \int_A y^2\,dA = \int_A y^2(2x)\,dy$$

$$= \int_{-a}^{a} y^2(2\sqrt{a^2 - y^2})\,dy = \frac{\pi a^4}{4} \qquad \text{Ans.}$$

SOLUTION II *(CASE 2)*

When the differential element is chosen as shown in Fig. 10–7b, the centroid for the element happens to lie on the x axis, and so, applying Eq. 10–3, noting that $dA = 2y\,dx$, we have

$$dI_x = \frac{1}{12}\,dx\,(2y)^3$$

$$= \frac{2}{3}y^3\,dx$$

Integrating with respect to x yields

$$I_x = \int_{-a}^{a} \frac{2}{3}(a^2 - x^2)^{3/2}\,dx = \frac{\pi a^4}{4} \qquad \text{Ans.}$$

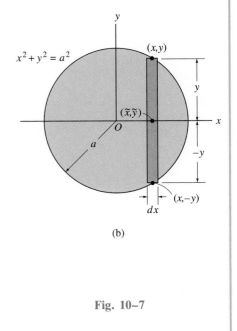

(b)

Fig. 10–7

Example 10–4

(a)

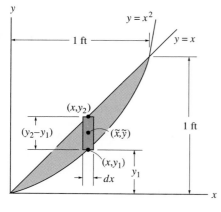

(b)

Fig. 10–8

Determine the moment of inertia of the shaded area shown in Fig. 10–8a about the x axis.

SOLUTION I (*CASE* 1)

The differential element of area parallel to the x axis is chosen for integration, Fig. 10–8a. The element intersects the curve at the *arbitrary points* (x_2, y) and (x_1, y). Consequently, its area is $dA = (x_1 - x_2)\, dy$. Since all parts of the element lie at the same distance y from the x axis, we have

$$I_x = \int_A y^2\, dA = \int_0^1 y^2(x_1 - x_2)\, dy = \int_0^1 y^2(\sqrt{y} - y)\, dy$$

$$I_x = \frac{2}{7}y^{7/2} - \frac{1}{4}y^4 \Big|_0^1 = 0.0357 \text{ ft}^4 \qquad\qquad \textit{Ans.}$$

SOLUTION II (*CASE* 2)

The differential element of area parallel to the y axis is shown in Fig. 10–8b. It intersects the curves at the *arbitrary points* (x, y_2) and (x, y_1). Since all parts of its entirety do *not* lie at the same distance from the x axis, we must first use the parallel-axis theorem to find the element's moment of inertia about the x axis, then integrate this result to determine I_x. Thus,

$$dI_x = d\bar{I}_{x'} + dA\,\widetilde{y}^2 = \frac{1}{12}\, dx\, (y_2 - y_1)^3 +$$

$$(y_2 - y_1)\, dx\left(y_1 + \frac{y_2 - y_1}{2}\right)^2$$

$$= \frac{1}{3}(y_2^3 - y_1^3)\, dx = \frac{1}{3}(x^3 - x^6)\, dx$$

$$I_x = \frac{1}{3}\int_0^1 (x^3 - x^6)\, dx = \frac{1}{12}x^4 - \frac{1}{21}x^7 \Big|_0^1 = 0.0357 \text{ ft}^4 \quad \textit{Ans.}$$

By comparison, Solution I requires much less computation. If an integral using a particular element appears difficult to evaluate, try solving the problem using an element oriented in the other direction.

PROBLEMS

10–1. The irregular area has a moment of inertia about the A–A axis of 500 in^4. If the total area is 12 in^2, determine the moment of inertia of the area about the B–B axis. The C–C axis passes through the centroid of the area.

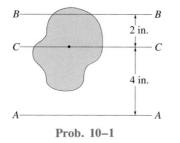

Prob. 10–1

10–2. Determine the moment of inertia of the triangular area about the x axis.

10–3. Determine the moment of inertia of the triangular area about the y axis.

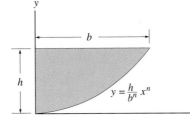

Probs. 10–2/10–3

***10–4.** Determine the moments of inertia I_x and I_y of the shaded area.

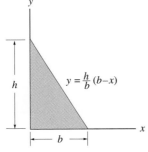

Prob. 10–4

10–5. Determine the moment of inertia about the x axis of the area which is bounded by the lines $x = 0$, $y = 1$ in., and the parabola $y = 3x^2$.

10–6. Determine the moments of inertia about the x and y axes of the area bounded by the curve $y^2 = 2x$ and the lines $x = 2$ in. and $y = 0$.

10–7. Determine the polar moments of inertia J_O for the cross-sectional area of a solid shaft having a radius of 20 mm and a tube having an outer radius of 20 mm and inner radius of 15 mm. What percentage of J_O is contributed by the tube to that of the solid shaft?

***10–8.** Determine the moment of inertia of the area about the x axis.

Prob. 10–8

10–9. Determine the moment of inertia of the shaded area about the x axis.

10–10. Determine the moment of inertia of the shaded area about the y axis.

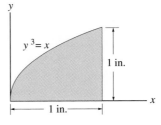

Probs. 10–9/10–10

10–11. Determine the moment of inertia of the shaded area about the x axis.

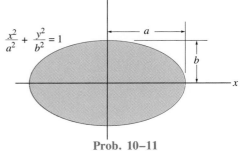

Prob. 10–11

*10–12. Determine the moments of inertia of the shaded area about the x and y axes.

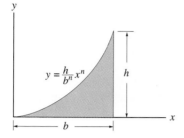

Prob. 10–12

10–13. Determine the moment of inertia of the quarter-circular area about the x axis.

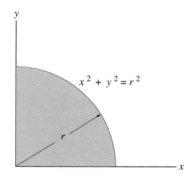

Prob. 10–13

10–14. Determine the radius of gyration of the shaded area about the y axis.

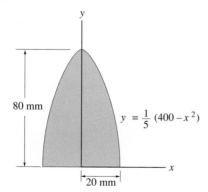

Prob. 10–14

10–15. Determine the moment of inertia of the shaded area about the x axis.

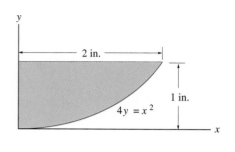

Prob. 10–15

*10–16. Determine the moments of inertia of the shaded area about the x and y axes.

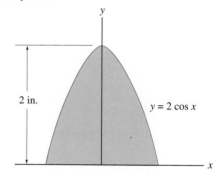

Prob. 10–16

10–17. Determine the moment of inertia of the shaded area about the y axis.

10–18. Determine the moment of inertia of the shaded area about the x axis.

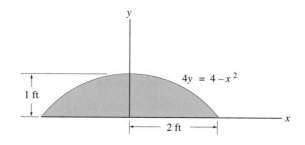

Probs. 10–17/10–18

10–19. Determine the moment of inertia of the shaded area about the x axis.

*10–20. Determine the moment of inertia of the shaded area about the y axis.

■10–21. Determine the moment of inertia of the shaded area about the y axis. Use Simpson's rule to evaluate the integral.

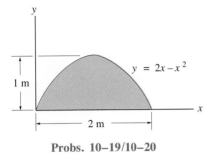

$y = 2x - x^2$

1 m

2 m

Probs. 10–19/10–20

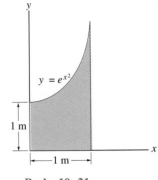

$y = e^{x^2}$

1 m

1 m

Prob. 10–21

10.5 Moments of Inertia for Composite Areas

A composite area consists of a series of connected "simpler" parts or shapes, such as semicircles, rectangles, and triangles. Provided the moment of inertia of each of these parts is known or can be determined about a common axis, then the moment of inertia of the composite area equals the *algebraic sum* of the moments of inertia of all its parts.

PROCEDURE FOR ANALYSIS

The following procedure provides a method for determining the moment of inertia of a composite area about a reference axis.

Composite Parts. Using a sketch, divide the area into its composite parts and indicate the perpendicular distance from the *centroid* of each part to the reference axis.

Parallel-Axis Theorem. The moment of inertia of each part should be determined about its centroidal axis, which is parallel to the reference axis. For the calculation use the table given on the inside back cover. If the centroidal axis does not coincide with the reference axis, the parallel-axis theorem, $I = \bar{I} + Ad^2$, should be used to determine the moment of inertia of the part about the reference axis.

Summation. The moment of inertia of the entire area about the reference axis is determined by summing the results of its composite parts. In particular, if a composite part has a "hole," its moment of inertia is found by "subtracting" the moment of inertia for the hole from the moment of inertia of the entire part including the hole.

Example 10–5

Compute the moment of inertia of the composite area shown in Fig. 10–9a about the x axis.

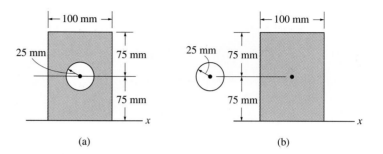

Fig. 10–9

SOLUTION

Composite Parts.   The composite area is obtained by *subtracting* the circle from the rectangle as shown in Fig. 10–9b. The centroid of each area is located in the figure.

Parallel-Axis Theorem.   The moments of inertia about the x axis are determined using the parallel-axis theorem and the data in the table on the inside back cover.

Circle

$$I_x = \bar{I}_{x'} + Ad_y^2$$

$$= \frac{1}{4}\pi(25)^4 + \pi(25)^2(75)^2 = 11.4(10^6) \text{ mm}^4$$

Rectangle

$$I_x = \bar{I}_{x'} + Ad_y^2$$

$$= \frac{1}{12}(100)(150)^3 + (100)(150)(75)^2 = 112.5(10^6) \text{ mm}^4$$

Summation.   The moment of inertia for the composite area is thus

$$I_x = -11.4(10^6) + 112.5(10^6)$$

$$= 101(10^6) \text{ mm}^4 \qquad \qquad Ans.$$

Example 10–6

Determine the moments of inertia of the beam's cross-sectional area shown in Fig. 10–10a about the x and y centroidal axes.

SOLUTION

Composite Parts.   The cross section can be considered as three composite rectangular areas A, B, and D shown in Fig. 10–10b. For the calculation, the centroid of each of these rectangles is located in the figure.

Parallel-Axis Theorem.   From the table on the inside back cover, or Example 10–1, the moment of inertia of a rectangle about its centroidal axis is $\bar{I} = \frac{1}{12}bh^3$. Hence, using the parallel-axis theorem for rectangles A and D, the calculations are as follows:

Rectangle A

$$I_x = \bar{I}_{x'} + Ad_y^2 = \frac{1}{12}(100)(300)^3 + (100)(300)(200)^2$$

$$= 1.425(10^9) \text{ mm}^4$$

$$I_y = \bar{I}_{y'} + Ad_x^2 = \frac{1}{12}(300)(100)^3 + (100)(300)(250)^2$$

$$= 1.90(10^9) \text{ mm}^4$$

Rectangle B

$$I_x = \frac{1}{12}(600)(100)^3 = 0.05(10^9) \text{ mm}^4$$

$$I_y = \frac{1}{12}(100)(600)^3 = 1.80(10^9) \text{ mm}^4$$

Rectangle D

$$I_x = \bar{I}_{x'} + Ad_y^2 = \frac{1}{12}(100)(300)^3 + (100)(300)(200)^2$$

$$= 1.425(10^9) \text{ mm}^4$$

$$I_y = \bar{I}_{y'} + Ad_x^2 = \frac{1}{12}(300)(100)^3 + (100)(300)(250)^2$$

$$= 1.90(10^9) \text{ mm}^4$$

Summation.   The moments of inertia for the entire cross section are thus

$$I_x = 1.425(10^9) + 0.05(10^9) + 1.425(10^9)$$
$$= 2.90(10^9) \text{ mm}^4 \qquad \qquad \textit{Ans.}$$
$$I_y = 1.90(10^9) + 1.80(10^9) + 1.90(10^9)$$
$$= 5.60(10^9) \text{ mm}^4 \qquad \qquad \textit{Ans.}$$

(a)

(b)

Fig. 10–10

PROBLEMS

10–22. The area has a moment of inertia about the $y-y$ axis of $25(10^6)$ mm^4. If the area is $15(10^3)$ mm^2, determine the moment of inertia of the area about the $y'-y'$ axis. The $\bar{y}-\bar{y}$ axis passes through the centroid C of the area.

Prob. 10–22

10–23. The beam is constructed from the two channels and two cover plates. If each channel has a cross-sectional area of $A_c = 11.8$ in^2 and a moment of inertia about a vertical axis passing through its own centroid, C_c, of $(\bar{I}_{\bar{y}})_{C_c} = 349$ in^4, determine the moment of inertia of the beam about the y axis.

Prob. 10–23

* **10–24.** Determine the distance $\bar{y}$ to the centroid of the beam constructed from the two channels and the cover plate. If each channel has a cross-sectional area of $A_c = 11.8$ in^2 and a moment of inertia about a horizontal axis passing through its own centroid, C_C, of $(\bar{I}_{\bar{x}})_{C_C} = 349$ in^4, determine the moment of inertia of the beam about the $x'-x'$ axis.

Prob. 10–24

10–25. Determine the radius of gyration k_x of the column's cross-sectional area.

10–26. Determine the radius of gyration k_y of the column's cross-sectional area.

Probs. 10–25/10–26

10–27. Determine the moment of inertia of the beam's cross-sectional area about the x axis.

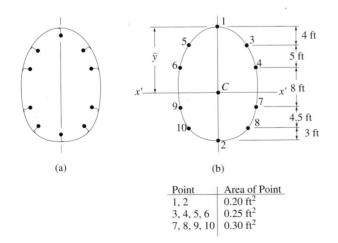

Prob. 10–27

*10–28. The fuselage of an airplane is made up of stringers covered with sheet metal as shown in *(a)*. If the combined areas of the angles and the sheet metal are tabulated and assumed to be concentrated at points 1 through 10 as shown in *(b)*, determine the distance $\bar{y}$ to the centroid and the moment of inertia about a horizontal $x'x'$ axis passing through the centroid C of the total cross section. *Hint:* Using the area elements, note that $\bar{y} = \Sigma \tilde{y} A / \Sigma A$ and $\bar{I}_{x'x'} = \Sigma \tilde{y}^2 A$.

Point	Area of Point
1, 2	0.20 ft^2
3, 4, 5, 6	0.25 ft^2
7, 8, 9, 10	0.30 ft^2

Prob. 10–28

10–29. Determine the moments of inertia of the shaded area with respect to the x and y centroidal axes.

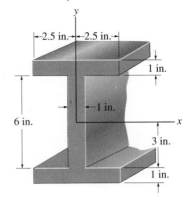

Prob. 10–29

10–30. The composite beam consists of a wide-flange beam and cover plates welded together as shown. Determine the moment of inertia of the cross-sectional area with respect to a horizontal axis passing through the beam's centroid.

Prob. 10–30

10–31. Determine the distance $\bar{y}$ to the centroid of the channel's cross-sectional area, and then determine the moment of inertia with respect to the $x'-x'$ axis passing through the centroid.

Prob. 10–31

*10–32. Determine the polar moment of inertia of the shaded area about the centroid, point O.

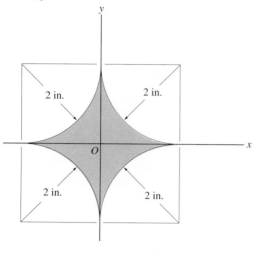

Prob. 10–32

10–33. Determine the moments of inertia of the "Z" section with respect to the x and y centroidal axes.

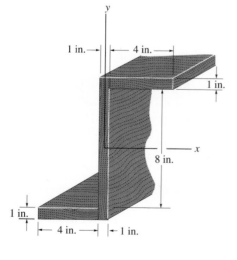

Prob. 10–33

10–34. Determine $\bar{y}$, which locates the centroid C of the *wing channel*, and then determine the moment of inertia $\bar{I}_{x'}$ about the centroidal x' axis. Neglect the effect of rounded corners. The material has a uniform thickness of 0.5 in.

Prob. 10–34

10–35. Determine the moment of inertia of the shaded area with respect to a horizontal axis passing through the centroid of the section.

Prob. 10–35

*10–36. Determine the moment of inertia of the beam's cross-sectional area about the x axis which passes through the centroid C.

10–37. Determine the moment of inertia of the beam's cross-sectional area about the y axis, which passes through the centroid C.

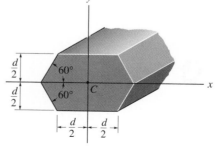

Prob. 10–36/10–37

10–38. Determine the moment of inertia of the shaded area about the *x* axis.

10–39. Determine the moment of inertia of the shaded area about the *y* axis.

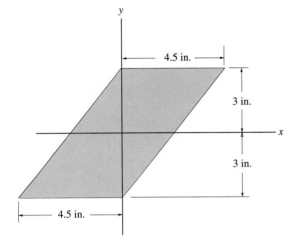

Probs. 10–38/10–39

*10–40. Determine the moments of inertia I_x and I_y of the "Z" section about the *x* and *y* axes, which pass through the centroid *C*.

Prob. 10–40

10–41. Determine the distance $\bar{y}$ to the centroid of the beam's cross-sectional area; then find the moment of inertia about the *x'* axis.

10–42. Determine the moment of inertia of the beam's cross-sectional area about the *y* axis.

Probs. 10–41/10–42

10–43. Determine the distance $\bar{y}$ to the centroid of the beam's cross-sectional area; then find the moment of inertia about the *x'* axis.

*10–44. Determine the moment of inertia of the beam's cross-sectional area about the *y* axis.

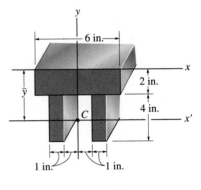

Probs. 10–43/10–44

10–45. Determine the location $(\bar{x}, \bar{y})$ of the centroid *C* of the cross-sectional area, and then compute the moments of inertia with respect to the *x'* and *y'* axes.

Prob. 10–45

10–46. Determine the moments of inertia of the triangular area about the x' and y' axes, which pass through the centroid C of the area.

10–47. Determine the moments of inertia of the parallelogram about the x' and y' axes, which pass through the centroid C of the area.

Prob. 10–46

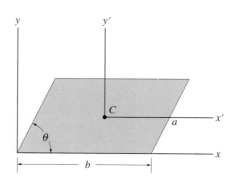

Prob. 10–47

*10.6 Product of Inertia for an Area

Fig. 10–11

In general, the moment of inertia for an area is different for every axis about which it is computed. In some applications of structural or mechanical design it is necessary to know the orientation of those axes which give, respectively, the maximum and minimum moments of inertia for the area. The method for determining this is discussed in Sec. 10.7. To use this method, however, one must first compute the product of inertia for the area as well as its moments of inertia for given x, y axes.

The product of inertia for an element of area located at point (x, y), Fig. 10–11, is defined as $dI_{xy} = xy\,dA$. Thus, for the entire area A, the *product of inertia* is

$$I_{xy} = \int_A xy\,dA \qquad (10\text{–}7)$$

If the element of area chosen has a differential size in two directions, as shown in Fig. 10–11, a double integration must be performed to evaluate I_{xy}. Most often, however, it is easier to choose an element having a differential size or thickness in only one direction, in which case the evaluation requires only a single integration (see Example 10–7).

Like the moment of inertia, the product of inertia has units of length raised to the fourth power, e.g., m⁴, mm⁴ or ft⁴, in⁴. However, since x or y may be a negative quantity, while the element of area is always positive, the product of inertia may be positive, negative, or zero, depending on the location and orientation of the coordinate axes. For example, the product of inertia I_{xy} for an area will be *zero* if either the x or y axis is an axis of *symmetry* for the area. To show this, consider the shaded area in Fig. 10–12, where for every element dA located at point (x, y) there is a corresponding element dA located at $(x, -y)$. Since the products of inertia for these elements are, respectively, $xy\ dA$ and $-xy\ dA$, the algebraic sum or integration of all the elements that are chosen in this way will cancel each other. Consequently, the product of inertia for the total area becomes zero. It also follows from the definition of I_{xy} that the "sign" of this quantity depends on the quadrant where the area is located. As shown in Fig. 10–13, if the area is rotated from one quadrant to another one, the sign of I_{xy} will change.

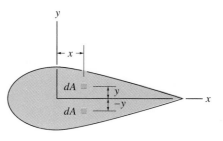

Fig. 10–12

Parallel-Axis Theorem. Consider the shaded area shown in Fig. 10–14, where x' and y' represent a set of axes passing through the *centroid* of the area, and x and y represent a corresponding set of parallel axes. Since the product of inertia of dA with respect to the x and y axes is $dI_{xy} = (x' + d_x)(y' + d_y)\ dA$, then for the entire area,

$$I_{xy} = \int_A (x' + d_x)(y' + d_y)\ dA$$

$$= \int_A x'y'\ dA + d_x \int_A y'\ dA + d_y \int_A x'\ dA + d_x d_y \int_A dA$$

The first term on the right represents the product of inertia of the area with respect to the centroidal axis, $\bar{I}_{x'y'}$. The integrals in the second and third terms are zero since the moments of the area are taken about the centroidal axis. Realizing that the fourth integral represents the total area A, the final result is therefore

$$\boxed{I_{xy} = \bar{I}_{x'y'} + Ad_x\ d_y} \qquad (10\text{--}8)$$

The similarity between this equation and the parallel-axis theorem for moments of inertia should be noted. In particular, it is important that the *algebraic signs* for d_x and d_y be maintained when applying Eq. 10–8. As illustrated in Example 10–8, the parallel-axis theorem finds important application in determining the product of inertia of a *composite area* with respect to a set of x, y axes.

Fig. 10–13

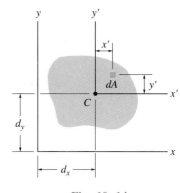

Fig. 10–14

Example 10–7

(a)

(b)

(c)

Fig. 10–15

Determine the product of inertia I_{xy} of the triangle shown in Fig. 10–15a.

SOLUTION I

A differential element that has a thickness dx, Fig. 10–15b, has an area $dA = y\ dx$. The product of inertia of the element about the x, y axes is determined using the parallel-axis theorem.

$$dI_{xy} = d\bar{I}_{x'y'} + dA\ \widetilde{x}\ \widetilde{y}$$

where $(\widetilde{x}, \widetilde{y})$ locates the *centroid* of the element or the origin of the x', y' axes. Since $dI_{x'y'} = 0$, due to symmetry, and $\widetilde{x} = x$, $\widetilde{y} = y/2$, then

$$dI_{xy} = 0 + (y\ dx)x\left(\frac{y}{2}\right) = \left(\frac{h}{b}x\ dx\right)x\left(\frac{h}{2b}x\right)$$

$$= \frac{h^2}{2b^2}x^3\ dx$$

Integrating with respect to x from $x = 0$ to $x = b$ yields

$$I_{xy} = \frac{h^2}{2b^2}\int_0^b x^3\ dx = \frac{b^2 h^2}{8} \qquad\qquad Ans.$$

SOLUTION II

The differential element that has a thickness dy, Fig. 10–15c, and area $dA = (b - x)\ dy$ can also be used. The *centroid* is located at point $\widetilde{x} = x + (b - x)/2 = (b + x)/2$, $\widetilde{y} = y$, so the product of inertia of the element becomes

$$dI_{xy} = d\bar{I}_{x'y'} + dA\ \widetilde{x}\ \widetilde{y}$$

$$= 0 + (b - x)\ dy\left(\frac{b + x}{2}\right)y$$

$$= \left(b - \frac{b}{h}y\right)dy\left[\frac{b + (b/h)\,y}{2}\right]y = \frac{1}{2}y\left(b^2 - \frac{b^2}{h^2}y^2\right)dy$$

Integrating with respect to y from $y = 0$ to $y = h$ yields

$$I_{xy} = \frac{1}{2}\int_0^h y\left(b^2 - \frac{b^2}{h^2}y^2\right)dy = \frac{b^2 h^2}{8} \qquad\qquad Ans.$$

Example 10–8

Compute the product of inertia of the beam's cross-sectional area, shown in Fig. 10–16a, about the x and y centroidal axes.

Fig. 10–16

(a)   (b)

SOLUTION

As in Example 10–6, the cross section can be considered as three composite rectangular areas A, B, and D, Fig. 10–16b. The coordinates for the centroid of each of these rectangles are shown in the figure. Due to symmetry, the product of inertia of *each rectangle* is *zero* about a set of x', y' axes that pass through the rectangle's centroid. Hence, application of the parallel-axis theorem to each of the rectangles yields

Rectangle A

$$I_{xy} = \bar{I}_{x'y'} + Ad_xd_y$$
$$= 0 + (300)(100)(-250)(200)$$
$$= -1.50(10^9) \text{ mm}^4$$

Rectangle B

$$I_{xy} = \bar{I}_{x'y'} + Ad_xd_y$$
$$= 0 + 0$$
$$= 0$$

Rectangle D

$$I_{xy} = \bar{I}_{x'y'} + Ad_xd_y$$
$$= 0 + (300)(100)(250)(-200)$$
$$= -1.50(10^9) \text{ mm}^4$$

The product of inertia for the entire cross section is therefore

$$I_{xy} = -1.50(10^9) + 0 - 1.50(10^9) = -3.00(10^9) \text{ mm}^4 \quad \textit{Ans.}$$

*10.7 Moments of Inertia for an Area About Inclined Axes

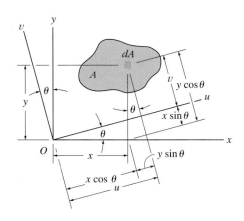

Fig. 10–17

In structural and mechanical design, it is sometimes necessary to calculate the moments and product of inertia I_u, I_v, and I_{uv} for an area with respect to a set of inclined u and v axes when the values for θ, I_x, I_y, and I_{xy} are *known*. To do this we will use *transformation equations* which relate the x, y and u, v coordinates. From Fig. 10–17, these equations are

$$u = x \cos \theta + y \sin \theta$$
$$v = y \cos \theta - x \sin \theta$$

Using these equations, the moments and product of inertia of dA about the u and v axes become

$$dI_u = v^2 \, dA = (y \cos \theta - x \sin \theta)^2 \, dA$$
$$dI_v = u^2 \, dA = (x \cos \theta + y \sin \theta)^2 \, dA$$
$$dI_{uv} = uv \, dA = (x \cos \theta + y \sin \theta)(y \cos \theta - x \sin \theta) \, dA$$

Expanding each expression and integrating, realizing that $I_x = \int y^2 \, dA$, $I_y = \int x^2 \, dA$, and $I_{xy} = \int xy \, dA$, we obtain

$$I_u = I_x \cos^2 \theta + I_y \sin^2 \theta - 2I_{xy} \sin \theta \cos \theta$$
$$I_v = I_x \sin^2 \theta + I_y \cos^2 \theta + 2I_{xy} \sin \theta \cos \theta$$
$$I_{uv} = I_x \sin \theta \cos \theta - I_y \sin \theta \cos \theta + I_{xy}(\cos^2 \theta - \sin^2 \theta)$$

These equations may be simplified by using the trigonometric identities $\sin 2\theta = 2 \sin \theta \cos \theta$ and $\cos 2\theta = \cos^2\theta - \sin^2\theta$, in which case

$$I_u = \frac{I_x + I_y}{2} + \frac{I_x - I_y}{2} \cos 2\theta - I_{xy} \sin 2\theta$$

$$I_v = \frac{I_x + I_y}{2} - \frac{I_x - I_y}{2} \cos 2\theta + I_{xy} \sin 2\theta \qquad (10\text{–}9)$$

$$I_{uv} = \frac{I_x - I_y}{2} \sin 2\theta + I_{xy} \cos 2\theta$$

If the first and second equations are added together, we can show that the polar moment of inertia about the z axis passing through point O is *independent* of the orientation of the u and v axes; i.e.,

$$J_O = I_u + I_v = I_x + I_y$$

Principal Moments of Inertia. From Eqs. 10–9, it may be seen that I_u, I_v, and I_{uv} depend on the angle of inclination, θ, of the u, v axes. We will now determine the orientation of the u, v axes about which the moments of inertia for the area, I_u and I_v, are maximum and minimum. This particular set of axes is called the *principal axes* of the area, and the corresponding moments of inertia with respect to these axes are called the *principal moments of* *inertia*. In general, there is a set of principal axes for every chosen origin O, although in structural and mechanical design, the area's centroid is an important location for O.

The angle $\theta = \theta_p$, which defines the orientation of the principal axes for the area, may be found by differentiating the first of Eqs. 10–9 with respect to θ and setting the result equal to zero. Thus,

$$\frac{dI_u}{d\theta} = -2\left(\frac{I_x - I_y}{2}\right)\sin 2\theta - 2I_{xy}\cos 2\theta = 0$$

Therefore, at $\theta = \theta_p$,

$$\tan 2\theta_p = \frac{-I_{xy}}{(I_x - I_y)/2} \qquad (10\text{–}10)$$

This equation has two roots, θ_{p_1} and θ_{p_2}, which are 90° apart and so specify the inclination of the principal axes. In order to substitute them into Eq. 10–9, we must first find the sine and cosine of $2\theta_{p_1}$ and $2\theta_{p_2}$. This can be done using the triangles shown in Fig. 10–18, which are based on Eq. 10–10.

For θ_{p_1},

$$\sin 2\theta_{p_1} = -I_{xy}\Big/\sqrt{\left(\frac{I_x - I_y}{2}\right)^2 + I_{xy}^2}$$

$$\cos 2\theta_{p_1} = \left(\frac{I_x - I_y}{2}\right)\Big/\sqrt{\left(\frac{I_x - I_y}{2}\right)^2 + I_{xy}^2}$$

For θ_{p_2},

$$\sin 2\theta_{p_2} = I_{xy}\Big/\sqrt{\left(\frac{I_x - I_y}{2}\right)^2 + I_{xy}^2}$$

$$\cos 2\theta_{p_2} = -\left(\frac{I_x - I_y}{2}\right)\Big/\sqrt{\left(\frac{I_x - I_y}{2}\right)^2 + I_{xy}^2}$$

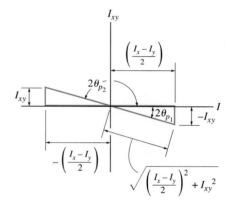

Fig. 10–18

Substituting these two sets of trigonometric relations into the first or second of Eqs. 10–9 and simplifying, we obtain

$$I_{\substack{max \\ min}} = \frac{I_x + I_y}{2} \pm \sqrt{\left(\frac{I_x - I_y}{2}\right)^2 + I_{xy}^2} \qquad (10\text{–}11)$$

Depending on the sign chosen, this result gives the maximum or minimum moment of inertia for the area. Furthermore, if the above trigonometric relations for θ_{p_1} and θ_{p_2} are substituted into the third of Eqs. 10–9, it can be shown that $I_{uv} = 0$; that is, the *product of inertia with respect to the principal axes is zero*. Since it was indicated in Sec. 10.6 that the product of inertia is zero with respect to any symmetrical axis, it therefore follows that *any symmetrical axis represents a principal axis of inertia for the area*.

■ Example 10–9

100 mm

400 mm

100 mm

C

400 mm

100 mm

600 mm

(a)

y

u

v

$\theta_{p_1} = 57.1°$

x

C

$\theta_{p_2} = -32.9°$

(b)

Fig. 10–19

Determine the principal moments of inertia for the beam's cross-sectional area shown in Fig. 10–19a with respect to an axis passing through the centroid.

SOLUTION

The moments and product of inertia of the cross section with respect to the x, y axes have been computed in Examples 10–6 and 10–8. The results are

$$I_x = 2.90(10^9) \text{ mm}^4 \qquad I_y = 5.60(10^9) \text{ mm}^4 \qquad I_{xy} = -3.00(10^9) \text{ mm}^4$$

Using Eq. 10–10, the angles of inclination of the principal axes u and v are

$$\tan 2\theta_p = \frac{-I_{xy}}{(I_x - I_y)/2} = \frac{3.00(10^9)}{[2.90(10^9) - 5.60(10^9)]/2} = -2.22$$

$$2\theta_{p_1} = -65.8° \qquad \text{and} \qquad 2\theta_{p_2} = 114.2°$$

Thus, as shown in Fig. 10–19b,

$$\theta_{p_1} = -32.9° \quad \text{and} \quad \theta_{p_2} = 57.1°$$

The principal moments of inertia with respect to the u and v axes are determined from Eq. 10–11. Hence,

$$I_{\substack{max \\ min}} = \frac{I_x + I_y}{2} \pm \sqrt{\left(\frac{I_x - I_y}{2}\right)^2 + I_{xy}^2}$$

$$= \frac{2.90(10^9) + 5.60(10^9)}{2}$$

$$\pm \sqrt{\left[\frac{2.90(10^9) - 5.60(10^9)}{2}\right]^2 + [-3.00(10^9)]^2}$$

$$I_{\substack{max \\ min}} = 4.25(10^9) \pm 3.29(10^9)$$

or

$$I_{max} = 7.54(10^9) \text{ mm}^4 \qquad I_{min} = 0.960(10^9) \text{ mm}^4 \qquad \textbf{\textit{Ans.}}$$

Specifically, the maximum moment of inertia, $I_{max} = 7.54(10^9) \text{ mm}^4$, occurs with respect to the selected u axis, since *by inspection* most of the cross-sectional area is farthest away from this axis. Or, stated in another manner, I_{max} occurs about the u axis since it is located within $\pm 45°$ of the y axis, which has the largest value of I ($I_y > I_x$). Also, this may be concluded mathematically by substituting the data with $\theta = 57.1°$ into the first of Eqs. 10–9.

*10.8 Mohr's Circle for Moments of Inertia

Equations 10–9 to 10–11 have a graphical solution that is convenient to use and generally easy to remember. Squaring the first and third of Eqs. 10–9 and adding, it is found that

$$\left(I_u - \frac{I_x + I_y}{2}\right)^2 + I_{uv}^2 = \left(\frac{I_x - I_y}{2}\right)^2 + I_{xy}^2 \qquad (10\text{--}12)$$

In a given problem, I_u and I_{uv} are *variables,* and I_x, I_y, and I_{xy} are *known constants.* Thus, Eq. 10–12 may be written in compact form as

$$(I_u - a)^2 + I_{uv}^2 = R^2$$

When this equation is plotted on a set of axes that represent the respective moment of inertia and the product of inertia, Fig. 10–20, the resulting graph represents a *circle* of radius

$$R = \sqrt{\left(\frac{I_x - I_y}{2}\right)^2 + I_{xy}^2}$$

having its center located at point $(a, 0)$, where $a = (I_x + I_y)/2$. The circle so constructed is called *Mohr's circle,* named after the German engineer Otto Mohr (1835–1918).

(a)

(b)

Fig. 10–20

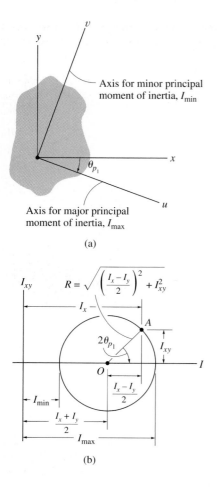

Axis for minor principal moment of inertia, I_{min}

Axis for major principal moment of inertia, I_{max}

(a)

(b)

Fig. 10–20

PROCEDURE FOR ANALYSIS

The main purpose in using Mohr's circle here is to have a convenient means for transforming I_x, I_y, and I_{xy} into the principal moments of inertia. The following procedure provides a method for doing this.

Determine I_x, I_y, I_{xy}. Establish the x, y axes for the area, with the origin located at the point P of interest, and determine I_x, I_y, and I_{xy}, Fig. 10–20a.

Construct the Circle. Construct a rectangular coordinate system such that the abscissa represents the moment of inertia I, and the ordinate represents the product of inertia I_{xy}, Fig. 10–20b. Determine the center of the circle, O, which is located at a distance $(I_x + I_y)/2$ from the origin, and plot the reference point A having coordinates (I_x, I_{xy}). By definition, I_x is always positive, whereas I_{xy} will be either positive or negative. Connect the reference point A with the center of the circle, and determine the distance OA by trigonometry. This distance represents the radius of the circle, Fig. 10–20b. Finally, draw the circle.

Principal Moments of Inertia. The points where the circle intersects the abscissa give the values of the principal moments of inertia I_{min} and I_{max}. Notice that the *product of inertia will be zero at these points*, Fig. 10–20b.

Principal Axes. To find the direction of the major principal axis, determine by trigonometry the angle $2\theta_{p_1}$, *measured from the radius OA to the positive I axis*, Fig. 10–20b. This angle represents *twice* the angle from the x axis of the area in question to the axis of maximum moment of inertia I_{max}, Fig. 10–20a. Both the angle on the circle, $2\theta_{p_1}$, and the angle to the axis on the area, θ_{p_1}, *must be measured in the same sense*, as shown in Fig. 10–20. The axis for minimum moment of inertia I_{min} is perpendicular to the axis for I_{max}.

Using trigonometry, the above procedure may be verified to be in accordance with the equations developed in Sec. 10.7.

■ Example 10–10 ■

Using Mohr's circle, determine the principal moments of inertia for the beam's cross-sectional area, shown in Fig. 10–21*a*, with respect to an axis passing through the centroid.

(a)

SOLUTION

Determine I_x, I_y, I_{xy}.   The moments of inertia and the product of inertia have been determined in Examples 10–6 and 10–8 with respect to the *x*, *y* axes shown in Fig. 10–21*a*. The results are $I_x = 2.90(10^9)$ mm^4, $I_y = 5.60(10^9)$ mm^4, and $I_{xy} = -3.00(10^9)$ mm^4.

Construct the Circle.   The I and I_{xy} axes are shown in Fig. 10–21*b*. The center of the circle, *O*, lies at a distance $(I_x + I_y)/2 = (2.90 + 5.60)/2 = 4.25$ from the origin. When the reference point $A(2.90, -3.00)$ is connected to point *O*, the radius *OA* is determined from the triangle *OBA* using the Pythagorean theorem.

$$OA = \sqrt{(1.35)^2 + (-3.00)^2} = 3.29$$

The circle is constructed in Fig. 10–21*c*.

Principal Moments of Inertia.   The circle intersects the I axis at points (7.54, 0) and (0.960, 0). Hence,

$$I_{max} = 7.54(10^9) \text{ mm}^4 \qquad\qquad Ans.$$
$$I_{min} = 0.960(10^9) \text{ mm}^4 \qquad\qquad Ans.$$

Principal Axes.   As shown in Fig. 10–21*c*, the angle $2\theta_{p_1}$ is determined from the circle by measuring counterclockwise from *OA* to the direction of the *positive I* axis. Hence,

$$2\theta_{p_1} = 180° - \sin^{-1}\left(\frac{|BA|}{|OA|}\right) = 180° - \sin^{-1}\left(\frac{3.00}{3.29}\right) = 114.2°$$

The principal axis for $I_{max} = 7.54(10^9)$ mm^4 is therefore oriented at an angle $\theta_{p_1} = 57.1°$, measured *counterclockwise*, from the *positive x* axis to the *positive u* axis. The *v* axis is perpendicular to this axis. The results are shown in Fig. 10–21*d*.

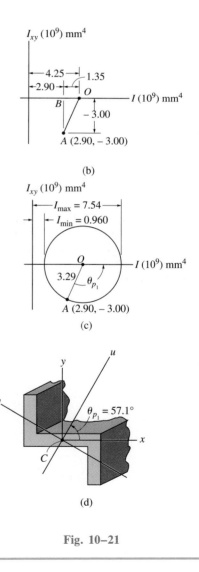

(b)

(c)

(d)

Fig. 10–21

PROBLEMS

*10–48.   Determine the product of inertia of the parabolic area with respect to the x and y axes.

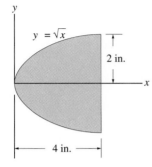

Prob. 10–48

10–49.   Determine the product of inertia of the shaded area of the ellipse with respect to the x and y axes.

Prob. 10–49

10–50.   Determine the product of inertia of the shaded area with respect to the x and y axes.

Prob. 10–50

10–51.   Determine the product of inertia of the quarter-circular area with respect to the x and y axes. Then use the parallel-axis theorem to determine the product of inertia about the centroidal x' and y' axes.

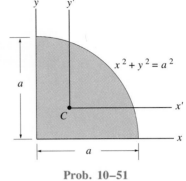

Prob. 10–51

*10–52.   Determine the product of inertia of the shaded area with respect to the x and y axes.

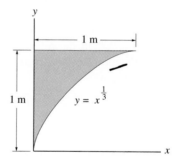

Prob. 10–52

10–53.   Determine the product of inertia of the shaded section of the ellipse with respect to the x and y axes.

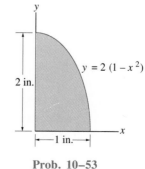

Prob. 10–53

10–54. Determine the product of inertia of the shaded area with respect to the x and y axes.

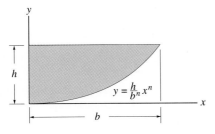

Prob. 10–54

10–55. Determine the product of inertia of the shaded area with respect to the x and y axes.

$$y = \frac{h}{b^n} x^n$$

Prob. 10–55

*10–56.** Determine the product of inertia of the shaded area with respect to the x and y axes.

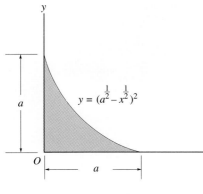

$$y = (a^{\frac{1}{2}} - x^{\frac{1}{2}})^2$$

Prob. 10–56

10–57. Determine the product of inertia of the parallelogram with respect to the x and y axes.

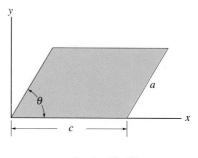

Prob. 10–57

10–58. Determine the location $(\bar{x}, \bar{y})$ of the centroid C of the cross-sectional area and then compute the product of inertia with respect to the x and y axes.

Prob. 10–58

10–59. Determine the product of inertia of the shaded area with respect to the x and y axes.

Prob. 10–59

*10–60. Determine the product of inertia of the beam's cross-sectional area with respect to the x and y axes that have their origin located at the centroid C.

Prob. 10–60

10–61. Determine the location $(\bar{x}, \bar{y})$ of the centroid C of the angle's cross-sectional area, and then compute the product of inertia with respect to the x and y axes. Assume that all corners are square.

Prob. 10–61

10–62. Determine the product of inertia for the angle with respect to the x and y axes passing through the centroid C. Assume all corners to be square.

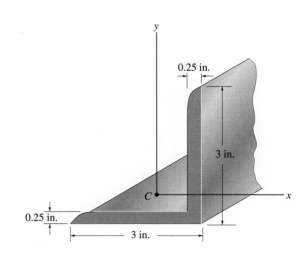

Prob. 10–62

10–63. Determine the product of inertia of the "Z" section with respect to the x and y axes that have their origin located at the centroid C.

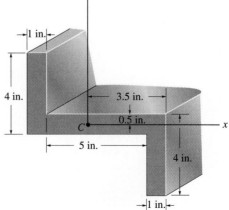

Prob. 10–63

*10–64. Determine the product of inertia of the area with respect to the *x* and *y* axes.

10–66. Determine the moments of inertia and the product of inertia for the rectangular area with respect to the *u* and *v* axes passing through the centroid *C*.

Prob. 10–64

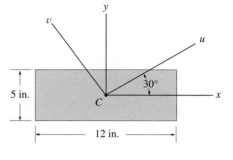

Prob. 10–66

10–65. Determine the product of inertia for the beam's cross-sectional area with respect to the *u* and *v* axes.

10–67. Determine the moments of inertia I_u and I_v and the product of inertia I_{uv} for the elliptical area. The *u* and *v* axes pass through the centroid *C*. *Hint:* See Prob. 10–11.

Prob. 10–65

Prob. 10–67

*10–68.   Determine the moments of inertia I_u, I_v, and I_{uv} of the shaded area.

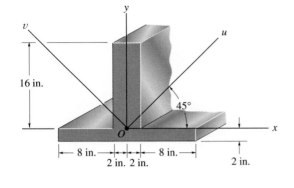

Prob. 10–68

10–69.   Determine the moments of inertia of the shaded area with respect to the u and v axes.

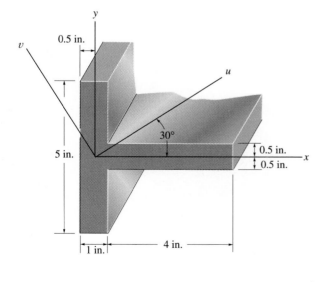

Prob. 10–69

10–70.   Determine the principal moments of inertia of the composite area with respect to a set of principal axes that have their origin located at the centroid C. Use the equations developed in Sec. 10.7. $I_{xy} = 15.84(10^6)$ mm⁴.

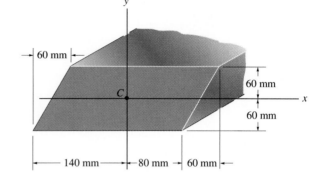

Prob. 10–70

10–71.   Determine the directions of the principal axes with origin located at point O, and the principal moments of inertia for the rectangular area about these axes.

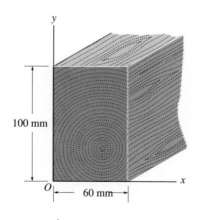

Probs. 10–71

*10–72. Determine the directions of the principal axes with origin located at point O, and the principal moments of inertia for the quarter-circular area about these axes. Use the results of Prob. 10–51.

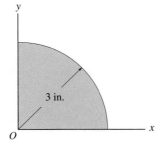

Prob. 10–72

10–73. Using Mohr's circle, determine the principal moments of inertia of the triangular area and the orientation of the principal axes of inertia having an origin at point O.

Prob. 10–73

10–74. Solve Prob. 10–70 using Mohr's circle.

10–75. Solve Prob. 10–71 using Mohr's circle.

*10–76. Solve Prob. 10–72 using Mohr's circle.

10.9 Mass Moment of Inertia

The mass moment of inertia of a body is a property that measures the resistance of the body to angular acceleration. Since it is used in dynamics to study rotational motion, methods for its calculation will now be discussed.

We define the *mass moment of inertia* as the integral of the "second moment" about an axis of all the elements of mass dm which compose the body.* For example, consider the rigid body shown in Fig. 10–22. The body's moment of inertia about the z axis is

$$I = \int_m r^2 \, dm \qquad (10\text{–}13)$$

Here the "moment arm" r is the perpendicular distance from the axis to the arbitrary element dm. Since the formulation involves r, the value of I is *unique* for each axis z about which it is computed. However, the axis which is generally chosen for analysis passes through the body's mass center G. The moment of inertia computed about this axis will be defined as I_G. Realize that because r is squared in Eq. 10–13, the mass moment of inertia is always a *positive quantity*. Common units used for its measurement are kg · m² or slug · ft².

Fig. 10–22

---

*Another property of the body which measures the symmetry of the body's mass with respect to a coordinate system is the mass product of inertia. This property most often applies to the three-dimensional motion of a body and is discussed in *Engineering Mechanics: Dynamics* (Chapter 21).

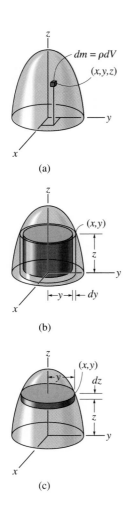

Fig. 10-23

PROCEDURE FOR ANALYSIS

For integration, we will consider only symmetric bodies having surfaces which are generated by revolving a curve about an axis. An example of such a body which is generated about the z axis is shown in Fig. 10–23.

If the body consists of material having a variable density, $\rho = \rho(x, y, z)$, the elemental mass dm of the body may be expressed in terms of its density and volume as $dm = \rho \, dV$. Substituting dm into Eq. 10–13, the body's moment of inertia is then computed using *volume elements* for integration; i.e.,

$$I = \int_V r^2 \rho \, dV \tag{10–14}$$

In the special case of ρ being a *constant,* this term may be factored out of the integral and the integration is then purely a function of geometry:

$$I = \rho \int_V r^2 \, dV \tag{10–15}$$

When the elemental volume chosen for integration has differential sizes in all three directions, e.g., $dV = dx \, dy \, dz$, Fig. 10–23a, the moment of inertia of the body must be determined using "triple integration." The integration process can, however, be simplified to a *single integration* provided the chosen elemental volume has a differential size or thickness in only *one direction*. Shell or disk elements are often used for this purpose.

Shell Element.   If a *shell element* having a height z, radius y, and thickness dy is chosen for integration, Fig. 10–23b, then the volume $dV = (2\pi y)(z) \, dy$. This element may be used in Eq. 10–14 or 10–15 for determining the moment of inertia I_z of the body about the z axis, since the *entire element,* due to its "thinness," lies at the *same* perpendicular distance $r = y$ from the z axis (see Example 10–11).

Disk Element.   If a disk element having a radius y and a thickness dz is chosen for integration, Fig. 10–23c, then the volume $dV = (\pi y^2) \, dz$. In this case, however, the element is *finite* in the radial direction, and consequently its parts *do not* all lie at the *same radial distance r* from the z axis. As a result, Eq. 10–14 or 10–15 *cannot* be used to determine I_z. Instead, to perform the integration using this element, it is first necessary to determine the moment of inertia *of the element* about the z axis and then integrate this result (see Example 10–12).

Example 10–11

Determine the moment of inertia of the cylinder shown in Fig. 10–24a about the z axis. The density ρ of the material is constant.

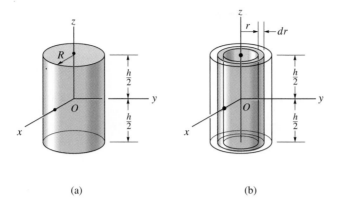

Fig. 10–24

SOLUTION

Shell Element.   This problem may be solved using the *shell element* in Fig. 10–24b and single integration. The volume of the element is $dV = (2\pi r)(h)\ dr$, so that its mass is $dm = \rho\ dV = \rho(2\pi hr\ dr)$. Since the *entire element* lies at the same distance r from the z axis, the moment of inertia *of the element* is

$$dI_z = r^2\ dm = \rho 2\pi hr^3\ dr$$

Integrating over the entire region of the cylinder yields

$$I_z = \int_m r^2\ dm = \rho 2\pi h \int_0^R r^3\ dr = \frac{\rho\pi}{2} R^4 h$$

The mass of the cylinder is

$$m = \int_m dm = \rho 2\pi h \int_0^R r\ dr = \rho\pi hR^2$$

so that

$$I_z = \frac{1}{2} mR^2 \qquad\qquad \textit{Ans.}$$

Example 10–12

A solid is formed by revolving the shaded area shown in Fig. 10–25a about the y axis. If the density of the material is 5 slug/ft³, determine the moment of inertia about the y axis.

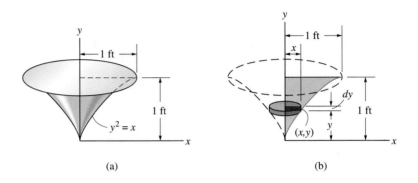

Fig. 10–25

SOLUTION

Disk Element. The moment of inertia will be determined using a *disk element,* as shown in Fig. 10–25b. Here the element intersects the curve at the arbitrary point (x, y) and has a mass

$$dm = \rho\, dV = \rho(\pi x^2)\, dy$$

Although all portions of the element are *not* located at the same distance from the y axis, it is still possible to determine the moment of inertia dI_y *of the element* about the y axis. In Example 10–11 it was shown that the moment of inertia of a cylinder about its longitudinal axis is $I = \frac{1}{2}mR^2$, where m and R are the mass and radius of the cylinder. Since the height of the cylinder is not involved in this formula, we can also use it for a disk. Thus, for the disk element in Fig. 10–25b, we have

$$dI_y = \frac{1}{2}(dm)x^2 = \frac{1}{2}[\rho(\pi x^2)\, dy]x^2$$

Substituting $x = y^2$, $\rho = 5$ slug/ft³, and integrating with respect to y, from $y = 0$ to $y = 1$ ft, yields the moment of inertia for the entire solid:

$$I_y = \frac{5\pi}{2}\int_0^1 x^4\, dy = \frac{5\pi}{2}\int_0^1 y^8\, dy = 0.873 \text{ slug} \cdot \text{ft}^2 \qquad Ans.$$

Parallel-Axis Theorem. If the moment of inertia of the body about an axis passing through the body's mass center is known, then the moment of inertia about any other *parallel axis* may be determined by using the *parallel-axis theorem*. This theorem can be derived by considering the body shown in Fig. 10–26. The z' axis passes through the mass center G, whereas the corresponding *parallel z axis* lies at a constant distance d away. Selecting the differential element of mass dm which is located at point (x', y') and using the Pythagorean theorem, $r^2 = (d + x')^2 + y'^2$, we can express the moment of inertia of the body about the z axis as

$$I = \int_m r^2 \, dm = \int_m [(d + x')^2 + y'^2] \, dm$$

$$= \int_m (x'^2 + y'^2) \, dm + 2d \int_m x' \, dm + d^2 \int_m dm$$

Since $r'^2 = x'^2 + y'^2$, the first integral represents I_G. The second integral equals *zero,* since the z' axis passes through the body's mass center, i.e., $\int x' \, dm = \bar{x} \int dm = 0$ since $\bar{x} = 0$. Finally, the third integral represents the total mass m of the body. Hence, the moment of inertia about the z axis can be written as

$$\boxed{I = I_G + md^2} \tag{10–16}$$

where I_G = moment of inertia about the z' axis passing through the mass center G

m = mass of the body

d = perpendicular distance between the parallel axes

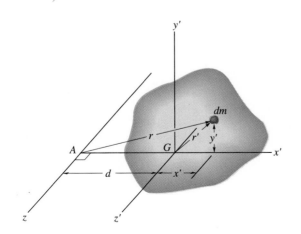

Fig. 10–26

Radius of Gyration. Occasionally, the moment of inertia of a body about a specified axis is reported in handbooks using the *radius of gyration, k.* This value has units of length, and when it and the body's mass m are known, the moment of inertia is determined from the equation

$$I = mk^2 \quad \text{or} \quad k = \sqrt{\frac{I}{m}} \qquad (10\text{–}17)$$

Note the *similarity* between the definition of k in this formula and r in the equation $dI = r^2 \, dm$, which defines the moment of inertia of an elemental mass dm of the body about an axis.

Composite Bodies. If a body is constructed from a number of simple shapes such as disks, spheres, and rods, the moment of inertia of the body about any axis z can be determined by adding algebraically the moments of inertia of all the composite shapes computed about the z axis. Algebraic addition is necessary since a composite part must be considered as a negative quantity if it has already been included within another part—for example, a "hole" subtracted from a solid plate. The parallel-axis theorem is needed for the calculations if the center of mass of each composite part does not lie on the z axis. For the calculation, then, $I = \Sigma(I_G + md^2)$, where I_G for each of the composite parts is computed by integration or can be determined from a table, such as the one given on the inside back cover.

Example 10–13

If the plate shown in Fig. 10–27a has a density of 8000 kg/m³ and a thickness of 10 mm, compute its moment of inertia about an axis directed perpendicular to the page and passing through point O.

Fig. 10–27

(a)   (b)

SOLUTION

The plate consists of two composite parts, the 250-mm-radius disk *minus* a 125-mm-radius disk, Fig. 10–27b. The moment of inertia about O can be determined by computing the moment of inertia of each of these parts about O and then *algebraically* adding the results. The computations are performed by using the parallel-axis theorem in conjunction with the data listed in the table on the inside back cover.

Disk.   The moment of inertia of a disk about an axis perpendicular to the plane of the disk is $I_G = \frac{1}{2}mr^2$. The mass center of the disk is located at a distance of 0.25 m from point O. Thus,

$$m_d = \rho_d V_d = 8000 \text{ kg/m}^3[\pi(0.25 \text{ m})^2(0.01 \text{ m})] = 15.71 \text{ kg}$$
$$(I_O)_d = \frac{1}{2}m_d r_d^2 + m_d d^2$$
$$= \frac{1}{2}(15.71 \text{ kg})(0.25 \text{ m})^2 + (15.71 \text{ kg})(0.25 \text{ m})^2$$
$$= 1.473 \text{ kg} \cdot \text{m}^2$$

Hole.   For the 125-mm-radius disk (hole), we have

$$m_h = \rho_h V_h = 8000 \text{ kg/m}^3[\pi(0.125 \text{ m})^2(0.01 \text{ m})] = 3.93 \text{ kg}$$
$$(I_O)_h = \frac{1}{2}m_h r_h^2 + m_h d^2$$
$$= \frac{1}{2}(3.93 \text{ kg})(0.125 \text{ m})^2 + (3.93 \text{ kg})(0.25 \text{ m})^2$$
$$= 0.276 \text{ kg} \cdot \text{m}^2$$

The moment of inertia of the plate about point O is therefore

$$I_O = (I_O)_d - (I_O)_h$$
$$= 1.473 \text{ kg} \cdot \text{m}^2 - 0.276 \text{ kg} \cdot \text{m}^2$$
$$= 1.20 \text{ kg} \cdot \text{m}^2 \qquad\qquad Ans.$$

Example 10–14

O

$\bar{y}$

2 ft

G

A

B   *C*

|←1 ft→|←1 ft→|

Fig. 10–28

The pendulum consists of two thin rods each having a weight of 10 lb and suspended from point *O* as shown in Fig. 10–28. Determine the pendulum's moment of inertia about an axis passing through (a) the pin at *O*, and (b) the mass center *G* of the pendulum.

SOLUTION

Part (a). Using the table on the inside back cover, the moment of inertia of rod *OA* about an axis perpendicular to the page and passing through the end point *O* of the rod is $I_O = \frac{1}{3}ml^2$. Hence,

$$(I_{OA})_O = \frac{1}{3}ml^2 = \frac{1}{3}\left(\frac{10}{32.2}\right)(2)^2 = 0.414 \text{ slug} \cdot \text{ft}^2$$

The same value may be computed using $I_G = \frac{1}{12}ml^2$ and the parallel-axis theorem; i.e.,

$$(I_{OA})_O = \frac{1}{12}ml^2 + md^2 = \frac{1}{12}\left(\frac{10}{32.2}\right)(2)^2 + \frac{10}{32.2}(1)^2$$
$$= 0.414 \text{ slug} \cdot \text{ft}^2$$

For rod *BC* we have

$$(I_{BC})_O = \frac{1}{12}ml^2 + md^2 = \frac{1}{12}\left(\frac{10}{32.2}\right)(2)^2 + \frac{10}{32.2}(2)^2$$
$$= 1.346 \text{ slug} \cdot \text{ft}^2$$

The moment of inertia of the pendulum about *O* is therefore

$$I_O = 0.414 + 1.346 = 1.76 \text{ slug} \cdot \text{ft}^2 \qquad \textit{Ans.}$$

Part (b). The mass center *G* will be located relative to the pin at *O*. Assuming this distance to be $\bar{y}$, Fig. 10–28, and using the formula for determining the mass center, we have

$$\bar{y} = \frac{\Sigma \tilde{y}m}{\Sigma m} = \frac{1(10/32.2) + 2(10/32.2)}{(10/32.2) + (10/32.2)} = 1.50 \text{ ft}$$

The moment of inertia I_G may be computed in the same manner as I_O, which requires successive applications of the parallel-axis theorem in order to transfer the moments of inertia of rods *OA* and *BC* to *G*. A more direct solution, however, involves applying the parallel-axis theorem using the result for I_O determined above; i.e.,

$$I_O = I_G + md^2; \qquad 1.76 = I_G + \left(\frac{20}{32.2}\right)(1.50)^2$$

$$I_G = 0.362 \text{ slug} \cdot \text{ft}^2 \qquad \textit{Ans.}$$

PROBLEMS

10–77.   Determine the moment of inertia of the slender rod with respect to the y axis. The rod's density ρ and cross-sectional area A are constant. Express the result in terms of the rod's total mass m.

Prob. 10–77

10–78.   The paraboloid is formed by revolving the shaded area around the x axis. Determine the radius of gyration k_x. The material has a constant density ρ.

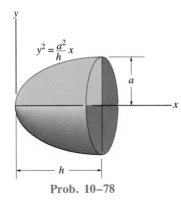

$$y^2 = \frac{a^2}{h} x$$

Prob. 10–78

10–79.   Determine the moment of inertia with respect to the x axis, and express the result in terms of the total mass m of the sphere. The sphere has a constant density ρ.

$$x^2 + y^2 = r^2$$

Prob. 10–79

*10–80.   Determine the moment of inertia of the thin ring about the z axis. The ring has a mass m.

Prob. 10–80

10–81.   The right circular cone is formed by revolving the shaded area around the x axis. Determine the moment of inertia I_x and express the result in terms of the total mass m of the cone. The cone has a constant density ρ.

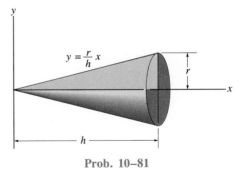

$$y = \frac{r}{h} x$$

Prob. 10–81

10–82.   The concrete solid is formed by rotating the shaded area about the y axis. Determine the moment of inertia I_y. The specific weight of concrete is $\gamma = 150$ lb/ft^3.

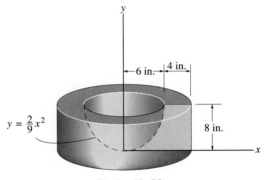

$$y = \frac{2}{9} x^2$$

6 in.   4 in.

8 in.

Prob. 10–82

10–83. The solid is formed by revolving the shaded area around the x axis. Determine the radius of gyration k_x. The mass density of the material is $\rho = 5$ Mg/m³.

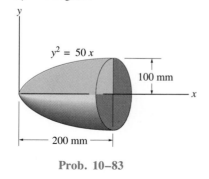

$y^2 = 50\,x$

100 mm

200 mm

Prob. 10–83

*10–84. The solid is formed by revolving the shaded area around the y axis. Determine the radius of gyration k_y. The specific weight of the material is $\gamma = 380$ lb/ft³.

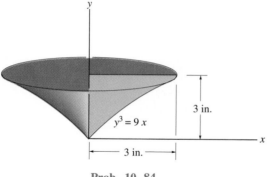

3 in.

$y^3 = 9\,x$

3 in.

Prob. 10–84

10–85. The frustum is formed by rotating the shaded area around the x axis. Determine the moment of inertia I_x and express the result in terms of the total mass m of the frustum. The frustrum has a constant density.

$y = \dfrac{b}{a}x + b$

$2b$

b

a

Prob. 10–85

10–86. The ellipsoid is formed by rotating the shaded area around the x axis. Determine the moment of inertia I_x and express the result in terms of the total mass m of the ellipsoid. The ellipsoid has a constant density.

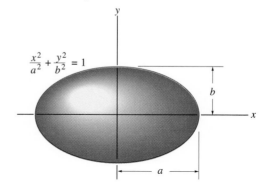

$\dfrac{x^2}{a^2} + \dfrac{y^2}{b^2} = 1$

b

a

Prob. 10–86

10–87. The slender rods have a weight of 3 lb/ft. Determine the moment of inertia of the assembly about an axis perpendicular to the page and passing through point A.

A

2 ft

1.5 ft

1.5 ft

Prob. 10–87

*10–88. The pendulum consists of a plate having a weight of 12 lb and a slender rod having a weight of 4 lb. Determine the radius of gyration of the pendulum about an axis perpendicular to the page and passing through point O.

O

1 ft

1 ft

3 ft

2 ft

Prob. 10–88

10–89. The pendulum consists of the 3-kg slender rod and the 5-kg thin plate. Determine the location $\bar{y}$ of the center of mass G of the pendulum; then calculate the moment of inertia of the pendulum about an axis perpendicular to the page and passing through G.

Prob. 10–89

10–91. Determine the moment of inertia of the assembly about an axis which is perpendicular to the page and passes through point O. The material has a specific weight of $\gamma = 90$ lb/ft^3.

Prob. 10–91

10–90. Determine the moment of inertia of the assembly about an axis which is perpendicular to the page and passes through the center of mass G. The material has a specific weight of $\gamma = 90$ lb/ft^3.

Prob. 10–90

*10–92.** Determine the moment of inertia of the thin plate about an axis perpendicular to the page and passing through the pin at O. The plate has a hole in its center. Its thickness is 50 mm and the material has a density of $\rho = 50$ kg/m^3.

Prob. 10–92

10–93. Each of the three rods has a mass m. Determine the moment of inertia of the assembly about an axis which is perpendicular to the page and passes through the center point O.

10–95. Determine the moment of inertia I_z of the frustum of the cone which has a conical depression. The material has a density of 200 kg/m³.

Prob. 10–93

Prob. 10–95

10–94. Determine the distance $\bar{y}$ to the center of mass G of the assembly, and then calculate the moment of inertia about an axis perpendicular to the page and passing through G. The block has a mass of 3 kg and the mass of the semicylinder is 5 kg.

*10–96. The pendulum consists of a disk having a mass of 2 kg and slender rods AB and DC which have a mass of 2 kg/m. Determine the length L of DC so that the center of mass is at the bearing O. What is the moment of inertia of the assembly about an axis perpendicular to the page and passing through point O?

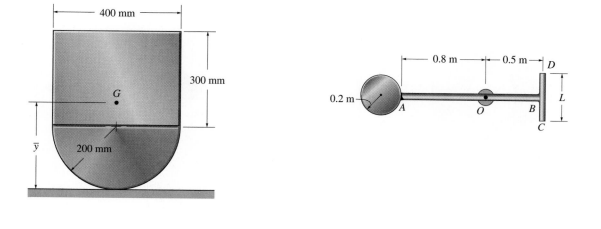

Prob. 10–94

Prob. 10–96

REVIEW PROBLEMS

10–97. Determine the product of inertia of the parabolic area with respect to the x and y axes.

10–99. Determine the product of inertia of the cross-sectional area with respect to the x and y axes that have their origin located at the centroid C.

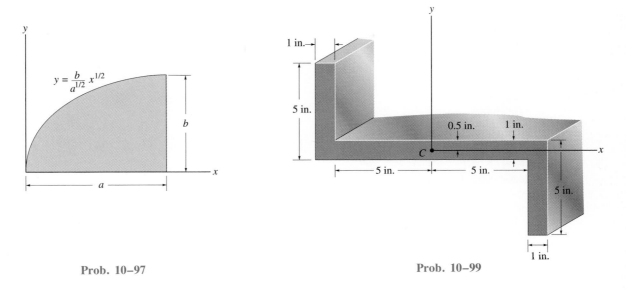

Prob. 10–97

Prob. 10–99

10–98. Determine the product of inertia of the shaded area with respect to the x and y axes.

*10–100. Determine the location $(\bar{x}, \bar{y})$ of the centroid C of the cross-sectional area for the angle. Then find the moments of inertia about the x' and y' centroidal axes.

Prob. 10–98

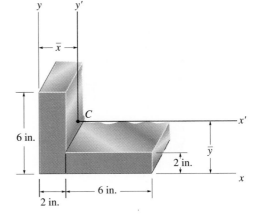

Prob. 10–100

10–101. Determine the principal moments of inertia for the angle with respect to the x and y axes passing through the centroid C. Use the equations developed in Sec. 10.7. Assume all corners to be square.

Prob. 10–101

10–102. Determine the moment of inertia of the shaded area about the x axis.

10–103. Determine the moment of inertia of the shaded area about the y axis.

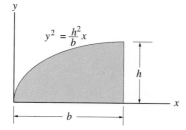

$$y^2 = \frac{h^2}{b}x$$

Probs. 10–102/10–103

*10–104. The pendulum consists of the slender rod OA, which has a mass of 3 kg/m. The thin plate has a mass of 12 kg/m². Determine the distance $\bar{y}$ to the center of mass G of the pendulum; then calculate the moment of inertia of the pendulum about an axis perpendicular to the page and passing through G.

Prob. 10–104

10–105. Determine the product of inertia for the angle's cross-sectional area with respect to the x' and y' axes having their origin located at the centroid C. Assume all corners to be square.

Prob. 10–105

10–106. Determine the moment of inertia of the solid steel assembly about the x–x axis. Steel has a specific weight of $\gamma_{st} = 490$ lb/ft³.

Prob. 10–106

11 Virtual Work

In this chapter we will use the principle of virtual work and the potential-energy method to determine the equilibrium position of a series of connected rigid bodies. Although application requires more mathematical sophistication than using the equations of equilibrium, it will be shown that, once the equation of virtual work or the potential-energy function is established, the solution may be obtained *directly*, without having to dismember the system in order to obtain relationships between forces occurring at the connections. Furthermore, by using the potential-energy method, we will be able to investigate the "type" of equilibrium or the stability of the configuration.

11.1 Definition of Work and Virtual Work

Work of a Force. In mechanics a force **F** does work only when it undergoes a displacement in the direction of the force. For example, consider the force **F** in Fig. 11–1, which is located on the path s specified by the position vector **r**. If the force moves along the path to a new position $\mathbf{r}' = \mathbf{r} + d\mathbf{r}$, the displacement is $d\mathbf{r}$ and therefore the work dU is a *scalar quantity*, defined by the dot product

$$dU = \mathbf{F} \cdot d\mathbf{r}$$

Because $d\mathbf{r}$ is infinitesimal, the magnitude of $d\mathbf{r}$ can be represented by ds,

Fig. 11–1

501

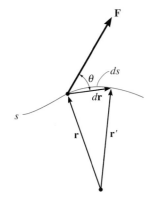

Fig. 11–1

the differential arc segment along the path. If the angle between the tails of $d\mathbf{r}$ and $\mathbf{F}$ is θ, Fig. 11–1, then by definition of the dot product, the above equation may also be written as

$$dU = F \, ds \, \cos \theta$$

Work expressed by this equation may be interpreted in one of two ways: either as the product of $\mathbf{F}$ and the component of displacement in the direction of the force, i.e., $ds \cos \theta$; or as the product of ds and the component of force in the direction of displacement, i.e., $F \cos \theta$. Note that if $0° \leq \theta < 90°$, then the force component and the displacement have the *same sense,* so that the work is *positive;* whereas if $90° < \theta \leq 180°$, these vectors have an *opposite sense,* and therefore the work is *negative.* Also, $dU = 0$ if the force is *perpendicular* to displacement, since $\cos 90° = 0$, or if the force is applied at a *fixed point,* in which case the displacement $ds = 0$.

The basic unit for work combines the units of force and displacement. In the SI system a *joule* (J) is equivalent to the work done by a force of 1 newton which moves 1 meter in the direction of the force (1 J = 1 N $\cdot$ m). In the FPS system, work is defined in units of ft $\cdot$ lb. The moment of a force has the same combination of units; however, the concepts of moment and work are in no way related. A moment is a vector quantity, whereas work is a scalar.

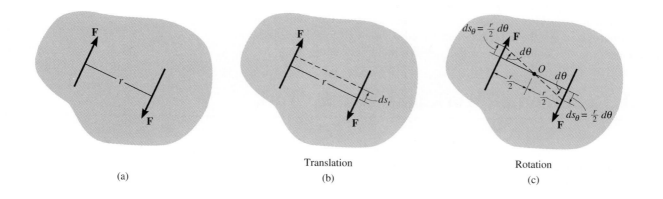

(a)

Translation
(b)

Rotation
(c)

Fig. 11–2

Work of a Couple.

Work of a Couple. The two forces of a couple do work when the couple *rotates* about an axis perpendicular to the plane of the couple. To show this, consider the body in Fig. 11–2a, which is subjected to a couple whose moment has a magnitude $M = Fr$. Any general differential displacement of the body can be considered as a combination of a translation and rotation. When the body *translates* such that the *component of displacement* along the line of action of each force is ds_t, clearly the "positive" work of one force $(F \, ds_t)$ *cancels* the "negative" work of the other $(-F \, ds_t)$, Fig. 11–2b. Consider now a differential *rotation* $d\theta$ of the body about an axis perpendicular to the plane of the couple, which intersects the plane at point O, Fig. 11–2c. (For the derivation, any other point in the plane may also be considered.) As shown, each force undergoes a displacement $ds_\theta = (r/2) \, d\theta$ in the direction of the force; hence, the work of both forces is

$$dU = F\left(\frac{r}{2} \, d\theta\right) + F\left(\frac{r}{2} \, d\theta\right) = (Fr) \, d\theta$$

or

$$dU = M \, d\theta$$

The resultant work is *positive* when the sense of **M** is the *same* as that of $d\boldsymbol{\theta}$, and negative when they have an opposite sense. As in the case of the moment vector, the *direction and sense* of $d\boldsymbol{\theta}$ are defined by the right-hand rule, where the fingers of the right hand follow the rotation or "curl" and the thumb indicates the direction of $d\boldsymbol{\theta}$. Hence, the line of action of $d\boldsymbol{\theta}$ will be *parallel* to the line of action of **M** if movement of the body occurs in the *same plane*. If the body rotates in space, however, the *component* of $d\boldsymbol{\theta}$ in the direction of **M** is required. Thus, in general, the work done by a couple is defined by the dot product, $dU = \mathbf{M} \cdot d\boldsymbol{\theta}$.

Virtual Work.

Virtual Work. The definitions of the work of a force and a couple have been presented in terms of *actual movements* expressed by differential displacements having magnitudes of ds and $d\theta$. Consider now an *imaginary* or *virtual movement*, which indicates a displacement or rotation that is *assumed* and *does not actually exist*. These movements are first-order differential quantities and will be denoted by the symbols δs and $\delta \theta$ (delta s and delta θ), respectively. The *virtual work* done by a force undergoing a virtual displacement δs is

$$\delta U = F \cos \theta \, \delta s \qquad (11\text{--}1)$$

Similarly, when a couple undergoes a virtual rotation $\delta \theta$ in the plane of the couple forces, the *virtual work* is

$$\delta U = M \, \delta \theta \qquad (11\text{--}2)$$

11.2 Principle of Virtual Work for a Particle and a Rigid Body

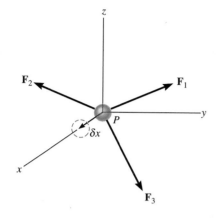

Fig. 11-3

If a particle is in equilibrium, the resultant of the force system acting on it must be equal to zero. Hence, if the particle undergoes an imaginary or virtual displacement in the x, y, or z direction, the virtual work (δU) done by the force system must be equal to zero since the components $\Sigma F_x = 0$, $\Sigma F_y = 0$, $\Sigma F_z = 0$. Alternatively, this may be expressed as

$$\delta U = 0$$

For example, if the particle in Fig. 11-3 is given a virtual displacement $\delta \mathbf{x}$, only the x components of the forces acting on the particle do work. (No work is done by the y and z components, since they are perpendicular to the displacement.) The virtual work equation is therefore

$$\delta U = 0; \qquad F_{1x}\,\delta x + F_{2x}\,\delta x + F_{3x}\,\delta x = 0$$

Factoring out δx, which is common to every term, yields

$$(F_{1x} + F_{2x} + F_{3x})\,\delta x = 0$$

Since $\delta x \neq 0$, this equation is satisfied only if the sum of the force components in the x direction is equal to zero, i.e., $\Sigma F_x = 0$. Two other virtual work equations can be written by assuming virtual displacements $\delta \mathbf{y}$ and $\delta \mathbf{z}$ in the y and z directions, respectively. Doing this, however, amounts to satisfying the equilibrium equations $\Sigma F_y = 0$ and $\Sigma F_z = 0$ for the particle.

In a similar manner, a rigid body that is subjected to a coplanar force system will be in equilibrium provided $\Sigma F_x = 0$, $\Sigma F_y = 0$, and $\Sigma M_O = 0$. We can also write a set of three virtual work equations for the body, each of which requires $\delta U = 0$. If these equations involve separate virtual translations in the x and y directions and a virtual rotation about an axis perpendicular to the x-y plane and passing through point O, then it can be shown that they will correspond to the above-mentioned three equilibrium equations. When writing these equations, it is *not necessary* to include the work done by the *internal forces* acting within the body, since a rigid body *does not deform* when subjected to an external loading, and furthermore, when the body moves through a virtual displacement, the internal forces occur in equal but opposite collinear pairs, so that the corresponding work done by each pair of forces *cancels*.

As in the case of a particle, however, no added advantage would be gained by solving rigid-body equilibrium problems using the principle of virtual work. This is because for each application of the virtual-work equation the virtual displacement, common to every term, factors out, leaving an equation that could have been obtained in a more *direct manner* by applying the equations of equilibrium.

11.3 Principle of Virtual Work for a System of Connected Rigid Bodies

The method of virtual work is most suitable for solving equilibrium problems that involve a system of several *connected* rigid bodies such as the ones shown in Fig. 11–4. Before we can apply the principle of virtual work to these systems, however, we must first specify the number of degrees of freedom for a system and establish coordinates that define the position of the system.

Degrees of Freedom. A system of connected bodies takes on a unique shape that can be specified provided we know the position of a number of specific points on the system. These positions are defined using *independent coordinates q,* which are measured from fixed reference points. For every coordinate established, the system will have a *degree of freedom* for displacement along the coordinate axis such that it is consistent with the constraining action of the supports. Thus, an *n*-degree-of-freedom system requires *n* independent coordinates q_n to specify the location of all its members. For example, the link and sliding-block arrangement shown in Fig. 11–4a is an example of a one-degree-of-freedom system. The independent coordinate $q = \theta$ may be used to specify the location of the two connecting links and the block. The coordinate *x* could also be used as the independent coordinate. However, since the block is constrained to move within the slot, *x* is not independent of θ; rather, it can be related to θ using the cosine law, $b^2 = a^2 + x^2 - 2ax \cos \theta$. The double-link arrangement, shown in Fig. 11–4b, is an example of a two-degree-of-freedom system. To specify the location of each link, the coordinate angles θ_1 and θ_2 must be known, since a rotation of one link is independent of a rotation of the other.

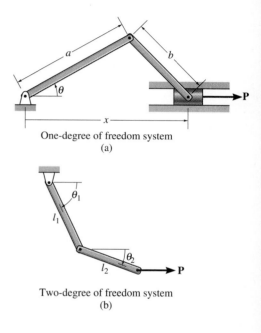

One-degree of freedom system
(a)

Two-degree of freedom system
(b)

Fig. 11–4

Principle of Virtual Work. The principle of virtual work for a system of rigid bodies whose connections are *frictionless* may be stated as follows: *A system of connected rigid bodies is in equilibrium provided the virtual work done by all the external forces and couples acting on the system is zero for each independent virtual displacement of the system.* Mathematically, this may be expressed as

$$\delta U = \qquad\qquad \text{(11–3)}$$

where δU represents the virtual work of all the external forces (and couples) acting on the system during any independent virtual displacement.

As stated above, if a system has *n* degrees of freedom it takes *n* independent coordinates q_n to completely specify the location of the system. Hence, for the system it is possible to write n independent virtual-work equations, one for every virtual displacement taken along each of the independent coordinate axes, while the remaining $n - 1$ independent coordinates are held *fixed*.*

PROCEDURE FOR ANALYSIS

The following procedure provides a method for applying the equation of virtual work to solve problems involving a system of frictionless connected rigid bodies having a single degree of freedom.

Free-Body Diagram.   Draw the free-body diagram of the entire system of connected bodies and define the *independent coordinate q*. Sketch the "deflected position" of the system on the free-body diagram when the system undergoes a *positive* virtual displacement δq. From this, specify the "active" forces and couples, that is, those that do work.

Virtual Displacements.   Indicate *position coordinates s_i*, measured from a *fixed point* on the free-body diagram to each of the i number of "active" forces and couples. Each coordinate axis should be parallel to the line of action of the "active" force to which it is directed, so that the virtual work along the coordinate axis can be calculated.

Relate each of the position coordinates s_i to the independent coordinate q; then *differentiate* these expressions in order to express the virtual displacements δs_i in terms of δq.

Virtual-Work Equation.   Write the *virtual-work equation* for the system assuming that, whether possible or not, all the position coordinates s_i undergo *positive* virtual displacements δs_i. Using the relations for δs_i, express the work of *each* "active" force and couple in the equation in terms of the single independent virtual displacement δq. By factoring out this common displacement, one is left with an equation that generally can be solved for an unknown force, couple, or equilibrium position.

If the system contains n degrees of freedom, n independent coordinates q_n must be specified. In this case, follow the above procedure and let *only one* of the independent coordinates undergo a virtual displacement, while the remaining $n - 1$ coordinates are held fixed. In this way, n virtual-work equations can be written, one for each independent coordinate.

The following examples should help to clarify application of this procedure.

---

*This method of applying the principle of virtual work is sometimes called the *method of virtual displacements*, since a virtual displacement is applied, resulting in the calculation of a real force. Although it is not to be used here, realize that we can also apply the principle of virtual work as a method of virtual forces. This method is often used to determine the displacements of points on deformable bodies. See R. C. Hibbeler, *Mechanics of Materials,* Macmillan Publishing Company, New York, 1991.

Example 11–1

Determine the angle θ for equilibrium of the two-member linkage shown in Fig. 11–5a. Each member has a mass of 10 kg.

SOLUTION

Free-Body Diagram.   The system has only one degree of freedom, since the location of both links may be specified by the single independent coordinate ($q =$) θ. As shown on the free-body diagram in Fig. 11–5b, when θ undergoes a *positive* (clockwise) virtual rotation $\delta\theta$, only the active forces, **F** and the two 98.1-N weights, do work. (The reactive forces $\mathbf{D}_x$ and $\mathbf{D}_y$ are fixed, and $\mathbf{B}_y$ does not move along its line of action.)

Virtual Displacements.   If the origin of coordinates is established at the *fixed* pin support D, the location of **F** and **W** may be specified by the *position coordinates* x_B and y_w, as shown in the figure. In order to determine the work, note that these coordinates are parallel to the lines of action of their associated forces.

Expressing the position coordinates in terms of the independent coordinate θ and taking the derivatives yields

$$x_B = 2(1 \cos \theta) \text{ m} \qquad \delta x_B = -2 \sin \theta \, \delta\theta \text{ m} \qquad (1)$$

$$y_w = \tfrac{1}{2}(1 \sin \theta) \text{ m} \qquad \delta y_w = 0.5 \cos \theta \, \delta\theta \text{ m} \qquad (2)$$

It is seen by the *signs* of these equations, and indicated in Fig. 11–5b, that an *increase* in θ (i.e., $\delta\theta$) causes a *decrease* in x_B and an *increase* in y_w.

Virtual-Work Equation.   If the virtual displacements δx_B and δy_w were *both positive,* then the forces **W** and **F** would do positive work since the forces and their corresponding displacements would have the same sense. Hence, the virtual-work equation for the displacement $\delta\theta$ is

$$\delta U = 0; \qquad W \, \delta y_w + W \, \delta y_w + F \, \delta x_B = 0 \qquad (3)$$

Substituting Eqs. 1 and 2 into Eq. 3 in order to relate the virtual displacements to the common virtual displacement $\delta\theta$ yields

$$98.1(0.5 \cos \theta \, \delta\theta) + 98.1(0.5 \cos \theta \, \delta\theta) + 25(-2 \sin \theta \, \delta\theta) = 0$$

Notice that the "negative work" done by **F** (force in the opposite sense to displacement) has been *accounted for* in the above equation by the "negative sign" of Eq. 1. Factoring out the *common displacement* $\delta\theta$ and solving for θ, noting that $\delta\theta \neq 0$, yields

$$(98.1 \cos \theta - 50 \sin \theta) \, \delta\theta = 0$$

$$\theta = \tan^{-1} \frac{98.1}{50} = 63.0° \qquad \qquad \textit{Ans.}$$

If this problem had been solved using the equations of equilibrium, it would have been necessary to dismember the links and apply three scalar equations to *each* link. The principle of virtual work, by means of calculus, has eliminated this task so that the answer is obtained directly.

(a)

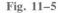

(b)

Fig. 11–5

Example 11-2

$M_1 = 0.5$ N · m

$P = 2$ N

(a)

(b)

Fig. 11-6

Using the principle of virtual work, determine the angle θ required to maintain equilibrium of the mechanism shown in Fig. 11–6a. Neglect the weight of the links. The spring is unstretched when $\theta = 0°$ and it maintains a horizontal position due to the roller.

SOLUTION

Free-Body Diagram. The mechanism has one degree of freedom, and therefore the location of each member may be specified using the independent coordinate θ. When θ undergoes a *positive* virtual displacement $\delta\theta$, as shown on the free-body diagram in Fig. 11–6b, links *AB* and *EC* rotate by the same amount since they have the same length, and link *BC* only translates. Since a couple moment does work *only* when it rotates, the work done by $\mathbf{M}_2$ is zero. The reactive forces at *A* and *E* do no work. Why?

Virtual Displacements. The position coordinates x_B and x_D are *parallel* to the lines of action of $\mathbf{P}$ and $\mathbf{F}$, and these coordinates locate these forces with respect to the *fixed points A* and *E*. From Fig. 11–6b,

$$x_B = 0.4 \sin \theta \text{ m}$$
$$x_D = 0.2 \sin \theta \text{ m}$$

Thus,

$$\delta x_B = 0.4 \cos \theta \, \delta\theta \text{ m}$$
$$\delta x_D = 0.2 \cos \theta \, \delta\theta \text{ m}$$

Virtual-Work Equation. Applying the equation of virtual work, noting that, for positive virtual displacements, $\mathbf{F}_s$ is opposite to δx_D and hence does negative work, we obtain

$$\delta U = 0; \qquad M_1 \, \delta\theta + P \, \delta x_B - F_s \, \delta x_D = 0$$

Relating each of the virtual displacements to the *common* virtual displacement $\delta\theta$ yields

$$0.5 \, \delta\theta + 2(0.4 \cos \theta \, \delta\theta) - F_s(0.2 \cos \theta \, \delta\theta) = 0$$
$$(0.5 + 0.8 \cos \theta - 0.2F_s \cos \theta) \, \delta\theta = 0 \qquad (1)$$

For the arbitrary angle θ, the spring is stretched a distance of $x_D = (0.2 \sin \theta)$ m; and therefore, $F_s = 60$ N/m$(0.2 \sin \theta)$ m $= (12 \sin \theta)$ N. Substituting into Eq. 1 and noting that $\delta\theta \neq 0$, we have

$$0.5 + 0.8 \cos \theta - 0.2(12 \sin \theta) \cos \theta = 0$$

Since $\sin 2\theta = 2 \sin \theta \cos \theta$, then

$$1 = 2.4 \sin 2\theta - 1.6 \cos \theta$$

Solving for θ by trial and error yields

$$\theta = 36.3° \qquad \qquad \textit{Ans.}$$

Example 11–3

Using the principle of virtual work, determine the horizontal force that the pin at C must exert in order to hold the mechanism shown in Fig. 11–7a in equilibrium when $\theta = 45°$. Neglect the weight of the members.

SOLUTION

Free-Body Diagram. The reaction C_x can be obtained by *releasing* the pin constraint at C in the x direction and allowing the frame to be displaced in this direction. The system then has only one degree of freedom, defined by the independent coordinate θ, Fig. 11–7b. When θ undergoes a *positive* virtual displacement $\delta\theta$, only C_x and the 200-N force do work.

Virtual Displacements. Forces C_x and 200 N are located from the fixed origin A using position coordinates y_B and x_C. From Fig. 11–7b, x_C can be related to θ by the "law of cosines." Hence,

$$(0.7)^2 = (0.6)^2 + x_C^2 - 2(0.6)x_C \cos \theta \tag{1}$$

$$0 = 0 + 2x_C \, \delta x_C - 1.2 \, \delta x_C \cos \theta + 1.2 x_C \sin \theta \, \delta\theta$$

$$\delta x_C = \frac{1.2 x_C \sin \theta}{1.2 \cos \theta - 2x_C} \, \delta\theta \tag{2}$$

Also,

$$y_B = 0.6 \sin \theta$$

$$\delta y_B = 0.6 \cos \theta \, \delta\theta \tag{3}$$

Virtual-Work Equation. When y_B and x_C undergo *positive* virtual displacements δy_B and δx_C, C_x and 200 N do *negative work,* since they both act in the opposite sense to δy_B and δx_C. Hence,

$$\delta U = 0; \qquad -200 \, \delta y_B - C_x \, \delta x_C = 0$$

Substituting Eqs. 2 and 3 into this equation, factoring out $\delta\theta$, and solving for C_x yields

$$-200(0.6 \cos \theta \, \delta\theta) - C_x \frac{1.2 x_C \sin \theta}{1.2 \cos \theta - 2x_C} \, \delta\theta = 0$$

$$C_x = \frac{-120 \cos \theta (1.2 \cos \theta - 2x_C)}{1.2 x_C \sin \theta} \tag{4}$$

At the required equilibrium position $\theta = 45°$, the corresponding value of x_C can be found by using Eq. 1, in which case

$$x_C^2 - 1.2 \cos 45° \, x_C - 0.13 = 0$$

Solving for the positive root yields

$$x_C = 0.981 \text{ m}$$

Thus, from Eq. 4,

$$C_x = 114 \text{ N} \qquad\qquad Ans.$$

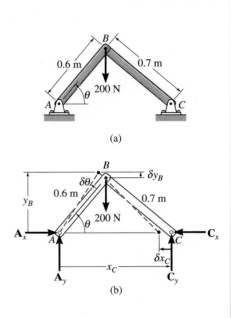

(a)

(b)

Fig. 11–7

Example 11–4

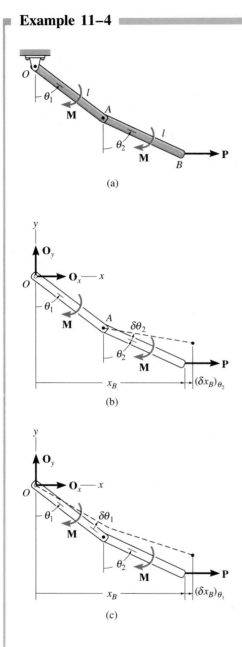

(a)

(b)

(c)

Fig. 11–8

Using the principle of virtual work, determine the equilibrium position of the two-bar linkage shown in Fig. 11–8a. Neglect the weight of the links.

SOLUTION

The system has two degrees of freedom, since the *independent coordinates* θ_1 and θ_2 must be known to locate the position of both links. The position coordinate x_B, measured from the fixed point O, is used to specify the location of **P**, Fig. 11–8b and c.

If θ_1 is held *fixed* and θ_2 varies by an amount $\delta\theta_2$, as shown in Fig. 11–8b, the virtual-work equation becomes

$$[\delta U = 0]_{\theta_2}; \qquad P(\delta x_B)_{\theta_2} - M\,\delta\theta_2 = 0 \tag{1}$$

Here P and M represent the magnitudes of the applied force and couple moment acting on link AB.

When θ_2 is held *fixed* and θ_1 varies by an amount $\delta\theta_1$, as shown in Fig. 11–8c, the virtual-work equation becomes

$$[\delta U = 0]_{\theta_1}; \qquad P(\delta x_B)_{\theta_1} - M\,\delta\theta_1 - M\,\delta\theta_1 = 0 \tag{2}$$

The *position coordinate* x_B may be related to the independent coordinates θ_1 and θ_2 by the equation

$$x_B = l \sin\theta_1 + l \sin\theta_2 \tag{3}$$

To obtain the variation of δx_B in terms of $\delta\theta_2$, it is necessary to take the *partial derivative* of x_B with respect to θ_2 since x_B is a function of both θ_1 and θ_2. Hence,

$$\frac{\partial x_B}{\partial \theta_2} = l \cos\theta_2 \qquad (\delta x_B)_{\theta_2} = l \cos\theta_2\,\delta\theta_2$$

Substituting into Eq. 1, we have

$$(Pl \cos\theta_2 - M)\,\delta\theta_2 = 0$$

Since $\delta\theta_2 \neq 0$, then

$$\theta_2 = \cos^{-1}\left(\frac{M}{Pl}\right) \qquad\qquad Ans.$$

Using Eq. 3 to obtain the variation of x_B with θ_1 yields

$$\frac{\partial x_B}{\partial \theta_1} = l \cos\theta_1 \qquad (\delta x_B)_{\theta_1} = l \cos\theta_1\,\delta\theta_1$$

Substituting into Eq. 2, we have

$$(Pl \cos\theta_1 - 2M)\,\delta\theta_1 = 0$$

Since $\delta\theta_1 \neq 0$, then

$$\theta_1 = \cos^{-1}\left(\frac{2M}{Pl}\right) \qquad\qquad Ans.$$

PROBLEMS

11–1. Determine the force developed in the spring required to keep the 6-kg rod in equilibrium when $\theta = 30°$.

Prob. 11–1

11–2. The punch press consists of the ram R, connecting rod AB, and a flywheel. If a torque of $M = 50$ N · m is applied to the flywheel, determine the force $\mathbf{F}$ applied at the ram to hold the rod in the position $\theta = 60°$.

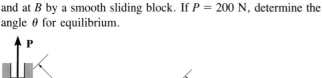

Prob. 11–2

11–3. The pin-connected mechanism is constrained at A by a pin and at B by a smooth sliding block. If $P = 200$ N, determine the angle θ for equilibrium.

Prob. 11–3

***11–4.** Determine the magnitude of the applied couple moments $\mathbf{M}$ needed to maintain equilibrium when $\theta = 20°$. The plate E has a weight of 50 lb. Neglect the weight of the links AB and CD.

Prob. 11–4

11–5. The scissors jack supports a load $\mathbf{P}$. Determine the axial force in the screw necessary for equilibrium when the jack is in the position θ. Each of the four links has a length L and is pin-connected at its center. Points B and D can move horizontally.

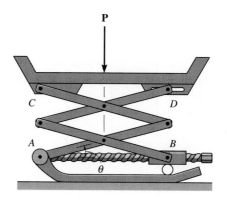

Prob. 11–5

11–6. The thin rod of weight W rests against the smooth wall and floor. Determine the magnitude of force **P** needed to hold it in equilibrium.

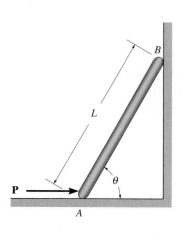

Prob. 11–6

*11–8.** The crankshaft is subjected to a torque of $M = 50$ lb · ft. Determine the vertical compressive force **F** applied to the piston for equilibrium when $\theta = 60°$.

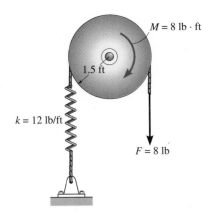

Prob. 11–8

11–7. Determine the force **F** acting on the cord which is required to maintain equilibrium of the horizontal 10-kg bar AB. *Hint:* Express the total *constant vertical length l* of the cord in terms of the position coordinates s_1 and s_2. The derivative of this equation yields a relation between δs_1 and δs_2.

Prob. 11–7

11–9. The disk has a weight of 10 lb and is subjected to a vertical force $F = 8$ lb and a couple moment $M = 8$ lb · ft. Determine the disk's rotation θ if the end of the spring wraps around the periphery of the disk as the disk turns. The spring is originally unstretched.

Prob. 11–9

11–10. If each of the three links of the mechanism has a weight of 20 lb, determine the angle θ for equilibrium. The spring, which always remains vertical, is unstretched when $\theta = 0°$. Set **P** = **0**.

11–11. The spring is unstretched when $\theta = 0°$. If $P = 100$ lb, determine the angle θ for equilibrium. Neglect the weight of the links.

Probs. 11–10/11–11

11–13. The two-bar linkage is subjected to a couple moment $M = 100$ N · m and horizontal force $P = 500$ N. Determine the angle θ for equilibrium. The spring is unstretched when $\theta = 45°$. Neglect the mass of each bar.

11–14. The two-bar linkage is subjected to a couple moment $M = 100$ N · m and horizontal force $P = 500$ N. Also, each bar is uniform and has a mass of 10 kg. Determine the angle θ for equilibrium. The spring is unstretched when $\theta = 45°$.

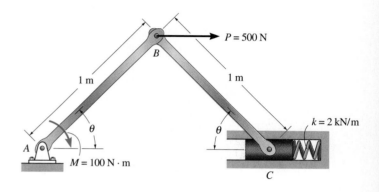

Probs. 11–13/11–14

11–15. The spring has an unstretched length of 0.3 m. Determine the angle θ for equilibrium if the uniform links each have a mass of 5 kg. Set **P** = **0**.

*11–12. The uniform rod AB weighs 10 lb. If the spring is unstretched when $\theta = 60°$, determine the angle θ for equilibrium.

*11–16. The spring has an unstretched length of 0.3 m. Determine the angle θ for equilibrium if $P = 50$ N. Neglect the weight of the links.

Prob. 11–12

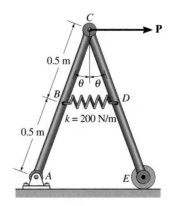

Probs. 11–15/11–16

11–17.   Each member of the pin-connected mechanism has a mass of 8 kg. If the spring is unstretched when $\theta = 0°$, determine the angle θ for equilibrium.

Prob. 11–17

11–19.   Determine the moment M required to maintain equilibrium of the slider mechanism at $\theta = 60°$.

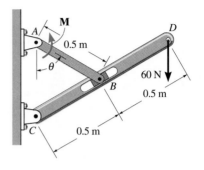

Prob. 11–19

11–18.   The vent plate is supported at B by a pin. If it weighs 15 lb and has a center of gravity at G, determine the stiffness k of the spring so that the plate remains in equilibrium at $\theta = 30°$. The spring is unstretched when $\theta = 0°$.

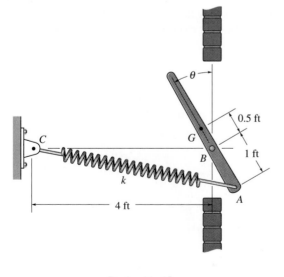

Prob. 11–18

*11–20.   Determine the force **P** that must be applied perpendicular to the handle in order to hold the mechanism in equilibrium for any angle θ of rod CD. There is a couple moment **M** applied to the link BA. A smooth collar attached at A slips over rod CD.

Prob. 11–20

11–21.   The bar is supported by a pin at A and a smooth collar at B. If the spring has an unstretched length of 1 ft, determine the angle θ for equilibrium when the vertical force acting on the bar is $F = 50$ lb.

11–22.   Solve Prob. 11–21 if the force **F** is applied *perpendicular* to the bar, i.e., so that it is directed down and to the right.

Probs. 11–21/11–22

*11–24.   A block having a weight W is attached to the end of the rod ABC. If the rod is supported by a smooth slider block at C and rod BD, determine the angle θ for equilibrium. Neglect the weights of the rods and the slider.

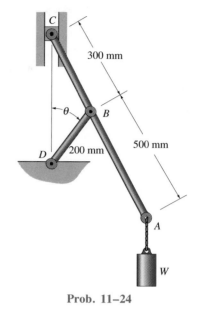

Prob. 11–24

11–23.   Determine the horizontal force $\mathbf{P}_1$ and vertical force $\mathbf{P}_2$ necessary to maintain the equilibrium position of the two uniform links. Each link weighs 20 lb.

11–25.   A weight W is suspended at one end of the parallel linkage mechanism shown. Determine the weights W_1 and W_2 that must be suspended from the other ends in order that the system remain in equilibrium for any angle θ or ϕ of the members. Neglect the weight of the members in the calculation.

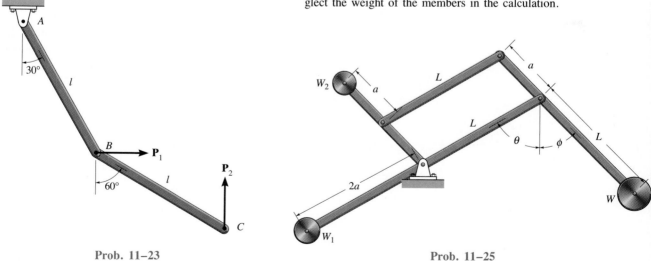

Prob. 11–23   Prob. 11–25

*11.4 Conservative Forces

The work done by a force when it undergoes a *differential displacement* has been defined as $dU = F \cos \theta \, ds$. If the force is displaced over a path that has a *finite length s,* the work is determined by integrating over the path; i.e.,

$$U = \int_s F \cos \theta \, ds$$

To evaluate the integral, it is necessary to obtain a relationship between F and the component of displacement $ds \cos \theta$. In some instances, however, the work done by a force will be *independent* of its path and, instead, will depend only on the initial and final locations of the force along the path. A force that has this property is called a *conservative force*.

Weight. Consider the body in Fig. 11–9, which is initially at P'. If the body is moved *down* along the *arbitrary path A* to the dashed position, then, for a given displacement ds along the path, the displacement component in the direction of **W** has a magnitude of $dy = ds \cos \theta$, as shown. Since both the force and displacement are in the same direction, the work is positive; hence,

$$U = \int_s W \cos \theta \, ds = \int_0^y W \, dy$$

or

$$U = Wy$$

In a similar manner, the work done by the weight when the body moves up a distance y back to P', along the arbitrary path A', is

$$U = -Wy$$

Why is the work negative?

The weight of a body is therefore a conservative force, since the work done by the weight depends *only* on the body's *vertical displacement* and is independent of the path along which the body moves.

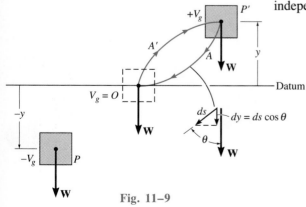

Fig. 11–9

Elastic Spring. The force developed by an elastic spring ($F_s = ks$) is also a conservative force. If the spring is attached to a body and the body is displaced along *any path*, such that it causes the spring to elongate or compress from a position s_1 to a further position s_2, the work will be negative, since the spring exerts a force $\mathbf{F}_s$ *on the body* that is opposite to the body's displacement ds, Fig. 11–10. For either extension or compression, the work is independent of the path and is simply

$$U = \int_{s_1}^{s_2} F_s \, ds = \int_{s_1}^{s_2} (-ks) \, ds$$
$$= -(\tfrac{1}{2}ks_2^2 - \tfrac{1}{2}ks_1^2)$$

Friction. In contrast to a conservative force, consider the force of *friction* exerted on a moving body by a fixed surface. The work done by the frictional force depends on the path; the longer the path, the greater the work. Consequently, frictional forces are *nonconservative*, and the work done is dissipated from the body in the form of heat.

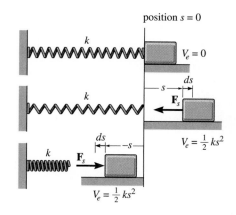

position $s = 0$

$V_e = 0$

$V_e = \frac{1}{2} ks^2$

$V_e = \frac{1}{2} ks^2$

Fig. 11–10

*11.5 Potential Energy

When a conservative force acts on a body, it gives the body the capacity to do work. This capacity, measured as *potential energy*, depends on the location of the body.

Gravitational Potential Energy. If a body is located a distance y *above* a fixed horizontal reference or datum, Fig. 11–9, the weight of the body has *positive* gravitational potential energy V_g since $\mathbf{W}$ has the capacity of doing positive work when the body is moved back down to the datum. Likewise, if the body is located a distance y *below* the datum, V_g is *negative* since the weight does negative work when the body is moved back up to the datum. At the datum, $V_g = 0$.

Measuring y as *positive upward,* the gravitational potential energy of the body's weight $\mathbf{W}$ is thus

$$\boxed{V_g = Wy} \qquad\qquad (11\text{–}4)$$

Elastic Potential Energy. The elastic potential energy V_e that a spring produces on an attached body, when the spring is elongated or compressed from an undeformed position ($s = 0$) to a final position s, is

$$\boxed{V_e = \tfrac{1}{2}ks^2} \qquad\qquad (11\text{–}5)$$

Here V_e is *always positive*, since in the deformed position the spring has the capacity of doing *positive work* in *returning* the body back to the spring's undeformed position, Fig. 11–10.

Potential Function.

In the general case, if a body is subjected to *both* gravitational and elastic forces, the *potential energy or potential function V* of the body can be expressed as the algebraic sum

$$V = V_g + V_e \qquad (11\text{--}6)$$

where measurement of V depends on the location of the body with respect to a selected datum in accordance with Eqs. 11–4 and 11–5.

In general, if a system of frictionless connected rigid bodies has a *single degree of freedom* such that its position from the datum is defined by the independent coordinate q, then the potential function for the system can be expressed as $V = V(q)$. The work done by all the conservative forces acting on the system in moving it from q_1 to q_2 is measured by the *difference* in V; i.e.,

$$U_{1-2} = V(q_1) - V(q_2) \qquad (11\text{--}7)$$

For example, the potential function for a system consisting of a block of weight **W** supported by a spring, Fig. 11–11*a*, can be expressed in terms of its independent coordinate $(q =)\ y$, measured from a fixed datum located at the unstretched length of the spring; we have

$$\begin{aligned} V &= V_g + V_e \\ &= -Wy + \tfrac{1}{2}ky^2 \end{aligned} \qquad (11\text{--}8)$$

If the block moves from y_1 to a farther downward position y_2, then the work of **W** and $\mathbf{F}_s$ is

$$U_{1-2} = V(y_1) - V(y_2) = -W[y_1 - y_2] + \tfrac{1}{2}ky_1^2 - \tfrac{1}{2}ky_2^2$$

(a)   (b)

Fig. 11–11

*11.6 Potential-Energy Criterion for Equilibrium

System Having One Degree of Freedom. When the displacement of a frictionless connected system is *infinitesimal*, i.e., from q to $q + dq$, Eq. 11–7 becomes

$$dU = V(q) - V(q + dq)$$

or

$$dU = -dV$$

Furthermore, if the system undergoes a *virtual displacement* δq, rather than an actual displacement dq, then $\delta U = -\delta V$. For equilibrium, the principle of virtual work requires that $\delta U = 0$ and therefore, provided the potential function for the system is known, this also requires that $\delta V = 0$. We can also express this requirement as

$$\boxed{\dfrac{dV}{dq} = 0} \tag{11–9}$$

Hence, *when a frictionless connected system of rigid bodies is in equilibrium, the first variation or change in V is zero*. This change is determined by taking the *first derivative* of the potential function and setting it equal to zero. For example, using Eq. 11–8 to determine the equilibrium position for the spring and block in Fig. 11–11a, we have

$$\frac{dV}{dy} = W - ky = 0$$

Hence, the equilibrium position $y = y_{eq}$ is

$$y_{eq} = \frac{W}{k}$$

Of course, the *same result* is obtained by applying $\Sigma F_y = 0$ to the forces acting on the free-body diagram of the block, Fig. 11–11b.

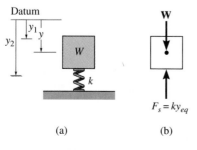

(a)   (b)

Fig. 11–11

System Having n Degrees of Freedom. When the system of connected bodies has n degrees of freedom, the total potential energy stored in the system will be a function of n independent coordinates q_n; i.e., $V = V(q_1, q_2, \ldots, q_n)$. In order to apply the equilibrium criterion $\delta V = 0$, it is necessary to determine the change in potential energy δV by using the "chain rule" of differential calculus; i.e.,

$$\delta V = \frac{\partial V}{\partial q_1} \delta q_1 + \frac{\partial V}{\partial q_2} \delta q_2 + \cdots + \frac{\partial V}{\partial q_n} \delta q_n = 0$$

Since the virtual displacements $\delta q_1, \delta q_2, \ldots, \delta q_n$ are independent of one another, the equation is satisfied provided

$$\frac{\partial V}{\partial q_1} = 0, \quad \frac{\partial V}{\partial q_2} = 0, \quad \ldots, \quad \frac{\partial V}{\partial q_n} = 0$$

Hence *it is possible to write n independent equations for a system having n degrees of freedom.*

*11.7 Stability of Equilibrium

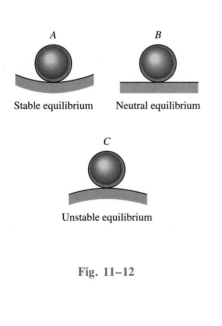

A

Stable equilibrium

B

Neutral equilibrium

C

Unstable equilibrium

Fig. 11–12

Once the equilibrium configuration for a body or system of connected bodies is defined, it is sometimes important to investigate the "type" of equilibrium or the stability of the configuration. For example, consider the position of a ball resting at a point on each of the three paths shown in Fig. 11–12. Each situation represents an equilibrium state for the ball. When the ball is at A, it is said to be in *stable equilibrium* because if it is given a small displacement up the hill, it will always *return* to its original, lowest, position. At A, its total potential energy is a *minimum*. When the ball is at B, it is in *neutral equilibrium*. A small displacement to either the left or right of B will not alter this condition. The ball *remains* in equilibrium in the displaced position, and therefore its potential energy is *constant*. When the ball is at C, it is in *unstable equilibrium*. Here a small displacement will cause the ball's potential energy to be *decreased*, and so it will roll farther *away* from its original, highest position. At C, the potential energy of the ball is a *maximum*.

Type of Equilibrium. The example just presented illustrates that one of three types of equilibrium positions can be specified for a body or system of connected bodies.

1. *Stable equilibrium* occurs when a small displacement of the system causes the system to return to its original position. In this case the original potential energy of the system is a minimum.

2. *Neutral equilibrium* occurs when a small displacement of the system causes the system to remain in its displaced state. In this case the potential energy of the system remains constant.
3. *Unstable equilibrium* occurs when a small displacement of the system causes the system to move farther away from its original position. In this case the original potential energy of the system is a maximum.

System Having One Degree of Freedom.

For *equilibrium* of a system having a single degree of freedom, defined by the independent coordinate q, it has been shown that the first derivative of the potential function for the system must be equal to zero; i.e., $dV/dq = 0$. If the potential function $V = V(q)$ is plotted, Fig. 11–13, the first derivative (equilibrium position) is represented as the slope dV/dq, which is zero when the function is maximum, minimum, or an inflection point.

If the *stability* of a body at the equilibrium position is to be investigated, it is necessary to determine the *second derivative* of V and evaluate it at the equilibrium position $q = q_{\text{eq}}$. As shown in Fig. 11–13a, if $V = V(q)$ is a *minimum,* then

$$\frac{dV}{dq} = 0, \quad \frac{d^2V}{dq^2} > 0 \quad \text{stable equilibrium} \qquad (11\text{–}10)$$

If $V = V(q)$ is a *maximum,* Fig. 11–13b, then

$$\frac{dV}{dq} = 0, \quad \frac{d^2V}{dq^2} < 0 \quad \text{unstable equilibrium} \qquad (11\text{–}11)$$

If the second derivative is zero, it will be necessary to investigate *higher-order* derivatives to determine the stability. In particular, stable equilibrium will occur if the order of the lowest remaining nonzero derivative is *even* and the sign of this nonzero derivative is positive when it is evaluated at $q = q_{\text{eq}}$; otherwise, it is unstable.

If the system is in neutral equilibrium, Fig. 11–13c, it is required that

$$\frac{dV}{dq} = \frac{d^2V}{dq^2} = \frac{d^3V}{dq^3} = \cdots = 0 \quad \text{neutral equilibrium} \qquad (11\text{–}12)$$

since then V must be constant at and around the "neighborhood" of q_{eq}.

System Having Two Degrees of Freedom.

A criterion for investigating stability becomes increasingly complex as the number of degrees of freedom for the system increases. For a system having two degrees of freedom, defined by independent coordinates (q_1, q_2), it may be verified (using the calculus of functions of two variables) that equilibrium and stability occur

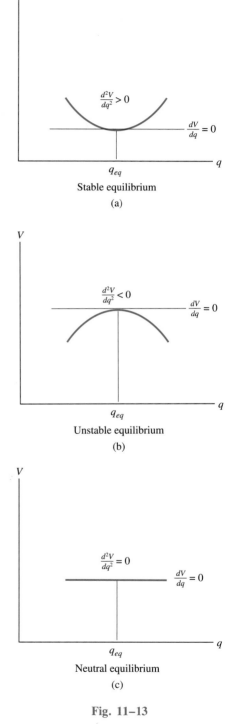

Stable equilibrium

(a)

Unstable equilibrium

(b)

Neutral equilibrium

(c)

Fig. 11–13

at a point (q_{1eq}, q_{2eq}) when

$$\frac{\partial V}{\partial q_1} = \frac{\partial V}{\partial q_2} = 0$$

$$\left[\left(\frac{\partial^2 V}{\partial q_1 \, \partial q_2}\right)^2 - \left(\frac{\partial^2 V}{\partial q_1^2}\right)\left(\frac{\partial^2 V}{\partial q_2^2}\right)\right] < 0$$

$$\left(\frac{\partial^2 V}{\partial q_1^2} + \frac{\partial^2 V}{\partial q_2^2}\right) > 0$$

Both equilibrium and instability occur when

$$\frac{\partial V}{\partial q_1} = \frac{\partial V}{\partial q_2} = 0$$

$$\left[\left(\frac{\partial^2 V}{\partial q_1 \, \partial q_2}\right)^2 - \left(\frac{\partial^2 V}{\partial q_1^2}\right)\left(\frac{\partial^2 V}{\partial q_2^2}\right)\right] < 0$$

$$\left(\frac{\partial^2 V}{\partial q_1^2} + \frac{\partial^2 V}{\partial q_2^2}\right) < 0$$

PROCEDURE FOR ANALYSIS

Using potential-energy methods, the equilibrium positions and the stability of a body or a system of connected bodies having a single degree of freedom can be obtained by applying the following procedure.

Potential Function. Formulate the potential function $V = V_g + V_e$ for the system. To do this, sketch the system so that it is located at some *arbitrary position* specified by the independent coordinate q. A horizontal *datum* is established through a *fixed point,** and the *gravitational potential energy* V_g is expressed in terms of the weight W of each member and its vertical distance y from the datum, $V_g = Wy$, Eq. 11–4. The elastic potential energy V_e of the system is expressed in terms of the stretch or compression, s, of any connecting spring and the spring's stiffness k, $V_e = \frac{1}{2}ks^2$, Eq. 11–5. Once V has been established, express the *position coordinates* y and s in terms of the independent coordinate q.

Equilibrium Position. The equilibrium position is determined by taking the first derivative of V and setting it equal to zero, $\delta V = 0$, Eq. 11–9.

Stability. Stability at the equilibrium position is determined by evaluating the second or higher-order derivatives of V as indicated by Eqs. 11–10 to 11–12.

The following examples illustrate this procedure numerically.

*The location of the datum is *arbitrary* since only the *changes* or differentials of V are required for investigation of the equilibrium position and its stability.

Example 11–5

The uniform link shown in Fig. 11–14a has a mass of 10 kg. The spring is unstretched when $\theta = 0°$. Determine the angle θ for equilibrium and investigate the stability at the equilibrium position.

SOLUTION

Potential Function. The datum is established at the top of the link when the *spring is unstretched,* Fig. 11–14b. When the link is located at the arbitrary position θ, the spring increases its potential energy by stretching and the weight decreases its potential energy. Hence,

$$V = V_e + V_g = \frac{1}{2}ks^2 - W\left(s + \frac{l}{2}\cos\theta - \frac{l}{2}\right)$$

Since $l = s + l\cos\theta$ or $s = l(1 - \cos\theta)$, then

$$V = \frac{1}{2}kl^2(1 - \cos\theta)^2 - \frac{Wl}{2}(1 - \cos\theta)$$

Equilibrium Position. The first derivative of V gives

$$\frac{dV}{d\theta} = kl^2(1 - \cos\theta)\sin\theta - \frac{Wl}{2}\sin\theta = 0$$

or

$$l\left[kl(1 - \cos\theta) - \frac{W}{2}\right]\sin\theta = 0$$

This equation is satisfied provided

$$\sin\theta = 0 \qquad \theta = 0° \qquad \textit{Ans.}$$

$$\theta = \cos^{-1}\left(1 - \frac{W}{2kl}\right) = \cos^{-1}\left[1 - \frac{10(9.81)}{2(200)(0.6)}\right] = 53.8° \quad \textit{Ans.}$$

Stability. Determining the second derivative of V gives

$$\frac{d^2V}{d\theta^2} = kl^2(1 - \cos\theta)\cos\theta + kl^2\sin\theta\sin\theta - \frac{Wl}{2}\cos\theta$$

$$= kl^2(\cos\theta - \cos 2\theta) - \frac{Wl}{2}\cos\theta$$

Substituting values for the constants, with $\theta = 0°$ and $\theta = 53.8°$, yields

$$\left.\frac{d^2V}{d\theta^2}\right|_{\theta=0°} = 200(0.6)^2(\cos 0° - \cos 0°) - \frac{10(9.81)(0.6)}{2}\cos 0°$$

$$= -29.4 < 0 \qquad \text{(unstable equilibrium at } \theta = 0°) \qquad \textit{Ans.}$$

$$\left.\frac{d^2V}{d\theta^2}\right|_{\theta=53.8°} = 200(0.6)^2(\cos 53.8° - \cos 107.6°) - \frac{10(9.81)(0.6)}{2}\cos 53.8°$$

$$= 46.9 > 0 \qquad \text{(stable equilibrium at } \theta = 53.8°) \qquad \textit{Ans.}$$

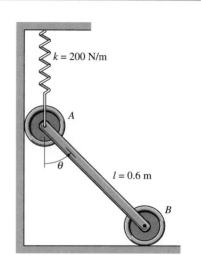

$k = 200$ N/m

A

θ

$l = 0.6$ m

B

(a)

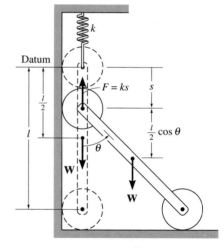

k

Datum

$F = ks$

s

$\frac{l}{2}$

l

$\frac{l}{2}\cos\theta$

θ

W

W

(b)

Fig. 11–14

Example 11-6

(a)

(b)

Fig. 11-15

Determine the mass m of the block required for equilibrium of the uniform 10-kg rod shown in Fig. 11-15a when $\theta = 20°$. Investigate the stability at the equilibrium position.

SOLUTION

Potential Function.   The datum is established through point A, Fig. 11-15b. When $\theta = 0°$, the block is assumed to be suspended $(y_W)_1$ below the datum. Hence, in the position θ,

$$V = V_e + V_g = 98.1\left(\frac{1.5 \sin \theta}{2}\right) - m(9.81)(\Delta y) \qquad (1)$$

The distance $\Delta y = (y_W)_2 - (y_W)_1$ may be related to the independent coordinate θ by measuring the difference in cord lengths $B'C$ and BC. Since

$$B'C = \sqrt{(1.5)^2 + (1.2)^2} = 1.92$$
$$BC = \sqrt{(1.5 \cos \theta)^2 + (1.2 - 1.5 \sin \theta)^2} = \sqrt{3.69 - 3.60 \sin \theta}$$

then

$$\Delta y = B'C - BC = 1.92 - \sqrt{3.69 - 3.60 \sin \theta}$$

Substituting the above result into Eq. 1 yields

$$V = 98.1\left(\frac{1.5 \sin \theta}{2}\right) - m(98.1)(1.92 - \sqrt{3.69 - 3.60 \sin \theta}) \quad (2)$$

Equilibrium Position

$$\frac{dV}{d\theta} = 73.6 \cos \theta - \left[\frac{m(98.1)}{2}\right]\left(\frac{3.60 \cos \theta}{\sqrt{3.69 - 3.60 \sin \theta}}\right) = 0$$

$$\left.\frac{dV}{d\theta}\right|_{\theta=20°} = 69.16 - 10.58m = 0$$

$$m = \frac{69.16}{10.58} = 6.54 \text{ kg} \qquad\qquad Ans.$$

Stability.   Taking the second derivative of Eq. 2, we obtain

$$\frac{d^2V}{d\theta^2} = -73.6 \sin \theta - \left[\frac{m(9.81)}{2}\right]\left(\frac{-1}{2}\right)\frac{(3.60 \cos \theta)^2}{(3.69 - 3.60 \sin \theta)^{3/2}}$$
$$- \frac{m(9.81)}{2}\left(\frac{-3.60 \sin \theta}{\sqrt{3.69 - 3.60 \sin \theta}}\right)$$

For the equilibrium position $\theta = 20°$, also $m = 6.54$ kg, so

$$\frac{d^2V}{d\theta^2} = 47.6 > 0 \qquad \text{(stable equilibrium at } \theta = 20°) \qquad Ans.$$

Example 11–7

The homogeneous block having a mass m rests on the top surface of the cylinder, Fig. 11–16a. Show that this is a condition of unstable equilibrium if $h > 2R$.

SOLUTION

Potential Function. The datum is established at the base of the cylinder, Fig. 11–16b. If the block is displaced by an amount θ from the equilibrium position, the potential function may be written in the form

$$V = V_e + V_g$$
$$= 0 + mgy$$

From Fig. 11–16b,

$$y = \left(R + \frac{h}{2}\right)\cos\theta + R\theta\sin\theta$$

Thus,

$$V = mg\left[\left(R + \frac{h}{2}\right)\cos\theta + R\theta\sin\theta\right]$$

(a)

(b)

Fig. 11–16

Equilibrium Position

$$\frac{dV}{d\theta} = mg\left[-\left(R + \frac{h}{2}\right)\sin\theta + R\sin\theta + R\theta\cos\theta\right] = 0$$

$$= mg\left(-\frac{h}{2}\sin\theta + R\theta\cos\theta\right) = 0$$

Obviously, $\theta = 0°$ is the equilibrium position that satisfies this equation.

Stability. Taking the second derivative of V yields

$$\frac{d^2V}{d\theta^2} = mg\left(-\frac{h}{2}\cos\theta + R\cos\theta - R\theta\sin\theta\right)$$

At $\theta = 0°$,

$$\left.\frac{d^2V}{d\theta^2}\right|_{\theta=0°} = -mg\left(\frac{h}{2} - R\right)$$

Since all the constants are positive, the block is in unstable equilibrium if $h > 2R$, for then $d^2V/d\theta^2 < 0$.

PROBLEMS

11–26. Solve Prob. 11–10 using the principle of potential energy, and investigate the stability at the equilibrium position.

11–27. Solve Prob. 11–15 using the principle of potential energy, and investigate the stability at the equilibrium position.

*11–28.** Solve Prob. 11–12 using the principle of potential energy, and investigate the stability at the equilibrium position.

11–29. Solve Prob. 11–18 using the principle of potential energy, and investigate the stability at the equilibrium position.

11–30. If the potential energy for a conservative two-degree-of-freedom system is expressed by the relation $V = (3y^2 + 2x^2)$ J, where y and x are given in meters, determine the equilibrium positions and investigate the stability at each position.

11–31. If the potential energy for a conservative one-degree-of-freedom system is expressed by the relation $V = (4x^3 - x^2 - 3x + 10)$ ft · lb, where x is given in feet, determine the equilibrium positions and investigate the stability at each position.

*11–32.** If the potential energy for a conservative one-degree-of-freedom system is expressed by the relation $V = (10 \cos 2\theta + 24 \sin \theta)$ ft · lb, $0 \le \theta \le 180°$, determine the equilibrium positions and investigate the stability at each position.

11–33. If the potential energy for a conservative two-degree-of-freedom system is expressed by the relation $V = (6y^2 + 2x^2)$ J, where x and y are given in meters, determine the equilibrium position and investigate the stability at this position.

11–34. The uniform rod OA weighs 20 lb, and when the rod is in the vertical position, the spring is unstretched. Determine the position θ for equilibrium. Investigate the stability at the equilibrium position.

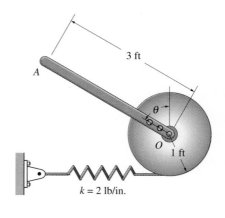

Prob. 11–34

11–35. Each of the three links of the mechanism has a weight of 10 lb. Determine the angle θ for equilibrium. The spring is unstretched when $\theta = 0°$. Investigate the stability at the equilibrium position.

Prob. 11–35

*11–36. The uniform members *AB* and *BC* each weigh 20 lb. Determine the angle θ for equilibrium. The spring has an unstretched length of 6 in. Investigate the stability at the equilibrium position.

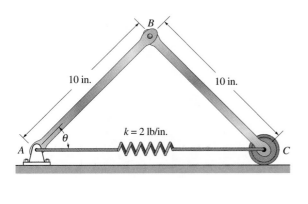

Prob. 11–36

11–37. The uniform beam has a weight *W*. Determine the angle θ for equilibrium. The spring is uncompressed when $\theta = 90°$. Neglect the weight of the rollers.

Prob. 11–37

11–38. The spring of the scale has an unstretched length *a*. Determine the angle θ for equilibrium when a weight *W* is supported on the platform. Neglect the weights of the members. What value of *W* would be required to keep the scale in neutral equilibrium when $\theta = 0°$?

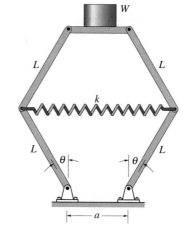

Prob. 11–38

11–39. The uniform bar *AB* weighs 10 lb. If the attached spring is unstretched when $\theta = 90°$, determine the angle θ for equilibrium. Investigate the stability of the bar when it is in the equilibrium position. Note that the spring always remains in the vertical position since the top end of the bar is attached to a roller which moves freely along the horizontal plane.

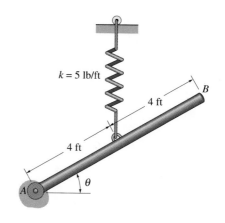

Prob. 11–39

*11–40.   Each of the two springs has an unstretched length of 500 mm. Determine the mass m needed to hold the springs in the equilibrium position shown, i.e., $y = 1$ m.

Prob. 11–40

11–42.   The spring attached to the mechanism has an unstretched length when $\theta = 90°$. Determine the position θ for equilibrium and investigate the stability of the mechanism at this position. Disk A is pin-connected to the frame at B and has a mass of 20 kg. Neglect the weight of the links.

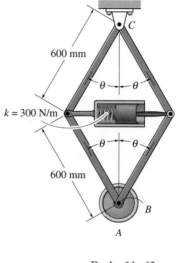

Prob. 11–42

11–41.   If each of the three links of the mechanism has a weight W, determine the angle θ for equilibrium. The spring, which always remains vertical, is unstretched when $\theta = 0°$.

11–43.   Determine the angle θ for equilibrium of the 10-lb weight. The spring is unstretched when $\theta = 0°$. Investigate the stability at the equilibrium position. Neglect the weight of the link.

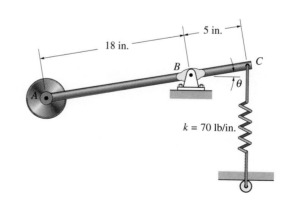

Prob. 11–41

Prob. 11–43

*11–44. The bar supports a weight of 500 lb at its end. If the springs are originally unstretched when the bar is vertical, and $k_1 = 300$ lb/ft, $k_2 = 500$ lb/ft, investigate the stability of the bar when it is in the vertical position.

11–46. The two bars each have a weight of 8 lb. Determine the angle θ for equilibrium and investigate the stability at the equilibrium position. The spring has an unstretched length of 1 ft.

3 ft

k_2

k_1 3 ft

3 ft

A

Prob. 11–44

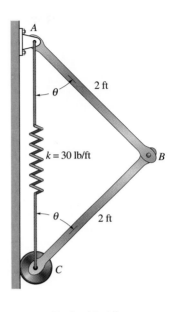

A

2 ft

θ

$k = 30$ lb/ft

θ 2 ft

B

C

Prob. 11–46

11–45. The bent rod has a weight of 5 lb/ft. A pivot of negligible size is attached at its center A and it is balanced as shown. Determine the length L of its vertical segments so that it remains in neutral equilibrium. Neglect the thickness of the rod.

11–47. Two uniform bars, each having a weight W, are pin-connected at their ends. If they are placed over a smooth cylindrical surface, show that the angle θ for equilibrium must satisfy the equation $\cos \theta / \sin^3 \theta = a/2r$.

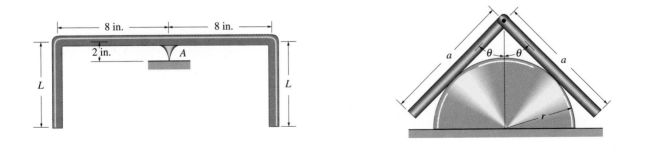

8 in. 8 in.

2 in. A

L L

Prob. 11–45

a θ θ a

r

Prob. 11–47

*11–48. The homogeneous cone has a conical cavity cut into it as shown. Determine the depth d of the cavity in terms of h so that the cone balances on the pivot and remains in neutral equilibrium.

Prob. 11–48

11–49. The hemispherical block has a conical cavity cut into it as shown. Determine the depth d of the cavity in terms of r so that the block balances on the pivot and remains in neutral equilibrium.

Prob. 11–49

11–50. A right circular cone is attached to the hemisphere. If both pieces have the same density ρ, determine the height h of the cone if the configuration is to be in neutral equilibrium at the position shown.

Prob. 11–50

11–51. A homogeneous block rests on top of the cylindrical surface. Derive the relationship between the radius of the cylinder, r, and the dimension of the block, b, for stable equilibrium. (*Hint:* Establish the potential energy function for a *small* angle θ; i.e., approximate $\sin \theta \approx 0$, and $\cos \theta \approx 1 - \theta^2/2$.)

Prob. 11–51

*11–52. The assembly consists of a semicylinder and a rectangular block. If the block weighs 8 lb and the semicylinder weighs 2 lb, investigate the stability when the assembly is resting in the equilibrium position. Set $h = 4$ in.

Prob. 11–52

11–53. The 2-lb semicylinder supports the block which has a specific weight of $\gamma = 80$ lb/ft^3. Determine the height h of the block which will produce neutral equilibrium in the position shown. The block has a length of 10 in.

11–54. The two blocks weigh 10 lb each and are suspended from the ends of springs as shown. Determine the stretch of the springs, x_1 and x_2, when the blocks are in equilibrium. The springs are unstretched when $x_1 = x_2 = 0$.

4 in.

Prob. 11–53

$k = 3$ lb/in.

x_1

$k = 8$ lb/in.

x_2

Prob. 11–54

REVIEW PROBLEMS

11–55. Determine the angle θ for equilibrium of the uniform 7-kg rod AC. Due to the roller guide at B, the spring remains vertical and is unstretched when $\theta = 0°$. Use the principle of virtual work.

*11–56. The assembly is used for exercise. It consists of four pin-connected bars, each of length L, and a spring of stiffness k and unstretched length a ($<2L$). If horizontal forces **P** and $-$**P** are applied to the handles so that θ is slowly decreased, determine the angle θ at which the magnitude of **P** becomes a maximum. Use the principle of virtual work.

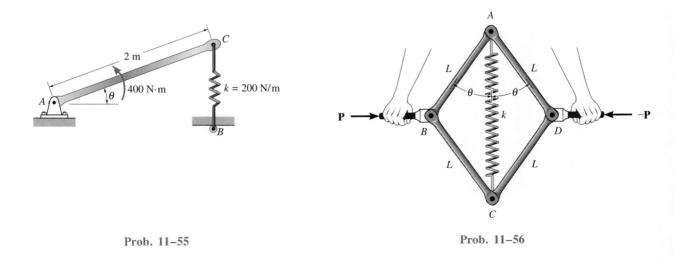

Prob. 11–55

Prob. 11–56

11–57.   If a force $P = 40$ N is applied perpendicular to the handle of the toggle press, determine the vertical compressive force developed at C; $\theta = 60°$. Use the principle of virtual work.

Prob. 11–57

11–59.   Determine the force developed in the spring required to keep the 10-lb rod AB in equilibrium when $\theta = 35°$. Use the principle of virtual work.

Prob. 11–59

11–58.   For the press shown, it takes n turns of the worm shaft at A to produce one turn of gear B. Determine the compression developed at C if a load P is applied to the handle. The pitch of the screw S is p. Use the principle of virtual work.

Prob. 11–58

*11–60.   The uniform rod AB has a weight of 10 lb. If the spring DC is unstretched when $\theta = 0°$, determine the angle θ for equilibrium using the principle of virtual work. The spring always remains in the horizontal position because of the roller guide at D.

11–61.   Solve Prob. 11–60 using the principle of potential energy. Investigate the stability of the rod when it is in the equilibrium position.

Probs. 11–60/11–61

11–62. The 4-ft members of the mechanism are pin-connected at their centers. If vertical forces act at C and E as shown, determine the angle θ for equilibrium. The spring is unstretched when $\theta = 45°$. Neglect the weight of the members. Use the principle of virtual work.

*11–64. A uniform can having a weight W is suspended by a cord. Assuming that the wall is smooth, determine the angle θ at which the cord hangs from the wall for equilibrium. Is the can in stable equilibrium?

Prob. 11–62

Prob. 11–64

11–63. If the uniform bar has a weight W, determine the angle θ for equilibrium. The spring is unstretched when $\theta = 0°$. Neglect the weight of the rollers. Use the principle of potential energy.

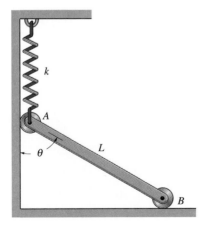

Prob. 11–63

A Mathematical Expressions

Quadratic Formula

If $ax^2 + bx + c = 0$, then $x = \dfrac{-b \pm \sqrt{b^2 - 4ac}}{2a}$

Hyperbolic Functions

$\sinh x = \dfrac{e^x - e^{-x}}{2}$, $\cosh x = \dfrac{e^x + e^{-x}}{2}$, $\tanh x = \dfrac{\sinh x}{\cosh x}$

Trigonometric Identities

$\sin^2 \theta + \cos^2 \theta = 1$

$\sin (\theta \pm \phi) = \sin \theta \cos \phi \pm \cos \theta \sin \phi$

$\sin 2\theta = 2 \sin \theta \cos \theta$

$\cos (\theta \pm \phi) = \cos \theta \cos \phi \mp \sin \theta \sin \phi$

$\cos 2\theta = \cos^2 \theta - \sin^2 \theta$

$\cos \theta = \pm\sqrt{\dfrac{1 + \cos 2\theta}{2}}$, $\sin \theta = \pm\sqrt{\dfrac{1 - \cos 2\theta}{2}}$

$\tan \theta = \dfrac{\sin \theta}{\cos \theta}$

$1 + \tan^2 \theta = \sec^2 \theta$ $\qquad\qquad$ $1 + \cot^2 \theta = \csc^2 \theta$

Power-Series Expansions

$\sin x = x - \dfrac{x^3}{3!} + \dfrac{x^5}{5!} - \dfrac{x^7}{7!} + \cdots$

$\cos x = 1 - \dfrac{x^2}{2!} + \dfrac{x^4}{4!} - \dfrac{x^6}{6!} + \cdots$

$\sinh x = x + \dfrac{x^3}{3!} + \dfrac{x^5}{5!} + \cdots$

$\cosh x = 1 + \dfrac{x^2}{2!} + \dfrac{x^4}{4!} + \cdots$

Derivatives

$\dfrac{d}{dx}(u^n) = nu^{n-1}\dfrac{du}{dx}$ $\qquad$ $\dfrac{d}{dx}(\sin u) = \cos u\,\dfrac{du}{dx}$ $\qquad$ $\dfrac{d}{dx}(\cot u) = -\csc^2 u\,\dfrac{du}{dx}$ $\qquad$ $\dfrac{d}{dx}(\sinh u) = \cosh u\,\dfrac{du}{dx}$

$\dfrac{d}{dx}(uv) = u\,\dfrac{dv}{dx} + v\,\dfrac{du}{dx}$ $\qquad$ $\dfrac{d}{dx}(\cos u) = -\sin u\,\dfrac{du}{dx}$ $\qquad$ $\dfrac{d}{dx}(\sec u) = \tan u \sec u\,\dfrac{du}{dx}$ $\qquad$ $\dfrac{d}{dx}(\cosh u) = \sinh u\,\dfrac{du}{dx}$

$\dfrac{d}{dx}\left(\dfrac{u}{v}\right) = \dfrac{v\,\dfrac{du}{dx} - u\,\dfrac{dv}{dx}}{v^2}$ $\qquad$ $\dfrac{d}{dx}(\tan u) = \sec^2 u\,\dfrac{du}{dx}$ $\qquad$ $\dfrac{d}{dx}(\csc u) = -\csc u \cot u\,\dfrac{du}{dx}$

Integrals

$$\int x^n \, dx = \frac{x^{n+1}}{n+1} + C, \; n \neq -1$$

$$\int \frac{dx}{a+bx} = \frac{1}{b} \ln(a+bx) + C$$

$$\int \frac{dx}{a+bx^2} = \frac{1}{2\sqrt{-ba}} \ln\left[\frac{\sqrt{a}+2\sqrt{-b}}{\sqrt{a}-x\sqrt{-b}}\right] + C,$$
$$a > 0, \; b < 0$$

$$\int \frac{x \, dx}{a+bx^2} = \frac{1}{2b} \ln(bx^2+a) + C$$

$$\int \frac{x^2 \, dx}{a+bx^2} = \frac{x}{b} - \frac{a}{b\sqrt{ab}} \tan^{-1} \frac{x\sqrt{ab}}{a} + C$$

$$\int \frac{dx}{a^2-x^2} = \frac{1}{2a} \ln\left[\frac{a+x}{a-x}\right] + C, \; a^2 > x^2$$

$$\int \sqrt{a+bx} \, dx = \frac{2}{3b} \sqrt{(a+bx)^3} + C$$

$$\int x\sqrt{a+bx} \, dx = \frac{-2(2a-3bx)\sqrt{(a+bx)^3}}{15b^2} + C$$

$$\int x^2\sqrt{a+bx} \, dx =$$
$$\frac{2(8a^2-12abx+15b^2x^2)\sqrt{(a+bx)^3}}{105b^3} + C$$

$$\int \sqrt{a^2-x^2} \, dx = \frac{1}{2}\left[x\sqrt{a^2-x^2} + a^2 \sin^{-1} \frac{x}{a}\right] + C,$$
$$a > 0$$

$$\int x\sqrt{a^2-x^2} \, dx = -\frac{1}{3}\sqrt{(a^2-x^2)^3} + C$$

$$\int x^2\sqrt{a^2-x^2} \, dx = -\frac{x}{4}\sqrt{(a^2-x^2)^3}$$
$$+ \frac{a^2}{8}\left(x\sqrt{a^2-x^2} + a^2 \sin^{-1} \frac{x}{a}\right) + C, \; a > 0$$

$$\int \sqrt{x^2 \pm a^2} \, dx =$$
$$\frac{1}{2}\left[x\sqrt{x^2 \pm a^2} \pm a^2 \ln(x + \sqrt{x^2 \pm a^2})\right] + C$$

$$\int x\sqrt{x^2 \pm a^2} \, dx = \frac{1}{3}\sqrt{(x^2 \pm a^2)^3} + C$$

$$\int x^2\sqrt{x^2 \pm a^2} \, dx = \frac{x}{4}\sqrt{(x^2 \pm a^2)^3}$$
$$\pm \frac{a^2}{8}x\sqrt{x^2 \pm a^2} - \frac{a^4}{8} \ln(x + \sqrt{x^2 \pm a^2}) + C$$

$$\int \frac{dx}{\sqrt{a+bx}} = \frac{2\sqrt{a+bx}}{b} + C$$

$$\int \frac{x \, dx}{\sqrt{x^2 \pm a^2}} = \sqrt{x^2 \pm a^2} + C$$

$$\int \frac{dx}{\sqrt{a+bx+cx^2}} = \frac{1}{\sqrt{c}} \ln\left[\sqrt{a+bx+cx^2} + \right.$$
$$\left. x\sqrt{c} + \frac{b}{2\sqrt{c}}\right] + C, \; c > 0$$
$$= \frac{1}{\sqrt{-c}} \sin^{-1}\left(\frac{-2cx-b}{\sqrt{b^2-4ac}}\right) + C, \; c < 0$$

$$\int \sin x \, dx = -\cos x + C$$

$$\int \cos x \, dx = \sin x + C$$

$$\int x \cos(ax) \, dx = \frac{1}{a^2} \cos(ax) + \frac{x}{a} \sin(ax) + C$$

$$\int x^2 \cos(ax) \, dx = \frac{2x}{a^2} \cos(ax) + \frac{a^2x^2-2}{a^3} \sin(ax) + C$$

$$\int e^{ax} \, dx = \frac{1}{a}e^{ax} + C$$

$$\int x \, e^{ax} \, dx = \frac{e^{ax}}{a^2}(ax-1) + C$$

$$\int \sinh x \, dx = \cosh x + C$$

$$\int \cosh x \, dx = \sinh x + C$$

B Numerical and Computer Analysis

Occasionally the application of the laws of mechanics will lead to a system of equations for which a closed-form solution is difficult or impossible to obtain. When confronted with this situation, engineers will often use a numerical method which in most cases can be programmed on a microcomputer or "programmable" pocket calculator. Here we will briefly present a computer program for solving a set of linear algebraic equations and three numerical methods which can be used to solve an algebraic or transcendental equation, evaluate a definite integral, and solve an ordinary differential equation. Application of each method will be explained by example, and an associated computer program written in Microsoft BASIC, which is designed to run on most personal computers, is provided.* A text on numerical analysis should be consulted for further discussion regarding a check of the accuracy of each method and the inherent errors that can develop from the methods.

B.1 Linear Algebraic Equations

Application of the equations of static equilibrium or the equations of motion sometimes requires solving a set of linear algebraic equations. The computer program listed in Fig. B–1 can be used for this purpose. It is based on the

---

*Similar types of programs can be written or purchased for programmable pocket calculators.

```
1 PRINT"Linear system of equations":PRINT
2 DIM A(10,11)
3 INPUT"Input number of equations : ",N
4 PRINT
5 PRINT"A  coefficients"
6 FOR I = 1 TO N
7 FOR J = 1 TO N
8 PRINT "A(";I;",";J;
9 INPUT")=",A(I,J)
10 NEXT J
11 NEXT I
12 PRINT
13 PRINT"B  coefficients"
14 FOR I = 1 TO N
15 PRINT "B(";I;
16 INPUT")=",A(I,N+1)
17 NEXT I
18 GOSUB 25
19 PRINT

20 PRINT"Unknowns"
21 FOR I = 1 TO N
22 PRINT "X(";I;")=";A(I,N+1)
23 NEXT I
24 END
25 REM Subroutine Guassian
26 FOR M=1 TO N
27 NP=M
28 BG=ABS(A(M,M))
29 FOR I = M TO N
30 IF ABS(A(I,M))<=BG THEN 33
31 BG=ABS(A(I,M))
32 NP=I
33 NEXT I
34 IF NP=M THEN 40
35 FOR I = M TO N+1
36 TE=A(M,I)
37 A(M,I)=A(NP,I)
38 A(NP,I)=TE

39 NEXT I
40 FOR I = M+1 TO N
41 FC=A(I,M)/A(M,M)
42 FOR J = M+1 TO N+1
43 A(I,J)=A(I,J)-FC*A(M,J)
44 NEXT J
45 NEXT I
46 NEXT M
47 A(N,N+1)=A(N,N+1)/A(N,N)
48 FOR I = N-1 TO 1 STEP -1
49 SM=0
50 FOR J=I+1 TO N
51 SM=SM+A(I,J)*A(J,N+1)
52 NEXT J
53 A(I,N+1)=(A(I,N+1)-SM)/A(I,I)
54 NEXT I
55 RETURN
```

Fig. B–1

method of a Gaussian elimination and can solve at most 10 equations with 10 unknowns. To do so, the equations should first be written in the following general format:

$$A_{11}x_1 + A_{12}x_2 + \cdots + A_{1n}x_n = B_1$$
$$A_{21}x_1 + A_{22}x_2 + \cdots + A_{2n}x_n = B_2$$
$$\cdot$$
$$\cdot$$
$$\cdot$$
$$A_{n1}x_1 + A_{n2}x_2 + \cdots + A_{nn}x_n = B_n$$

The "A" and "B" coefficients are "called" for when running the program. The output presents the unknowns $x_1, \ldots, x_n$.

Example B–1

Solve the two equations

$$3x_1 + x_2 = 4$$
$$2x_1 - x_2 = 10$$

SOLUTION

When the program begins to run, it first calls for the number of equations (2); then the A coefficients in the sequence $A_{11} = 3$, $A_{12} = 1$, $A_{21} = 2$, $A_{22} = -1$; and finally the B coefficients $B_1 = 4$, $B_2 = 10$. The output appears as

Unknowns

$X(1) = 2.8$ *Ans.*

$X(2) = -4.4$ *Ans.*

B.2 Simpson's Rule

Simpson's rule is a numerical method that can be used to determine the area under a curve given as a graph or as an explicit function $y = f(x)$. Likewise, it can be used to compute the value of a definite integral which involves the function $y = f(x)$. To do so, the area must be subdivided into an *even number* of strips or intervals having a width h. The curve between three consecutive ordinates is approximated by a parabola, and the entire area or definite integral is then determined from the formula

$$\int_{x_0}^{x_n} f(x)\ dx \simeq \frac{h}{3}[y_0 + 4(y_1 + y_3 + \cdots + y_{n-1})$$
$$+ 2(y_2 + y_4 + \cdots + y_{n-2}) + y_n]\quad \text{(B–1)}$$

The computer program for this equation is given in Fig. B–2. For its use, we must first specify the function (on line 6 of the program). The upper and lower limits of the integral and the number of intervals are called for when the program is executed. The value of the integral is then given as the output.

```
1 PRINT"Simpson's rule":PRINT
2 PRINT" To execute this program :":PRINT
3 PRINT"    1- Modify right-hand side of the equation given below,
4 PRINT"       then press RETURN key"
5 PRINT"    2- Type  RUN 6":PRINT:EDIT 6
6 DEF FNF(X)=LOG(X)
7 PRINT:INPUT" Enter Lower Limit = ",A
8 INPUT" Enter Upper Limit = ",B
9 INPUT" Enter Number (even) of Intervals = ",N%
10 H=(B-A)/N%:AR=FNF(A):X=A+H
11 FOR J%=2 TO N%
12 K=2*(2-J%+2*INT(J%/2))
13 AR=AR+K*FNF(X)
14 X=X+H:NEXT J%
15 AR=H*(AR+FNF(B))/3
16 PRINT" Integral = ",AR
17 END
```

Fig. B–2

Example B–2

Evaluate the definite integral

$$\int_2^5 \ln x \, dx$$

SOLUTION

The interval $x_0 = 2$ to $x_6 = 5$ will be divided into six equal parts ($n = 6$), each having a width $h = (5 - 2)/6 = 0.5$. We then compute $y = f(x) = \ln x$ at each point of subdivision.

n	x_n	y_n
0	2	0.693
1	2.5	0.916
2	3	1.099
3	3.5	1.253
4	4	1.386
5	4.5	1.504
6	5	1.609

Thus, Eq. B–1 becomes

$$\int_2^5 \ln x \, dx \simeq \frac{0.5}{3} [0.693 + 4(0.916 + 1.253 + 1.504)$$
$$+ 2(1.099 + 1.386) + 1.609]$$
$$\simeq 3.66 \qquad \qquad Ans.$$

This answer is equivalent to the exact answer to three significant figures. Obviously, accuracy to a greater number of significant figures can be improved by selecting a smaller interval h (or larger n).

Using the computer program, we first specify the function $\ln x$, line 6 in Fig. B–2. During execution, the program input requires the upper and lower limits 2 and 5, and the number of intervals $n = 6$. The output appears as

$$\text{Integral} = 3.66082 \qquad \qquad Ans.$$

B.3 The Secant Method

The secant method is used to find the real roots of an algebraic or transcendental equation $f(x) = 0$. The method derives its name from the fact that the formula used is established from the slope of the secant line to the graph $y = f(x)$. This slope is $[f(x_n) - f(x_{n-1})]/(x_n - x_{n-1})$, and the secant formula is

$$x_{n+1} = x_n - f(x_n)\left[\frac{x_n - x_{n-1}}{f(x_n) - f(x_{n-1})}\right] \qquad \text{(B-2)}$$

For application it is necessary to provide two initial guesses, x_0 and x_1, and thereby evaluate x_2 from Eq. B–2 ($n = 1$). One then proceeds to reapply Eq. B–2 with x_1 and the calculated value of x_2 and obtain x_3 ($n = 2$), etc., until the value $x_{n+1} \simeq x_n$. One can see this will occur if x_n is approaching the root of the function $f(x) = 0$, since the correction term on the right of Eq. B–2 will tend toward zero. In particular, the larger the slope, the smaller the correction to x_n, and the faster the root will be found. On the other hand, if the slope is very small in the neighborhood of the root, the method leads to large corrections for x_n, and convergence to the root is slow and may even lead to a failure to find it. In such cases other numerical techniques must be used for solution.

A computer program based on Eq. B–2 is listed in Fig. B–3. We must first specify the function on line 7 of the program. When the program is executed, two initial guesses, x_0 and x_1, must be entered in order to approximate the solution. The output specifies the value of the root. If it cannot be determined, this is so stated.

```
1 PRINT"Secant method":PRINT
2 PRINT" To execute this program :":PRINT
3 PRINT"    1) Modify right hand side of the equation given below,"
4 PRINT"       then press RETURN key."
5 PRINT"    2) Type  RUN 7"
6 PRINT:EDIT 7
7 DEF FNF(X)=.5*SIN(X)-2*COS(X)+1.3
8 INPUT"Enter point #1 =",X
9 INPUT"Enter point #2 =",X1
10 IF X=X1 THEN 14
11 EP=.00001:TL=2E-20
12 FP=(FNF(X1)-FNF(X))/(X1-X)
13 IF ABS(FP)>TL THEN 15
14 PRINT"Root can not be found.":END
15 DX=FNF(X1)/FP
16 IF ABS(DX)>EP THEN 19
17 PRINT "Root = ";X1;"      Function evaluated at this root = ";FNF(X1)
18 END
19 X=X1:X1=X1-DX
20 GOTO 12
```

Fig. B–3

Example B–3

Determine the root of the equation

$$f(x) = 0.5 \sin x - 2 \cos x + 1.30 = 0$$

SOLUTION

Guesses of the initial roots will be $x_0 = 45°$ and $x_1 = 30°$. Applying Eq. B–2,

$$x_2 = 30° - (-0.1821)\frac{(30° - 45°)}{(-0.1821 - 0.2393)} = 36.48°$$

Using this value in Eq. B–2, along with $x_1 = 30°$, we have

$$x_3 = 36.48° - (-0.0108)\frac{36.48° - 30°}{(-0.0108 + 0.1821)} = 36.89°$$

Repeating the process with this value and $x_2 = 36.48°$ yields

$$x_4 = 36.89° - (0.0005)\left[\frac{36.89° - 36.48°}{(0.0005 + 0.0108)}\right] = 36.87°$$

Thus $x = 36.9°$ is appropriate to three significant figures.

If the problem is solved using the computer program, first we specify the function, line 7 in Fig. B–3. During execution, the first and second guesses must be entered in radians. Choosing these to be 0.8 rad and 0.5 rad, the result appears as

Root = 0.6435022.

Function evaluated at this root = 1.66893E–06.

This result converted from radians to degrees is therefore

$$x = 36.9° \qquad\qquad Ans.$$

B.4 The Runge-Kutta Method

The Runge-Kutta method is used to solve an ordinary differential equation. It consists of applying a set of formulas which are used to find specific values of y for corresponding incremental values h in x. The formulas given in general form are as follows:

First-Order Equation. To integrate $\dot{x} = f(t, x)$ step by step, use

$$x_{i+1} = x_i + \frac{1}{6}(k_1 + 2k_2 + 2k_3 + k_4) \qquad \text{(B–3)}$$

where

$$k_1 = hf(t_i, x_i)$$
$$k_2 = hf\left(t_i + \frac{h}{2}, x_i + \frac{k_1}{2}\right)$$
$$k_3 = hf\left(t_i + \frac{h}{2}, x_i + \frac{k_2}{2}\right) \qquad \text{(B–4)}$$
$$k_4 = hf(t_i + h, x_i + k_3)$$

Second-Order Equation. To integrate $\ddot{x} = f(t, x, \dot{x})$, use

$$x_{i+1} = x_i + h\left[\dot{x}_i + \frac{1}{6}(k_1 + k_2 + k_3)\right]$$
$$\dot{x}_{i+1} = \dot{x}_i + \frac{1}{6}(k_1 + 2k_2 + 3k_3 + k_4) \qquad \text{(B–5)}$$

where

$$k_1 = hf(t_i, x_i, \dot{x}_i)$$
$$k_2 = hf\left(t_i + \frac{h}{2}, x_i + \frac{h}{2}\dot{x}_i, \dot{x}_i + \frac{k_1}{2}\right)$$
$$k_3 = hf\left(t_i + \frac{h}{2}, x_i + \frac{h}{2}\dot{x}_i + \frac{h}{4}k_1, \dot{x}_i + \frac{k_2}{2}\right) \qquad \text{(B–6)}$$
$$k_4 = hf\left(t_i + h, x_i + h\dot{x}_i + \frac{h}{2}k_2, \dot{x}_i + k_3\right)$$

To apply these equations, one starts with initial values $t_i = t_0$ and $x_i = x_0$. Also, $\dot{x}_i = \dot{x}_0$ for the second-order equation. Choosing an increment h for t_0, the four constants k are computed and these results are substituted into Eq. B–3 or B–5 in order to compute $x_{i+1} = x_1$, $\dot{x}_{i+1} = x_1$, corresponding to $t_{i+1} = t_1 = t_0 + h$. Repeating this process using t_1, x_1, $\dot{x}_1$ and h, values for x_2, $\dot{x}_2$ and $t_2 = t_1 + h$ are then computed, etc.

Computer programs which solve first and second-order differential equations by this method are listed in Figs. B–4 and B–5, respectively. In order to use these programs, we must specify the function $\dot{x} = f(t, x)$ or $\ddot{x} = f(t, x, \dot{x})$ on line 7, and during the program execution, input the initial values t_0, x_0, $\dot{x}_0$ (for second-order equation), the final time t_n, and the step size h. The output gives the values of t, x, and $\dot{x}$ for each time increment until t_n is reached.

```
1 PRINT"Runge-Kutta Method for 1-st order Differential Equation":PRINT
2 PRINT" To execute this program :":PRINT
3 PRINT"   1) Modify right hand side of the equation given below,"
4 PRINT"      then Press RETURN key"
5 PRINT"   2) Type  RUN 7"
6 PRINT:EDIT 7
7 DEF FNF(T,X)=5*T+X
8 CLS:PRINT" Initial Conditions":PRINT
9 INPUT"Input  t   = ",T
10 INPUT"       x   = ",X
11 INPUT"Final  t   = ",T1
12 INPUT"step size  = ",H:PRINT
13 PRINT"       t              x"
14 IF T>=T1+H THEN 23
15 PRINT USING"######.#####";T;X
16 K1=H*FNF(T,X)
17 K2=H*FNF(T+.5*H,X+.5*K1)
18 K3=H*FNF(T+.5*H,X+.5*K2)
19 K4=H*FNF(T+H,X+K3)
20 T=T+H
21 X=X+(K1+K2+K2+K3+K3+K4)/6
22 GOTO 14
23 END
```

Fig. B–4

```
1 PRINT"Runge-Kutta Method for 2-nd order Differential Equation":PRINT
2 PRINT" To execute this program :":PRINT
3 PRINT"   1) Modify right hand side of the equation given below,"
4 PRINT"      then Press RETURN key"
5 PRINT"   2) Type  RUN 7"
6 PRINT:EDIT 7
7 DEF FNF(T,X,XD)=
8 INPUT"Input  t   = ",T
9 INPUT"       x   = ",X
10 INPUT"     dx/dt = ",XD
11 INPUT"Final  t   = ",T1
12 INPUT"step size  = ",H:PRINT
13 PRINT"       t              x         dx/dt"
14 IF T>=T1+H THEN 24
15 PRINT USING"######.#####";T;X;XD
16 K1=H*FNF(T,X,XD)
17 K2=H*FNF(T+.5*H,X+.5*H*XD,XD+.5*K1)
18 K3=H*FNF(T+.5*H,X+(.5*H)*(XD+.5*K1),XD+.5*K2)
19 K4=H*FNF(T+H,X+H*XD+.5*H*K2,XD+K3)
20 T=T+H
21 X=X+H*XD+H*(K1+K2+K3)/6
22 XD=XD+(K1+K2+K2+K3+K3+K4)/6
23 GOTO 14
24 END
```

Fig. B–5

Example B–4

Solve the differential equation $\dot{x} = 5t + x$. Obtain the results for two steps using time increments of $h = 0.02$ s. At $t_0 = 0$, $x_0 = 0$.

SOLUTION

This is a first-order equation, so Eqs. B–3 and B–4 apply. Thus, for $t_0 = 0$, $x_0 = 0$, $h = 0.02$, we have

$$k_1 = 0.02[0 + 0] = 0$$
$$k_2 = 0.02[5(0.1) + 0] = 0.001$$
$$k_3 = 0.02[5(0.01) + 0.0005] = 0.00101$$
$$k_4 = 0.02[5(0.02) + 0.00101] = 0.00202$$
$$x_1 = 0 + \tfrac{1}{6}[0 + 2(0.001) + 2(0.00101) + 0.00202] = 0.00101$$

Using the values $t_1 = 0 + 0.02 = 0.02$ and $x_1 = 0.00101$ with $h = 0.02$, the value for x_2 is now computed from Eqs. B–3 and B–4.

$$k_1 = 0.02[5(0.02) + 0.00101] = 0.00202$$
$$k_2 = 0.02[5(0.03) + 0.00202] = 0.00304$$
$$k_3 = 0.02[5(0.03) + 0.00253] = 0.00305$$
$$k_4 = 0.02[5(0.04) + 0.00406] = 0.00408$$
$$x_2 = 0.001 + \tfrac{1}{6}[0.00202 + 2(0.00304) + 2(0.00305) + 0.00408]$$
$$x_2 = 0.00405 \qquad\qquad\qquad Ans.$$

To solve this problem using the computer program in Fig. B–4, the function is first specified on line 7, then during the execution of the program, the data $t_0 = 0$, $x_0 = 0$, $t_n = 0.04$, and $h = 0.02$ is entered. The results appear as

t	x
0.00000	0.00000
0.02000	0.00101
0.04000	0.00405

Ans.

Answers

2–35. 222 N, 4.26°

2–37. $F_A = 1.41$ kN, $F_R = 2.73$ kN$\downarrow$

2–38. $F_C = 4.51$ kN, $F_B = 5.96$ kN

2–39. 40 cosec θ lb

2–41. 84.3 lb, 37.2°

2–42. 407 lb, 207°

2–43. 103°, 88.1 lb

2–45. 63.7°, $F_3 = 1.20$ F_1

2–46. 68.6°, 960 N

2–47. 839 N, 14.8°

2–49. $F_B = 311$ N, $F_A = 439$ N

2–50. 60°, $F_A = 520$ N, $F_B = 300$ N

2–51. 11.3 kN

2–53. $\mathbf{F}_1 = \{30.6\mathbf{i} - 25.7\mathbf{j}\}$ N,
$\mathbf{F}_2 = \{-23.0\mathbf{i} + 19.3\mathbf{j} + 52.0\mathbf{k}\}$ N,
$\mathbf{F}_R = \{7.66\mathbf{i} - 6.43\mathbf{j} + 52.0\mathbf{k}\}$ N, $F_R = 52.9$ N,
$\alpha = 81.7°$, $\beta = 97.0°$, $\gamma = 10.9°$

2–54. $\mathbf{F}_1 = \{-25.0\mathbf{i} + 43.3\mathbf{j}\}$ lb, $\mathbf{F}_2 = \{-90.0\mathbf{j} + 120\mathbf{k}\}$ lb,
$\mathbf{F}_3 = \{74.5\mathbf{j} - 43.0\mathbf{k}\}$ lb,
$\mathbf{F}_R = \{-25.0\mathbf{i} + 27.8\mathbf{j} + 77.0\mathbf{k}\}$ lb, $F_R = 85.6$ lb,
$\alpha = 107°$, $\beta = 71.1°$, $\gamma = 25.9°$

2–55. $\mathbf{F} = \{289\mathbf{i} + 225\mathbf{j} - 261\mathbf{k}\}$ lb

2–57. 581 N, $\alpha = 72.5°$, $\beta = 83.4°$, $\gamma = 18.8°$

2–58. 166 N, $\alpha = 97.5°$, $\beta = 63.7°$, $\gamma = 27.5°$

2–59. $\alpha_1 = 36.9°$, $\beta_2 = 90°$, $\gamma_3 = 53.1°$, $\alpha_R = 69.3°$,
$\beta_R = 52.2°$, $\gamma_R = 45°$

2–61. $\alpha = 52.2°$, $\beta = 52.2°$, $\gamma = 120°$

2–62. $\mathbf{F} = \{13.7\mathbf{i} + 37.6\mathbf{j} + 30\mathbf{k}\}$ N

2–63. 50 N, $\alpha = 74.1°$, $\beta = 41.3°$, $\gamma = 53.1°$

2–65. $\mathbf{F} = \{200\mathbf{i} + 283\mathbf{j} + 200\mathbf{k}\}$lb

2–66. $F_x = 6$ kN, $F_y = 6$ kN, $F_z = 8.49$ kN

2–67. $F = 2.02$ kN, $F_y = 0.523$ kN

2–69. 180 N, $\alpha_2 = 147°$, $\beta_2 = 119°$, $\gamma_2 = 75.0°$

2–70. $\alpha = 73.4°$, $\beta = 64.6°$, $\gamma = 31.0°$

2–71. $\mathbf{r}_{AB} = \{2\mathbf{i} - 7\mathbf{j} - 5\mathbf{k}\}$ m, $r_{AB} = 8.83$ m, $\alpha = 76.9°$,
$\beta = 142°$, $\gamma = 124°$

2–73. 7.55 m, $\alpha = 105°$, $\beta = 22.0°$, $\gamma = 74.6°$

2–74. $\mathbf{r} = \{4\mathbf{i} - 2\mathbf{j} + 3\mathbf{k}\}$ m, $r = 5.39$ m, $\alpha = 42.0°$,
$\beta = 112°$, $\gamma = 56.1°$

2–75. $\mathbf{r} = \{9.48\mathbf{i} - 1.24\mathbf{j} + 2.30\mathbf{k}\}$ m, $r = 9.83$ m,
$\alpha = 15.4°$, $\beta = 97.2°$, $\gamma = 76.5°$

2–77. 6.40 ft

2–78. 19.0 in.

2–79. $y = 16.7$ ft, $z = 50.2$ ft

2–81. 284 ft

2–82. $r_{AD} = 1.5$ m, $r_{BD} = 1.5$ m, $r_{CD} = 1.73$ m

2–83. $\mathbf{F} = \{-2\mathbf{i} + 1\mathbf{j} + 2\mathbf{k}\}$ kN, $\alpha = 132°$, $\beta = 70.5°$,
$\gamma = 48.2°$

2–85. $\mathbf{F} = \{366\mathbf{i} + 294\mathbf{j} - 172\mathbf{k}\}$ lb, $\alpha = 43.0°$, $\beta = 53.9°$,
$\gamma = 110°$

2–87. $\mathbf{F} = \{29.7\mathbf{i} + 35.4\mathbf{j} - 80.0\mathbf{k}\}$lb, $\alpha = 71.3°$, $\beta = 67.5°$,
$\gamma = 150°$

2–89. $\mathbf{F}_1 = \{75.5\mathbf{i} - 43.6\mathbf{j} - 122\mathbf{k}\}$ lb,
$\mathbf{F}_2 = \{26.8\mathbf{i} + 33.5\mathbf{j} - 90.4\mathbf{k}\}$ lb, $F_R = 236$ lb,
$\alpha = 64.3°$, $\beta = 92.5°$, $\gamma = 154°$

2–90. 10.0 ft, $\mathbf{F} = \{-19.1\mathbf{i} - 14.9\mathbf{j} + 43.7\mathbf{k}\}$ lb, $\alpha = 112°$,
$\beta = 107°$, $\gamma = 29.0°$

2–91. $\mathbf{F}_1 = \{-3.79\mathbf{i} + 11.4\mathbf{k}\}$ lb,
$\mathbf{F}_2 = \{-6.65\mathbf{i} - 11.8\mathbf{j} + 11.8\mathbf{k}\}$ lb, $F_R = 28.1$ lb,
$\alpha = 112°$, $\beta = 115°$, $\gamma = 34.2°$

2–93. $\mathbf{F}_C = \{-324\mathbf{i} - 130\mathbf{j} + 195\mathbf{k}\}$ N,
$\mathbf{F}_B = \{-324\mathbf{i} + 130\mathbf{j} + 195\mathbf{k}\}$ N,
$\mathbf{F}_E = \{-194\mathbf{i} + 291\mathbf{k}\}$ N

2–94. $\mathbf{F}_A = \{-1.46\mathbf{i} + 5.82\mathbf{k}\}$ kN,
$\mathbf{F}_C = \{0.857\mathbf{i} + 0.857\mathbf{j} + 4.85\mathbf{k}\}$ kN,
$\mathbf{F}_B = \{0.970\mathbf{i} - 1.68\mathbf{j} + 7.76\mathbf{k}\}$ kN, $F_R = 18.5$ kN,
$\alpha = 88.8°$, $\beta = 92.6°$, $\gamma = 2.81°$

2–95. $x = 1.96$ m, $y = 2.34$ m, 704 N

2–98. 85.9°

2–99. $\mathbf{r}_2 = \{0.143\mathbf{i} + 0.214\mathbf{j} - 0.0714\mathbf{k}\}$ m

2–101. $\mathbf{r}_2 = \{2.58\mathbf{i} - 1.61\mathbf{j} - 0.645\mathbf{k}\}$ ft

2–102. 2.67 m

2–103. 3.87 m

2–105. $F_{AB} = 49.0$ N, $F_{per} = 63.2$ N

2–106. $F_1 = 18.3$ lb, $F_{per} = 35.6$ lb

2–107. $F_{AC} = 26.5$ lb, $F_{BA} = 66.4$ lb

2–109. 143°

2–110. 34.2°

2–111. $F_x = 47.8$ lb, $F_{AC} = 45.5$ lb

2–113. $(F_1)_{AO} = 18.5$ N, $(F_2)_{AO} = 21.3$ N

2–114. 5.43 lb

2–115. 100°

2–117. $\mathbf{F}_A = \{285\mathbf{j} - 93.0\mathbf{k}\}$ N,
$\mathbf{F}_C = \{159\mathbf{i} + 183\mathbf{j} - 59.7\mathbf{k}\}$ N

2–118. $\mathbf{F}_1 = \{20\mathbf{i} + 34.6\mathbf{j}\}$ lb, $\mathbf{F}_2 = \{-42.4\mathbf{i} + 42.4\mathbf{j}\}$ lb,
$F_R = 80.3$ lb, $\theta = 106°$

2–119. 80.3 lb, 106°

2–121. $x = 7.65$ ft, $y = 4.24$ ft, $z = 3.76$ ft

2–122. $x = 8.67$ ft, $y = 1.89$ ft

2–123. 0.667 kN

2–125. 50.3°, 369 N

2–126. 90°

Chapter 3

3–1. $F_1 = 259$ N, $F_2 = 366$ N

3–2. 21.8°, 3.23 kN

3–3. 73.9°, 180 lb

3–5. $F_1 = 3.59$ kN, $F_2 = 0.536$ kN

3–6. $F_D = 4.90$ kN, $F_B = 3.46$ kN

3–7. 968 lb

3–9. $F_{AD} = 46.7$ lb, $F_{AC} = 30.5$ lb

3–10. $F_{AB} = F_{AC} = F = 1.32$ kip

3–11. $F_{AC} = 267$ N, $F_{AB} = 98.6$ N

3–13. 106 lb

3–14. 240 lb

3–15. 43.0°

3–17. $x_{AB} = 0.467$ m, $x_{AC} = 0.793$ m, $x_{AD} = 0.490$ m

3–18. 3.40 ft

3–19. $F_{AF} = 589$ N, $F_{AC} = 736$ N, $F_{AB} = 441$ N, $F_{CE} = 589$ N, $F_{CD} = 441$ N

3–21. $T_B = 0.0104$ lb, $W = 0.012$ lb, $\theta = 19.1°$, $T_A = 0.0159$ lb

3–23. $T = 0.5W \csc \theta$

3–25. 53.1 lb

3–26. 9.30 lb

3–27. $F_{BD} = 171$ N, $F_{BC} = 145$ N

3–29. 2.45 m

3–30. $\theta = 90°$, $(F_{AC})_{max} = 160$ lb, $\theta = 120°$, $(F_{AB})_{max} = 160$ lb

3–31. $F_1 = 0$, $F_3 = 0$, $F_2 = 200$ lb

3–33. $F_1 = 2.83$ kN, $F_2 = 1.10$ kN, $F_3 = 3.46$ kN

3–34. $F_1 = 400$ lb, $F_2 = 280$ lb, $F_3 = 357$ lb

3–35. $F_1 = 5.10$ kN, $F_2 = 11.8$ kN, $F_3 = 3.92$ kN

3–37. $F_1 = 0$, $F_2 = 311$ lb, $F_3 = 238$ lb

3–38. $F_{AD} = 118$ N, $F_{AB} = 78.5$ N, $F_{AC} = 39.2$ N

3–39. $F_{CD} = 625$ lb, $F_{CB} = F_{CA} = 198$ lb

3–41. $F_{AD} = F_{AC} = 104$ N, $F_{AB} = 220$ N

3–42. $F_{AB} = F_{AC} = 16.6$ kN, $F_{AD} = 55.2$ kN

3–43. $F_{AD} = 510$ N, $F_{AC} = 130$ N, $F = 1.60$ kN

3–45. $F_{AB} = 0.980$ kN, $F_{AC} = 0.463$ kN, $F_{AD} = 1.55$ kN

3–46. $F_{AB} = 1.21$ kN, $F_{AC} = 0.278$ kN, $F_{AD} = 1.64$ kN

3–47. $T = 1.79(10^{-3})$ lb

3–49. $F_{AB} = 359$ lb, $F_{AC} = F_{AD} = 254$ lb

3–50. $F_{AB} = 469$ lb, $F_{AC} = F_{AD} = 331$ lb

3–51. 88.8 lb

3–53. $F_{DA} = 21.5$ lb, $F_{DB} = 14.0$ lb, $F_{DC} = 17.6$ lb

3–54. 40.2°

3–55. $F_2 = 9.60$ kN, $F_1 = 1.83$ kN

3–57. $F = 0.850(10^{-3})$ N

Chapter 4

4–5. 3.15 kN · m ↺

4–6. 858 lb · in. ↻

4–7. 1.17 kip · in. ↻

4–9. 4.47 kN · m ↺

4–10. 14.5 lb · ft ↻

4–11. 39.2°

4–13. 11.2 kip · ft ↻

4–14. 6.16 kip · ft ↺

4–15. 66.1 N

4–17. $M_A = 80 \cos \theta + 320 \sin \theta$

4–18. 1.90 N · m ↺

4–19. $M_P = \{538 \cos \theta + 75 \sin \theta\}$ lb · ft

4–21. $\mathbf{M}_O = \{67.8\mathbf{i} - 17.0\mathbf{j} + 84.8\mathbf{k}\}$ lb · ft

4–22. $\mathbf{M}_P = \{153\mathbf{i} - 29.7\mathbf{j} + 174\mathbf{k}\}$ lb · ft

4–23. $(\mathbf{M}_R)_O = \{-160\mathbf{i} + 200\mathbf{j} + 510\mathbf{k}\}$ N · m

4–25. $\mathbf{M}_A = \{360\mathbf{i} + 240\mathbf{j}\}$ N · m

4–26. $\mathbf{M}_1 = \{-52\mathbf{i} + 24\mathbf{j} - 36\mathbf{k}\}$ N · m, $\mathbf{M}_2 = \{-34\mathbf{i} + 34\mathbf{j} - 68\mathbf{k}\}$ N · m, 147 N · m, $\alpha = 126°$, $\beta = 66.7°$, $\gamma = 135°$

4–27. $\mathbf{M}_1 = \{-36\mathbf{i} - 36\mathbf{k}\}$ N · m, $\mathbf{M}_2 = \{-42\mathbf{i} - 14\mathbf{j} - 68\mathbf{k}\}$ N · m, 131 N · m, $\alpha = 127°$, $\beta = 96.1°$, $\gamma = 143°$

4–29. $\mathbf{M}_A = \{-16.0\mathbf{i} - 32.1\mathbf{k}\}$ N · m

4–30. $(\mathbf{M}_R)_A = \{36\mathbf{i} - 78\mathbf{j}\}$ lb · in., $(\mathbf{M}_R)_B = \{36\mathbf{i} + 192\mathbf{j} + 90\mathbf{k}\}$ lb · in., $(\mathbf{M}_R)_C = \{-396\mathbf{i} + 192\mathbf{j} - 222\mathbf{k}\}$ lb · in.

4–31. 165 lb · ft, $\alpha = 91.3°$, $\beta = 62.5°$, $\gamma = 152°$

4–33. $\mathbf{M}_C = \{-226\mathbf{i} - 226\mathbf{j} - 226\mathbf{k}\}$ lb · ft

4–34. $\mathbf{M}_O = \{-1.77\mathbf{i} - 1.10\mathbf{j}\}$ kN · m

4–35. $\mathbf{M}_A = \{-5.39\mathbf{i} + 13.1\mathbf{j} + 11.4\mathbf{k}\}$ N · m

4–37. $c = 2$ in., $b = 5$ in.

4–38. $z = 3$ m, $x = 2$ m

4–39. $y = 2$ m, $x = 3$ m, $d = 3.63$ m

4–41. $z = 3$ m, $y = 1$ m, $d = 1.15$ m

4–42. $\mathbf{M}_{Oa} = \{218\mathbf{j} + 163\mathbf{k}\}$ N · m

4–43. $(\mathbf{M}_R)_{aa} = \{-41.3\mathbf{i} - 44.7\mathbf{j} + 13.8\mathbf{k}\}$ kN · m

4–45. $(\mathbf{M}_R)_{Oa} = \{26.1\mathbf{i} - 15.1\mathbf{j}\}$ lb · ft

4–46. $M_x = -72$ N · m, $M_y = 12$ N · m, $M_z = -52$ N · m

4–47. $\mathbf{M}_{OA} = \{-17.3\mathbf{i} - 13.0\mathbf{j}\}$ N · m

4–49. 62 lb · in.

4–50. 44.4 lb · ft

4–51. 3.75 N · m

4–53. $\mathbf{M}_y = \{0.828\mathbf{j}\}$ N · m

4–54. $\mathbf{M}_y = \{87.4\mathbf{j}\}$ N · m

4–55. $\mathbf{M}_z = \{-15.9\mathbf{k}\}$ N · m

4–57. 19.8 lb

4–58. 14.8 N · m

4–59. 20.2 N

4–61. 21.9 kN · m ↲

4–62. 3.12 kip · ft ↲

4–63. 67.5 lb

4–65. 27.7 lb

4–66. 28.9 lb · ft

4–67. 0.909 kip

4–69. 139 lb

4–70. 830 N

4–71. 5.54 ft

4–73. $M_R = 39.6$ lb · ft ↰

4–74. 9.69 kN · m ↓

4–75. $\mathbf{M}_C = \{56\mathbf{i} - 15\mathbf{j} + 57\mathbf{k}\}$ kN · m

4–77. 78.1 N · m, $\alpha = 48.3°$, $\beta = 50.2°$, $\gamma = 67.4°$

4–78. $\mathbf{M}_C = \{-50\mathbf{i} + 60\mathbf{j}\}$ lb · ft, $M_C = 78.1$ lb · ft

4–79. $\mathbf{M}_R = \{-12.1\mathbf{i} - 10\mathbf{j} - 17.3\mathbf{k}\}$ N · m

4–81. $F = 3$ lb, $P = 4$ lb

4–82. $22.6° \angle\theta$, 204 lb · ft ↓

4–83. $30° \angle\theta$, 2.46 kN · m ↓

4–85. 1.19 kN, $85.9° \theta\angle$, 21.6 kN · m ↲

4–86. 17.3 N, $60° \angle\theta$, 42.0 N · m ↲

4–87. 17.3 N, $60° \angle\theta$, 14.0 N · m ↱

4–89. 52.1 lb, $20.9° \angle\theta$, 60.9 ft

4–90. 52.1 lb, $20.9° \angle\theta$, 23.3 ft

4–91. 73.8 lb, $59.0° \angle\theta$, 5.66 ft

4–93. 60 N

4–94. 29.9 lb, $78.4° \angle\theta$, 214 lb · in. ↲

4–95. 26.4 lb, $85.7° \angle\theta$, 205 lb · in. ↲

4–97. 342 N, $43.0° \theta\angle$, 99.5 N · m ↓

4–98. 464 lb, $45.1° \angle\theta$, 1595 lb · ft ↓

4–99. 464 lb, $45.1° \angle\theta$, 832 lb · ft ↓

4–101. 110 lb, $67.5° \theta\angle$, 5.95 ft

4–102. 110 lb, $67.5° \theta\angle$, 705 lb · ft ↰

4–103. $26.6°$, 8.94 kN, 4.5 m

4–105. 219 lb, $79.5° \theta\angle$, 1.19 kip · ft ↲

4–106. 219 lb, $79.5° \theta\angle$, 1.39 kip · ft ↓

4–107. 98.5 lb, $24.0° \angle\theta$, 390 lb · ft ↓

4–109. 98.5 lb, $24.0° \angle\theta$, 6.0 ft

4–110. 4.5 kN ↓, 2.22 m

4–111. $\mathbf{F} = \{50\mathbf{i} - 30\mathbf{j} + 80\mathbf{k}\}$ N, $\mathbf{M}_{RO} = \{410\mathbf{i} + 230\mathbf{j} - 170\mathbf{k}\}$ N · m

4–113. $\mathbf{F} = \{8\mathbf{i} + 6\mathbf{j} + 8\mathbf{k}\}$ kN, $\mathbf{M}_{RO} = \{-10\mathbf{i} + 18\mathbf{j} - 56\mathbf{k}\}$ kN · m

4–114. $\mathbf{F} = \{8\mathbf{i} + 6\mathbf{j} + 8\mathbf{k}\}$ kN, $\mathbf{M}_{RP} = \{-46\mathbf{i} + 66\mathbf{j} - 56\mathbf{k}\}$ kN · m

4–115. $\mathbf{F}_R = \{-80\mathbf{i} - 80\mathbf{j} + 40\mathbf{k}\}$ lb, $\mathbf{M}_{RP} = \{-240\mathbf{i} + 720\mathbf{j} + 960\mathbf{k}\}$ lb · ft

4–117. $F_C = 600$ N, $F_D = 500$ N

4–118. 140 kN ↓, $y = 7.14$ m, $x = 5.71$ m

4–119. 40 lb ↑, 0.5 ft

4–121. $F_R = 495$ N, $M_R = 1.54$ kN · m, $z = 1.16$ m, $y = 2.06$ m

4–122. $F_R = 108$ lb, $M_R = 624$ lb · ft, $x = 8.69$ ft, $y = 0.414$ ft

4–123. $\mathbf{F}_R = \{141\mathbf{i} + 100\mathbf{j} + 159\mathbf{k}\}$ N, $\mathbf{M}_{RO} = \{122\mathbf{i} - 183\mathbf{k}\}$ N · m

4–125. 1.95 kip · ft ↓

4–126. 27 kN, 2.11 m

4–127. 13.25 lb, 0.340 ft to the right

4–129. 13.5 kip, 6.0 ft

4–130. 3.25 kip, $67.2° \theta\angle$, 3.86 ft

4–131. 3.90 kip ↑, 11.3 ft

4–133. 18.0 kip ↓, 11.7 ft

4–134. 162 lb ↓, 6.80 ft

4–135. $b = 9.0$ ft, $a = 7.50$ ft

4–137. 30 kN ↓, 4.10 m

4–138. 667 lb ↓, 1.67 kip · ft

4–139. 107 kN ←, 1.60 m

4–141. 1.87 kip, 3.66 ft

4–142. 195 lb, $60.6° \angle$, 3.21 ft

4–143. 46.9 kN, 7.11 m

4–145. $(M_1)_B = 3.96$ kip · ft $\downarrow$, $(M_2)_B = 4.80$ kip · ft $\downarrow$, $(M_3)_B = 1.30$ kip · ft $\downarrow$, $M_{RB} = 10.1$ kip · ft $\downarrow$

4–146. $\mathbf{M}_{BA} = \{-3.39\mathbf{i} + 2.54\mathbf{j} - 2.54\mathbf{k}\}$ N · m

4–147. $\mathbf{M}_{OC} = \{-4.61\mathbf{i} - 3.46\mathbf{j}\}$N · m

4–149. $\mathbf{M}_O = \{-128\mathbf{i} + 128\mathbf{j} - 257\mathbf{k}\}$ N · m

4–150. $\mathbf{M}_B = \{-37.6\mathbf{i} + 90.7\mathbf{j} - 155\mathbf{k}\}$ N · m

4–151. 10.75 kip $\downarrow$, 99.5 kip · ft $\downarrow$

4–153. $\mathbf{F}_R = \{-40\mathbf{j} - 40\mathbf{k}\}$ N, $(\mathbf{M}_R)_A = \{-12\mathbf{j} + 12\mathbf{k}\}$ N · m

4–154. $\mathbf{F}_R = \{-28.3\mathbf{j} - 68.3\mathbf{k}\}$N, $(\mathbf{M}_R)_A = \{-20.5\mathbf{j} + 8.49\mathbf{k}\}$ N · m

Chapter 5

5–11. $N_B = 592$ N, $F_A = 92.3$ N

5–13. $T = 830$ N, $A_x = 498$ N, $A_y = 709$ N

5–14. $F_B = 100$ lb, $A_x = 100$ lb, $A_y = 20$ lb

5–15. $N_B = 2.14$ kip, $A_x = 1.29$ kip, $A_y = 1.49$ kip

5–17. $N_A = 23.7$ lb, $N_B = 12.2$ lb, $N_C = 5.77$ lb

5–18. $A_x = 0$, $A_y = 3.83$ kN, $B_y = 11.2$ kN

5–19. $A_x = 0$, $A_y = 200$ N, $B_y = 400$ N

5–21. $A_x = 42$ N, $A_y = 10.5$ N, $B_y = 10.5$ N

5–22. $A_x = 0$, $A_y = 6.75$ kN, $M_A = 46.25$ kN · m

5–23. $F_A = 105$ lb, $B_x = 97.4$ lb, $B_y = 269$ lb

5–25. 1.69 kN, 1.69 kN

5–26. $T_{BC} = 16.4$ kN, $A_x = 13.1$ kN, $A_y = 14.6$ kN

5–27. $F_{BD} = 41.6$ kN, $A_x = 41.6$ kN, $A_y = 8$ kN

5–29. $N_B = 110$ kip, $A_x = 95.3$ kip, $A_y = 5.0$ kip

5–30. $A_x = 4.00$ kN, $A_y = 8.71$ kN, $N_B = 78.2$ kN

5–31. $N_A = N_B = 49.1$ N, $N_C = N_D = 42.5$ N

5–33. (a) $N_E = 1.16$ kip, $N_D = 2.19$ kip, (b) 4.74 kip

5–34. $N_A = 41.6$ kip, $N_B = 105$ kip

5–35. $T_{BD} = 76.8$ kN, $A_x = 62.9$ kN, $A_y = 65.6$ kN

5–37. $A_x = 33.3$ lb, $B_x = 33.3$ lb, $B_y = 100$ lb

5–38. $F_B = 533$ N, $N_A = 667$ N

5–39. (a) 1.64 kN, (b) 213 N

5–41. $T = 8.83$ kN, $A_x = 4.41$ kN, $A_y = 9.61$ kN

5–42. $F_{BC} = 3.76$ kip, $A_x = 3.26$ kip, $A_y = 3.12$ kip

5–43. $T_{BC} = 1.83$ kN, $A_x = 1.10$ kN, $A_y = 314$ N

5–46. $A_x = 4.61$ lb, $A_y = B_y = 43.3$ lb

5–47. $F_A = 432$ lb, $F_B = 0$, $F_C = 432$ lb

5–49. $\theta = \cos^{-1}[(L + \sqrt{L^2 + 128r^2})/16r]$

5–50. $w_1 = 3.22$ kN/m, $w_2 = 1.63$ kN/m

5–51. $P = F_A = 0.01$ N

5–53. $T_{BC} = 342$ lb, $F_A = 223$ lb

5–54. 12.8°

5–55. 6.70 lb

5–57. $d = 3a/4$

5–58. $\theta = \tan^{-1} \frac{1}{2}(\cot \psi - \cot \phi)$

5–59. $F_A = (W/2)\sqrt{(4l^2 - 3d^2)/(l^2 - d^2)}$

5–61. 11.2 lb/ft

5–62. 20.4°

5–63. 24.1°

5–65. $F_A = A_{z'} = 125$ lb, $M_A = 206$ lb · ft

5–66. $A_x = -8$ lb, $A_y = 4$ lb, $A_z = 0$, $(M_A)_y = 44$ lb · ft, $(M_A)_z = -32$ lb · ft

5–67. $N_C = 375$ lb, $N_A = N_B = 187.5$ lb

5–69. $F_{DE} = 21.3$ kip, $F_{BC} = F_{AC} = 8.89$ kip

5–70. $F_C = 147$ N, $F_B = 81.8$ N, $F_A = 164$ N

5–71. $A_x = 0$, $A_y = 5$ kN, $A_z = 16.7$ kN, $F_B = 16.7$ kN

5–73. $A_y = 0$, $F = 34.6$ N, $B_z = 129$ N, $A_z = -66.0$ N, $B_x = 95.0$ N, $A_x = 75.0$ N

5–74. $A_x = 25.5$ lb, $A_y = 18.75$ lb, $B_x = 22.6$ lb, $B_y = 18.75$ lb, $B_z = 50$ lb, $C_x = 32.1$ lb

5–75. $P = 75$ lb, $B_z = 75$ lb, $B_x = 112$ lb, $A_y = 0$, $A_x = 37.5$ lb, $A_z = 75$ lb

5–77. $A_y = 168$ lb, $A_z = 368$ lb, $B_x = -358$ lb, $B_y = -168$ lb, $C_x = 358$ lb, $C_z = -168$ lb

5–78. $A_x = -56.6$ lb, $A_z = 28.3$ lb, $B_y = -28.3$ lb, $B_z = 72.4$ lb, $C_y = 28.3$ lb, $C_z = 15.9$ lb

5–79. $T_{CD} = 115$ N, $A_y = 80$ N, $A_z = 0$, $B_x = -30$ N, $B_y = -40$ N, $B_z = 0$

5–81. $A_x = -1933$ N, $A_y = -867$ N, $B_y = 467$ N, $B_z = 667$ N, $C_x = 2433$ N, $C_z = -867$ N

5–82. $F_{BD} = F_{CD} = 70.2$ lb, $A_x = 37.5$ lb, $A_y = 0$, $A_z = 37.5$ lb

5–83. $F_{BC} = 175$ lb, $A_x = 130$ lb, $A_y = -10$ lb, $M_{Ax} = -300$ lb · ft, $M_{Ay} = 0$, $M_{Az} = -720$ lb · ft

5–85. $F_{BC} = 47.5$ lb, $A_x = -12.5$ lb, $A_y = 5.0$ lb, $A_z = 0$, $M_{Ay} = -60$ lb · ft, $M_{Az} = -37.5$ lb · ft

5–86. $F_{BC} = 209$ lb, $A_y = 80$ lb, $A_z = -130$ lb, $M_{Ax} = -780$ lb · ft, $M_{Ay} = -480$ lb · ft, $M_{Az} = -360$ lb · ft

5–87. $T_{DE} = 721$ lb, $T_{BC} = 2.16$ kip, $A_x = -309$ lb, $A_y = 1.55$ kip, $A_z = -1.21$ kip

5–89. 1.02°

5–90. 2.50 kN/m

5–91. $P = 100$ lb, $B_z = 40$ lb, $B_x = -35.7$ lb, $A_x = 136$ lb, $B_y = 0$, $A_z = 40$ lb

5–93. $w_1 = 62.7$ lb/ft, $w_2 = 91.1$ lb/ft

5–94. $A_x = 6.0$ kip, $A_y = 5.92$ kip, $B_y = 3.58$ kip

5–95. $A_y = 7.36$ kip, $B_x = 0.5$ kip, $B_y = 16.6$ kip

5–97. $T = 1.01$ kN, $F_D = 982$ N

5–98. $F_{BC} = 186$ N, $A_x = 161$ N, $A_y = 457$ N

5–99. $B_y = 1.28$ kN, $A_x = 0$, $A_y = 1.82$ kN

Chapter 6

6–1. $F_{BC} = 374$ lb (C), $F_{BA} = 143$ lb (T), $F_{CA} = 264$ lb (T)

6–2. $F_{BA} = 1.18$ kip (C), $F_{AC} = 833$ lb (T), $F_{BC} = 373$(C)

6–3. $F_{DE} = 0$, $F_{DC} = 16.5$ kip (C), $F_{AB} = 4.21$ kip (C), $F_{AE} = 4.33$ kip (T), $F_{CE} = 7.81$ kip (T), $F_{CB} = 2.33$ kip (C), $F_{BE} = 3.50$ kip (T)

6–5. $F_{CB} = 3.0$ kN (T), $F_{CD} = 2.60$ kN (C), $F_{DE} = 2.60$ kN (C), $F_{DB} = 2$ kN (T), $F_{BE} = 2$ kN (C), $F_{BA} = 5$ kN (T)

6–6. $F_{CB} = 507$ N (T), $F_{CD} = 439$ N (C), $F_{DE} = 439$ N (C), $F_{DB} = 303$ N (T), $F_{BE} = 826$ N (C), $F_{BA} = 1.36$ kN (T)

6–7. $F_{EF} = 0$, $F_{ED} = 13.1$ kN (C), $F_{DF} = 5.21$ kN (T), $F_{DC} = 4.17$ kN (C), $F_{BC} = 3$ kN (C), $F_{BA} = 8$ kN (C), $F_{CA} = 1.46$ kN (C), $F_{CF} = 3.12$ kN (C), $F_{FA} = 4.17$ kN (T)

6–9. $F_{EC} = 9.01$ kN (C), $F_{ED} = 7.50$ kN (T), $F_{DC} = 5$ kN (C), $F_{DA} = 7.50$ kN (T), $F_{CA} = 4.51$ kN (T), $F_{CB} = 13.5$ kN (C)

6–10. $F_{GC} = F_{FE} = 0$, $F_{CD} = F_{ED} = 0$, $F_{GB} = F_{FA} = 4.24$ kN (C), $F_{AB} = 3.0$ kN (T), $F_{GD} = F_{FD} = 4.24$ kN (C), $F_{CB} = F_{EA} = 4$ kN (C)

6–11. $F_{CD} = 400$ lb (T), $F_{CB} = 346$ lb (C), $F_{BD} = 0$, $F_{BA} = 346$ lb (C), $F_{DA} = 400$ lb (C), $F_{DE} = 800$ lb (T), $F_{AE} = 200$ lb (T)

6–13. $F_{AB} = F_{ED} = 667$ lb (C), $F_{AG} = F_{EF} = 533$ lb (T), $F_{BG} = F_{DF} = 0$, $F_{BC} = F_{DC} = 667$ lb (C), $F_{GC} = F_{FC} = 0$, $F_{GF} = 533$ lb (T)

6–14. $F_{GB} = 0$, $F_{BA} = 4.47$ kN (C), $F_{BC} = 2.0$ kN (C), $F_{GA} = F_{GC} = 5.66$ kN (C)

6–15. $F_{EC} = 1.20$ P (T), $F_{ED} = 0$, $F_{AB} = F_{AD} = F = 0.373$ P (C), $F_{DC} = 0.373$ P (C), $F_{DB} = 0.333$ P (T), $F_{BC} = 0.373$ P (C)

6–17. $F_{AB} = 10.6$ kN (C), $F_{AH} = 7.50$ kN (T), $F_{EF} = 10.6$ kN (T), $F_{ED} = 7.50$ kN (C), $F_{BC} = 7.50$ kN (C), $F_{BH} = 7.50$ kN (T),

$F_{FG} = 7.50$ kN (T), $F_{FD} = 2.50$ kN (C), $F_{HC} = 3.54$ kN (C), $F_{HG} = 10.0$ kN (T), $F_{DG} = 3.54$ kN (T), $F_{DC} = 10.0$ kN (C), $F_{CG} = 2.50$ kN (T)

6–18. $F_{AI} = 4.04$ kip (C), $F_{AB} = 3.75$ kip (T), $F_{EF} = 12.1$ kip (C), $F_{ED} = 7.75$ kip (T), $F_{BI} = 0$, $F_{BC} = 3.75$ kip (T), $F_{DF} = 0$, $F_{DC} = 7.75$ kip (T), $F_{FC} = 4.04$ kip (C), $F_{FG} = 8.08$ kip (C), $F_{HG} = 2.15$ kip (C), $F_{HI} = 0.8$ kip (T), $F_{CI} = 0.269$ kip (T), $F_{CG} = 1.40$ kip (T), $F_{GI} = 5.92$ kip (C)

6–19. $F_{FG} = 10.1$ kip (C), $F_{FE} = 8.53$ kip (T), $F_{AJ} = 7.26$ kip (C), $F_{AB} = 4.53$ kip (T), $F_{GH} = 10.1$ kip (C), $F_{GE} = 3$ kip (C), $F_{IC} = 0$, $F_{JI} = 7.26$ kip (C), $F_{JB} = 2$ kip (C), $F_{EH} = 3.84$ kip (T), $F_{ED} = 6.13$ kip (T), $F_{BI} = 2.15$ kip (T), $F_{BC} = 3.73$ kip (T), $F_{HD} = 4.50$ kip (C), $F_{HI} = 7.23$ kip (C), $F_{DI} = 5.10$ kip (T), $F_{DC} = 3.73$ kip (T)

6–21. $F_{FC} = 1.39$ kip (T), $F_{GF} = 19.0$ kip (C), $F_{CD} = 16.7$ kip (T)

6–22. $F_{HG} = 15$ kN (C), $F_{HC} = 7.07$ kN (T), $F_{GC} = 0$

6–23. $F_{CF} = 10$ kN (C), $F_{GC} = 7.89$ kN (T)

6–25. $F_{GF} = 12.5$ kN (C), $F_{CD} = 6.67$ kN (T), $F_{GC} = 0$

6–26. $F_{DF} = 2.26$ kN (T), $F_{CE} = 4.53$ kN (C), $F_{ED} = 400$ N (C)

6–27. $F_{FE} = 4.50$ kip (C), $F_{CD} = 1.41$ kip (C), $F_{CE} = 2.0$ kip (T)

6–29. $F_{BC} = 3.25$ kN (C), $F_{CH} = 1.92$ kN (T)

6–30. $F_{FE} = 1.0$ kip (T), $F_{BC} = 2.24$ kip (C), $F_{FB} = 0$

6–31. $F_{CD} = 671$ N (C), $F_{KJ} = 800$ N (T), $F_{CJ} = 283$ N (C)

6–33. $F_{BC} = F_{AB} = F_{CD} = F_{DE} = 0$, $F_{IC} = 6.88$ kN (C), $F_{CG} = 11.0$ kN (T)

6–34. $F_{BC} = F_{AB} = F_{CD} = F_{DE} = 0$, $F_{JE} = 10.62$ kN (C), $F_{GF} = 6.38$ kN (T)

6–35. $F_{BC} = 10.4$ kN (C), $F_{HC} = 2.24$ kN (T), $F_{HG} = 9.15$ kN (T)

6–37. $F_{CD} = 2.62$ kip (T), $F_{KJ} = 3.02$ kip (C), $F_{JN} = 0$

6–38. $F_{GF} = 7.83$ kN (C), $F_{GD} = 1.80$ kN (C)

6–39. $F_{DE} = 3.38$ kip (C), $F_{JI} = 3.38$ kip (T), $F_{DO} = 0.944$ kip (C)

6–41. $F_{AB} = 6.46$ kN (T), $F_{AC} = F_{AD} = 1.50$ kN (C), $F_{BC} = F_{BD} = F = 3.70$ kN (C), $F_{BE} = 4.80$ kN (T)

6–42. $F_{BD} = 731$ lb (C), $F_{BC} = 250$ lb (T), $F_{BA} = 167$ lb (T), $F_{AD} = 786$ lb (T), $F_{AC} = 391$ lb (C), $F_{CD} = 0$

6–43. $F_{AC} = F_{BC} = 400$ N (C), $F_{CD} = 566$ N (T), $F_{BD} = 283$ N (T), $F_{AB} = 0$, $F_{AD} = 283$ N (T)

6–45. $F_{BA} = 175$ N (C), $F_{BC} = 79.1$ N (T), $F_{BD} = 25$ N (T),
$F_{DA} = 175$ N (C), $F_{DC} = 79.1$ N (T), $F_{CA} = 335$ N (C)

6–46. $F_{FA} = 0$, $F_{FD} = F_{FE} = F = 1.73$ kN (T),
$F_{EA} = 9.37$ kN (C), $F_{EB} = 7.79$ kN (T),
$F_{ED} = 1.73$ kN (T), $F_{DA} = 9.37$ kN (C),
$F_{DC} = 7.79$ kN (T), $F_{AC} = F_{AB} = 5.20$ kN (T),
$F_{BC} = 2.60$ kN (C)

6–47. $F_z = 2$ kN, $E_y = 6.93$ kN, $F_y = 3.46$ kN,
$A_x = 0$, $A_y = 3.46$ kN, $A_z = 2$ kN, $F_{CD} = 2.31$ kN (T),
$F_{CF} = 0$, $F_{CB} = 1.15$ kN (C), $F_{DF} = 4.16$ kN (C),
$F_{DB} = 0$, $F_{DE} = 3.46$ kN (T), $F_{BE} = 4.16$ kN (T),
$F_{BA} = 3.46$ kN (C), $F_{EA} = 2.31$ kN (C),
$F_{EF} = 0$, $F_{FA} = 1.15$ kN (T)

6–49. $F_C = 220$ lb, $F_A = 434$ lb, $\theta = 64.0° \measuredangle \theta$

6–50. $A_x = 300$ N, $A_y = 300$ N, $C_x = 300$ N, $C_y = 300$ N

6–51. $B_y = -24.0$ lb, $B_x = 74.0$ lb,
$A_x = 74.0$ lb, $A_y = 124$ lb, $C_x = 74.0$ lb, $C_y = 76.0$ lb

6–53. $F_{BC} = 576$ N, $F_{AB} = 360$ N, $D_x = 360$ N,
$D_y = 450$ N, $M_D = 450$ N · m

6–54. $F_{BC} = 297$ lb, $D_y = 210$ lb, $D_x = 60$ lb

6–55. $A_y = 2$ kN, $B_x = 0$, $B_y = 12$ kN, $M_B = 32$ kN · m

6–57. $B_y = 1.34$ kip, $A_y = 1.02$ kip, $E_x = 0$, $E_y = 240$ lb,
$M_E = 1.20$ kip · ft

6–58. $T_{AB} = 357$ lb, $N_D = 70$ lb, $C_x = C_y = 0$

6–59. $F_{AB} = 8.60$ MN, $F_C = 9.78$ MN

6–61. $C_x = 1600$ lb, $C_y = 800$ lb, $D_x = 800$ lb, $D_y = 800$ lb

6–62. $B_x = 1600$ lb, $B_y = 800$ lb, $D_x = 800$ lb, $D_y = 800$ lb

6–63. 667 N

6–65. $F_{DH} = 10.2$ kN, $E_x = 6.13$ kN, $E_y = 13.1$ kN,
$F_{AB} = 4.60$ kN, $C_x = 3.98$ kN, $C_y = 7.21$ kN

6–66. $N = 10$ lb

6–67. 36.0 lb

6–69. 5.45 lb

6–70. $C_y = 285$ lb, $C_x = 326$ lb, $A_x = 176$ lb, $A_y = 135$ lb

6–71. $D_x = 50$ lb, $D_y = 50$ lb, $B_x = 112$ lb, $A_x = 62.5$ lb,
$B_y = 117$ lb, $A_y = 167$ lb

6–73. $C_x = 75$ lb, $C_y = 100$ lb

6–74. $D_x = 300$ lb, $D_y = 300$ lb, $E_y = 600$ lb, $B_y = 300$ lb,
$E_x = 225$ lb, $B_x = 75$ lb

6–75. $F_{AB} = 1.71$ kip, $F_C = 2.28$ kip

6–77. $B_x = 2.06$ kN, $A_y = 0$, $A_x = 2.06$ kN, $F_C = 2.43$ kN

6–78. $F_A = 258$ lb, $F_B = 142$ lb

6–79. $C_x = 19.2$ lb, $C_y = 25.6$ lb, $B_x = 305$ lb, $B_y = 0$,
$A_x = 206$ lb, $A_y = 25.6$ lb

6–81. $W_C = 0.812W$, $N_C = 0.544W$

6–82. $T = 147$ N, $\theta = 14.6°$

6–83. $N_B = 40$ lb, $N_C = 11.6$ lb, $D_x = 11.6$ lb, $D_y = 40$ lb

6–85. 170 lb · in.

6–86. $F_{DE} = W \cot \theta$

6–87. 850 N

6–89. $M = [(2PL \sin 2\theta)/(\sin \phi)][\cos(\phi - \theta)]$

6–90. $M = [(4PL \sin^2 \theta)/(\sin \phi)][\cos(\phi - \theta)]$

6–91. $G_x = 333$ lb, $G_y = 250$ lb, $F_x = G_x = 333$ lb,
$F_y = G_y = 250$ lb

6–93. (a) $\theta = \tan^{-1}(0.5 \cot \phi)$, (b) $16.1°$

6–94. $F_{AB} = 1.56$ kN, $(M_E)_y = 0$, $E_x = 0$, $E_y = 0$,
$(M_E)_x = 0.5$ kN · m

6–95. $A_z = 0$, $A_x = 172$ N, $A_y = 115$ N, $C_x = 47.3$ N,
$C_y = 61.9$ N, $C_z = 125$ N, $(M_C)_y = -429$ N · m,
$(M_C)_z = 0$

6–97. $F_{DE} = 270$ lb, $B_x = -30$ lb, $B_y = -13.3$ lb, $B_z = 0$

6–98. $F_{DF} = F_{EF} = F_{DE} = 22.2$ lb

6–99. $A_x = 1.5$ kN, $A_z = 0$, $F_{DE} = 5.30$ kN, $A_y = 0$,
$C_x = -2.25$ kN, $C_y = 3.0$ kN, $C_z = 2.25$ kN

6–101. $P = [kL/(2 \tan \theta \sin \theta)](2 - \csc \theta)$

6–102. $F_{AD} = 990$ lb (C), $F_{AB} = 700$ lb (T), $F_{DB} = 495$ lb (C),
$F_{DC} = 1.48$ kip (C), $F_{CB} = 1.05$ kip (T)

6–103. $F_{HD} = 7.07$ kN (C), $F_{CD} = 50$ kN (T), $F_{GD} = 5$ kN (T)

6–105. $A_y = 250$ N, $B_x = 1.40$ kN, $C_y = 1.70$ kN,
$C_x = 500$ N, $A_x = 1.40$ kN

6–106. $F_{AD} = 2.47$ kip (T), $F_{AC} = F_{AB} = F = 1.22$ kip (C)

6–107. $B_y = 449$ lb, $A_x = 92.3$ lb, $A_y = 186$ lb,
$M_A = 359$ lb · ft

6–109. $P = 25$ lb, $P = 33.3$ lb, $T = 33.33$ lb, $P = 11.1$ lb

6–110. $F_{BC} = 17.9$ kN (C), $F_{CD} = 8$ kN (T), $F_{BD} = F_{BE} = 0$,
$F_{AB} = 17.9$ kN (C), $F_{DE} = 8$ kN (T),
$F_{AE} = 8$ kN (T)

Chapter 7

7–1. $N_C = 0$, $V_C = 1.0$ kip, $M_C = 56$ kip · ft, $N_D = 0$,
$V_D = 1.0$ kip, $M_D = 48$ kip · ft

7–2. $N_B = 0$, $V_B = 100$ lb, $M_B = 750$ lb · in., $N_D = 0$,
$V_D = 75$ lb, $M_D = 750$ lb · in.

7–3. $T_D = 0$, $T_C = 200$ N · m, $T_B = 200$ N · m,
$T_A = 100$ N · m

7–5. $N_B = 0$, $V_B = 288$ lb, $M_B = 1.15$ kip · ft

7–6. $N_C = 0$, $V_C = 11.8$ kN, $M_C = 35.4$ kN · m

7–7. $N_D = 750$ N, $V_D = 250$ N, $M_D = 125$ N · m

7–9. $N_C = 0$, $V_C = 12$ kN, $M_C = 24$ kN · m

7–10. $N_C = 80$ lb, $V_C = 0$, $M_C = 480$ lb $\cdot$ in.

7–11. $N_C = 45.0$ kip, $V_C = 0$, $M_C = 9.00$ kip $\cdot$ ft

7–13. $N_C = 0$, $V_C = 2.01$ kip, $M_C = 15.0$ kip $\cdot$ ft, $N_D = 0$,
$V_D = 1.11$ kip, $M_D = 3.77$ kip $\cdot$ ft

7–14. $N_C = 20$ kN, $V_C = 70.6$ kN, $M_C = 292$ kN $\cdot$ m

7–15. $N_C = 4$ kip, $V_C = 1.0$ kip, $M_C = 14.0$ kip $\cdot$ ft,
$N_D = 4$ kip, $V_D = 1.75$ kip, $M_D = 5.0$ kip $\cdot$ ft

7–17. $V_A = 0$, $N_A = 5$ kN, $M_A = 0.8$ kN $\cdot$ m, $V_B = 0$,
$N_B = 7.5$ kN, $M_B = 0.2$ kN $\cdot$ m

7–18. $N_C = 2.49$ kip, $V_C = 2.49$ kip, $M_C = 4.97$ kip $\cdot$ ft,
$N_D = 0$, $V_D = 2.49$ kip, $M_D = 16.5$ kip $\cdot$ ft

7–19. $N_D = 220$ N, $V_D = 220$ N; $M_D = 54.9$ N $\cdot$ m,
$N_E = 80.4$ N, $V_E = 0$, $M_E = 112$ N $\cdot$ m

7–21. $N_C = 280$ lb, $V_C = 50$ lb, $M_C = 150$ lb $\cdot$ ft

7–22. $N_D = 800$ N, $V_D = 0$, $M_D = 1200$ N $\cdot$ m

7–23. $N_F = 0$, $V_F = 150$ lb, $M_F = 600$ lb $\cdot$ ft, $N_E = 0$,
$V_E = 230$ lb, $M_E = 387$ lb $\cdot$ ft

7–25. $V_B = 496$ lb, $N_B = 59.8$ lb, $M_B = 480$ lb $\cdot$ ft,
$N_C = 495$ lb, $V_C = 70.7$ lb, $M_C = 1.59$ kip $\cdot$ ft

7–26. $N = P \sin \theta$, $V = P \cos \theta$, $M = Pr \sin \theta$

7–27. 1.69 kip, 10.0 kip $\cdot$ ft

7–29. $(M_C)_x = -900$ lb $\cdot$ ft, $(M_C)_y = -260$ lb $\cdot$ ft,
$(M_C)_z = 1000$ lb $\cdot$ ft, $(A_C)_x = -50$ lb, $(V_C)_y = 500$ lb,
$(V_C)_z = -170$ lb

7–30. $(M_C)_x = 20$ lb $\cdot$ ft, $(M_C)_y = 72$ lb $\cdot$ ft,
$(M_C)_z = -178$ lb $\cdot$ ft, $(V_C)_x = 104$ lb, $(A_C)_y = 0$,
$(V_C)_z = 10$ lb

7–31. $(M_D)_x = 49.2$ kN $\cdot$ m, $(M_D)_y = 87.0$ kN $\cdot$ m,
$(M_D)_z = 26.2$ kN $\cdot$ m, $(V_D)_x = 116$ kN,
$(A_D)_y = -65.6$ kN, $(V_D)_z = 0$

7–33. -300 lb, $\{-300x\}$ lb $\cdot$ ft, 300 lb,
$M = \{300x - 900\}$ lb $\cdot$ ft

7–34. 1200 lb, $\{1200x\}$ lb $\cdot$ ft, 400 lb,
$M = \{400x + 9600\}$ lb $\cdot$ ft, -400 lb,
$M = \{28,800 - 400x\}$ lb $\cdot$ ft, -1200 lb,
$M = \{57,600 - 1200x\}$ lb $\cdot$ ft

7–35. $V = -2M_O/L$, $M = -(2M_O/L)x$, $V = -2M_O/L$,
$M = M_O - (2M_O/L)x$, $V = -2M_O/L$,
$M = 2M_O - (2M_O/L)x$

7–37. 0.75 kN, $\{0.75x\}$ kN $\cdot$ m, $V = \{3.75 - 1.5x\}$ kN,
$M = \{3.75x - 0.75x^2 - 3\}$ kN $\cdot$ m

7–38. $V = \{-2x + 2.5\}$ kN, $M = \{2.5x - x^2\}$ kN $\cdot$ m,
$V = \{-7.5\}$ kN, $M = \{-7.5x + 75\}$ kN $\cdot$ m

7–39. $V = \{250 - 300x\}$ N, $M = \{250x - 150x^2\}$ N $\cdot$ m,
$V = \{550 - 600x\}$ N,
$M = \{-300x^2 + 550x - 150\}$ N $\cdot$ m

7–41. $V = (w/2)(3L - 2x)$, $M = (w/2)(3Lx - x^2 - 2L^2)$

7–42. $V = \{360 - 3.33x^2\}$ lb, $M = \{360x - 1.11x^3\}$ lb $\cdot$ ft

7–43. $V = \{-50x^2 - 650\}$ N, $M = \{-16.7x^3 - 650x\}$ N $\cdot$ m,
$V = \{-300x + 2100\}$ N,
$M = \{-150x^2 + 2100x - 7350\}$ N $\cdot$ m

7–45. $V = \{3000 - 41.7x^2\}$ lb,
$M = \{-13.9x^3 + 3000x - 18000\}$ lb $\cdot$ ft,
$V = \{41.7x^2 - 1000x + 6000\}$ lb,
$M = \{-13.9(12 - x)^3\}$ lb $\cdot$ ft

7–46. $V = \{-0.417x^2\}$ kip, $M = \{-0.139x^3 - 15\}$ kip $\cdot$ ft,
$V = \{-5x + 55\}$ kip,
$M = \{-2.5x^2 + 55x - 285\}$ kip $\cdot$ ft,
$V = \{0.417(22 - x)^2\}$ kip,
$M = \{-0.139(22 - x)^3 - 15\}$ kip $\cdot$ ft

7–47. $V = (\gamma ht/2d)x^2$, $M = -(\gamma ht/6d)x^3$

7–49. $a = L/3$

7–50. $a = L/2$

7–51. $a = 0.414L$

7–53. See Prob. 7–34.

7–54. See Prob. 7–37.

7–55. See Prob. 7–38.

7–57. See Prob. 7–43.

7–58. See Prob. 7–45.

7–59. $x = 0$, $V = 8$, $M = 0$; $x = 10$, $V = 0$, $M = 43$

7–61. $x = 0$, $V = -20$, $M = 0$; $x = 15$, $V = -115$,
$M = -1062.5$

7–62. $x = 0$, $V = 50$, $M = -660$; $x = 12$, $V = 50$, $M = 0$

7–63. $x = 0$, $V = 358$, $M = -1458$; $x = 25$, $V = -41.7$,
$M = -500$

7–65. $x = 0$, $V = 0$, $M = 0$; $x = 12$, $V = 0$, $M = 2700$

7–66. $x = 0$, $V = 48$, $M = 0$; $x = 5$, $V = -12$, $M = 60$

7–67. $x = 0$, $V = 1.52$, $M = 0$; $x = 800$, $V = -1.98$,
$M = 0.395$

7–69. $x = 0$, $V = 0$, $M = 0$; $x = 6^+$, $V = 800$, $M = -1200$

7–70. $x = 0$, $V = 0$, $M = 0$; $x = 6^+$, $V = 9$, $M = -36$

7–71. $x = 0$, $V = wL/3$, $M = 0$; $x = L^+$, $V = wL/2$,
$M = -wL^2/6$

7–73. $x = 0$, $V = 1.10$, $M = -10.4$; $x = 12$, $V = 0.4$, $M = 0$

7–74. $x = 0$, $V = 4$, $M = 0$; $x = 10.97$, $V = 0$, $M = 25.2$

7–75. $x = 0$, $V = -0.45$, $M = 0$; $x = 6$, $V = -0.75$,
$M = -3.30$

7–77. 5.65 ft

7–78. $y_B = 8.67$ ft, $y_D = 7.04$ ft

7–79. 658 lb, 6.44 ft

7–81. 84.0 lb

7–82. 2.43 m, $T_{max} = T_{CD} = 157$ N

7–83. $T_{AB} = 83.0$ lb, $T_{CD} = 88.1$ lb, $T_{BC} = 46.7$ lb, 20.2 ft

7–85. 184 kN

7–86. 1.36 kN/m

7–87. $y = \{0.0385x^2 + 0.577x\}$ m, $T_{max} = 5.20$ kN

7–89. $T_A = 61.7$ kip, $T_B = 36.5$ kip, $T_C = 50.7$ kip

7–90. $h/L = 0.141$

7–91. 82.1 m, $T_{max} = 3.22$ kN

7–93. 55.6 ft, $h = 10.6$ ft

7–95. 46.3 ft, 75.2 lb

7–97. 292 lb

7–98. 326 ft

7–99. 5.14 ft

7–101. 76.7 lb

7–102. 10.4 ft

7–103. 8.07 m

7–105. $l = 238$ ft, $h = 93.8$ ft

7–106. $x = 0$, $V = 0$, $M = 24$; $x = 3$, $V = -12$, $M = 12$

7–107. $N_C = 0$, $V_C = 108$ lb, $M_C = 433$ lb · ft

7–109. $x = 0$, $V = 52.8$, $M = 0$; $x = 4.36$, $V = 0$, $M = 153$

7–110. $x = 0$, $V = 0$, $M = 0$; $x = 18$, $V = 0$, $M = 0$

7–111. $N_D = 6.08$ kN, $V_D = 2.60$ kN, $M_D = 13.0$ kN · m

7–113. $V_D = M_D = 0$, $N_D = F_{CD} = 86.6$ lb, $N_E = 0$,
　　　　$V_E = 28.9$ lb, $M_E = 86.6$ lb · ft

7–114. $x = 0$, $V = 2.5$, $M = 0$; $x = 1.25$, $V = 0$, $M = 1.56$

Chapter 8

8–1. 344 lb, $\mu_s \geq 0.70$

8–2. $P = 224$ lb

8–3. 76.4 lb $\leq P \leq 144$ lb

8–5. the spool will not remain in equilibrium

8–6. $\mu = 0.127$

8–7. 6.67 in.

8–9. 34.8 lb, 112 lb

8–10. 45°, 1

8–11. $\theta = \tan^{-1}[(1 - \mu^2)/2\mu]$

8–13. $\theta = \tan^{-1}\mu$

8–14. 21.7 lb

8–15. 20 lb

8–17. 6.97 kN, 15.3 kN

8–18. 1.25 kN, 6.89 kN

8–19. 27.0 lb, 0.171

8–21. 2.21 kip · ft

8–22. he can push cabinet to the left

8–23. he can pull cabinet to the right

8–25. 377 lb

8–26. 33.8°

8–27. 1.10 ft

8–29. 1.17 kN

8–30. $\mu = 0.415$

8–31. he cannot move the crate

8–33. 21.8°, $P = 72.0$ lb, $N_A = 769.4$ lb

8–34. the bicycle can move forward

8–35. 26.6°

8–37. 318 lb, 360 lb

8–38. 33.4°

8–39. 33.4°

8–41. slipping at B

8–42. 40.2 N

8–43. $\theta = \tan^{-1}[(\mu L/a) - \mu]$

8–45. 46.4°

8–46. 79.3 N

8–47. 0.576 in.

8–49. $P = W \sin(\alpha + \theta)$, $\phi = \theta$

8–50. 0.344

8–51. 0.601

8–53. 32.3 kip

8–54. 1.90 kN

8–55. 4.05 kip

8–57. $P' = 4.96$ kip, $P = 8.16$ kip

8–58. wedges are self-locking

8–59. 7.32 kN

8–61. 22.6°

8–62. 25.3°

8–63. 4.83 kN

8–65. 29.7 lb · in., 112 lb

8–66. 1.98 kN

8–67. 2.02 N · m

8–69. 1.35 N

8–70. 21.8 kN

8–71. 76.1 lb · ft

8–73. 31.0 N · m

8–74. 5.69 lb · in.

8–75. 880 N, 352 N · m

8–77. 4.53 N · m

8–78. 72.7 N

8–79. $N = (T_1/\mu)(e^{\mu\theta} - 1) + N_1$

8–81. $T_A = 1.95$ kN, $T_B = 0.759$ kN

8–82. 1.31 kN, 0.372 kN

8–83. (a) 4.60 kN, (b) 16.2 kN

8–85. Possible for B to pull A towards him

8–86. 22.9 N $\leq P \leq$ 151 N

8–87. 8.28 ft

8–89. 2.49 kN

8–90. 12.7 lb

8–91. $F_{AB} = F_B = 15.4$ lb, $F_{AC} = F_C = 5.47$ lb

8–93. 77.8 lb

8–94. 300 N

8–95. 63.5°

8–97. 38.2°

8–98. 78.7 lb

8–99. 15.8 lb · ft

8–101. 36.3 lb · ft

8–102. 1.62 kip

8–103. $M = 2\mu PR/3 \cos \theta$

8–105. $M = \frac{1}{2}\mu PR$

8–106. $M = \frac{8}{15}\mu PR$

8–107. 905 lb · in.

8–109. 0.0455 N · m

8–110. 826 N

8–111. 814 N

8–113. 0.0407

8–114. 13.8 lb

8–115. 29.0 lb

8–117. 0.129

8–118. 18.9 N

8–119. 1.35 mm

8–121. 486 lb

8–123. 131 N

8–125. 2.25 kN

8–126. 78.8 lb

8–127. 90.9 lb

8–129. 0.277

8–130. 36.6 N

8–131. 12 lb · ft

8–133. 1.88 lb

8–134. 35.0°

8–135. 0.833

8–137. 32.5°

Chapter 9

9–1. 1.6 m

9–2. $\bar{x} = \dfrac{a \sin \theta}{\theta}$, $\bar{x} = (2\sqrt{2}/\pi)a$, $\bar{x} = (2/\pi)a$

9–3. $\bar{x} = 0$, $\bar{y} = 1.82$ ft

9–5. 1.15 m

9–6. 0.546 m, $O_x = 0$, $O_y = 7.06$ N, 3.85 N · m

9–7. 0.620 ft, $A_x = 0$, $A_y = 1.16$ lb, 0.720 lb · ft

9–9 $\bar{x} = (2/3)a$, $\bar{y} = (m/3)a$

9–10. $\bar{x} = (2/7)a$, $\bar{y} = (2/5)a$

9–11. $\bar{x} = [(1 + n)/2(2 + n)]a$, $\bar{y} = [(1 + n)/(1 + 2n)]h$

9–13. $\bar{x} = 1.61$ in., $\bar{y} = 1.33$ in.

9–14. $\bar{x} = 1.08$ in., $\bar{y} = 0.541$ in.

9–15. $\bar{x} = (5/8)a$, $\bar{y} = (2k/5)a$

9–17. $\bar{x} = (2a \sin \alpha/(3\alpha, y = 0)$

9–18. $\bar{x} = (4/3\pi)a$, $\bar{y} = (4/3\pi)b$

9–19. $\bar{x} = (\pi/2)a$, $\bar{y} = (\pi/8)a$

9–21. $\bar{x} = 0.4$ ft, $\bar{y} = 1.0$ ft

9–22. $\bar{x} = 0.914$ m, $\bar{y} = 0.357$ m

9–23. $\bar{x} = 1.26$ m, $\bar{y} = 0.143$ m

9–25. $(4/5)h$

9–26. 0.833 ft

9–27. 1.33 ft

9–29. 84.7 mm

9–30. 2.5 ft

9–31. $\bar{z} = \frac{3}{8}a$

9–33. $\bar{z} = \frac{1}{5}a$

9–34. $\bar{z} = \frac{27}{40}a$

9–35. 2 kg, 5 m

9–37. $\bar{x} = 1.29$ ft, $\bar{y} = 2.14$ ft, $\bar{z} = -0.286$ ft

9–38. $\bar{x} = 0$, $\bar{y} = 1.63$ in.

9–39. $\bar{x} = 124$ mm, $\bar{y} = 0$

9–41. $\bar{x} = 2.85$ in., $\bar{y} = 5.33$ in.

9–42. $\bar{x} = 2.43$ m, $\bar{y} = 1.31$ m

9–43. 179 mm

9–45. 1.55 in.

9–46. 95 mm

9–47. 87.5 mm

9–49. $-0.262a$

9–50. $\bar{x} = 2.5$ in., $\bar{y} = 4$ in.

9–51. $\bar{y} = (b^2c + 3abc + 3a^2c)/3(a + b)(2a + b)$,
$\bar{x} = b(b + 3a)/3(2a + b)$

9–53. $\bar{y} = h(2a + b)/3(a + b)$

9–54. $\bar{x} = 1.82$ m, $\bar{y} = 1.28$ m

9–55. 142 mm

9–57. $\bar{x} = 2.11$ in., $\bar{y} = 2.11$ in.

9–58. $\bar{x} = 2.73$ in., $\bar{y} = 1.42$ in.

9–59. 0.206 m

9–61. 7.18 ft, $B_y = 2633$ lb, $A_y = 667$ lb

9–62. 4.28 ft, $B_y = 620$ lb, $A_y = 830$ lb

9–63. 2.48 ft, 38.9°

9–65. $\bar{z} = 52.5$ mm, $\bar{x} = \bar{y} = 0$

9–66. 2.95 in.

9–67. $\bar{y} = 58.1$ mm, $\bar{x} = \bar{z} = 0$

9–69. $\bar{z} = \frac{1}{2}a$

9–70. 323 mm

9–71. $h = (a^3 \pm a^2\sqrt{a^2 - \pi r^2})/\pi r^2$

9–73. 4.71 m³

9–74. 6.77 lb

9–75. $A = 8\pi ac$, $V = 2\pi ca^2$

9–77. 3.56(10³) ft², 22.1(10)³ ft³

9–78. 29.3 kip

9–79. 2.26

9–81. 1.04 lb

9–82. 0.377 lb

9–83. 22.7(10⁻³) m³

9–85. 71.6(10⁻³) m²

9–86. 274 m²

9–87. $A = 3.33$ ft², $\bar{y} = 1.2$ ft, $V = 25.1$ ft³

9–89. $A = 0.941$ m², $\bar{y} = 0.569$ m, $V = 3.36$ m³

9–90. 1.41 MN, 4 m

9–91. 157 kN, 235 kN, 4.22 m

9–93. $A_y = 2.51$ MN, $B_x = 2.20$ MN, $B_y = 859$ kN

9–94. $F_h = 488$ kip, $F_v = 260$ kip

9–95. 4.92 kip, 24.0° θ

9–97. 4.0 kip

9–98. 13.8 kip, 61.4 kip

9–99. 40.0 kip, 8.0 ft

9–101. 2.02 kip

9–102. 133 lb, $\bar{y} = 7.50$ ft, $\bar{x} = 0$

9–103. 42.7 kN, $\bar{y} = 2.40$ m, $\bar{x} = 0$, $C_z = B_z = 12.8$ kN,
$A_x = 0$, $A_y = 0$

9–105. 14.8 lb/ft²

9–106. $\pi w_0 r$

9–107. $O_x = wr$, $O_y = wr$, $M_O = wr^2$

9–109. $F_{Rx} = (1/2)w_0 r$, $F_{Ry} = (\pi/4)w_0 r$, $M_{RO} = (1/2)w_0 r^2$

9–110. $F_{Rx} = (\pi/4)w_0 r$, $F_{Ry} = (1/2)w_0 r$, $M_{RO} = (\pi/4)w_0 r^2$

9–111. $F = (4ab/\pi^2)p_0$, $\bar{x} = a/2$, $\bar{y} = b/2$

9–113. at $a = 0.5$ ft, $F = 2.94$ lb; at $a = 1$ ft, $F = 16.6$ lb

9–114. 6.93 kN, -0.125 m

9–115. 90.5 mm

9–117. $\bar{x} = \frac{3}{8}b$, $\bar{y} = \frac{3}{8}h$

9–118. $\bar{z} = 0.422$ m, $\bar{x} = \bar{y} = 0$

9–119. $\bar{x} = 1.30$ m, $\bar{y} = 2.30$ m

9–121. $\bar{x} = 112$ mm, $\bar{y} = 112$ mm, $\bar{z} = 136$ mm

9–122. $\bar{x} = 3.30$ in., $\bar{y} = 3.30$ in.

9–123. $\bar{x} = -0.262$ in., $\bar{y} = 0.262$ in.

Chapter 10

10–1. 356 in⁴

10–2. $I_x = \frac{1}{12}bh^3$

10–3. $I_y = \frac{1}{12}hb^3$

10–5. 0.165 in⁴

10–6. $I_x = 2.13$ in⁴, $I_y = 4.57$ in⁴

10–7. $(J_O)_s = 251(10)^3$ mm⁴, $(J_O)_t = 172(10)^3$ mm⁴, 68.4%

10–9. 0.167 in⁴

10–10. 0.3 in⁴

10–11. $I_x = \pi ab^3/4$

10–13. $I_x = \pi r^4/16$

10–14. 8.94 mm

10–15. 0.571 in⁴

10–17. 2.13 ft⁴

10–18. 0.610 ft⁴

10–19. 0.305 m⁴

10–21. 0.628 m⁴

10–22. 15.4(10)⁶ mm⁴

10–23. 3.35(10³) in⁴

10–25. 74.7 mm

10–26. 59.4 mm

10–27. $I_x = b^3d^3/6(b^2 + d^2)$

10–29. $I_x = 141$ in⁴, $I_y = 21.3$ in⁴

10–30. 147(10)⁶ mm⁴

10–31. $\bar{y} = 2$ in., $I_{\bar{x}\bar{x}} = 128$ in^4

10–33. $I_x = 246$ in^4, $I_y = 61.5$ in^4

10–34. $\bar{y} = 1.29$ in., $I_{\bar{x}\bar{x}} = 6.74$ in^4

10–35. 86 in^4

10–37. $0.187d^4$

10–38. 81 in^4

10–39. 91.1 in^4

10–41. $\bar{y} = 170$ mm, $I_{x'} = 722(10)^6$ mm^4

10–42. $91.7(10)^6$ mm^4

10–43. $\bar{y} = 2.20$ in., $I_{x'} = 57.9$ in^4

10–45. $I_{x'} = 45.1(10)^6$ mm^4, $I_{y'} = 7.60(10)^6$ mm^4

10–46. $I_{x'} = \frac{1}{36}bh^3$, $I_{y'} = \frac{1}{36}hb(b^2 - ab + a^2)$

10–47. $I_{x'} = \frac{1}{12}a^3b \sin^3 \theta$, $I_{y'} = [(ab \sin \theta)/12](b^2 + a^2 \cos^2 \theta)$

10–49. 8 in^4

10–50. $I_{xy} = \frac{3}{16}b^2h^2$

10–51. $I_{xy} = \frac{1}{8}a^4$, $\bar{I}_{x'y'} = -0.0165a^4$

10–53. 0.333 in^4

10–54. 1.33 m^4

10–55. $I_{xy} = nb^2h^2/2(2 + 2n)$

10–57. $I_{xy} = (a^2c \sin^2 \theta/12)(4a \cos \theta + 3c)$

10–58. $\bar{x} = 85$ mm, $\bar{y} = 35$ mm, $I_{xy} = -7.5(10)^6$ mm^4

10–59. $98.4(10)^6$ mm^4

10–61. $\bar{x} = 44.1$ mm, $\bar{y} = 44.1$ mm, $I_{xy} = -6.26(10)^6$ mm^4

10–62. 0.740 in^4

10–63. -36 in^4

10–65. $135(10)^6$ mm^4

10–66. $I_u = 274$ in^4, $I_v = 572$ in^4, $I_{uv} = -258$ in^4

10–67. $I_u = 101$ in^4, $I_v = 134$ in^4, $I_{uv} = -28.6$ in^4

10–69. $I_u = 15.8$ in^4, $I_v = 25.8$ in^4

10–70. $10.5°$, $I_{max} = 117(10^6)$ mm^4, $I_{min} = 28.8(10^6)$ mm^4

10–71. $-27.3°$, $I_{max} = 24.6(10)^6$ mm^4, $I_{min} = 2.56(10)^6$ mm^4

10–73. $I_{max} = 194(10)^3$ mm^4, $I_{min} = 55.5(10)^3$ mm^4, $29.9°$

10–74. $I_{max} = 117(10)^6$ mm^4, $I_{min} = 28.8(10)^6$ mm^4, $10.5°$

10–75. $I_{max} = 24.6(10)^6$ mm^4, $I_{min} = 2.56(10)^6$ mm^4, $27.3°$

10–77. $I_y = \frac{1}{3} ml^2$

10–78. $k_x = a/\sqrt{3}$

10–79. $I_x = \frac{2}{5}mr^2$

10–81. $I_x = \frac{3}{10} mr^2$

10–82. 2.25 slug $\cdot$ ft^2

10–83. 57.7 mm

10–85. $I_x = \frac{93}{70} mb^2$

10–86. $I_x = \frac{2}{5} mb^2$

10–87. 1.58 slug $\cdot$ ft^2

10–89. 1.78 m, 4.45 kg $\cdot$ m^2

10–90. 118 slug $\cdot$ ft^2

10–91. 282 slug $\cdot$ ft^2

10–93. $I_O = \frac{1}{2} ma^2$

10–94. 0.203 m, 0.230 kg $\cdot$ m^2

10–95. 34.2 kg $\cdot$ m^2

10–97. $I_{xy} = \frac{1}{6}a^2b^2$

10–98. 0.667 in^4

10–99. -110 in^4

10–101. $I_{max} = 0.752$ in^4, $I_{min} = 0.206$ in^4

10–102. $I_x = \frac{2}{15} bh^3$

10–103. $I_y = \frac{2}{7} b^3h$

10–105. $17.0(10)^6$ mm^4

10–106. 5.64 slug $\cdot$ ft^2

Chapter 11

11–1. 51.0 N

11–2. 512 N

11–3. $75.1°$

11–5. $F = 2P \cot \theta$

11–6. $P = W/2 \cot \theta$

11–7. 24.5 N

11–9. $42.4°$

11–10. virtual work, $\theta = 90°$; spring force, $\theta = 11.5°$

11–11. $0°$, $60°$

11–13. $37.8°$

11–14. $36.1°$

11–15. $24.2°$

11–17. $27.4°$

11–18. 10.8 lb/ft

11–19. 26.0 N $\cdot$ m

11–21. $90°$, $17.0°$

11–22. $17.1°$

11–23. $P_1 = 11.5$ lb, $P_2 = 10$ lb

11–25. $W_1 = W_2 = WL/2a$

11–26. $90°$, $11.5°$

11–27. $24.2°$

11–29. 10.8 lb/ft

11–30. $x = 0$, $y = 0$

11–31. 0.590 ft and -0.424 ft

11–33. $x = 0$, $y = 0$

11–34. 64.8°

11–35. 0°, 60°

11–37. 90°, $\theta = \sin^{-1}(W/2kl)$

11–38. 0°, $\theta = \cos^{-1}(W/2kL)$, $W = 2kL$

11–39. 90°, 30°

11–41. 90°, $\theta = \sin^{-1}(4W/ka)$

11–42. 36.6°

11–43. 90°, 5.90°

11–45. 8 in.

11–46. 0°, 71.5°

11–49. $d = 0.451r$

11–50. 5.20 ft

11–51. $r > \dfrac{b}{2}$

11–53. 1.35 in.

11–54. $x_1 = 6.67$ in., $x_2 = 1.25$ in.

11–55. 29.1°

11–57. 427 N

11–58. $F_C = \dfrac{2n\pi l}{p}P$

11–59. 10.0 lb

11–61. 0°, 72.5°

11–62. 16.6°

11–63. 0°, $\theta = \cos^{-1}[1 - (W/2kl)]$

Index

Geometric Properties of Line and Area Elements

| Centroid Location | Centroid Location | Area Moment of Inertia |

Centroid Location

Circular arc segment

Quarter and semicircular arcs

Trapezoidal area
$$A = \tfrac{1}{2}h(a+b)$$
$$\tfrac{1}{3}\left(\tfrac{2a+b}{a+b}\right)h$$

Semiparabolic area
$$A = \tfrac{2}{3}ab$$

Exparabolic area
$$A = \tfrac{ab}{3}$$

Parabolic area
$$A = \tfrac{4}{3}ab$$

Centroid Location

Circular sector area
$$A = \theta r^2$$

Quarter circular area
$$A = \tfrac{\pi r^2}{4}$$

Semicircular area
$$A = \tfrac{\pi r^2}{2}$$

Circular area
$$A = \pi r^2$$

Rectangular area
$$A = bh$$

Triangular area
$$A = \tfrac{1}{2}bh$$

Area Moment of Inertia

$$I_x = \tfrac{1}{4}r^4(\theta - \tfrac{1}{2}\sin 2\theta)$$
$$I_y = \tfrac{1}{4}r^4(\theta + \tfrac{1}{2}\sin 2\theta)$$

$$I_x = \tfrac{1}{16}\pi r^4$$
$$I_y = \tfrac{1}{16}\pi r^4$$

$$I_x = \tfrac{1}{8}\pi r^4$$
$$I_y = \tfrac{1}{8}\pi r^4$$

$$I_x = \tfrac{1}{4}\pi r^4$$
$$I_y = \tfrac{1}{4}\pi r^4$$

$$I_x = \tfrac{1}{12}bh^3$$
$$I_y = \tfrac{1}{12}hb^3$$

$$I_x = \tfrac{1}{36}bh^3$$